Automotive Mechatronics: Operational and Practical Issues

International Series on
INTELLIGENT SYSTEMS, CONTROL AND AUTOMATION:
SCIENCE AND ENGINEERING

VOLUME 52

Editor

Professor S. G. Tzafestas, National Technical University of Athens, Greece

Editorial Advisory Board
Professor P. Antsaklis, University of Notre Dame, Notre Dame, IN, USA
Professor P. Borne, Ecole Centrale de Lille, Lille, France
Professor D.G. Caldwell, University of Salford, Salford, UK
Professor C.S. Chen, University of Akron, Akron, Ohio, USA
Professor T. Fukuda, Nagoya University, Nagoya, Japan
Professor S. Monaco, University La Sapienza, Rome, Italy
Professor G. Schmidt, Technical University of Munich, Munich, Germany
Professor S.G. Tzafestas, National Technical University of Athens, Athens, Greece
Professor F. Harashima, University of Tokyo, Tokyo, Japan
Professor N.K. Sinha, McMaster University, Hamilton, Ontario, Canada
Professor D. Tabak, George Mason University, Fairfax, Virginia, USA
Professor K. Valavanis, University of Denver, Denver, Colorado, USA

For other titles published in this series, go to
www.springer.com/series/6259

B.T. Fijalkowski

Automotive Mechatronics: Operational and Practical Issues

Volume II

B.T. Fijalkowski
Department of Mechanical Engineering
Cracow University of Technology
Al. Jana Pawla II 37
31-864 Krakow
Poland
pmfijalk@cyf-kr.edu.pl

ISBN 978-94-007-3572-9 ISBN 978-94-007-1183-9 (eBook)
DOI 10.1007/978-94-007-1183-9
Springer Dordrecht Heidelberg London New York

© Springer Science+Business Media B.V. 2011
Softcover reprint of the hardcover 1st edition 2011
No part of this work may be reproduced, stored in a retrieval system, or transmitted in any form or by any means, electronic, mechanical, photocopying, microfilming, recording or otherwise, without written permission from the Publisher, with the exception of any material supplied specifically for the purpose of being entered and executed on a computer system, for exclusive use by the purchaser of the work.

Cover design: SPi Publisher Services

Printed on acid-free paper

Springer is part of Springer Science+Business Media (www.springer.com)

For my daughter Madeleine

Preface

The purpose of this book is to present operational and practical issues of automotive mechatronics with special emphasis on the heterogeneous automotive vehicle systems approach.

The book is intended as a graduate text as well as a reference for scientists and engineers involved in the design of automotive mechatronic control systems.

As the complexity of automotive vehicles increases, so does the dearth of high competence, multi-disciplined automotive scientists and engineers. This book provides a discussion of the type of mechatronic control systems found in modern vehicles and the skills required by automotive scientists and engineers working in this environment.

Divided into two volumes and five parts, **Automotive Mechatronics** aims at improving automotive mechatronics education and emphasises the training of students' experimental hands-on abilities. The author hopes that this can stimulate and promote the education programme in students' experimental hands-on experience among high education institutes and produce more automotive mechatronics and automation engineers.

Contents

❖ VOLUME I
- ➢ Part 1 - RBW or XBW unibody or chassis-motion mechatronic control hypersystems;
  Part 2 - DBW AWD propulsion mechatronic control systems;
- ➢ Part 3 - BBW AWB dispulsion mechatronic control systems;

❖ VOLUME II
- ➢ Part 4 - SBW AWS conversion mechatronic control systems;
- ➢ Part 5 - ABW AWA suspension mechatronic control systems.

The book was developed for undergraduate and postgraduate students as well as for professionals involved in all disciplines related to the design or research and development of automotive vehicle dynamics, powertrains, brakes, steering, and shock absorbers (dampers). A basic knowledge of college mathematics, college physics, and knowledge of the functionality of automotive vehicle basic propulsion, dispulsion, conversion and suspension systems is required.

Individuals new to the subject matter of RBW or XBW unibody, space-chassis, skateboard-chassis or body-over-chassis motion mechatronic control systems, will benefit most from the material. This manual is not compulsory for individuals with a basic background in, or knowledge of DBW AWD propulsion, BBW AWB dispulsion, SBW AWS conversion and ABW AWA suspension mechatronic control systems. Into the bargain, please notice that because of proprietary considerations, this book does not present details of algorithm design, algorithm performance, or algorithm application.

I am the sole author of the book and all text contained herein is of my own conception unless otherwise indicated. Any text, figures, theories, results, or designs that are not of my own devising are appropriately referenced in order to give acknowledgement to the original authors. All sources of assistance have been assigned due acknowledgement.

All information in this book has been obtained and presented in accordance with academic rules and ethical conduct. I also wish to state declare that, as required by these rules and conduct, I have fully cited and referenced all material and results that are not original to this book.

I wish to express my sincere gratitude to Professor Spiros Tzafestas for his interest in the preparation of this book in the Intelligent Systems, Control, Automation, Science and Engineering book series (ISCA). My gratitude is also due to Ms Nathalie Jacobs and Ms Johanna F. A. Pot of Springer for their persistence in making this book a reality.

I am grateful to the many authors referenced in this book from whom, during the course of writing, I learned so much on the subjects which appear in the book. I am also indebted to my national and international colleagues who indirectly contributed to this book.

Most of all I wish to express thanks the following consortia and institutions: ABIresearch, ABResearch, ACURE Dynamics, ADAMS, ADVISOR, ADwin, AEG, Air Force Research Lab. (AFRL), AIRMATIC, AirRock, AKA Bose Corp., AMESim, AMI Semiconductor, AMT, AMTIAC, AR&C, AROQ Ltd., Audi AG, AUDIAG, AutoPro, AUTOSAR, AUTOTECH, Avio Pro, AVL, Bertone, BizWire, Bobbs-Merril Co., BMW, BOSCH GmbH, Bridgestone Corp., Cadillac, CAFS, California Linear Devices, Carnegie Mellon, Centro Richerche FIAT, CFC, Challenge Bibendum, Chalmers University of Technology, Climatronic, Cleveland State University, Cracow University of Technology, Continental TEVES Inc., Cosc/Psych, Cracow University of Technology, CRL, Chrysler, Daimler-Benz, DaimlerChrysler AG, D&R, DAS, DECOMSYS, Delco Electronics, Delco-Remy, Delphi, Delft Center for Systems and Control, DJH, DLR RoboDrive, DRDC, dSPACE GmbH, Dynamic Structures & Materials LLC, Energen Inc., ERFD, ER Fluid Developments Ltd. UK, eSTOP GmbH, FAA US DoT, FACE International Corp., FHWA-MC Fiat, Fichtel & Sachs, FlexRay Consortium, FMA, FortuneCity, FPDA, US DoT, Ford Europe, Ford Motor Co., Ford SRL, Freescale Semiconductor Inc., FUJI Microelectronics Inc. (FMA), FUJITSU, GM Chevrolet, GM Opel, General Motors Corp., German Aerospace

Centre (DLE e.V.), Gothorum Carolinae Sigillium Universita, Graz University of Technology (TUG), Haskell, Hitachi Co., Honda, How Stuff Works, Hunter, I-CAR, IEC, IEEE, IMechE, Intel, Institute of Robotics and Mechatronics, Intelligent Transportation Society (ITS), ISO, IPC website, IPG Automotive GmbH, Istanbul Technical University, Jäger GmbH, JB, JUST-AUTO.COM, Kalmar, Kinetic Suspension Technology, Lexus, Kungl. Tekniska Högskolan (KTH), Land Rover, Lord, Lotus Engineering, Lund Institute of Technology, Magneti Marelli, Magnet Motor, Mazda, McCormick, Mechanical Dynamics, Inc., Mecel, Messier-Bugatti, MICHELIN, MILLENWORKS, MIT Hatsoupulos Microfluid Lab., Mitsubishi Corporate, MOST Net-services,, MOTOROLA, NI, NASA Langley Research Center, National Highway Traffic Administration (NHTSA), Nissan, Office of Naval Research (ONR), Norwegian University of Science and Technology, Oldhams Ltd., OSEK-WORKS, Packard, PACIFICA Group Technologies Pty Ltd., PEIT, PHILIPS, PITechnology, Polski FIAT, Porsche, PSA PEUGEOT CITROËN , Purdue School of Engineering and Technology, SAAB, SAE, Scania, Sensormag, Siemens VDO Automotive, SKF, Star, Studebaker, Subaru, Radatec Inc., Southwest Research Institute (SwRI), Racelogic, Radatec Inc., Renault, Research Team for Technology (CARAMELS), Ricardo, RMSV, Robert Bosch GmbH, Rodmillen, SCANIA, Seoul National University, TACOM TARDEC, Technische Universität Darmstadt, Universität Koblenz, Universität Regensburg, TENNECO Automotive, The Motor Industry Research Association (MIRA), The New York Times, The University of Michigan, Toyota, TRIDEC, Triumph, TRW Automotive Inc., TTPbuild, TTPnode, TTTech Computertechnik AG, Universita 'di Bologna, UniversitÄt Salzburg, University of California Berkeley, University of Leicester, University of Limerick, University of Pennsylvania, University of Queensland, University of Sussex, University of Texas at Austin, University of York (UK), Uppsala University, US Army Research Office, US DLA, US DoD, US DoE, UT-CEM, Valentin Technologies Inc., Valeo, Van Doorne Transmissie BV, VCT, Vienna Institute of Technology, VOLKSWAGEN (VW), VOLVO, Wongkwang University, ZF Sachs AG, and XILINX for their text, figures, or designs included in this book in order to give them due credit and acknowledgement as well as to present their contemporary achievements in automotive mechatronics.

The book is full of advanced statements and information on technology developments of the automotive industry. These statements can be written and may be recognizable by terms such as *'may be'*, *'will'*, *'estimates'*, *'intends'*, *'anticipated, 'expects'* or terms with analogous sense. These statements are derived from presuppositions with reference to the developments of the technology of Europe, the Americas and Asia-Pacific countries, and in particular of their automotive industry, which I have prepared based on information accessible to me and which I think to be realistic at the time of going to press.

The estimates specified implicate a degree of risk, and the actual development may differ from those forecasts.

If the presuppositions underlying any of these statements prove incorrect, the actual results may noticeably differ from those expressed by or embedded in such statements.

I do not update advanced statements retrospectively. Such statements are of most value on the date of publication and can be superseded.

Anyone who has attempted to write such a book in their spare time knows how many weekends and vacation days go into it. I dedicate this book to my family for their continual encouragement, constant care, and assistance and infinite patience in making the writing of this book possible, as well as the generous understanding they have always shown me.

Cracow, November 2010 *BOGDAN THADDEUS FIJALKOWSKI*

Contents

PART 4

4 SBW AWS Conversion Mechatronic Control System

4.1 Introduction

In this part of the book, the interested readers may consider the lateral motion in the *y*-axis of the automotive vehicle. By lateral motion of the automotive vehicle, is meant how the vehicle responds to steering input.

A **human driver** (HD) controls the vehicle's lateral dynamics by indirectly affecting the forces generated by the wheel-tyres of the vehicle. Approximately, these forces are influenced by various systems:

- The steering system;
- The braking system (especially with differential braking);
- The absorbing (damping) system;
- The drivetrain system.

In fact, the response of the automotive vehicle to steering input is predominantly influenced by a **steer-by-wire** (SBW) **all-wheel-steered** (AWS) conversion mechatronic control system. On the other hand, the **brake-by-wire** (BBW) **all-wheel-braked** (AWB) dispulsion, **absorb-by-wire** (ABW) **all-wheel-absorbed** (AWA) suspension and **drive-by-wire** (DBW) **all-wheel-driven** (AWD) propulsion mechatronic control systems may also be used to influence the steering capabilities of the vehicle; it is therefore not surprising that research on controlling the lateral motions of a vehicle has recently concentrated on integrating these systems into a **ride-by-wire** (RBW) or **x-by-wire** (XBW) integrated unibody, space-chassis, skateboard-chassis, or body-over-chassis motion mechatronic control hypersystem.

Mechatronic control of lateral dynamics through SBW AWS conversion mechatronic control systems - Conventionally, vehicle steering systems have been used to control the lateral motion of the vehicle. Approximately, **research-and-development** (R&D) works on this subject may be broken down along the following lines; R&D work on active **front-wheel steering** (FWS), active **rear-wheel steering** (RWS) and **all-wheel steering** (AWS) systems. The SBW **four-wheel-steered** (4WS) conversion mechatronic controller may influence the wheel-tyres's direction in different modes, as represented in Figure 4.1 [VILLEGAS AND SHORTEN 2005]. An FWS controller alters the direction of the front wheels as a function of the driver's input with or without a mechanical link. RWS controllers on the other hand do not influence the front steering angle (this task is left to the driver) but rather affect the vehicle dynamics by adjusting the steering angles of the rear wheels.

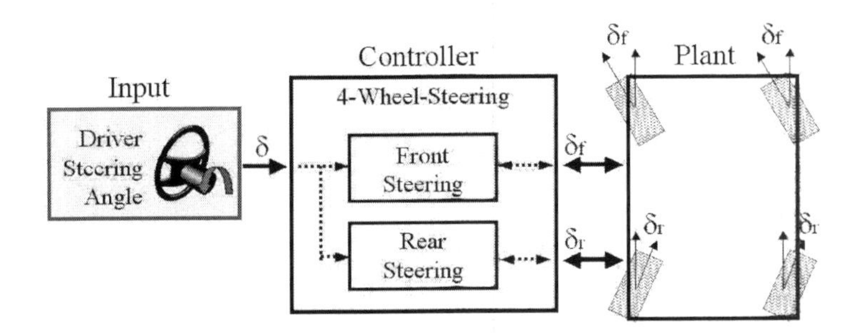

Fig. 4.1 Active SBW 4WS conversion mechatronic control system [VILLEGAS AND SHORTEN 2005].

Four-wheel steering (4WS) systems control both front and rear steering angles as a function of driver input and vehicle dynamics. An FWS system without a mechanical link between the steering **hand wheel** (HW) and the steering angle (i.e. an active FWS SBW conversion mechatronic control system) is presented in [MOCHIZUKI AND NAKANO 2000]. The control structure uses feedback and feedforward control as a function of the steering HW input, the measured velocity (speed), yaw-rate, and lateral acceleration measurements to control the vehicle's yaw-rate. The resulting controller is shown to have better vehicle stability during a lane-change manoeuvre over packed snow than one with brake and drive force distribution through **direct yaw-moment control** (DYC). In addition, the resulting system feeds back a torque signal to the driver by means of the steering HW. Other proposals for active steering remain a direct mechanical link from the steering HW but enhance a supplementary steering angle through the use of an **electro-mechanical** (E-M) motor introduced in the steering system. This category of system may, in principle, provide steering in circumstances of the failure of the mechatronic control system and makes available enhanced safety functionality. This category of structure is used in [ACKERMANN 1998; and KOBO ET AL. 2002].
In [ACKERMANN 1998], the driver's task of keeping in the correct lane and the automated yaw stabilisation are separated by means of a yaw-rate feedback control system correcting the steering angle. The test results (given in this reference) show the robustness against crosswind perturbations as well as against μ-split-braking. Another approach is presented in [VILA LANA 2004]. In this reference, a SBW 4WS conversion mechatronic control system and a conventional steering system are integrated into a single steering system. This construction permits the introduction of a safety management system that reverts to normal steering during the failure of the SBW 4WS function. Active RWS has been studied by various authors for controlling automotive vehicle lateral dynamics. Most of these structures rely on the use of gain-scheduled feedforward control to command the rear steering angle [FURUKAWA ET AL. 1989]. In such control structures, some of which have already been implemented in the manufacture of passenger vehicles, the rear steering angle is computed as a function of the front steering angle command from the driver's input to the steering HW.

Various different control laws have been proposed for controlling such systems (usually related to the improvement of the manoeuvrability and cornering stability of the vehicle).

In [INOE AND SUGASAWA 2002] and, some years later, in [HIRANO AND FUKATANI 1996], the authors combine feedforward and feedback control to command the rear steering angle, while the front steering angle remains under the direct control of the driver. The control objective here is to follow a predefined physical model of automotive vehicle dynamics.

In order to obtain a satisfactory degree of robustness, the feedback controller is designed using μ synthesis [HIRANO AND FUKATANI 1996]. The results of this demonstrate improved handling and stability in a variety of experimental circumstances. In some circumstances, it is desirable to control both side-slip angle and vehicle yaw-rate [NIETHAMMER 2000].

The mechatronic control of both of these dynamics simultaneously is not possible using only active RWS, or only active FWS. The mechatronic control of both of these signals requires at least two control inputs.

In automotive vehicles equipped with the **electronic stability program** (ESP) or DYC, this may be realised through the brake and drive force (one input) and by using the front steering angle as a second input. In fact, using ESP to control lateral dynamics is not desirable in all circumstances.

For vehicles operating under normal operation circumstances, controlling lateral dynamics using a SBW 4WS conversion mechatronic control system is clearly desirable; here the front and rear steering angles are the two control inputs. Under this circumstance, even undesired side-effects from ESP or DYC may be counteracted. The ability to control side-slip and yaw-rate independently of each other is depicted in Figure 4.2 [VILLEGAS AND SHORTEN 2005]; when the front and rear steering angles are in opposite directions, a yaw-rate without side-slip may be obtained so a constant radii curve could be managed: when both angles have the same sense of direction, a sideslip angle without a yaw-rate may be performed to change lanes.

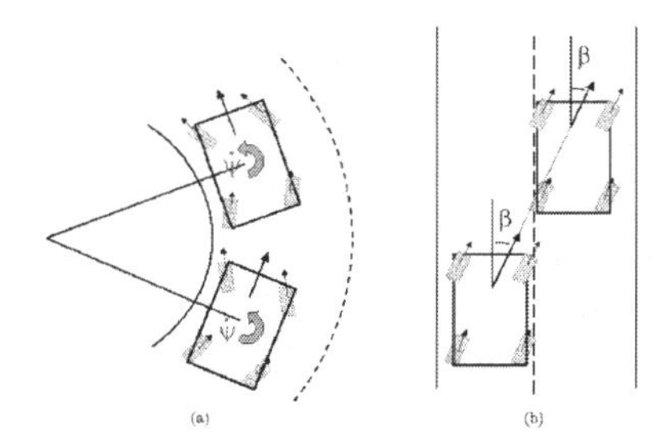

Fig. 4.2 Independent yaw-rate (a) and independent side-slip [VILLEGAS AND SHORTEN 2005].

A substantial outline of R&D works on the mechatronic control of SBW 4WS vehicles exists and a variety of control structures have been considered in the Bibliography and References that accompany this part of the book.

In [VILAPLANA 2004], an innovative feedback steering controller capable of modifying the lateral dynamics of SBW 4WS automotive vehicles to follow a given reference physical model is presented. The proposed steering controller commands the front and rear steering angles with the objective of tracking the reference sideslip angle and yaw rate signals obtained online from the driver's inputs to steering wheel and pedals. These reference signals describe the lateral motion that would result if the driver's input is applied to a physical model with the desired dynamics. Besides, the steering controller automatically rejects any disturbances in sideslip and yaw rate.

The R&D work described in [GIANONE ET AL. 1995] proposes a feedback control structure based on **virtual model following control** (VMFC) and robust **linear-quadratic regulator** (LQR) design. The physical model to be followed corresponds to the FWS automotive vehicle. An example of a steering controller specifically designed for vehicles equipped with SBW 4WS is presented in [ACKERMANN 1994]. The controller structure given in this reference is based on the cross-feedback of the measured yaw rate to the front steering angle. This structure decouples the control of the lateral acceleration from the control of the yaw rate. Two outer feedback loops are utilised so that FWS is used to track the desired lateral acceleration and RWS is used to regulate the damping of the resulting yaw dynamics.

In the Bibliography and References that accompany this chapter, a small number of the listed references to the automotive-vehicle emulation problem and only in [LEE 1995] results for an emulator are presented.

In [LEE 1995], a *'midsized'* vehicle with an active RWS algorithm is used to emulate lateral dynamics of *'small'*, *'compact'*, and *'full-sized'* vehicles. The emulation test results for *'small'* vehicles are good but not for *'compact'* and *'full-sized'* ones.

Mechatronic control of lateral dynamics through BBW AWB dispulsion and ABW AWA suspension mechatronic control systems – It has already been mentioned that lateral vehicle motion may be influenced by mechatronic control systems other than the SBW AWS conversion mechatronic control system; for example, both BBW AWB dispulsion and ABW AWA suspension mechatronic control systems have been used in the works of various authors to control these lateral motions.

The majority of this R&D work has been motivated by an automotive vehicle safety study. For instance, in certain critical circumstances, when the vehicle wheel-tyre limits are reached, the mode in which the vehicle responds to normal driving inputs such as steering angle alterations. In these circumstances, we may compensate for this effect using the ABW AWA suspension and/or the BBW AWB dispulsion mechatronic control systems. This is an essential concept that has been developed recently; the principal approaches used are individual-wheel **anti-lock braking systems** (ABS), **roll moment distribution** (RMD) and **external combustion engine** (ECE) or **internal combustion engine** (ICE) **torque distribution** (TD).

It should be observed that active ABW AWA suspension and individual-wheel BBW AWB dispulsion have also an effect on lateral dynamics in normal driving circumstances; these interactions, while mostly neglected in the literature, were discovered during R&D work on the CEMACS project [VILLEGAS AND SHORTEN 2005].

Anti-lock braking systems (ABS) and traction control systems (TCS) – Individual-wheel ABS and **traction control systems** (TCS) are found in various conventional automotive vehicles. Systems of this category are referred to by a variety of names; ESP or **vehicle dynamics control** (VDC).

In [ZANTEN VON 1996] an in-production VDC from Bosch is described. It uses an ABS and TCS infrastructure to maintain vehicle stability -- during dangerous manoeuvring and to enhance the use of friction potential. The controlled forces are the longitudinal wheel-tyre forces; these are controlled by individual-wheel brake pressures and an ECE or ICE torque. While this system maintains stability, it has been recognised by various authors that the use of individual-wheel ABS to control the yaw rate is not only uncomfortable for the driver, but also requires the vehicle to develop very large wheel-tyre forces (as illustrated in [ACKERMANN ET AL. 1999]). In fact, the use of such large forces to control the vehicle is, under certain circumstances, undesirable.

In [ACKERMANN ET AL. 1999] the authors present simulation results that show that a combination of active FWS, together with individual-wheel ABS during dangerous manoeuvres is more stable and, in addition, prevents vehicle roll-over (due to input from the driver). The use of individual-wheel ABS together with an active SBW 4WS conversion mechatronic control system is presented in [MOKHIAMAR AND ABE 2002]. The stability performance of this system is shown to be better than the use of mechatronic control systems operating independently of each other as it makes full use of the vehicle wheel-tyre forces. A mechatronic control system that uses both TD and active RWS is presented in [HIRANO ET AL. 1992]. Here the authors exploit the fact that the vehicle cornering force decreases as driving torque increases. The authors demonstrated, in a series of publications [HIRANO 1994], that the use of TD provides more vehicle stability that active RWS in slippery road surfaces.

All-wheel-absorbed (AWA) absorb-by-wire (ABW) suspension mechatronic control systems – It is well known that variations in the vertical load of each wheel influence the total force that may be developed by a wheel-tyre. Consequently, the on-off roadway inertial loads that are transferred to the ABW AWA suspension mechatronic control system may be modified using active and semi-active ABW AWA suspension mechatronic control systems with a view to affecting the lateral behaviour of the vehicle.

An analysis of the influence of VDC on improving vehicle stability in near-limit circumstances is presented in [ABE 1991]. Simulation results show VDC potential to keep the vehicle out of lateral/directional instability under hard manoeuvring circumstances. A semi-active approach that involves the use of varying damping coefficients is presented in [BODIE AND HAC 2002] (with **magneto-rheological** (MR) shock absorbers). In this reference, the yaw-rate is controlled through the distribution the damping forces between the front and rear axles thereby improving the vehicle response and reducing the driver's attempts at steering. It is also shown that variations in absorbing (damping) force have a great corrective yaw moment potential at high lateral accelerations.

VDC is tested experimentally in [WILLIAMS AND HADDAD 1995] using an active suspension with a non-linear controller. This controller shows good yaw-rate tracking performance when cornering. Apart from active suspensions, vehicles equipped with active stabilising bars at every axle canal influence the suspension forces, distributing their stiffness between the front and the rear axles.

In [ÖTTGEN AND BERTRAM 2001], it is shown that the understeer gradient (that defines the lateral behaviour of the vehicle when cornering) may be altered more by distributing the roll stiffness between the rear and front than by the addition of more roll stiffness. By doing so, the lateral force potential of the wheel-tyres may be fully exhausted because the wasted potential at the front axle of the vehicle is used.

A **proportional-integral-derivative** (PID) roll moment control together with a **proportional-integral** (PI) roll moment distribution controller are implemented with active stabilisers in order to reduce any undesired roll and maintain stability. Vehicle pitch and lift have also an influence on lateral dynamics as is shown by [HARADA AND HARADA 1999]. Using a frequency characterisation of on/off road disturbances, the authors deduce that the slip angle is correlated to the bounce velocity, while the yaw rate is related to the pitch velocity, and these relationships appear to be independent of the distribution of suspension spring stiffness between the front and rear axles. The authors develop a control algorithm based on LQR theory for bounce and pitch showing a reduction of maximum disturbance influence on sideslip and yaw rate, respectively, in simulation.

Integrated anti-lock braking systems (ABS), traction control systems (TCS), and absorb-by-wire (ABW) all-wheel-absorbed suspension mechatronic control systems – The influence of both suspension and longitudinal forces acting as a cohesive unit to affect vehicle lateral motion has also been investigated by a number of authors. Recent work in this area is summarised extensively in [GORDON ET AL. 2003]. Approximately, R&D work in this area has proceeded along two lines of enquiry. The first approach is concerned with the design of independently designed mechatronic control systems whose action is coordinated through the application of a supervisory mechanism. For instance, the design and analysis of a mechatronic control system constructed in such an approach analysed in [HAC AND BODIE 2002]; here, a control algorithm is implemented to coordinate active mechatronic control of brakes with MR ABW AWA suspension that significantly reduces the brake intervention required to maintain stability compared to using only BBW AWB dispulsion. Apart from ABS, TD may be controlled to provide stability in dangerous circumstances.

The effect of ECE or ICE torque on lateral vehicle motions and the effect of interaction between ECE or ICE torque and the ABW AWA suspension mechatronic control system are presented in [COOPER ET AL. 2004]. In this reference, an active ABW AWA suspension, together with **variable traction distribution** (VTD), is used to enhance handling and stability.

Even if VDC with an active ABW AWA suspension has a better yaw-rate tracking performance, the integrated controller is shown to have better stability and handling than those with the respective mechatronic control systems acting independently. All mechatronic control systems, active ABW AWA suspension, SBW 4WS conversion and VDC, were tested independently and coordinated by KITAJIMA AND HUEI [2000] using an H_1 control algorithm. The integrated control provides (in simulation) a significantly different performance and more stability than the uncoordinated controllers, showing the importance of interactions. Better performance of an integrated approach over a single approach is demonstrated using a test vehicle and the results are presented in [TRÄCHTLER 2004]. Here ESP is integrated with other controllers like **active front steering** (AFS), and the resulting mechatronic control system shows a significant improvement in lateral dynamic handling and is also shown to require less driver intervention and less braking pressure which is usually required to maintain stability. In addition, the results also show that these interactions may also be used to affect longitudinal dynamics; for example, to decrease braking distances. The AFS is integrated into the steering rod. Its main parts are an E-M actuator and an integrated planet-gear that operates the angle sensor and activates the E-M actuator, thus providing the power steering.

An upgraded AFS version no longer relies on mechanical steering commands and may implement a real SBW AWS conversion mechatronic control system. A steering controller sends commands to an E-M motor that responds by moving the steering gear and setting the front wheels into the right position. This feature may also permit programmed steering characteristics to be used as input (overlay) for automatic cruise control (load change in curve/yaw-rate regulation). For instance, using the SBW 4WS controller [VILAPLANA ET AL. 2004], the influence of vertical dynamics on lateral dynamics was tested by simulating a step in the steering HW with and without an **active suspension controller** (ASC) being activated. The results presented in Figure 4.3 [VILLEGAS AND SHORTEN 2005] depict the influence of roll vertical dynamics on the steering system's tracking performance. The roll dynamics influence principally the side-slip angle and induce an oscillation in the yaw-rate.

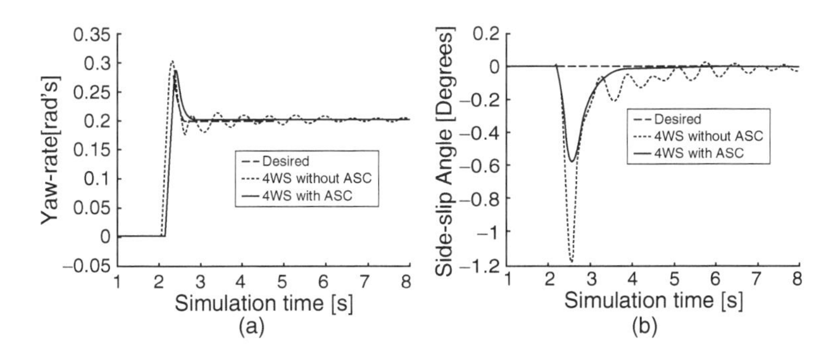

Fig. 4.3 Vertical dynamics influence on (a) yaw-rate and (b) side-slip response of 4WS controller [VILLEGAS AND SHORTEN 2005].

SBW AWS conversion mechatronic control systems function by including an additional SBW conversion mechatronic control system at the middle and/or rear of the vehicle. Fully developed versions were controlled mechanically. In both cases, when the front wheels are directionally pointed at certain angles, the middle and/or rear wheels are equally adjusted to angled positions, to provide superior platform control during handling and turning manoeuvres. Industrial three-wheeled vehicles have traditionally used RWS that may turn sharply in tighter spaces. The downside of these vehicle platforms is that they lack platform stability and require extensive driver training and experience. For instance, with SBW 4WS conversion mechatronic control systems, the rear wheels are turned in the same or opposite senses of direction as the front wheels, depending on vehicle velocity or the angle at which the steering HW is turned. All of this occurs as soon as the SBW 4WS mechatronic control system is used, allowing a tighter sight and turning line into the corners. At a predetermined vehicle velocity, when the steering HW is turned to the desired angle, the rear wheels may turn in the same sense of direction as the front wheels. At that steering angle or vehicle velocity, the rears may either move to a straight-line position or turn in the opposite sense of direction. These modern SBW 4WS conversion mechatronic control systems are either controlled by a set of drive shafts or mechatronically, depending on the type of vehicle platform. Obviously, additional components may be added to the underside of these platforms to monitor and control the RWS mechanical components. Usually, an additional steering gearbox that is similar to the FWS unit may be used to control the predetermined rear-wheel angles. A safety device or fail-safe unit may lock the alignment of the rear wheels in the conventional straight-ahead mode if a problem develops. With regard to consumer truck and **smart-utility vehicles** (SUV) platforms, scientists and engineers have designed and developed a SBW 4WS conversion mechatronic control system that may be offered on select vehicles, as well as full-size pickup platforms. The considerable function certainly is that this improvement allows larger vehicle platforms a shorter turning radius for tight manoeuvring and better road handling manners, especially under loaded and towing circumstances [SENEFSKY 2003]. Interfacing the driver and a single-chip microprocessor-controlled RBW or XBW integrated unibody, space-chassis, skateboard-chassis, or body-over-chassis motion mechatronic control hypersystem may result in less mass, more design flexibility, decreased manufacturing complexity, and higher reliability. The automotive vehicle's steering column as scientists and engineers already know may become obsolete within a few years, replaced by advanced SBW AWS automotive technology.

Automotive scientists and engineers expect that the first vehicles with very advanced complete SBW AWS technology may go into production in 2015. In the meantime, the automotive industry is developing a cost-effective interim vehicle at enhance vehicle functionality while leaving the steering column in place. SBW AWS means the end of the mechanical linkage between the steering HW and the road wheels, with mechatronically-controlled actuators setting the steering angles as well as mechatronically-controlled force feedback to the driver's steering device.

SBW AWS may enable fully integrated vehicle stability mechatronic control systems for collision avoidance systems and one day, potentially, even for autonomous driving.

The automotive industry is working to develop a mechatronic control system that may deliver advanced functionality enabled through an SBW AWS conversion mechatronic control system, such as a variable steering ratio according to vehicle velocity, but at a cost viable for the mass vehicle market. The leap forward for very advanced SBW AWS automotive technology development is that vehicle designers may no longer be constrained by the steering column – the field is open for console-style steering or side-shift steering. SBW AWS may also provide benefits in terms of **noise, vibration, and harshness** (NVH) and crashworthiness, due to the loss of the steering column, as well as opening up the tremendous possibilities of vehicle stability control. Coupled with the introduction of DBW AWD propulsion, BBW AWB dispulsion and ABW AWA suspension mechatronic control systems, the design of the integrated chassis may be greatly simplified. Ultimately, there may be four equal corner modules that may be significantly lighter and easy to fit. Recent advances in a dependable, very advanced embedded-system's automotive technology, as well as continuing demand for improved handling and passive and active safety improvements, have led vehicle manufacturers and suppliers to work towards developing a computer controlled, RBW or XBW integrated unibody, or body-over-chassis motion mechatronic control hypersystems with no mechanical link to the driver. These include SBW AWS, DBW AWD, BBW AWB, as well as ABW AWA, and are composed of mechanically decoupled sets of actuators and controllers connected through multiplexed in-vehicle computer networks. Actually, the success of such projects is highly dependent on the concurrent design of vehicle mechanics and mechatronic control architecture and the mechatronic integrated approach is an essential feature for the development of effective realisations; therefore, the availability of computer-aided modelling tools has been a basic requirement to assess beforehand a system's behaviour ever since the first design stages through dynamic simulation and virtual prototyping. An SBW AWS conversion mechatronic control system for even greater safety and comfort is better than the conventional steering systems. It may allow the wheel-tyres to move freely and it controls the vehicle velocity in accordance with the steering because the steering system and the actuator that operates the wheel-tyres are separate. The aim of the SBW AWS conversion mechatronic control system is to improve safety and comfort in vehicle dynamics by reducing the number of steering corrections and by simplifying the steering system communication structure. This SBW AWS conversion mechatronic control system may offer the additional advantage of flexibility in the interior design around the driver's seat. Automotive scientists and engineers are investigating optimal structures for the aforementioned actuator that may be the core of these SBW AWS conversion mechatronic control systems and may also be conducting R&D work for even greater safety and comfort through joint mechatronic control activities with BBW AWB dispulsion mechatronic control systems to evolve SBW AWS conversion mechatronic control systems into the RBW or XBW integrated unibody, space-chassis, skateboard-chassis, or body -over-chassis mechatronic control hypersystem. AWA SBW conversion mechatronic control systems underwent substantial changes from the onset. They began as a direct link between the steering column and the two front wheels. With automotive vehicles increasing in size and mass, the forces required to turn the wheels (especially while stopped) became unreasonable. The next step was to produce a favourable gearing ratio. One such method used to create the necessary mechanical advantage is the **parallelogram** (PG) **rack-and-pinion** (R&P) displayed in Figure 4.4 [NICE 2002]. Having maximised the reasonable gearing ratio, **fluido-mechanical** (F-M) or even an **electro-mechanical** E-M) assist was added - - known as **power steering** (PS). This is a contemporary mechatronic control system on most automotive vehicles; however, it is very complicated and may require excessive energy.

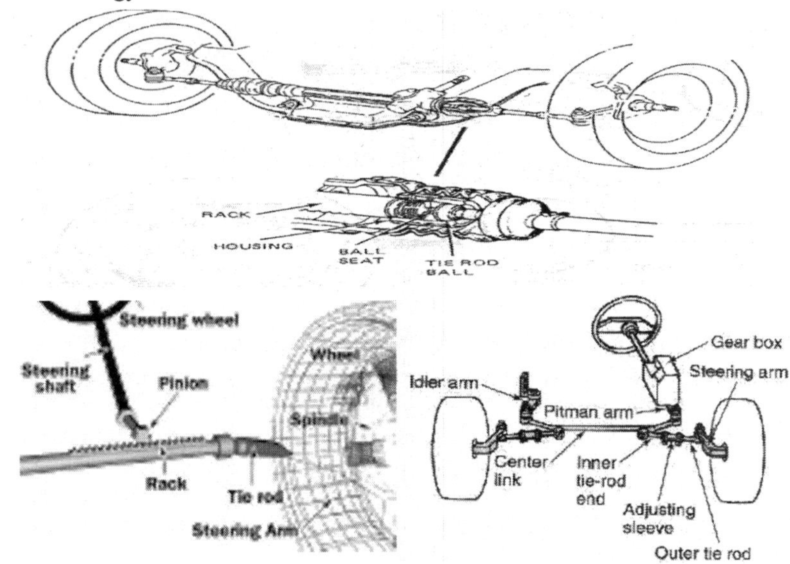

Fig. 4.4 Parallelogram (PG) and rack and pinion (R&P) [HOW STUFF WORKS—Bottom middle image].

The ECE or ICE drives, by means of rubber belts, a **mechano-fluidical** (M-F) pump or M-F compressor or even **mechano-electrical** (M-E) generator. This pressurised oily fluid or gas (air) or even induced electrical current, respectively, is used to assist the steering shaft in the direction it is turned. One of the main problems with F-M or E-M assist is one of available timing. The M-F pump or M-F compressor, or even M-E generator, respectively, must generate enough oily-fluid or gas (air) pressure or even electrical current to turn the vehicle while it is stopped (the hardest time to turn) when the ECE or ICE is running at idle values of its rotational velocity. When a value of the ECE or ICE rotational velocity rises and the vehicle velocity is increased, the pumping power being drawn from the ECE or ICE may be excessive. F-M or E-M mechatronic control systems also draw continuous power, even when PS is not indispensable.

The next logical advancement would be what is termed **electro-fluido-mechanical** (E-F-M) steering. This is where the **electro-mechano-fluidical** (E-M-F) pump or E-M-F compressor driven by an **electro-mechanical** (E-M) motor allows the E-M-F pump or E-M-F compressor, respectively, to provide only the required oily-fluid, or gas (air) pressure to aid the turning. E-F-M steering substantially reduces power consumption, however, due to the nature of the fluidics, oily-fluid, or gas (air) pressure that must be maintained for continual power use. An innovative advancement currently finding its way onto markets is referred to as E-M assist steering. In this mechatronic control system, the fluidics may be replaced by mechatronically controlled actuators and sensors. The sensors determine the manner in which the wheel is turning and cause the actuators to provide additional force in that direction to assist the mechanical linkage. With this configuration, energy is not wasted as with fluidics. Energy is only used when the wheels are turned [MOGHBELL 1992]. The mechanical linkage provides continual control in the case where electrical energy is lost. Substantial frictional losses and time constraints still exist from the mechanical connection. Removing the mechanical linkage between the steering HW and the wheel-tyres would simplify the design, increase efficiency, enhance performance, and improve overall safety. This completely E-M steering mechatronic control system is also termed SBW AWS. The E-M SBW 2WS conversion mechatronic control system is powered by the vehicle's **chemo-electrical/electro-chemical** (CH-E/E-CH) storage battery, not the ECE or ICE that adds to fuel-consumption efficiency and gives the vehicle manufacturer more options in configuring the steering system (Fig. 4.5) [UEKI ET AL. 2004; WILWERT ET AL. 2005].

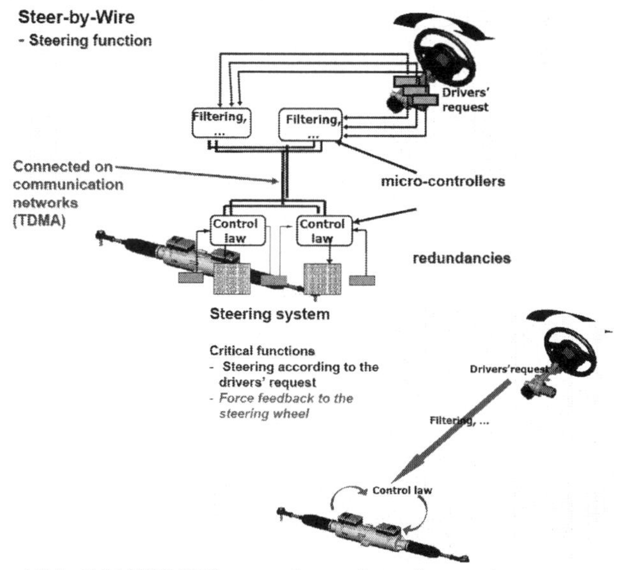

Fig. 4.5 An E-M SBW 2WS conversion mechatronic control system [Xilinx (photo); WILWERT ET AL. 2005].

Replacing a conventional M-F-M steering mechatronic control system with an E-M motor, as shown here, reduces the system's size and mass.

The E-M SBW 2WS conversion mechatronic control system also improves fuel efficiency and may give the vehicle manufacturer more design flexibility.

An SBW AWS conversion mechatronic control system would consist of a position sensor on the steering HW, an E-M motor providing forced feedback to the driver, actuators on all the wheels, and an **electronic control unit** (ECU) to coordinate the process. These four components would mark a substantial reduction in the number of parts as well as the size and mass of the overall mechatronic control system. The process is very efficient because power is only consumed when the wheels are turned. The ECU may receive data from other mechatronic control systems so, when the wheel-tyres are turned, only the necessary force is applied based on the instantaneous circumstances derived there from. Performance is enhanced because the steering ratio (how far the steering HW turns relative to how far the wheel-tyres turn) is fully variable, based on the vehicle velocity, traction control, and other pertinent variables. Safety is improved in the case of an accident.

The driver is most at risk in an accident because the steering HW is forced into the driver by the steering column. By removing the steering column, many of the injuries and fatalities experienced in automotive collisions could be eliminated. An SBW AWS conversion mechatronic control system is currently being used by military ground units and large construction companies, and in forklift trucks. For instance, Delphi Corporation has announced an innovative system which is scheduled to reach consumers in the 2015s and is termed *'QUADRA-STEER™'* [NICE 2002]. It is an SBW 4WS conversion mechatronic control system using RWS on the two rear wheels. This system is designed to assist large trucks and SUVs while manoeuvring in small spaces as well as to increase stability by highway values of vehicle velocity [NICE 2002].

An SBW 4WS conversion mechatronic control system has many advantages. By being mechatronically controlled, the rear wheel-tyres may change the way they turn based on driving parameters. When the vehicle is moving slowly, the rear wheel-tyres turn in the opposite sense of direction from the front wheel-tyres to aid in manoeuvring; at high values of vehicle velocity, the rear wheel-tyres turn in the same sense of direction as the front wheel-tyres to reduce yaw and improve stability [MURRAY 2002]. If the SBW on the rear wheels were to fail, the FWS mechatronic control system would allow the driver to maintain control. The advantages of an SBW AWS conversion mechatronic control system are clear. However, reliability is currently slow in its implementation into FWS mechatronic control systems.

This section describes the architecture of a RBW or XBW automotive integrated chassis mechatronic controller that lets a vehicle obtain good dynamic performances through a *'joystick'* **human-machine** (HM) interface; the independent motorisation of the front and rear wheels as well as a full-state sensorisation have been suitably used to enhance the standard capabilities of automotive vehicles.

A RBW or XBW automotive integrated chassis mechatronic control hypersystem consists of sensors and control elements connected by a redundant data bus: the driving dynamics controller plays a key role, since it tries to optimise the vehicle's handling behaviour and to allow safe cruising under all possible occurrences, actively taking over when the driver loses control of the vehicle; as a matter of fact, usually all components have a backup system to ensure maximum safety. To achieve such determined aspirations, the **fluido-mechanical** (F-M) and **mechano-mechanical** (M-M) transmissions are disconnected and replaced by servo E-M motors and switching elements, usually controlled by a fault-tolerant computer system. The latter receives its data not only from the driver, who issues commands to the system, but also from sensors that continually monitor the vehicle's status. As a matter of fact, the sensing system is a key part of the whole mechatronic control architecture and complex equipment is indispensable for the most performing realisations: typical implementations are based on sensing all four wheels' angular velocity, the yaw angular velocity, the steering angle, and the lateral acceleration.

SBW AWS conversion mechatronic control systems could eventually use joystick-like mechatronic controls that would eradicate the necessity for a steering HW as well as accelerator and brake pedals, freeing up room in the interior for other potential innovative advancements. In place of the steering HW and floor acceleration (gas) and brake pedals innovative mechatronic control systems may be utilised. The steering, accelerator, gear shifting and clutch actuator, as well as brakes may be mechatronically controlled by the SBW technology.

The SBW AWS conversion mechatronic control systems may be mechatronically controlled by the proof-of-concept driver interface shown in Figure 4.6 [SAE 2004; CITROËN 2005].

(a)

(b)

Fig. 4.6 Principle layout of the driver's mechatronic control system, consists of left and right steering control yokes for mechatronic control of steering (turning), throttling (acceleration), gear shifting and clutch actuation, as well as braking (deceleration) [SKF & Bertone - *Guida-Filo* (a); Citroën 2005 - Citroën *C5* (b)].

The driver interface's left and right steering control yokes are linked mechanically and have a full travel of just 20 deg.

Movement of the vehicle's front wheels is assisted by full-logic mechatronic control, with feedback to the driver being provided by a high-torque E-M motor. Driver *'feel'* is programmable, as is the relationship between yoke and front-wheel movement.

A next-generation steering actuator fits easily into the front subframe assembly of the latest automotive vehicle platforms.

The vehicle of the future may give the impression of being just similar to this: but it has no ECE or ICE, no steering column, and no brake pedal.

It does not run on petrol, emits no pollution (only a little water vapour) and hitherto handles in a similar way to a high-performance vehicle. It could act like an ecologist's dream.

In place of an ICE, for example, the vehicle of the future may be energised by **fuel cells** (FC) similar to those used in the orbiting space station [HAMILTON 2002]. Electrical energy is generated by an **electro-chemical** (E-CH) reaction of hydrogen and oxygen that emits only thermal energy (heat) and water (H_2O) as its side-effect. No smelly exhaust, no smog, no greenhouse gases. Replaced too are the cables and mechanical links that have held together vehicles since the beginning of the automobile age a century ago. As a substitute, the steering and braking are fully mechatronic; using techniques originating in **fly-by-wire** (FBW) aircraft cockpits. Instead of the steering column there is a miniature colour screen and two handgrips, as shown in Figure 4.7 [THIESEN 2003, SCHMIDT 2004].

Fig. 4.7 Miniature colour screen and two handgrips
[GM Opel's *Hy-Wire* FC HEV; THIESEN, 2003; SCHMIDT, 2004]

Without a steering column, vehicle designers may locate the mechatronic controls anywhere in the vehicle for maximum comfort and safety, even maybe on the backseat.

What appears to be a kind of video game to a conventional driver is actually a unique innovation based on a revolutionary concept known as *'SBW AWS'*. This new system not only offers improved safety, comfort, and ergonomics, but also provides extra advantages in terms of vehicle design and production.

It's all made possible by a mechatronic control system that replaces the mechanical and fluidical connections linking the steering wheel and pedals to the steering, drive, and brakes.

Designed so that it may only be moved to the left or right, a unique sidestick (see Fig. 4.8) enables drivers to steer with high precision [DAIMLERCHRYSLER 2004].

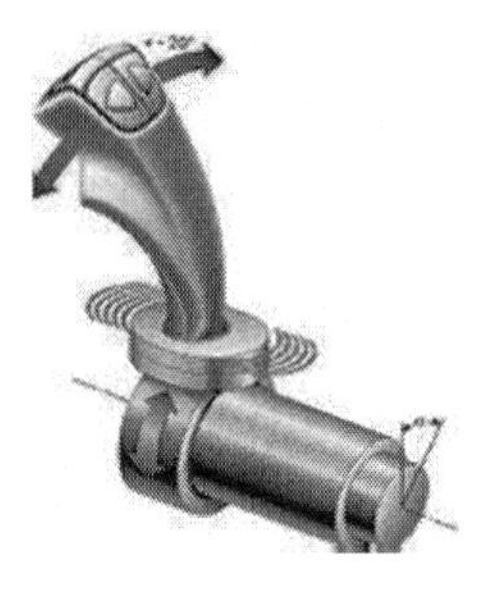

Fig. 4.8 Principle layout of a unique sidestick (joystick) [DAIMLERCHRYSLER 2004].

At the same time, an integrated E-M motor gives the driver a more realistic feeling of steering resistance. A two-dimensional force-measuring sensor that reacts to for- or backward hand pressure and registers commands to accelerate or brake.

The SBW AWS conversion mechatronic control system takes over control of the ECE or ICE plus braking and steering functions. In this manner, it may control the vehicle, as the driver would wish, even in a situation where the driver might not be able to react in time. Today, there are in normal driving operations *'SBW AWS'* vehicles of the future without steering HW, acceleration, or brake pedal that are steered only by a sidestick, as shown in Figure 4.9 [DAIMLER CHRYSLER 2004].

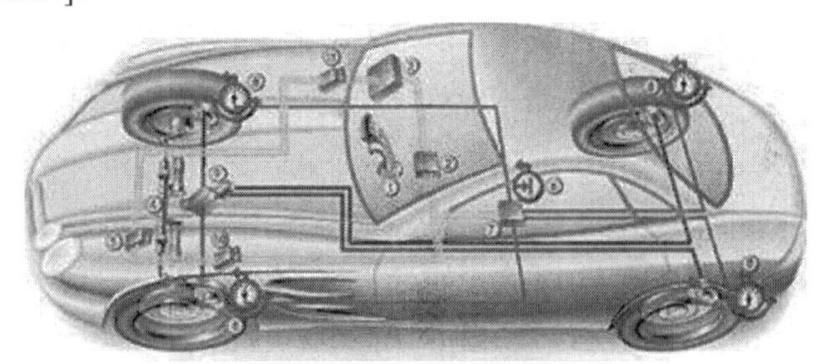

Fig. 4.9 Principle layout of the proof-of-concept automotive vehicle [DAILMLERCHRYSLER 2004]

Appearances are deceiving:

- The easy-to-use sidestick is based on a complex mechatronic system with redundant safeguards;
- An SBW AWS conversion mechatronic control system consists of sensors and control elements connected by a redundant data bus (black);
- The driving dynamics controller plays an important role here, actively taking over when the driver loses control of the vehicle;
- All mechatronic components have a backup system to ensure maximum safety.

To achieve this ambitious objective, automotive scientists and engineers disconnected the fluidical (hydraulical and/or pneumatical) and mechanical connections and replaced them with servo E-M motors and electronic switching elements.
Both types of mechatronic components are controlled by a fault-tolerant microcomputer system. The latter receives its data not only from the driver, who issues commands to the system, but also from sensors that continually monitor the vehicle's status.
'SBW AWS' automotive vehicles of the future may offer numerous benefits:

- Their safety systems react automatically to potentially dangerous driving situations within fractions of a second;
- The push of a button on a sidestick (side-mounted joystick) may be sufficient to make parking and other difficult manoeuvres child's play;
- The advanced concept may also enable designers to completely revamp automobile interiors.

High values of vehicle velocity, tight curves, and wet cobblestones -- even experienced racing car drivers would struggle with the steering under such conditions.

As far as automotive scientists and engineers are concerned, two fingers are sufficient to control the vehicle; the driver is driving with a hand-sized sidestick, or side-mounted joystick. The driver's left elbow may be supported against the centrifugal force by an arm console in the door and the right elbow may rest on the centre console. The driver literally has a handle on the vehicle.

The most important driving operations -- steering (turning), accelerating (driving), decelerating (braking), signalling, and honking the horn -- are integrated, for instance, into two sidesticks in the vehicle's armrests, as shown in Figure 4.10 [DAIMLERCHRYSLER 2004].

Fig. 4.10 View of two sidesticks in the vehicle's armrests [DAIMLERCHRYSLER 2004].

Much like a modern jet fighter, the automotive vehicle may be accelerated by lightly pushing forward a compact joystick.

Once the vehicle is en route, the integrated **adaptive cruise control** (ACC) and anti-skid **adaptive traction control** (ATC) automatically maintains its velocity. When the driver wants to brake, he simply pulls back the sidestick (see Fig. 4.11) [DAIMLERCHRYSLER 2004].

Fig. 4.11 The SBW vehicle may be accelerated by lightly pushing forward compact joysticks that are linked electronically [DAIMLERCHRYSLER 2004].

A significant safety feature of SBW AWS is that, unlike the ESP currently in use, it may be extended to act on the steering as well as on the brakes (See Fig. 4.12) [DAIMLERCHRYSLER 2004].

Fig. 4.12 View of the sidesticks' comfortable location [DAIMLERCHRYSLER 2004].

A test procedure in which strong winds are directed at the side of a vehicle also demonstrates the effectiveness of the stabilising algorithms. When a wind corridor is reached, a conventional vehicle immediately swerves off course and the driver must steer accordingly to counteract its effect. With SBW AWS, drivers hardly notice the wind at all. This is because the sensors immediately register the deviation it causes, while the computer already *'knows'* direction the driver wants to take due to the position of the sidestick.

As a result, the wheels are automatically turned in the right direction to offset the effect of side wind. With the stabilising algorithm of SBW AWS, however, sensors, computers, and actuators react so quickly that the vehicle neither skids nor swerves out of the lane. Instead, it maintains the desired course.
Passive Safety - Integrating driving functions into a sidestick offers additional passive safety benefits, too. It is well known that; if there's no steering column, then there's also no danger of chest injuries often caused in an accident and if the driver's foot can no longer get caught in one of the pedals during a collision, then the number of foot injuries may also be reduced.
Braking Speed - But the sidestick concept's ace in the hole is braking speed. In order to brake, drivers of conventional vehicles require an average 0.2 s to move their foot from the acceleration pedal to the brake pedal. At a speed of circa 50 km/h (30 mi/h), this translates into an additional braking distance of roughly 2.9 m (9.5 feet). The quicker reaction time of the sidestick system could therefore prevent many collisions [DAIMLERCHRYSLER, 1998-2004].
Automatic Vehicle Velocity Adjustment -- SBW AWS, whether via sidestick or mechatronic steering wheel -- also offers a great advantage. Because there is no longer a direct mechanical or fluidical (hydraulical) connection between the driver and the wheel, the steering ratio may be automatically adjusted to the vehicle's velocity. Summing up, SBW AWS conversion mechatronic control systems have the following advantages:

- Mechanical elements such as steering column removed;
- Position and torque sensors, plus an E-M motor for generating road feedback signals used on the driver's side;
- E-M actuators used to turning wheels;
- Mass set aside and an enhanced design flexibility effect;
- Systems such as active steering may be implemented entirely in mechatronic control software;
- Potential to extend the concept to automatic mechatronic control of rear wheels for enhanced turn radius and vehicle handling and stability.

Of course, **hybrid-electric** (HE) or **all-electric** (AE) traction presents very interesting features that well integrate with RBW or XBW very advanced automotive technology in fact, beyond the obvious considerations on environmental compatibility, motion distribution, and traction control may be more easily attained, especially in the case of independently powered wheels and/or **four-wheel steering** (4WS) capabilities. It must also be supposed that, though currently the driver is always in the loop, such systems are highly automated and therefore well establish for future extensions that may let autonomous guidance or even lead to the development of mobile telerobotic systems. To this purpose, this section describes a driving dynamics controller for a hypothetical vehicle '*Poly-Supercar*' (see Fig. 4.13): such a design uses the independent wheel motorisation to maximise the SBW AWS command [FIJALKOWSKI 1995].

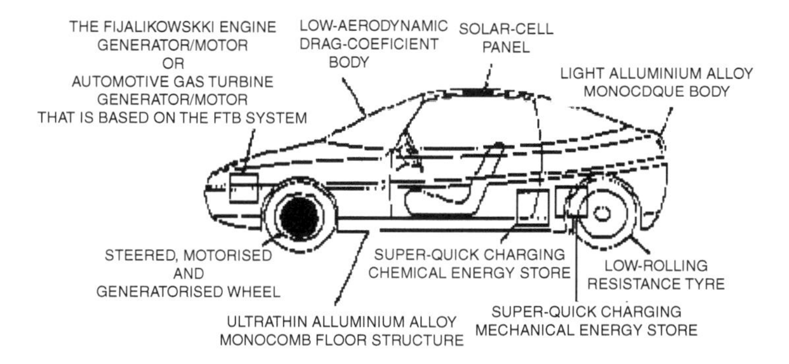

Fig. 4.13 Principle layout of a high-performance, all-round energy-efficient tri-mode automotive vehicle termed '*Poly-Supercar*' [FIJALKOWSKI 1995].

A particular physical model of the vehicle is fit out for motion-impaired drivers who should be able to use a conventional wheel-chair joystick to drive the vehicle along the desired path: the upper level of the mentioned controller deals with this kind of command interfacing the driver.

The hypothetical vehicle, termed '*Poly-Supercar*', has been studied by developing an **eight-degrees-of-freedom** (8-DoF) physical model to assess the performance of a RBW or XBW automotive unified chassis mechatronic controller by computer simulation, according to the different required performances or available sensing data [FIJALKOWSKI 1995].

The application of mechatronic means that a vehicle's mechano-fluido-mechanical (M-F-M) fluidically powered steering (FPS) electro-mechano-fluido-mechanical (E-M-F-M) electro-fluidically powered steering (EFPS), and electro-mechanical (E-M) electrically powered steering (EPS) SBW AWS conversion, as well as DBW AWD propulsion mechatronic control systems are, so far, increasing. The construction and normal action of such SBW AWS conversion as well as DBW AWD propulsion mechatronic control systems have not yet been clearly explained. Therefore, the data that proceeds is planned originally for modern SBW AWS conversion mechatronic control systems. These systems have formerly been announced and directed fully at the mechatronic means of those SBW AWS conversion mechatronic control systems and do not contain discourses that describe the fundamental function of the M-F-M FPS SBW AWS conversion mechatronic control systems.

In an oversteer situation, the turning round of a vehicle exceeds the intended steering angle and the vehicle spins out of control.

In the so-termed SBW AWS conversion mechatronic control system the vehicle's microcomputer-based ECU is able to correct over- or understeer by using the ABS or BBW AWB dispulsion mechatronic control system to apply the brake to a single particular wheel.

Therefore, an opposite torque is applied to the vehicle to '*twist*' it back in the right direction.

The onset of an over- or understeer event is detected by comparing the vehicle yaw angle with the driver's intended steering angle. This operation is carried out many times per second under microcomputer-based ECU.

Summing up, there are five types of PS SBW AWS conversion mechatronic control systems. Currently, FPS SBW AWS conversion mechatronic control systems are widely used, but EPS SBW AWS conversion mechatronic control systems are rapidly coming into use, primarily in small passenger vehicles.

Automotive vehicle manufacturers are ready to launch an innovative E-M steering mechatronic control system that would replace conventional M-F-M FPS SBW and/or 4WS conversion mechatronic control systems. Sensors measure vehicle velocity and driver torque on the steering shaft.

An ECU analyses the data and determines the sense of direction and amount of steering assistance required. Then, the **neuro-fuzzy** (NF) microcontroller generates a command to a variable-velocity, 14 V_{DC} or 42 V_{DC} steer-actuator E-M motor that drives a gear mechanism to provide the required assistance. For instance, the E-M EPS reduces **specific fuel consumption** (SFC) by eliminating the drain of M-F-M FPS -- power steering **mechano-fluidical** (M-F) pumps or **mechano-pneumatical** (M-P) compressors, hoses, oily-fluids or gas (air), drive belt and pulley -- on the ECE or ICE.

It is also easier to locate in the ECE or ICE compartment, simplifies tuning, and cuts assembly time by up to 3.5 min compared with conventional M-F-M FPS SBW 2WS conversion mechatronic control systems.

The E-M EPS SBW 2WS conversion mechatronic control system is more expensive but easier to assembly and easier to package in the ECE or ICE compartment. Therefore, it is a cost-saving, fuel-efficient device.

It is now about 10 -- 20% more expensive than M-F-M FPS SBW 2WS and/or 4WS conversion mechatronic control systems, but the price may soon be equivalent.

Vehicle manufacturers think that E-M EPS may overcome M-F-M FPS. E-M EPS is standard on several versions of the very advanced vehicles and optional on the remainder. This alternative to E-M FPS was developed and is up to 5 kg lighter, easier to install, and consumes 2 -- 5% less fuel. It only consumes power when necessary, unlike an M-F-M FPS that continuously powers its M-F pump.

The objective of this section is to impart user-prescribed dynamics upon a vehicle travelling at high velocity values under normal operation.

Emergency manoeuvres and abrupt alterations in vehicle velocity are not taken into consideration, thus a relatively simple physical model of an automotive vehicle has been used.

The physical model chosen, a **two-degrees-of-freedom** (2-DoF) one, commonly referred to as the single-track half-vehicle (bicycle) physical model, encompasses the dominant dynamics of the vehicle under prescribed conditions [LYNCH 2000].

The vehicle states are yaw rate ($\dot{\psi}$ or r) and lateral velocity (V). Figure 4.14 illustrates the vehicle states associated with a single-track half-vehicle (bicycle) physical model [LYNCH 2000].

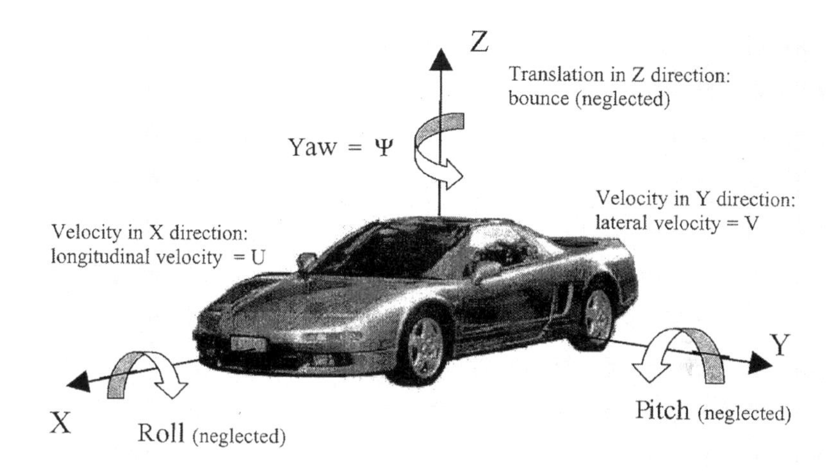

Fig. 4.14 Description of an automotive vehicle states [LYNCH 2000].

Here we present an approach to control the lateral steering and longitudinal propelling (tractive) forces in wheel-tyre/terrain contacts with a minimally required frictional coefficient.

Making use of examples drawn from SBW 2WS or SBW 4WS conversion as well as DBW 2WD or 4WD propulsion, the author demonstrates the various features of the optimisation. This study may form a useful basis for an on-board RBW or XBW automotive integrated unibody, space-chassis, skateboard-chassis or body-over-chassis mechatronic control hypersystem that complements the driver of an automotive vehicle to perform the manoeuvre so as to require minimal friction.

Automotive responsibilities necessitate a variety of steering activity: one is highway driving with minor turnings for hundreds of kilometres; another is urban handling that requires agile turning.

This section explains the responsibility of steering and propulsion (traction) for a variety of steering activities.

The knowledge gained of steering forces and torques (moments of force) as termed by the **Society of Automotive Engineers** (SAE), endows us with an introduction to the terminology and the forces and torques required on an automotive vehicle during a turn.

Even if a standard automotive integrated chassis is intended for much higher values of vehicle velocity than any off-road vehicles, many of the forces and torques are identical.

Vehicle velocity is the time rate of change of position of a vehicle; it is a vector quantity having a sense of direction as well as magnitude.

Vehicle speed is the time of change of position of a vehicle without regard to a sense of direction, in other words, the magnitude of the vehicle-velocity vector.

For a **wheeled vehicle** (WV) the forces and torques enforced on the steering mechanisms follow from those created at the wheel-tyre-ground interface, as depicted in Figure 4.15 [WONG 1993].

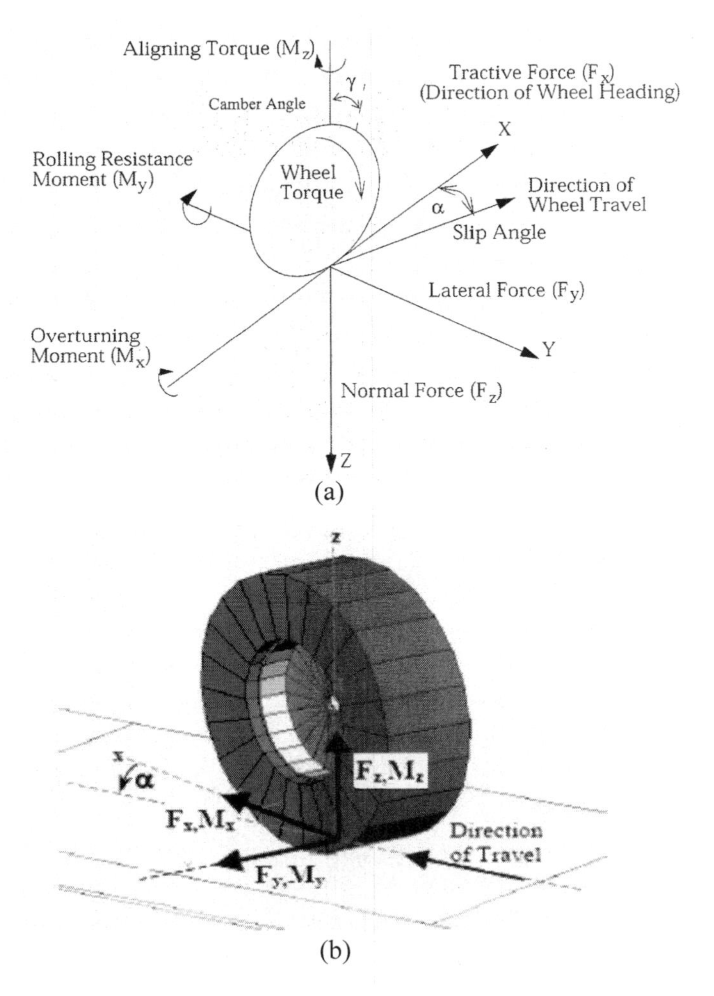

Fig. 4.15 Wheel forces and torques (moments of force)
[WONG 1993 (a); CHEN 2006 (b)].

The coordinate system is found at the base of the wheel where the *X* coordinate is in the direction of wheel travel.

The Y coordinate is parallel to the axis of the wheel's rotation and the Z coordinate is perpendicular to the terrain.

The wheel torque is created around the axis of rotation and is opposed by the rolling resistance moment of rotation.

The aligning torque opposes any alteration in the heading of the wheel around the Z axis.

The overturning moment of rotation opposes any lateral forces created as the wheel slides in the Y direction during a turn.

The slip angle α is termed as the difference between the direction in which the wheel is heading and the direction of wheel travel.

When a propelling (driving) torque affects a wheel, the distance that the wheel-tyre travels is less than that travelled by a wheel-tyre rotating in an unloaded and free rolling situation. This observable fact is known as longitudinal slip and is expressed by the following equation [WONG.93]:

$$i = \left(1 - \frac{V}{r\ \omega}\right). \tag{4.1}$$

Extreme or *1* longitudinal slip takes place when a wheel rotates without any translational motion. A universal theory to perfectly term the correlation between the propelling (driving) torque and the longitudinal slip does not exist.

Steering Configurations - One study of steering configurations, carried out in relation to WVs, concluded that no one optimum steering exists for all applications. The optimal steering may be set up as a function of operating circumstances, particular automotive responsibilities, service life, as well as cost constraints for manufacture.

The innovative tendency in optimising steering is the application of load sensors that optimise the fluidical or pneumatical pressure or electric current indispensable to actuate the wheels.

Manoeuvrability is dependent not only on the steering mechanism but also on steering mechatronic control [DUDZINSKI 1989].

The objective of the steering mechanism is to assure manoeuvrability and vehicle stability.

The major features of the evaluation are found in manoeuvrability, stability, propulsion (traction), and design complication. Energy draw is not a decisive metric for the automotive industry as fuel costs are low. Precise positioning for a manned vehicle is not indispensable because the driver is able to account for the vehicle's fuzziness. For an unmanned vehicle, energy draw and positioning become most decisive factors.

Steering Configuration Kinematics -- The analysis of the kinematics of different steering configurations lets the properties of the different steering modes be pragmatic in terms of various performance metrics. For instance, the unavailable volume, indispensable for accepting different steering can be considered by kinematics analysis.

On the other hand, any kinematic study is an idealised analysis in view of the fact that the wheel-tyre/ground interaction is not taken into consideration. Straight driving helps as a basis for evaluation with steady-state turning. A force diagram for straight driving, as is shown in Figure 4.16, depicts the force indispensable in turning a wheel which is equal and opposite to the resistance delivered by the terrain [BEKKER 1964].

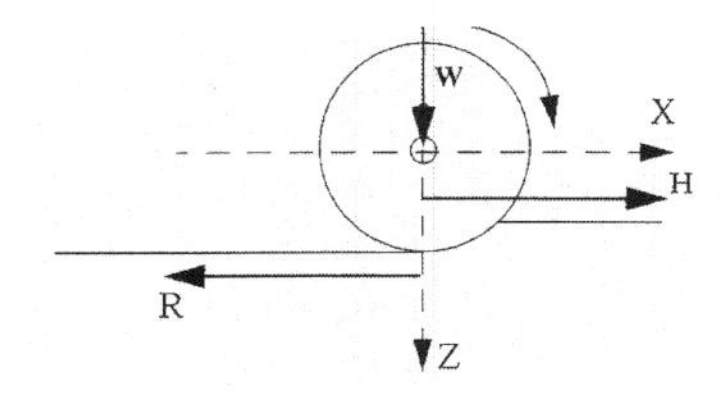

Fig. 4.16 A force diagram for straight driving [BEKKER.1964].

The thrust indispensable to vehicle automation is derived between the wheel-tyre and the terrain. The terrain delivers the resistance indispensable for locomotion until the terrain (soil) shears. The shearing conclusively creates a slip circumstance in which the wheel spins without delivering any vehicle thrust. The resolution of resistance forces is reliant on a physical and geometrical relationship between terrain and the vehicle [BEKKER 1964].

A double-track full-vehicle physical model for straight driving is depicted in Figure 4.17 [SHAMAH 1999].

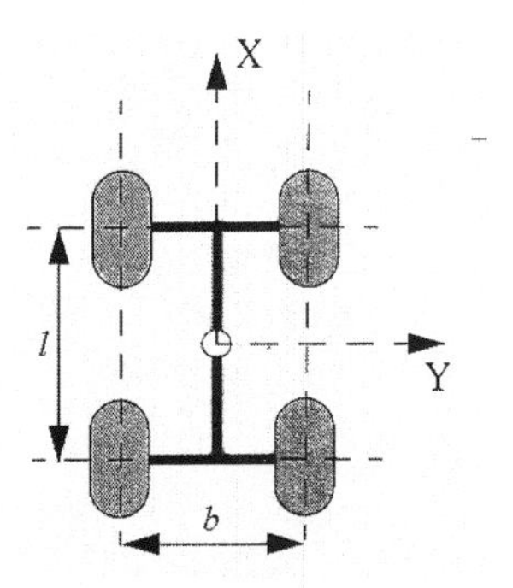

Fig. 4.17 A double-track full-vehicle physical model for straight driving [SHAMAH 1999].

The relationship has been resolved with the implementation of vehicle and terrain parameters. The terrain parameters can be resolved by means of a standardised set of tests.

Single Axle SBW 2WS Conversion Kinematics -- At present, for conventional automotive vehicles, the most common steering configuration is a single axle SBW 2WS conversion mechatronic control system in which two wheels are pivoted; that is, the coordinated *Ackerman* 2WS that mechanically coordinates the angle of the front two wheels.

In order to continue all wheels in a pure rolling situation during turning, the wheels necessitate following curved paths with different radii originating from a common centre.

A double-track full-vehicle physical model of the single axle SBW 2WS conversion is depicted in Figure 4.18 [SHAMAH 1999].

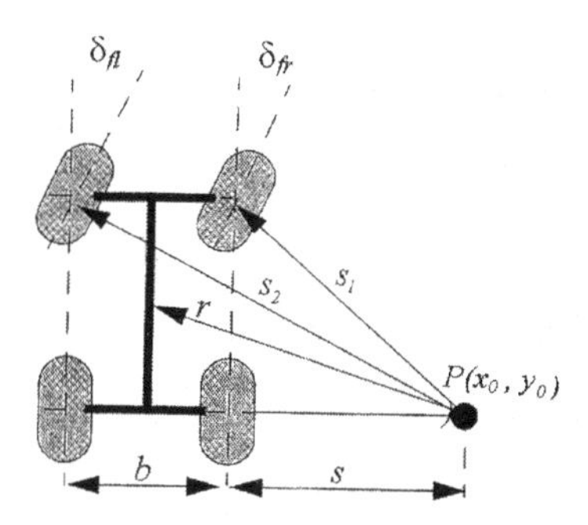

Fig. 4.18 A physical model of the single axle SBW 2WS conversion [SHAMAH 1999].

In order to minimise lateral forces on the wheel-tyres during the turn, all wheels should be in a rolling condition. The wheels must follow curved paths with different radii originating from a common centre.

The relation between the steering angle of the inside and the outside front wheels may be obtained using physics and geometry as follows [SHAMAH 1999]:

$$s = \sqrt{r^2 - \left(\frac{l}{2}\right)^2 - \frac{b}{2}}, \tag{4.2}$$

$$\delta_f = \tan\left(\frac{l}{s}\right), \tag{4.3}$$

$$\delta_r = \tan\left(\frac{l}{b+s}\right). \tag{4.4}$$

Since the outer wheels travel a longer path distance than the inner wheels, the velocity components must be distributed to match the path lengths.

<u>Double Axle SBW 4WS Conversion Kinematics</u> – A smart SBW 4WS conversion mechatronic control system offers greater manoeuvrability than single axle SBW 2WS conversion by moving the turn centre closer to the centre of the vehicle. A SBW 4WS vehicle accomplishes half the turn radius of a SBW 2WS vehicle for the same alteration in wheel heading.

A double-track, full vehicle physical model of the double axle SBW 4WS conversion is depicted in Figure 4.19 [SHAMAH 1999].

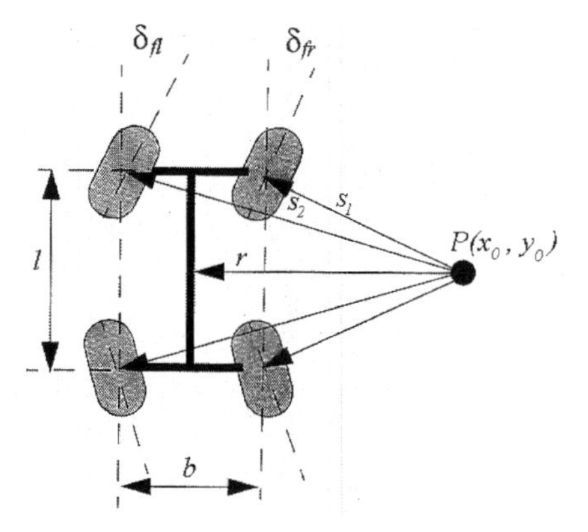

Fig. 4.19 A double-track, full-vehicle physical model of the double axle SBW 4WS conversion [SHAMAH 1999].

The relation between the steering angle of the inside front or rear wheels and the outside front or rear wheels can be obtained using geometry as follows [SHAMAH 1999]:

$$\delta_f = \tan\left(\frac{l/2}{r - b/2}\right), \tag{4.5}$$

$$\delta_r = \tan\left(\frac{l/2}{r + b/2}\right). \tag{4.6}$$

Skid SBW 4WS Conversion Kinematics – A skid SBW 4WS conversion mechatronic control system can be compact, light, require few parts, and exhibit agility from point turning to line driving using only the motions, components, and swept volume needed for straight driving. Skid SBW 4WS is achieved by creating a differential thrust between the left and right sides of the vehicle, thus causing an alteration in heading. Thus, skid SBW 4WS is accomplished by creating differential angular velocities between the inner and outer wheels, as is shown in Figure 4.20 [SHAMAH 1999].

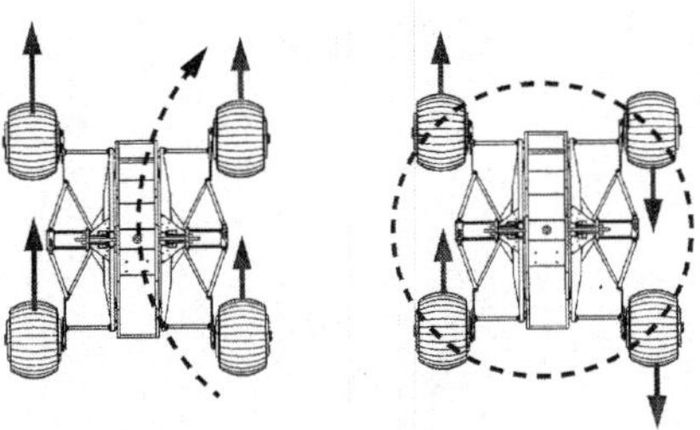

Fig. 4.20 Principle layout of skid SBW 4WS conversion [SHAMAH 1999 – *Nomad*].

The downside is that skidding causes unpredictable mechanical energy requirements because of terrain irregularities and non-linear tyre-terrain intersection. Skid SBW 4WS also fails to achieve the most aggressive steering that can be achieved with independent explicit SBW 4WS because the maximum forward thrust is not maintained during a turn. For skid SBW 4WS, the motion of the wheels is limited to rotation about one axis. Therefore a centralised drive can pass the propelling (tractive) torques directly to each wheel. Hence, skid SBW 4WS doesn't work for vehicles that are longer than they are wide.

The kinematics analysis of the skid SBW 4WS conversion mechatronic control system allows a preliminary determination of wheel angular velocities given the vehicle's dimensions, the desired radius, and the desired turn rate.

However, as in the previous kinematics physical models, no forces are studied. Therefore, the slippage that is more prevalent in skid SBW 4WS conversion is not accounted for; thus the kinematics double-track, full-vehicle physical model is even less accurate, as is shown in Figure 4.21 [SHAMAH 1999].

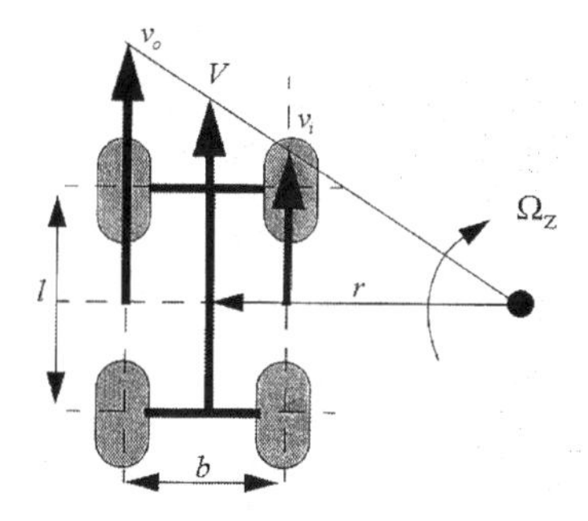

Fig. 4.21 A double-track, full-vehicle physical model of the skid SBW 4WS conversion [SHAMAH 1999].

The radius of the turn may be calculated from similar triangles [WONG 1993] as follows:

$$\frac{v_o}{v_i} = \frac{r + b/2}{r - b/2}, \tag{4.7}$$

$$r = \frac{b}{2}\frac{(v_o/v_i + 1)}{(v_o/v_i - 1)} = \frac{b}{2}\left(\frac{v_o + v_i}{v_o - v_i}\right). \tag{4.8}$$

However, this radius may only be achieved if no slippage occurs between the wheel and the ground. In order to account for the slippage of the outer wheels i_o and the inner wheels i_i, this radius may be

$$r' = \frac{b}{2}\left[\frac{v_o(1-i_o)+v_i(1-i_i)}{v_o(1-i_o)-v_i(1-i_i)}\right] \tag{4.9}$$

The turn rate or yaw angular velocity can be found from the following:

$$\Omega_z = \frac{v_o + v_i}{2r} = v_i \frac{(v_o/v_i - 1)}{b} \,. \tag{4.10}$$

Again, in order to account for the slippage, it may be

$$\Omega_z = \frac{v_o(1-i_o)+v_i(1-i_i)}{2r'} = v_i \frac{(1-i_o)v_o/v_i - (1-i_i)}{b} \,. \tag{4.11}$$

Given an accurate slippage physical model, the kinematics physical model can be used to provide accurate results. Without a longitudinal slip physical model, wheel angular velocities and turn radius can only be assumed to be estimates.
Independent explicit SBW 4WS Conversion Kinematics – An independent SBW 4WS conversion mechatronic control system explicitly articulates each of the wheels to the desired heading. Thus, explicit SBW 4WS is accomplished by changing the heading of the wheels to cause a change in heading of the vehicle, as is shown in Figure 4.22 [SHAMAH 1999].

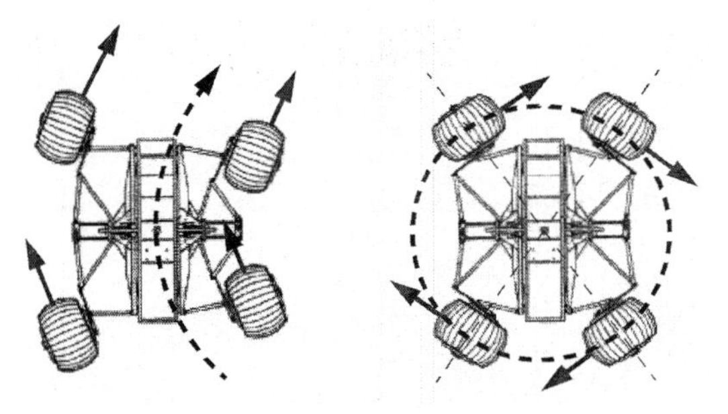

Fig. 4.22 Principle layout of independent explicit SBW 4WS conversion [SHAMAH 1999 – *Nomad*]

Apart from the issues of actuation complexity and accuracy of coordination mechatronic control, this steering configuration provides advantages to the manoeuvrability of all-terrain vehicles, especially those operating in very rough terrains.

A common variation of independent SBW 4WS conversion, not attainable by the other steering configurations, is crab steering in which all wheels turn by the same amount in the same direction. As a result, the all-terrain vehicle may move in a sideways fashion.

Coordination of SBW 4WS and DBW 4WD allows efficient manoeuvring and reduces the affect of internal losses due to actuator fighting.

Advantages of an explicit SBW 4WS conversion mechatronic control system include more aggressive steering with better dead reckoning (due to less slip of the wheels) and lower power consumption.

The downside of an explicit SBW 4WS conversion is a higher actuator count, part count, and the necessary swept volume.

Independent explicit SBW 4WS is more difficult since the wheels move the torque transmission about two axes.

If a centralised drive is used, the propelling (traction) torque must pass through universal joints and drive shafts that are inefficient.

Another approach for explicit SBW 4WS is to use individual drive FM, PM or EM motors inside each wheel with the necessary gearing.

Frame Articulated SBW 4WS Conversion Kinematics – A frame articulated SBW 4WS conversion mechatronic control system is prevalent in all-terrain vehicles. The heading of the vehicle alters by folding the hinged chassis units.

For all-terrain vehicles, frame articulated SBW 4WS has the advantage of allowing the vehicle to be significantly more manoeuvrable than a vehicle with coordinated SBW 2WS (*Ackerman modus operandi*).

Frame articulated SBW 4WS conversion has the advantage over skid SBW 4WS conversion in that, during a turn, the maximum value of thrust provided by the traction elements is maintained.

The primary advantages of frame articulated SBW 4WS conversion are greater manoeuvrability in limited space and minimal gross energy necessary to turn on soft ground in comparison to a knuckle SBW 2WS (*Ackerman modus operandi*) automotive vehicle. However, these articulated vehicles are commonly equipped with wheels.

Articulated WVs conversions have been recently developed in order to improve steering performance of skid SBW 4WS WVs.

A full-vehicle physical model, as depicted in Figure 4.23, shows the coordinate system and kinematics of the articulated WVs on level ground [DUDZINSKI 1999; WATANABE ET AL. 2000].

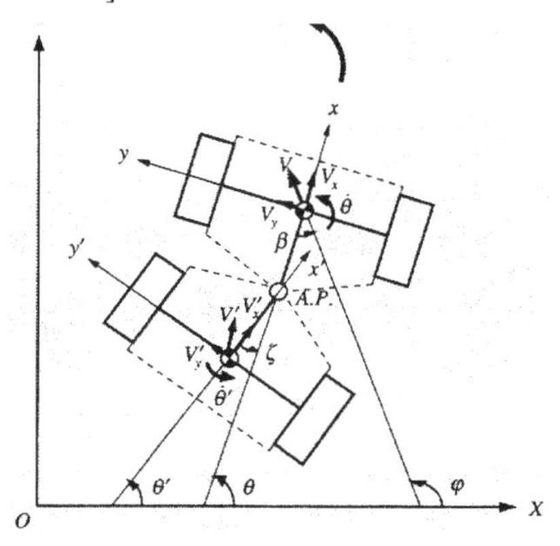

Fig. 4.23 A full-vehicle physical model of the frame articulated SBW 4WS conversion [DUDZINSKI 1999; WATANABE ET AL. 2000].

The reference coordinate system (X, Y, Z) is fixed on the ground and moving rectangular coordinate systems (x, y, z) and (x', y', z') are attached to the mass centre of the front and rear vehicle-frames.

The turning conditions, such as the running vehicle velocity, the steering angle, and the steering time affect the steering torque required to steer the WV. However, the required steering mechanical energy is relatively small compared with the required driving mechanical energy of the WV.

The turning radius of the WV is not affected very much by the mass ratio of front to rear vehicle frames. However, the relative tractive efforts (forces) of each wheel are influenced considerably.

Axle Articulated SBW 2WS Conversion Kinematics -- This mechatronic control system is performed by adding a free pivot to one of the vehicle axles, as is shown in Figure 4.24 [SHAMAH ET AL. 2001]. This steering configuration is common in wagons and carts.

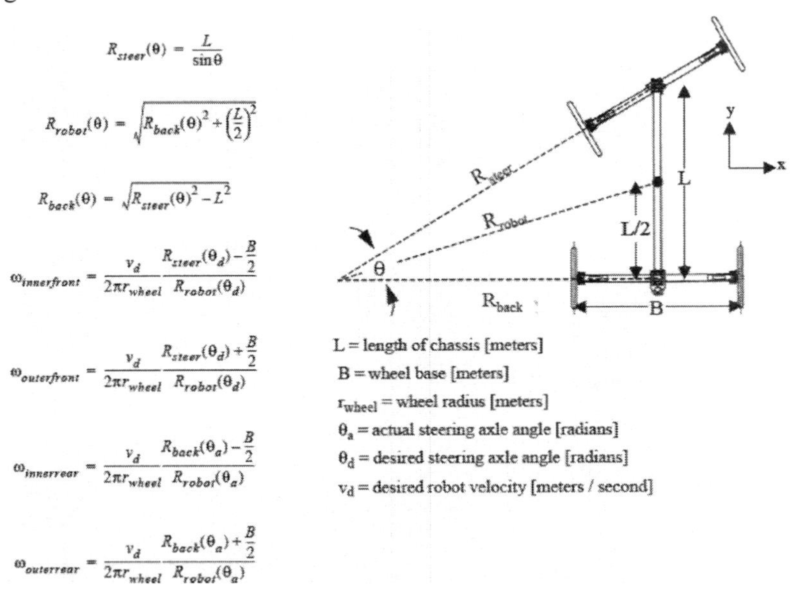

Fig. 4.24 Steering kinematics and equations of a full-vehicle physical model of the axle articulated SBW 2WS automotive vehicle [SHAMAH ET AL. 2001].

One disadvantage of single axle SBW 2WS is that the wheels run on separate tracks when going around curves. Under difficult ground conditions, this requires increased drive propulsion as each wheel is running over fresh terrain [DUDZINSKI 1989]. The advantages include mechanical simplicity, relatively low steering mechanical energy, and moderate manoeuvrability. The innovation of the design implemented on an experimental vehicle is to mechatronically control the angular velocity of the front wheels to maintain a desired angle of the front axle [SHAMAH ET AL. 2001].

An added benefit to the steered front axle is the ability to point the perception sensors (including: stereo IR thermovision cameras, a laser finder, and a panoramic camera) with the front axle. This sensor pointing increases the effective horizontal field of view [KELLY AND STENTZ 1998] resulting in more robust autonomous navigation. The axle articulated SBW 2WS joint is composed of two free rotations, as is shown in Figure 4.25 [SHAMAH ET AL. 2001]. The first is about the vertical axis that allows the alteration in the heading of the front axle. The second rotation allows a roll motion of the front axle that is necessary to enable all four wheels to contact the ground over rough terrain. When a vehicle travels at relatively low velocity values, then a spring suspension may be not used.

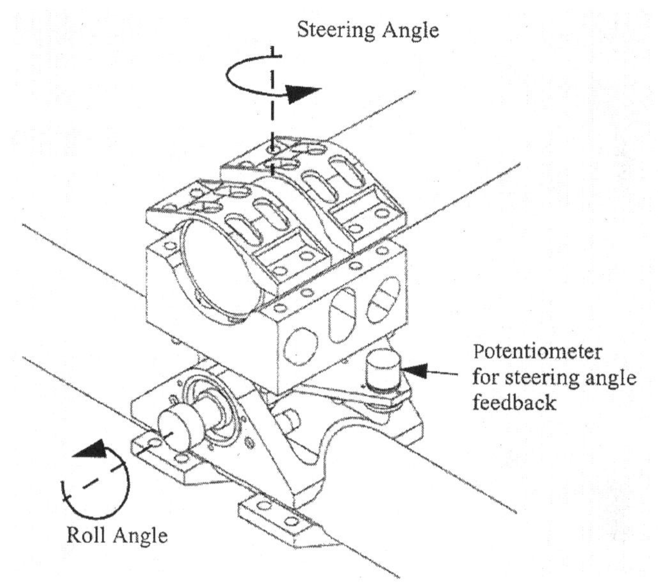

Fig. 4.25 Layout of the front axle steer and roll pivot for the axle-articulated SBW 2WS joint [SHAMAH ET AL. 2001].

This steering configuration offers the mechanical simplicity of conventional skid SBW 4WS while maintaining the steering performance of a frame-articulated vehicle. The absence of a steering actuator further reduces mass. However, the trade-off in creating such a mechanically simple steering mechanism lies in the added complexity of the mechatronic control system.

In comparison with the skid SBW 4WS WV, the required driving forces of wheels and wheel-tyre slippage during steering are significantly decreased in articulated WVs, resulting in great enhancement of the steerability, especially on soft ground. Consequently, articulated WVs significantly gain in mobility on soft terrain that would be impossible for the skid SBW 4WS WVs.

<u>SBW 6WS Conversion Kinematics</u> - The SBW AWS mechanism also may be used to steer multiple axles in close proximity to another, specifically multi-axle trucks [DJH 2003].

In SBW 6WS WVs (similar as in having three axles that all steer, instead of two axles) a single actuator may be used to steer the rear two axles (Fig. 4.26).

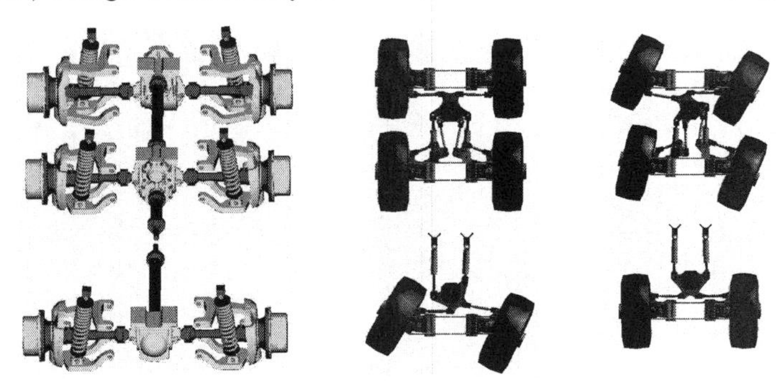

Fig. 4.26 A full-vehicle physical model of the 6WS SBW conversion [DJH 2003].

Reducing the system to a single actuator for the rear two axles greatly simplifies the SBW 6WS conversion mechatronic control system design. For contra-steering (as shown in Fig. 4.26), the rear wheels do not have the same steer angle to create a common turn centre. The relationship between them may be derived through geometric studies. If automotive designers implement a fixed relationship between the two rear axles in a turn, the in-phase steer would be compromised, as the angles would need to be equal.

SBW 8WS Conversion Kinematics -- The SBW 8WS conversion consists of eight independent steering systems, each of them all **electro-mechano-fluido-mechanical** (E-M-F-M) position servo systems. It has six kinds of working stations, and is shown in Figure 4.27 [YUNHUA ET AL. 2006]. Among of them, the *V*-axis steering is shown in Figure 4.28 [YUNHUA ET AL. 2006].

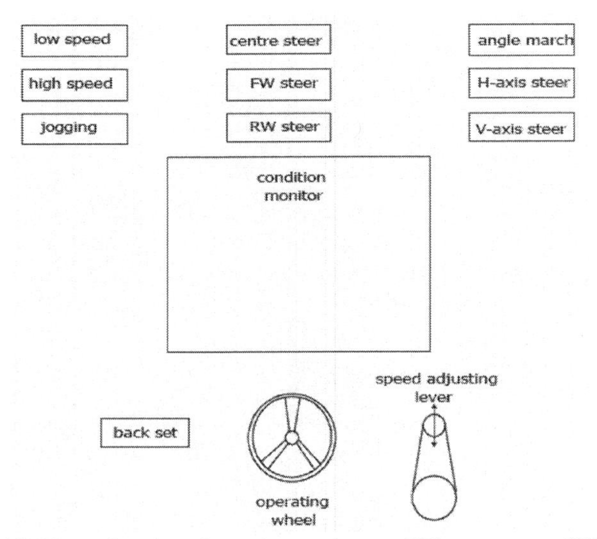

Fig. 4.27 Human interface-electronic operating panel [YUNHUA ET AL. 2006].

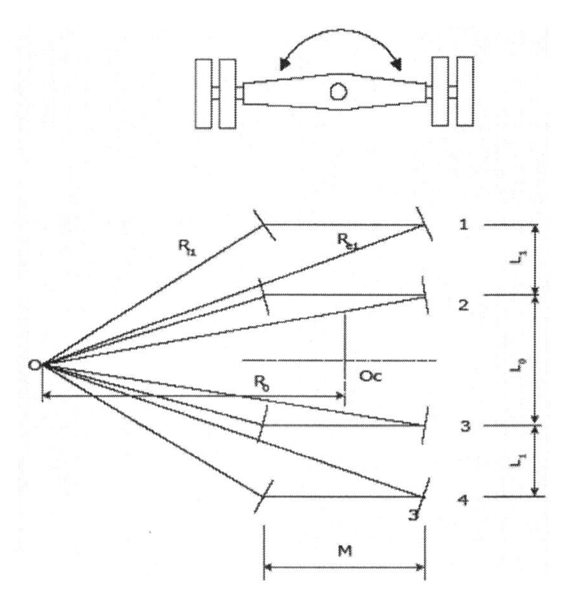

Fig. 4.28 A full-vehicle physical model of the 8WS SBW conversion [YUNHUA ET AL. 2006].

When a driver operates the 8WS truck to steer, the button of the steering station is first pressed and then the operating wheel is rotated. When the 8WS truck is steered, a coordinated motion control to the travelling system and steering system is necessary.

Strictly speaking, this is an MIMO mechatronic control system. Here, for acquiring the simple controller construction and the fast processing time, the compound control law with order-compensating and internal loop feedback is used [YUNHUA ET AL. 2006].

Kinematics of Major Steering Configurations - One way to reduce vehicle mass is to limit the number of actuated motions. During the configuration stage, various steering and drive modes ought to be studied to determine the best combination of mobility and simplicity required to traverse the desired terrain. The kinematics of the most important steering configurations is shown in Fig. 4.29 [SHAMAH ET AL. 2001].

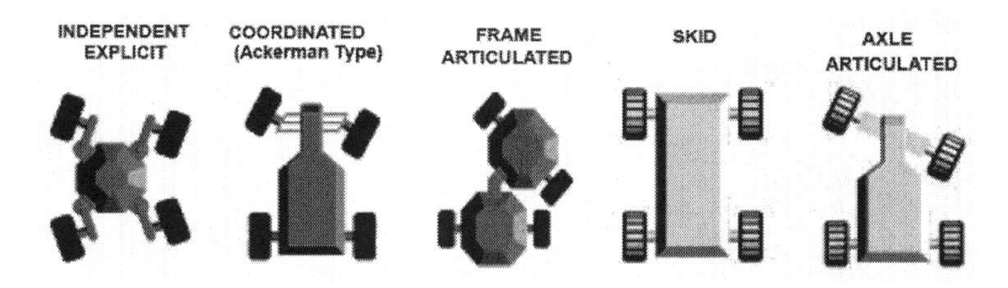

Fig. 4.29 Kinematics of major steering configurations [JEONG ET AL. 1994; SHAMAH ET AL. 2001]

Each steering configuration has advantages in certain areas, as shown in Table 4.1.

Table 4.1 Steering Configuration Evaluation for SBW 4WS and 2WS [SHAMAH ET AL. 2001]

	Manoeuvrability	Mechanical complexity	Control complexity	Propulsion power during steering manoeuvre	Number of actuated joints for steering
INDEPENDENT EXPLICIT SBW 4WS	Medium/High	Medium	Low	Medium	4
COORDINATED ACKERMAN SBW 2WS	Medium	Medium/High	Medium/Low	Medium/Low	1
FRAME ARTICULATED SBW 4WS	Medium	Low	Medium	Medium	1
SKID SBW 4WS	High	Low	Low	High	0
AXLE ARTICULATED SBW 2WS	Medium	Low	Medium/High	Low	0

What Everyone Should Know About Wheel Alignment **-** On the automotive vehicle shown in Figure 4.30, the front wheels are not aligned to the rear thrust line [HUNTER 2005]. This may happen from normal wear and stress, whether it has adjustable or non–adjustable rear suspension. To steer straight ahead, the driver would have to steer the front wheels slightly to the right.

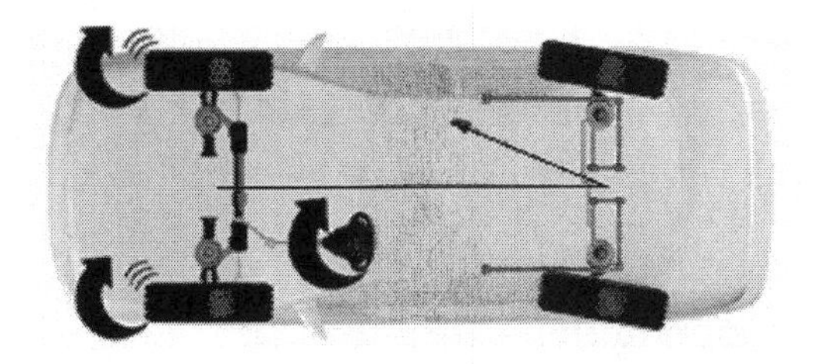

Fig. 4.30 Automotive vehicle with non-aligned front wheels [HUNTER 2005].

A common result would be that the vehicle would *'dog track'* and possibly *'pull'* to the side. Of course, the angles are exaggerated so the driver may more easily see the condition. But it takes only a small misalignment to create problems.

<u>*Why Four-Wheel Alignment?*</u>

- *Reduced Wheel-Tyre Wear* -- Improper alignment is a major cause of premature wheel-tyre wear. Over the years, a properly aligned vehicle may add thousands of kilometres (miles) to wheel-tyre life;
- *Better Liquid-Fuel Kilometreage (Mileage)* -- Liquid-fuel mileage increasees as rolling resistance decreases. Total alignment sets all four wheels parallel so that, along with proper inflation, it minimises rolling resistance;

- *Improved Vehicle Handling* -- Does the vehicle pull to one side? Does the steering HW vibrate? Does the driver constantly have to move the steering HW to keep the vehicle travelling straight ahead? Many handling problems may be corrected by total alignment. With all the system components aligned properly, on/off road shock is more efficiently absorbed for a smoother ride;
- *Safer Vehicle Driving* -- A vehicular suspension inspection is part of the alignment procedure. This allows customers to spot worn parts before they cause costly problems.

Here's what Automotive Designers do:

For vehicles with non–adjustable rear suspensions (See Fig. 4.31)

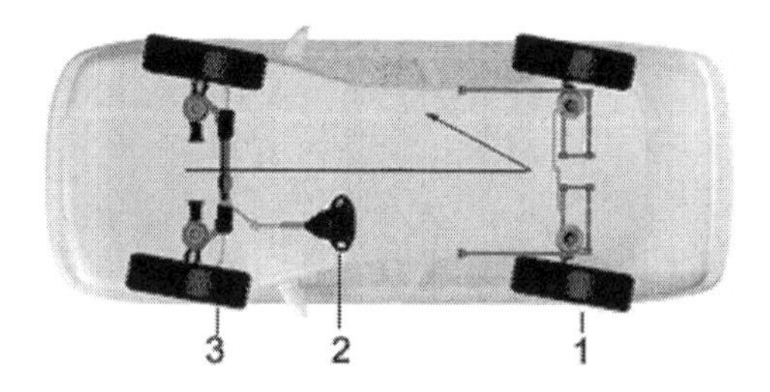

Fig. 4.31 Automotive vehicle with non-adjustable rear suspension: 1 - angle readings are measured at all four wheels; 2 - the steering HW is centred; 3- front wheels are referenced to the rear thrust line and set to specifications [HUNTER 2005].

Result: All four wheels are parallel and the steering HW is centred.

For automotive vehicles with adjustable rear suspensions (See Fig. 4.32)

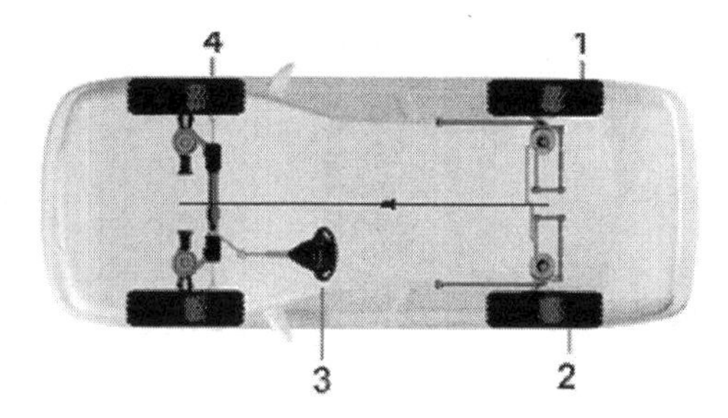

Fig. 4.32 Automotive vehicle with adjustable rear suspension: 1 -angle readings are measured at all four wheels; 2 - rear wheels are set to specification *(rear thrust line corresponds to the vehicle's centre line.); 3- steering HW is centred; 4 - front wheels are referenced to the rear thrust line and set to specification. [HUNTER 2005].

Result: all four wheels are positioned straight ahead and parallel, and the steering HW is centred.

4.2 Variable-Assist SBW 2WS Conversion Mechatronic Control Systems

4.2.1 Essentials of SBW 2WS Conversion Mechatronic Control Systems

An automotive M-F-M FPS, E-M-F-M EFPS or E-M EPS SBW 2WS conversion enhances steering performance and betters the feel of the steering and power-saving efficiency. It does so with conversion mechatronic control mechanisms that decrease the steering effort. An automotive FPS, EFPS or EPS SBW 2WS conversion mechatronic control system is joined to the M-F-M, E-M-F-M or E-M booster actuator, respectively. The intention of FPS, EFPS and EPS SBW 2WS conversion mechatronic control systems was originally, to reduce the steering effort when driving at low values of vehicle velocity and to provide a feedback loop for the proper steering reaction force when driving at high value of vehicle velocity. To reach those objectives, automotive velometers or speedometers are employed as vehicle velocity sensors to measure vehicle velocity according to any variations in the steering assist rate under circumstances differing between certain limits from steering manoeuvres at zero value of vehicle velocity to those at high values of vehicle velocity. Nevertheless, as vehicles became supplied with E-M booster actuators and smart E-M EPS SBW 2WS conversion mechatronic control systems, the affirmation for these systems initiated the diminution in power demands and superior performance. The important normal actions needed for SBW 2WS conversion mechatronic-control systems are listed in Table 4.2 [SATO 1995].

Table 4.2 Requirements for SBW 2WS Conversion Mechatronic-Control System

Attenuation of driver's nuisance when turning the hand steering wheel and betterment in the steering feel	Foolproof
➢ Decrease in steering effort ➢ Ease of steering operation ➢ Feedback of correct reaction steering forces ➢ Decrease of kickback [SATO 1991] ➢ Betterment in convergence [NISSAN 1991] ➢ Energy saving ➢ Formation of other innovative functions	Sustaining of manual steering function in the case of any malfunctions

Single Axle SBW 2WS Conversion Mechanism Normally, the aim of the single axle SBW 2WS conversion mechatronic control system is simply that of making available a mechatronic means whereby the driver may locate the vehicle as exactly as possible where the driver wants it to be situated on on/off road, for the selection of the course needed to steer round corners, and so that the driver can avoid other on- and/or off-road users and obstacles.

On the other hand, it must also keep the automotive vehicle stably on course irrespective of deformities in the surface over which the vehicle is roving. For the achievement of these basic aims, the first requirement is that, when the vehicle is moving very slowly, all the wheels should roll correctly, that is, without any lateral slip.

In Figure 4.33, motion of the wheel along YY is rolling; along XX it is slipping [NEWTON ET AL. 1989].

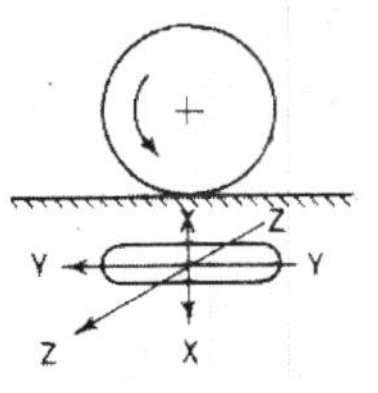

Fig. 4.33 Motion of the wheel [NEWTON ET AL. 1989].

In view of the fact that for all the wheels on a vehicle to roll in reality, they must all move in the same direction perpendicular to their lateral axes. These axes must all intersect at a common point.

If the vehicle is on a straight course, this point will be '*ad infinitum*', that is, the axes will be parallel. However, if the vehicle is turning a corner, this point will be, for all time, situated in the centre of the vehicle as the whole is turning, and the tighter the turn the closer it will be to the vehicle. If not both the front and rear wheels are to be steered --impractical on difficult terrain, except in particular conditions, for example on vehicles having more than eight wheels, where it may be practically unavoidable -- the common centre must be positioned everywhere along the lines of the axis created from the fixed rear axle. As may be seen from Figure 4.34 [NEWTON ET AL. 1989], this indicates that when the front wheels are steered, their axes must be turned through different angles and consequently the point $P(x_0, y_0)$ of their intersection is created each time on that axis.

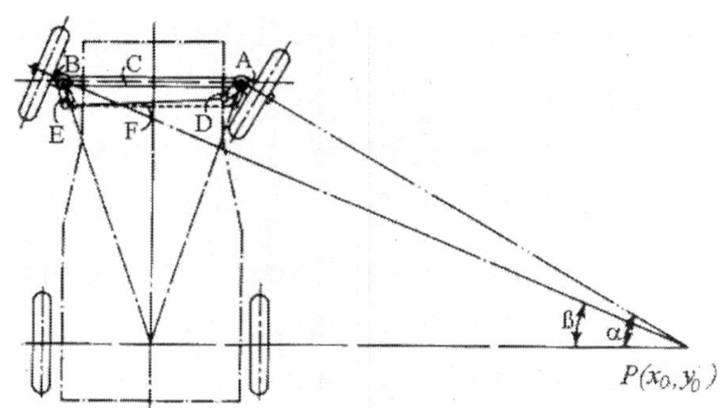

Fig. 4.34 A double-track full vehicle physical model of the coordinated *Ackerman* SBW 2WS principle [NEWTON ET AL. 1989].

A beam axle pivoting the whole axle assembly about a vertical axis midway between its ends can do this. On the other hand, such an arrangement is impractical for all but very slow vehicles. Normally, the wheels are carried on stub axles A and B in Figure 4.35 [NEWTON ET AL. 1989]. Except with autonomous suspension, these stub axles are pivoted on the ends of the axle beam C that, since it is linked by the road springs to the chassis frame, continues in reality parallel to the rear axle, as displayed in the illustration.

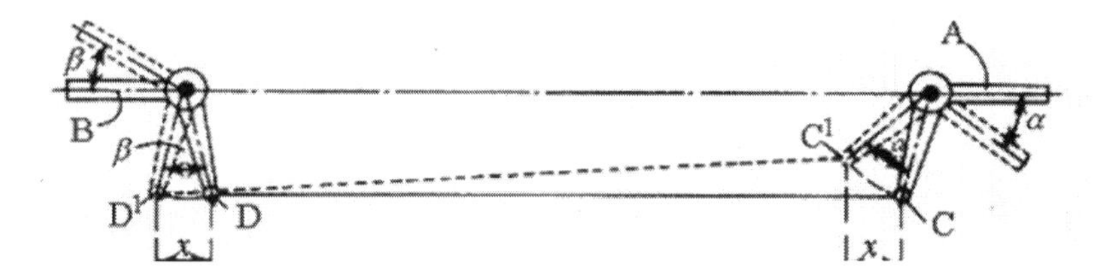

Fig. 4.35 Stub axles steered differentially by *Ackerman* linkage [NEWTON ET AL. 1989].

With autonomous suspension, the rule continues in the same way, albeit the mechanism is completely different. The arms D and E together with their coupled stub axles form what amounts to bell crank levers pivoted on the kingpins and are used for coupling the two wheels so that they move together when they are steered. These arms are called the '*track arms*' and are interconnected by the '*track rod*'. The actual steering is generally caused by a connecting link, termed a '*drag link*', between the steering gear and either what is called the '*steering arm*' on the adjoining stub axle assembly or, in some instances, part of the track rod system.

Ackerman Linkage - From Figure 4.34 it may be seen that there is a difference between the angles α and β in that the wheels on the inside and outside, respectively, of the curvature have to be turned. In reality, this difference is realised by setting the arms D and E at angles such that, in the straight ahead position, exposed dotted lines are drawn through the centres of the two pivots on each intersection near the centre of the rear axle. The precise position of this intersection points rely on the correlation between the wheelbase and track, and other factors.

From Figure 4.35 [NEWTON ET AL. 1989] it may be seen how the stub axles are steered differentially by this linkage, the full lines representing the straight ahead and the dotted lines a steered circumstance.

In the latter, the stub axle B has turned through an angle β and the end D of its track arm has moved to D^1, a distance x parallel to the axle beam. Ignoring the slight angle of inclination of the track rod, it follows that the end C of the other track arm must move the same distance x, parallel to the axle beam. This, on the other hand, entails movement of arm C through a greater angle than D, because the latter is swinging across the dead centre of the base, as depicted in the illustration, while the former is moving further from its related lowest point.

Even if, for realistic reasons, these arms may have to be curved, possibly to clear some other part of the wheel or brake assembly, the effective arm continues that of a straight line joining the centres of the kingpin and the pivot at the contrary end.

Figures 4.34 and 4.35 illustrate the track rod at the rear of the axle, but exceptionally it is in front, again with precisely inclined arms.

A feature of inserting it to the rear is the protection provided by the axle beam, but it is then loaded in compression and, as a result, must be of inflexible structure.

Conversely, when it is in front, complications is normally arise when giving clearance between its ball joints and the wheels.

With *Ackerman* SBW 2WS, the wheels really roll in no more than three positions – straight ahead or when turned through a particularly selected angle to the right and left. However, in the last two positions, rolling takes place only at low values of vehicle velocity. At all other angles, the axes of the front wheels do not intersect on that of the rear wheels, while at the higher values of vehicle velocity, the slip angles of the front and rear tyres normally alter and definitely those of the tyres on the outside may always alter from those of the inside of the curvature.

In all cases, the slip angle on both the front and rear wheels have the consequence of turning their real axes forward.

Linkages providing almost precise static steering geometry on all curls have been invented; they are complicated and in reality have not shown to be satisfactory because they cannot consider the difference in slip angle.

The *Ackerman* rule, derived from the best realistic compromise – as a rule slip angles are understood to be equal on all four wheels – is acceptable in reality, perhaps for the reason that flexing of the tyres contains the errors.

The Steering Hand-Wheel (HW) Angle--Torque Characteristics -- According to SHIMOMURA ET AL. [1991], in a mechanical steering system, the steering HW angle-torque (d_h - T_h) characteristic is clearly one of the most important items that influence the driving feeling. The d_h - T_h characteristic is influenced by the vehicle velocity, the wheel-tyre properties, the suspension parameters, and the on/off road surface, namely:

$$T_h = f(d_h, V, \textit{Wheel-Tyre Properties, Suspension Geometry, On/Off Road Surface})$$

In a SBW 2WS conversion mechatronic control system, different steering feelings may be obtained by modifying the d_h - T_h characteristic (Fig. 4.36) [SHIMOMURA ET AL. 1991; DOMINGUEZ-GARCIA AND KASSAKIAN 2003].

A. Vehicle Velocity Influence

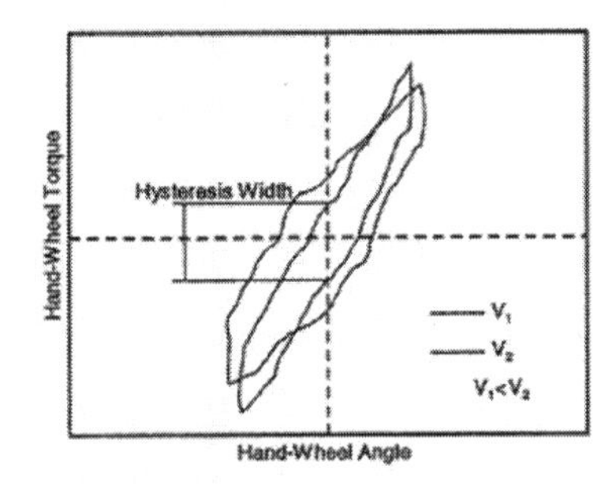

•At higher **vehicle velocity** (V), the gradient T_s/δ_s increases.

•At higher V, the hysteresis width of the hand-wheel (HW) torque for zero HW angle becomes narrower.

B. Wheel-Tyre Properties Influence

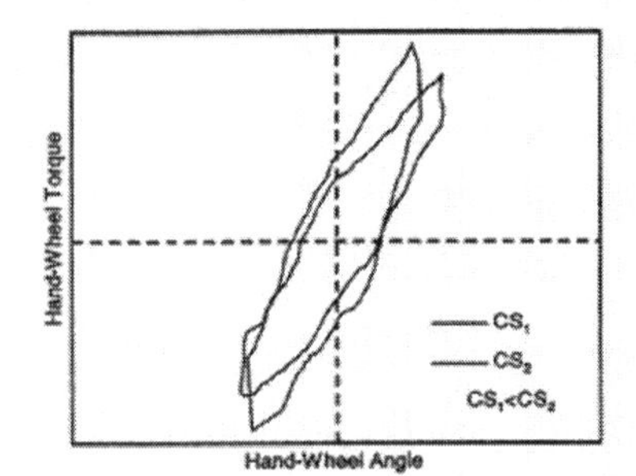

•As the **cornering stiffness** (CS) of the wheel-tyre increases, the gradient T_s/δ_s increases.

•As the CS, the hysteresis width of the hand-wheel (HW) torque for zero HW angle increases.

C. Suspension Geometry Influence

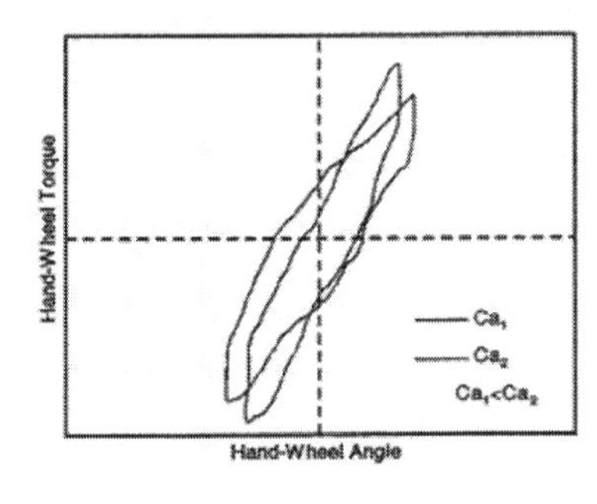

•As the **caster** (Ca) increases, the gradient T_s/δ_s increases.

•As the Ca increases, the hysteresis width of the hand-wheel (HW) torque for zero HW angle decreases.

D. On/Off Road Surface Influence

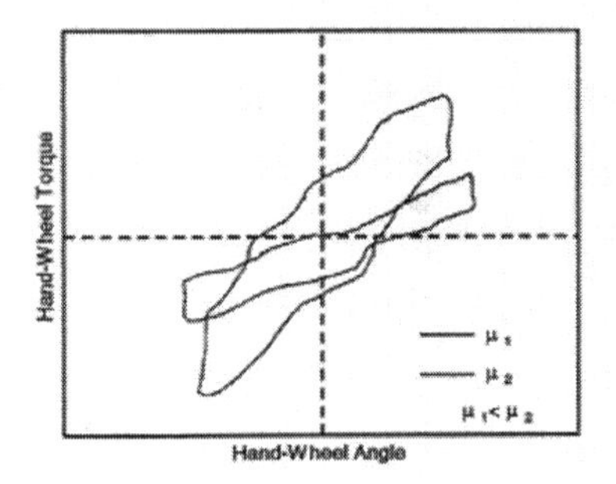

•As the **friction coefficient** (μ) between the wheel-tyre and the road surface increases, the gradient T_s/δ_s decreases.

•As the μ between the wheel-tyre and the road increases, the hysteresis width of the hand-wheel (HW) torque for zero HW angle becomes narrower.

Fig. 4.36 The hand-wheel angle vs.torque characteristics:
vehicle velocity influence (A); wheel-tyre properties influence (B);
suspension geometry influence (C); on/off road surface influence (D).
[SHIMOMURA ET AL. 1991; DOMINGUEZ--GARCIA AND KASSAKIAN 2003].

4.2.2 Categories of the SBW 2WS Conversion Mechatronic Control Systems

FPS, EFPS and EPS SBW 2WS conversion mechatronic control systems that may be soon enter production may be arranged systematically, in order of their fundamental structure and principles, into four types: full M-F-M, E-M-F-M and E-M ones, as displayed in Table 4.3. Particular explanations of these SBW 2WS conversion mechatronic control systems are also given in the table [SATO 1995].

Table 4.3 Categories of SBW 2WS Conversion Mechatronic-Control Systems [SATO 1995].

Fundamental structure		M-F-M FPS SBW 2WS conversion mechatronic control system				E-M-F-M EFPS SBW 2WS conversion mechatronic control system	E-M EPS SBW 2WS conversion mechatronic control system	
Mechatronic control methods		Oily-fluid or air flow	Power F-M cylinder bypass	Fluidic valve characteristics	Fluidic reaction force mechatronic control	Oily-fluid or air flow	E-M motor voltage	E-M motor current
Mechatronic control objects		Oily-fluid or air flow supply to power F-M cylinder	Effective actuation oily-fluid or air pressure given to power F-M cylinder	Oily-fluid or air pressure generated at mechatronic control value	Oily-fluid or air pressure acting on the fluidic reaction force mechanism	Oily-fluid or air flow supply to power F-M cylinder	E-M motor power	E-M motor torque
Automotive vehicle sensors	Vehicle velocity	☺	☺	☺	☺	☺	☺	☺
	Angular velocity					☺	☺	
	Steering torque				☺		☺	☺
	Electric current	☺	☺	☺		☺	☺	☺
Actuator		Solenoid fluidic valve	Solenoid fluidic valve	Solenoid fluidic valve	Solenoid fluidic valve	Linear or rotary E-M motor	Linear or rotary E-M motor	Linear or rotary E-M motor
Major effects	Steering force responsive to vehicle velocity	☺	☺	☺	☺	☺	☺	☺
	Energy saving					☺	☺	☺

4.2.3 Description of SBW 2WS Conversion Mechatronic Control Systems

Current SBW options available include E-M, M-F-M or E-M-F-M actuation. Sensors may be required for steering position and velocity both at the steering HW and the road wheels.

Torque measurements may be required for the road wheels as well as for the steering HW if force feedback is required.

Feedback may not be required for early prototypes but would certainly be required in a production situation in order to give the driver meaningful information about what is happening at the road wheels.

Other sensors and information may also be required, such as wheel speeds, lateral acceleration, yaw rate, and so on for the system's self-calibration and other features.

Steering actuation may be achieved through several methods. An easy way is to replace the input from the steering column into the steering rack with that from an E-M motor (Fig. 4.37), whilst retaining the PS [JB 2004]. This may require a reduction through a gearbox to achieve the required torque and speed, but the E-M motor may be relatively small and run off the standard 12 V_{DC} power supply.

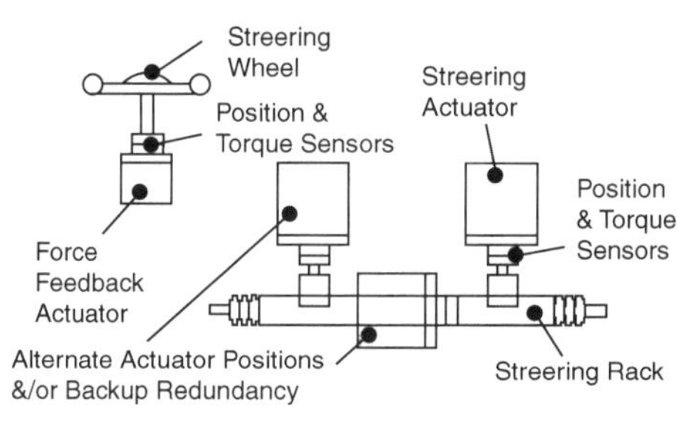

Fig. 4.37 Electro-mechano-fluido-mechanical (E-M-F-M) actuation [JB 2004].

This solution provides an easy method of modifying a current vehicle to SBW, provided there is space for the E-M motor, but still suffers from the package, mass, and fuel economy that the fluidics (hydraulics) of a PS system bring.

The fluidics only method makes use of the power assisted steering rack to provide the steering actuation with a series of fluidical valves to provide control over the rack position (Fig. 4.38) [JB 2004].

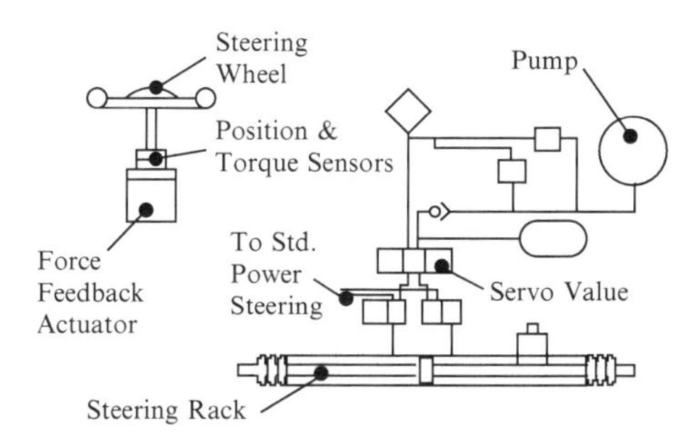

Fig. 4.38 Mechano-fluido-mechanical (M-F-M) actuation [JB 2004].

Again this solution provides an easy method of modifying a current automotive vehicle to SBW but still suffers from the package, mass, and fuel economy that the fluidics (hydraulics) of a PS system bring. The next step is to completely remove the M-F-M system and replace it with a direct drive from an E-M motor of sufficient power. This may be done with one E-M motor controlling both wheels through a steering rack or with one E-M motor per wheel providing independent control of each road wheel (Fig. 4.39) [JB 2004].

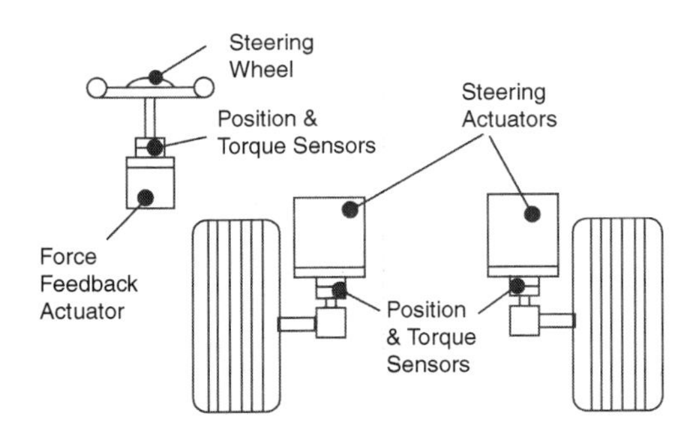

Fig. 4.39 Electro-mechanical (E-M) actuation [JB 2004].

With the need for redundancy, the ability to have twin actuation for the steering imposes additional restrictions on the above architectures.

The M-F-M only system, whilst acceptable for a prototype, becomes complicated for production implementation with the need for two M-F pumps, two racks or fluidical chambers within the rack, and two sets of control fluidical valves.

Packaging, cost, and complexity become a large issue. For the E-M-F-M system, the issue is compounded by the fact that the PS fluidics (hydraulics) and the E-M motor would be unnecessary.

If the PS was to fail in a standard vehicle, it is considered to be reasonable that the driver would still have sufficient energy to maintain a level of control over the vehicle. Here the E-M motor be designed to complement the PS and would therefore be insufficient to control the steering without assistance. The E-M motor could be sized to cope with this but then the system would be almost equivalent to the E-M implementation [JB 2004].

Mechatronic Control Architecture -- Once again the safety case dictates that the mechatronic control is fail safe and therefore must be somewhat unnecessary. As developed from the *Markov* model [HAMMETT ET AL. 2003], this could even be a dual-dual or triplex mechatronic control system (Fig. 4.40) [JB 2004]. An important factor in determining the mechatronic control system is not only the ability to detect failures, but also to determine what the failure is. For example if there are two position sensors in the system and if one fails, it could be obvious that they are different but how do you tell which one is correct?

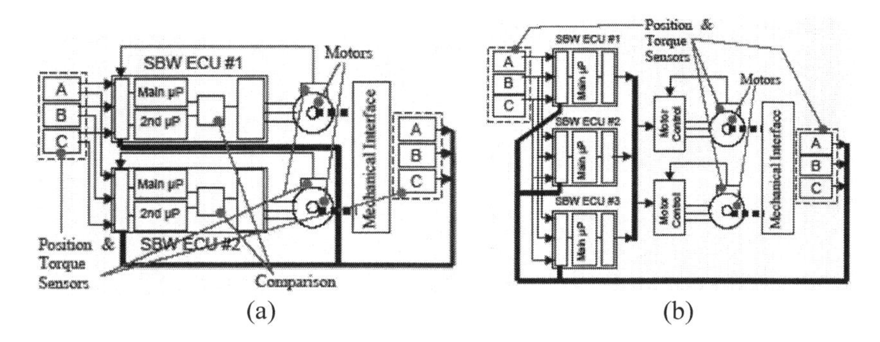

Fig. 4.40 Mechatronic control architecture: (a) -- dual-dual SBW 2WS conversion mechatronic control system; (b) -- triplex SBW 2WS conversion mechatronic control system [JB 2004].

Sensors -- The SBW 2WS conversion mechatronic control system requires measurements of angular position, angular velocity, and torque from both the steering wheel (that is, the driver's input) and the road wheels in order to provide suitable information to the controller for correct operation.

Other data such as yaw rate and lateral acceleration may allow additional features such as stability control and steering correction. These signals must be accurate and have good resolution to allow the system to perform well. In addition, for SBW at least one sensor at the steering HW and road wheels should be capable of measuring the absolute angle over the complete steering range.

These sensors must also be dependable (not affected by their age or environment), reliable, and conform to the usual package and cost considerations, and so on.

Suitable sensors are likely to be of the on-contact variety with optical and magnetoresistive sensors providing good performance and reliability.

The exact sensor and technology though is likely to rely on many factors dependent on the specific application [JB 2003].

Actuators -- The SBW 2WS conversion mechatronic control system requires the use of actuators for two purposes. The first is to provide the correct positioning and sufficient force to control the steering rack. The second is to provide meaningful feedback to the driver through the steering HW. Both require good performance in both power/torque and accuracy. This is currently best achieved through the use of brushless DC-AC macrocommutator IPM magneto-electrically-excited steering-actuator motors, likely to be of a higher voltage than the standard 12 V_{DC} for **energy-and-information networks** (E&IN) (e.g. 42 V_{DC}). Although this would not necessarily be required for feedback, it certainly would for the steering actuation. This could be achieved through a number of solutions including twin E-M motors connected via a gearbox or twin wound E-M motors sharing the same package. These E-M motors may need to be powerful enough to achieve sufficient steering force. Again, specific application may determine the exact requirements for the actuator [JB 2003].

Power Supply -- There are two main requirements for a SBW power supply: sufficient power for the steering actuators and redundancy. This may require dual power systems (two M-E generators, two storage batteries, two looms, and so on).

The main power system may have to be of a voltage higher than the standard 12 V_{DC} **electrical energy distribution** (EED) systems in order to provide the required performance (e.g. 24 V_{DC}, 42 V_{DC}, or higher).

With the need for a 12 V_{DC} for the main EED systems, this could then provide the required redundancy (albeit at a reduced level of performance).

The power wiring may have to be protected against single-point failure. This may require multiple wires from the power supplies to the critical components, wires routed separately, independent protection for the separate wires, and switching to the use of separate wires [JB 2004].

M-F-M FPS SBW 2WS Conversion Mechatronic Control System -- This mechatronic control system is composed of a linear electro-magneto-mechanically operated solenoid fluidical valve, a vehicle velocity sensor, and other automotive mechatronic devices situated in part of the fluidical circuit of the M-F-M FPS SBW 2WS conversion mechatronic control system (see Fig. 4.41) [HITACHI 2004].

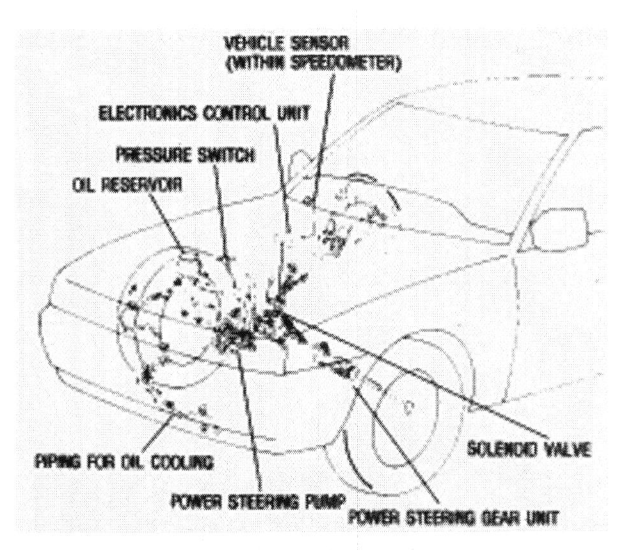

Fig. 4.41 Oily-fluid pressure M-F-M FPS SBW 2WS conversion mechatronic control system [Hitachi Co; HITACHI 2004].

The flow of oily fluid to the F-M cylinder is decreased when driving at high values of vehicle velocity, with the intention that for this mechatronic control system, the magnitude of the steering response rate and the steering reaction force are equal in value at the point of balance. FPS SBW 2WS conversion mechatronic control systems are capable of mechatronically adjusting the steering force according to vehicle velocity, and are used primarily in luxury vehicles.

For instance, a **vehicle-velocity responsive power steering** (VRPS) SBW AWS conversion mechatronic control system may be capable of adjusting steering force by controlling the oily-fluid pressure using fluidical valves according to the vehicle velocity (Fig. 4.42) [HITACHI 2004].

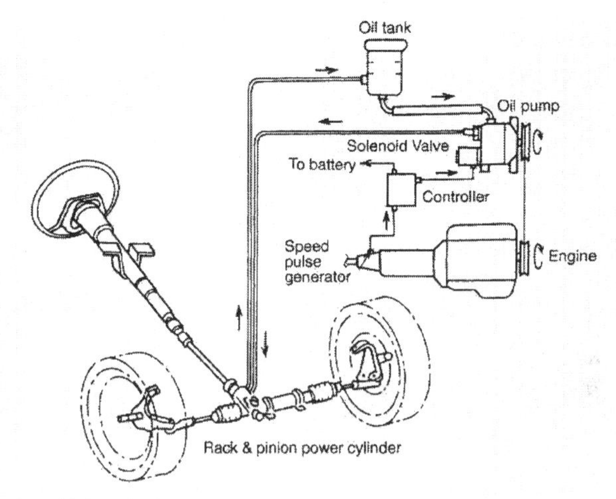

Fig. 4.42 Vehicle velocity-responsive M-F pump discharge oily-fluid–flow volume mechatronic control system [SATO 1991]

This, as shown in Fig. 4.43, requires a relatively simple fluidical valve structure and makes possible highly responsive steering because it converts the oily-fluid flow supplied by an M-F pump to a pressure that efficiently operates an F-M cylinder (a linear F-M motor) [HITACHI 2004].

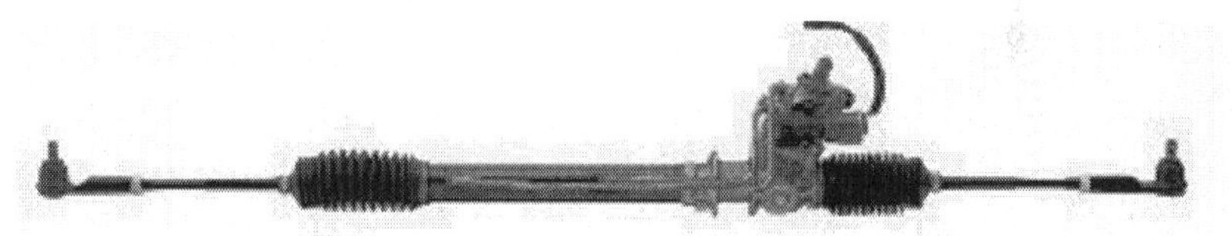

Fig. 4.43 Fluidical power steering (FPS) – vehicle velocity responsive type [Hitachi Co.; HITACHI 2004].

Power F-M Cylinder Bypass M-F-M FPS SBW 2WS Conversion Mechatronic-Control System -- In this mechatronic control system, a fluidical valve's solenoid and a bypass line are situated between both fluidical chambers of the power F-M cylinder [NISSAN 1991B]. The turning on of the fluidical valve's solenoid is expanded by the ECU in conformity with gains in vehicle velocity, thus weakening the pressure in the power F-M cylinder and increasing the steering effort. Similar to the oily-fluid flow mechatronic control system, this system may also try to reach the equilibrium point for the steering response rate and the steering reaction force [SATO 1995].

Rotary Fluidical Valve Characteristics M-M FPS SBW 2WS Conversion Mechatronic Control System -- In this mechatronic control system, the oily-fluid-pressure control limitations of the rotary fluidical-valve mechanism that controls the volume and pressure of the oily-fluid provided to the power F-M cylinder, are separated into second and third parts. A fourth part, controlled by means of the vehicle-velocity signal, is supplied in the fluidical line between the second and third parts, fulfilling variable control of the fourth part to vary the assistance ratio controls of the steering effort. Since the structure is unsophisticated and the oily-fluid flow from the M-F pump to the power F-M cylinder is provided effectively without dissipation, this mechatronic control system demonstrates a satisfactory response rate. For example, when a value of the electric current is about 300 mA, the solenoid fluidical valve is fully turned on and this demonstrates the high value of vehicle-velocity driving [SATO 1995].

Fluidical Reaction-Force M-F-M FPS SBW 2WS Conversion Mechatronic-Control System -- In this mechatronic control system, the steering effort is controlled with the aid of a fluidical reaction-force mechanism that is situated at the control rotary fluidical valve. A control rotary fluidical valve increases the reaction oily-fluid pressure leading to the reaction force fluidical chamber in uniformity with intensities in vehicle velocity. The stringency of the fluidical reaction force mechanism (equivalent spring constant) is variably controlled to directly control the steering effort. This system requires a reactive-force mechanism that makes the structure of the control fluidical valve more sophisticated but radically increases the cost. On the other hand, because the stiffness of the reaction-force mechanism rises in conformity with escalations in vehicle velocity, there is no difference in the steering feel in the section around the basic steering position. Since this mechatronic control system gives the steering reaction force without regard to the volume of oily-fluid provided to the power F-M cylinder, the magnitude of the steering reaction force may be adjusted independently without requiring substitution any of the steering response rates [SATO 1995].

4.2.4 Hybrid E-M-F-M EPFS SBW 2WS Conversion Mechatronic Control System

For SBW 2WS conversion mechatronic control systems, the most likely approach may be to implement the E-M-F-M EFPS option (Fig. 4.44) [JB 2003]. In actual fact, the PS on the automotive vehicle is likely to be mechatronic so the implementation may follow the philosophy discussed for the E-M-F-M architecture but may in fact have no fluidics (as power assist is electrical). These options allow the implementation of SBW on the vehicle without the need for major modifications to the vehicle's current steering from the rack down. The lack of FPS on the vehicle also prevents an M-F-M-only solution.

An additional benefit of using the E-M-F-M approach for the SBW 2WS mechatronic control system is that it may have a failsafe option.

This approach also avoids the need to source expensive and/or hard to obtain parts such as high-power E-M motors for the steering, whilst proving the concept and developing a model for potential production.

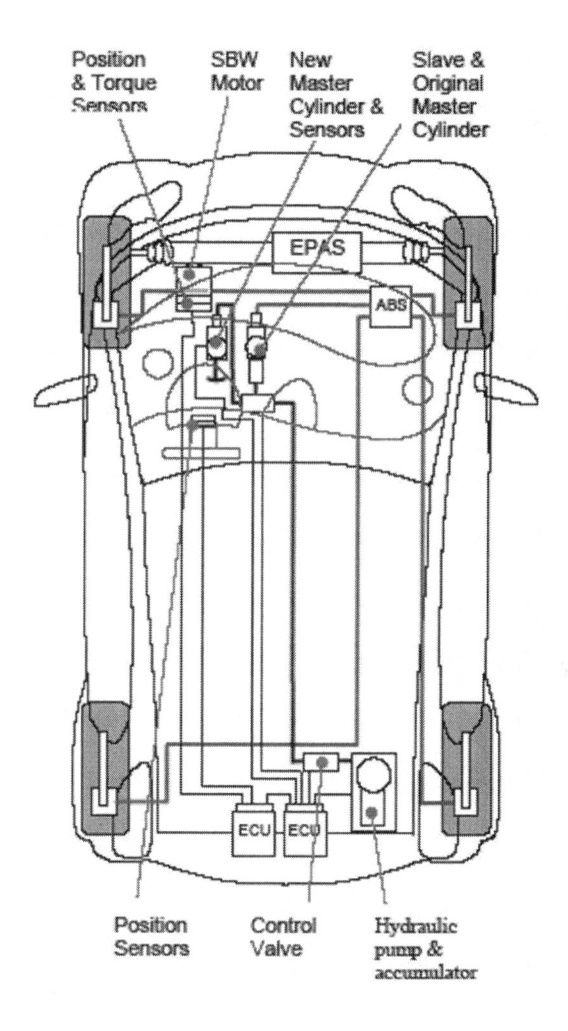

Fig. 4.44 Proposed Pininfarina *Autosicura* prototype implementation [JB 2003].

The SBW 2WS conversion mechatronic control system would consist of the M-F-M actuation shown in Figure 4.37 in the form of an E-M implementation (as the power assist may be electrical).

At least dual position sensors would be required for both pinion and steering HW angle although this would be confirmed by a safety case analysis. In addition, a torque sensor at the pinion or a rack force sensor would be required.

For the initial implementation it may not be necessary for force feedback of the steering, therefore not requiring an actuator or torque sensor at the driver interface. This functionality could be added at a later date though.

With the SBW 2WS conversion mechatronic control systems requiring a control ECU of some kind, there is the possibility of using each ECU as the other's backup, or one ECU as the master for SBW and the second as a safety monitor.

There are several suitable rapid prototype ECUs capable of performing this function, for example, PiTechnology's *Open* ECU [JB 2003].

The main power supply for the SBW 2WS conversion mechatronic control system would come from the vehicle's 12 V_{DC} supply.

As the steering main actuators would be F-M, there would be no need for a secondary power supply system; although a small secondary storage battery could be installed for the SBW ECUs using a split charge system with the vehicle's current AC-DC macrocommutator electromagnetically-excited generator. This again would be determined by safety and vehicle review.

Below (Fig. 4.45) is a high-level overview of the control strategies intended for the suggested SBW 2WS conversion mechatronic control systems [JB 2003].

This structural and functional block diagram is meant to give a basic understanding of the control strategy and therefore does not include safety monitors, diagnostics, fallback mechanisms, as well as redundancy and so on.

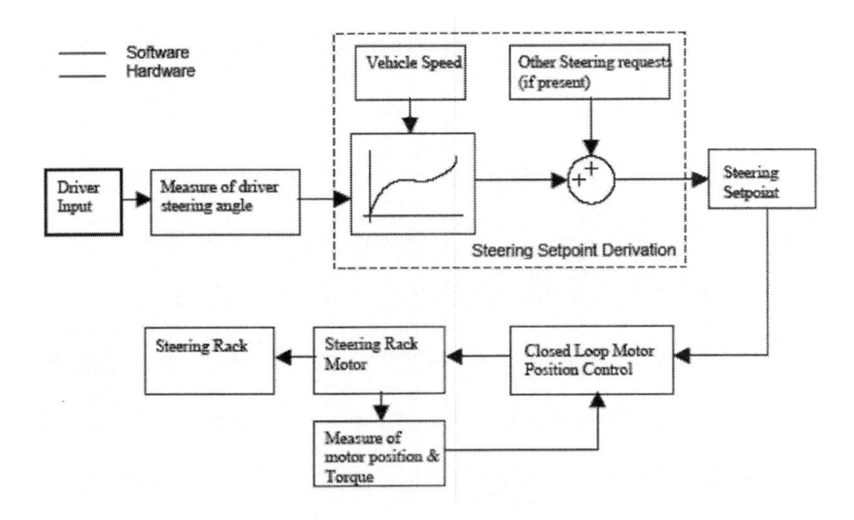

Fig. 4.45 Structural and functional block diagram of the SBW 2WS conversion mechatronic control system [JB 2003].

Hybrid E-M-F-M EFPS SBW 2WS conversion mechatronic control system for new passenger vehicles (Fig. 4.46) [TRW 2003] combines an electrically driven E-M-F pump with conventional R&P steering to give the most precise handling and steering assistance using minimal energy consumption. This mechatronic control system uses an oily-fluid-flow control method in which the power steering E-M-F pump is driven by an E-M motor. The steering effort is controlled by regulation of the angular velocity of the E-M-F pump's rotor, that is, the discharge.

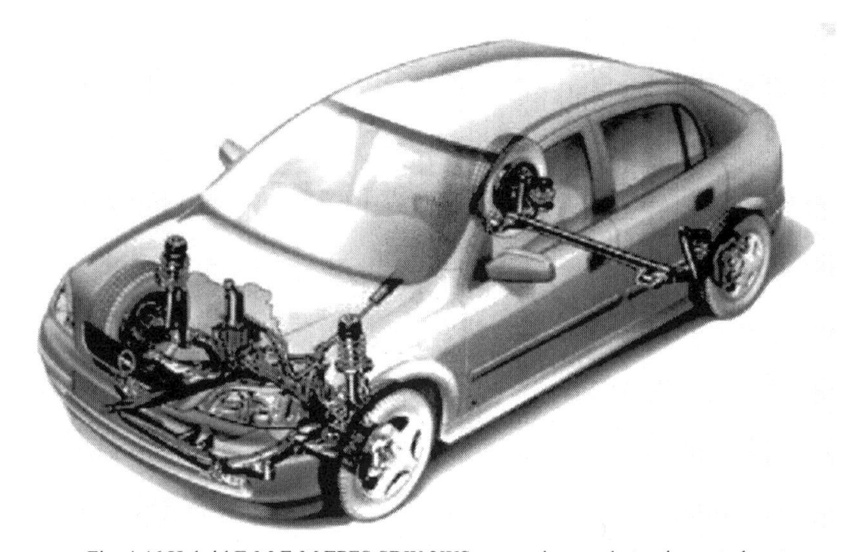

Fig. 4.46 Hybrid E-M-F-M EPFS SBW 2WS conversion mechatronic control system for the new Opel *Astra* [TRW 2003].

The driving efficiency of the M-E generator and E-M motor are related to that of the M-F pump which is driven by the vehicle's ECE or ICE and/or F-M, P-M or E-M motor(s).

Nevertheless, because any residual flow is not discharged, the power loss is lower than that of the ECE or ICE M-F pump when driving at high velocities.

For the reason that the vehicle's ECE or ICE and/or F-M, P-M or E-M motor(s) does not drive the E-M-F pump, there is also a great degree of freedom in the choice of mounting places for the E-M-F pump.

Driving Mode Responsive Hybrid EFPS SBW 2WS Conversion Mechatronic Control System -- In this mechatronic control system, it is composed of a vehicle-velocity sensor, steering-angular-velocity sensor, an ECU, and an E-M motor driven E-M-F pump, as illustrated in Figure 4.47 [IGA ET AL. 1988].

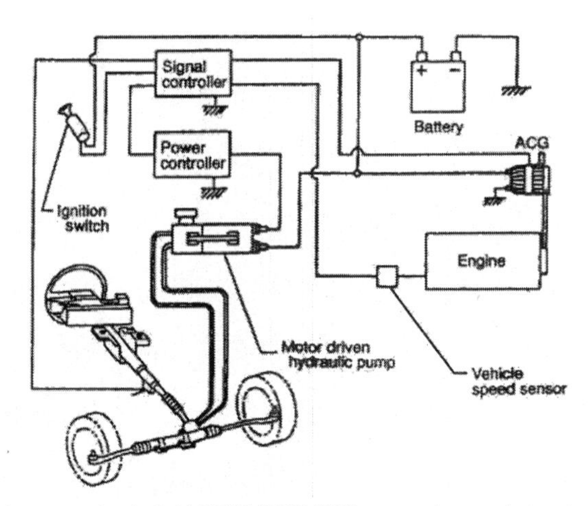

Fig. 4.47 Driving mode-responsive hybrid EPFS SBW 2WS conversion mechatronic control system [IGA ET AL. 1988]

Driving situations; such as driving in urban areas, country areas, winding regions, or highways; are instinctively deduced, and the E-M-F pump flow rate is controlled uniformly in order to provide proper steering effort for driving situations. Sensitive control adjustments are achieved with the aid of this mechatronic control system when related to the previously mentioned vehicle-velocity responses [SATO 1995].

Steering Wheel Angular-Velocity Responsive EFPS SBW 2WS Conversion Mechatronic Control System -- This system contains component parts such as vehicle-velocity sensor, steering wheel angular-velocity sensor, an ECU, and an E-M motor-driven E-F-M pump, as illustrated in Figure 4.48 [HONDA 1991].

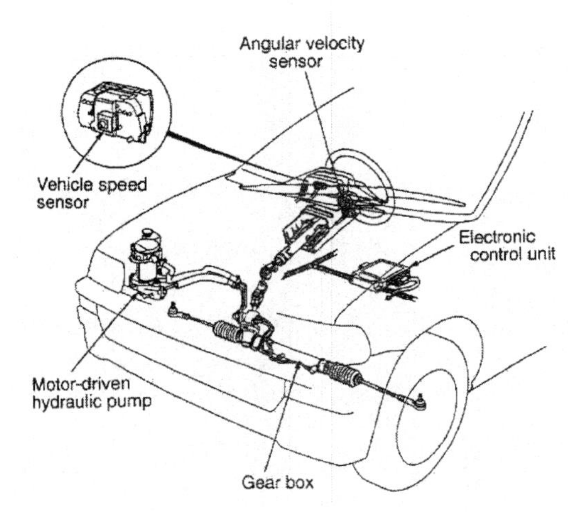

Fig. 4.48 Steering velocity-responsive hybrid EPFS SBW 2WS conversion mechatronic control system [HONDA 1991].

As mentioned above, the discharge flow volume of the E-M-F pump decreases and the steering response decreases when the vehicle is driven at high velocity. Consequently, in this mechatronic control system, the angular velocity of the E-M motor's rotor becomes more uniform with the sensed angular velocity of the steering HW according to the increasing discharge flow volume.

Correspondingly, power loss resulting from the dispersion of residual oily-fluid flow within the system are maintained to a minimum satisfactory level, and the magnitude of the reaction force may be controlled independently without neglecting any of the steering response ratios [SATO 1995].

4.2.5 E-M EPS SBW 2WS Conversion Mechatronic Control System

During the last few years, the automotive industry has focused efforts on the development of E-M EPS SBW 2WS conversion mechatronic control systems. Among the main advantages of such a system, the most important are the enhancement of passive safety systems and the introduction of lateral active safety systems.

An E-M EPS SBW 2WS conversion mechatronic control system is the key to vehicle performance and safety. Problems with existing F-M FPS SBW 2WS conversion mechatronic control systems [CHEN 2006]:

- Complex – high cost;
- *'On'* all the time and powered by ECE- or ICE-low fuel economy and safety concern;
- Hard to integrate advanced technologies.

Supplementary advantages with E-M EPS SBW 2WS conversion mechatronic control systems [CHEN 2006]:

- Reduce maintenance and recycling cost;
- Reduced assembly time (up to 4 min);
- Up to 4% improvement in fuel economy;
- Enhanced safety (still operate even if ECE or ICE is stalled);
- Active tuning, chassis mechatronic control integration.

One of the objectives in the development of such control systems is to equip the driver with tactile feedback from the on/off road surface.

The importance of force feedback or *'road feeling'* has been well understood in both the automotive field and in telemanipulator systems.

Force feedback is one of the most valuable parameters in providing the driver with accurate control of the vehicle.

In this context, it is very important to provide a tuneable realistic steering feel that ensures comfortable driving.

The next generation of E-M EPS SBW 2WS conversion mechatronic control systems [CHEN 2006]:

- Further cost reduction for design, testing, manufacturing/assembly, and warrantee service;
- Advanced functions to enhance reliability and safety: fault-detection and tolerant, parts erosion compensation, and so on;
- Higher torque generation and optimised power flow management.

Current R&D ought to be focusing on [CHEN 2006]:

- Low cost motor with high torque yield, for example, a brushed motor with an easy drive and performance control set-up (in contrast to brushless motors) with the aim of applying to bigger vehicles;
- Optimised torque/energy distribution control with robustness to improve the driver's steering feeling, attenuate undesired disturbance and noise from both tyres and rough road surfaces;
- On-board diagnostics, tolerance and compensation of non-critical faults with fault-tolerant control strategies to extend the life of parts or components.

The principle layout of a column-mounted E-M EPS SBW 2WS conversion mechatronic control system's module is shown in Figure 4.49 [CHEN 2006].

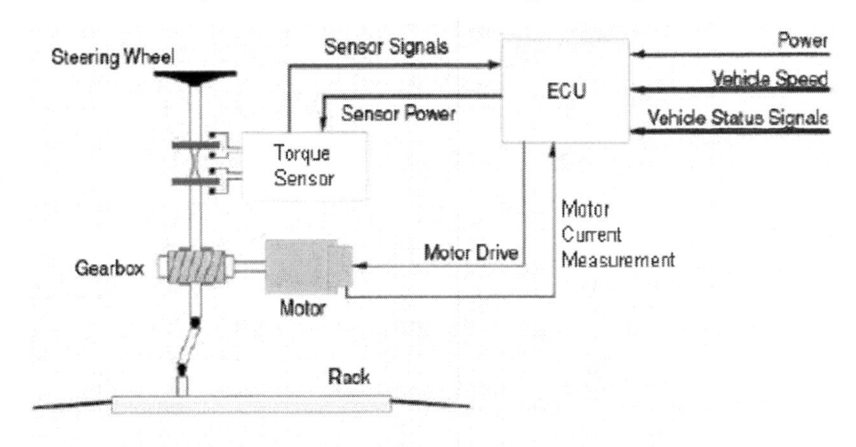

Fig. 4.49 The principle layout of a column mounted E-M EPS SBW 2WS conversion mechatronic control system's module [CHEN 2006].

Figure 4.50 shows a structural and functional block diagram of a column-mounted E-M EPS SBW 2WS conversion mechatronic control system's module for analysis and design [CHEN 2006].

A structural and functional block diagram of an E-M EPS SBW 2WS conversion mechatronic control system's module with road/surface dynamics using *CAR SIM*^{TM} is shown in Figure 4.51 CHEN 2006].

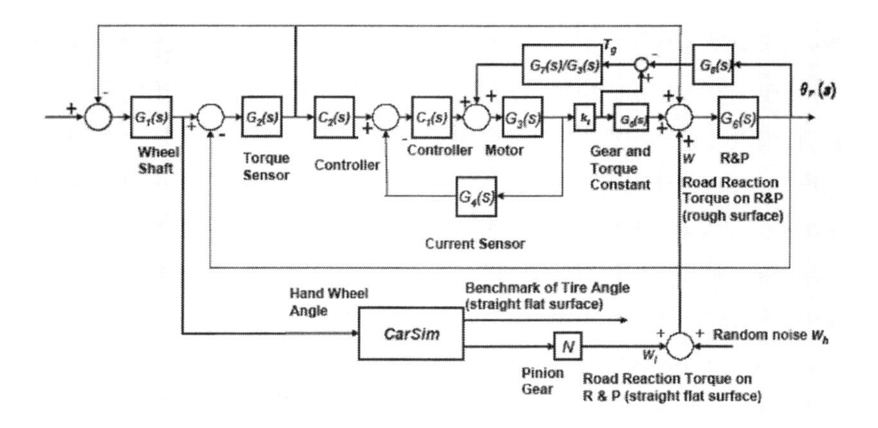

Fig. 4.50 Structural and functional block diagram of a column-mounted E-M EPS SBW 2WS conversion mechatronic control system's module [CHEN 2006].

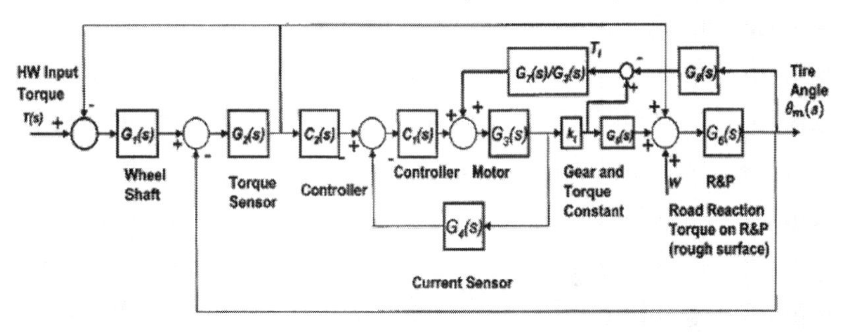

Fig. 4.51 Structural and functional block diagram of an E-M EPS SBW 2WS conversion mechatronic control system's module with road/surface dynamics using *CARSIM*[TM] [CHEN 2006].

A generic description of an E-M SBW 2WS conversion mechatronic control system [DOMINGUEZ-GARCIA AND KASSAKIAN 2003]:

- No longer a mechanical connection between the driver and the steering system;
- Significant similarities with telemanipulator systems;
- The control of both the steering wheel and steering system rely on a unique controller.

An E-M EPS SBW 2WS conversion mechatronic control system replaces conventional M-M linkage between the steering HW and the road-wheel actuator (e.g. an R&P steering) with an electronic connection (see Fig. 4.52) [GÜVENC 2005]. This allows flexibility in the packaging and modularity of the design. Since it removes the direct kinematics relationship between the steering and road wheels, it enables control algorithms to help enhance driver input.

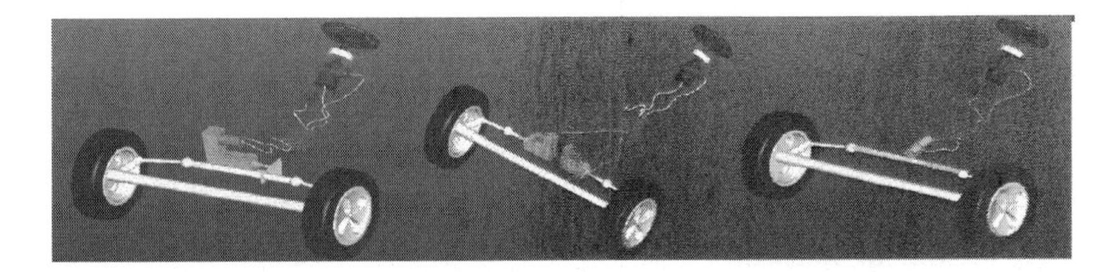

Fig. 4.52 E-M EPS SBW 2WS conversion concepts [GÜVENC 2005].

Figure 4.53 shows a principle layout for the E-M EPS SBW 2WS conversion mechatronic control system's conceptual design [AMBERKAR ET AL. 2003]. It may be subdivided into three major parts: an ECU system (Controller), a steering **hand-wheel** (HW) system, and a road-wheel system. The steering HW system contains sensors to provide information about driver steering input. This information is sent to the ECU hyposystem (Controller) that employs knowledge of the vehicle's current state to command the desired road wheel angle.

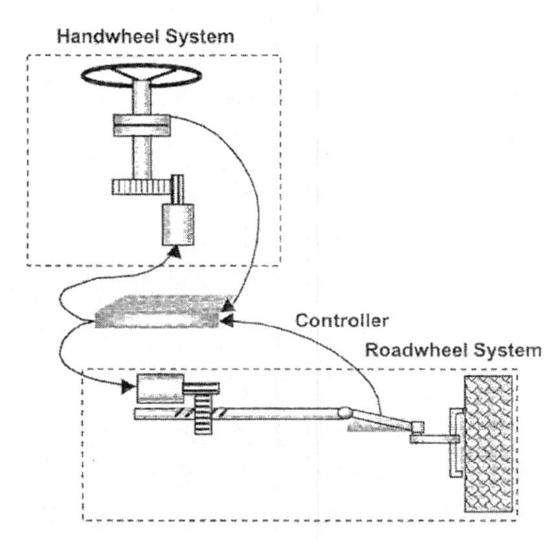

Fig. 4.53 Principle layout of the E-M EPS SBW 2WS conversion conceptual design [AMBERKAR ET AL. 2003].

The road-wheel system contains actuators to position the wheels. An actuator in the steering HW system provides road feedback to the driver. This also is commanded by the ECU system and based on information provided by the sensors in the road-wheel system. A place where the mechanical-to-electrical transformation is taking place, albeit slowly, is in the vehicle's PS.

Some smaller vehicles appeared with EPS back in 1998, with volumes increasing steadily as vehicle manufacturers adapted very advanced technology. By using an E-M motor to replace the conventional combination of M-F pump, oily-fluid, hoses, and oily-fluid reservoirs, drivers have a SBW 2WS conversion mechatronic control system that, while not significantly less expensive to produce, is smaller and lighter.

Summing up, the M-F-M SBW 2WS conversion mechatronic control system of recent automotive vehicles is equipped with a PS that reduces the steering effort of the driver, while allowing for manual steering in the case of a decrease in the oily fluid pressure.

However, it requires the use of fluidics (hydraulics) and large amounts of energy for driving the boost M-F pump, especially at low velocity [TREVETT 2002].

When a vehicle is idling or running at a low velocity, the rotary vane M-F pump must provide sufficient flow fluid to facilitate the turning of the wheel. Low velocities usually occur when turning takes place, so a good M-F-M SBW 2WS conversion mechatronic control system must operate well at these velocities. As a direct result of this, the M-F pump moves a much larger amount of oily fluid than is needed when the ECE or ICE speeds up [LUKIC AND EMADI 2003]. However, at high velocities, the steering wheel is barely turned at all. Therefore, there are huge losses involved with the current FPS. Also, the steering column that is used to couple the steering wheel with the drivetrain is a major source of injury in front-end collisions [TREVETT 2002]. In addition, the steering assembly places constraints on the design of the vehicle. The discovery of the E-M SBW 2WS conversion mechatronic control system opens the way to creating a steering system that is safer, cheaper, and more efficient.

In the M-F-M SBW 2WS conversion mechatronic control system, as shown in Figure 4.54, the steering column may be eliminated completely [LUKIC AND EMADI 2003]. As a replacement, there would be a purely mechatronic control system. This form of steering would contain sensors that would send signals to the actuators that make the wheels turn in the desired direction. It is even possible for the driver to know what the vehicle is experiencing.

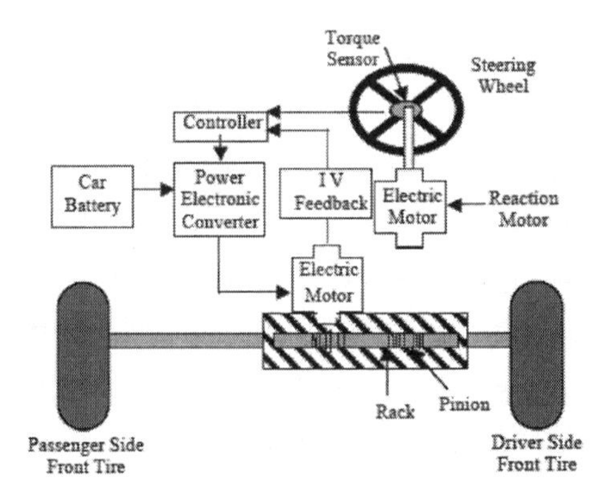

Fig. 4.54 E-M SBW 2WS conversion mechatronic control system [LUKIC AND EMADI 2003].

Energy may be consumed only when the steering wheel is being turned, leading to huge energy savings [LUKIC AND EMADI 2003].

The major issue with the E-M SBW 2WS conversion mechatronic control system is safety. If the FPS fails, there is a mechanical backup provided by the steering column. When the system is used, there is no physical connection between the steering shaft and the steering wheel; so there is no opportunity for mechanical backup. However, there are advances in mechatronic control that make the SBW failsafe [LEONI AND HEFFERMAN 2002].

The pinion assist type EPS SBW 2WS conversion mechatronic control systems consist of [UEKI ET AL. 2004]:

- ❖ A torque sensor that detects steering force;
- ❖ An electronic control unit (ECU) that calculates signals from the torque sensor and supplies the necessary electrical energy to the E-M motor;
- ❖ An E-M motor that conveys an assist force to a pinion shaft through a reduction gear mechanism;
- ❖ A rack and pinion (R&P) type steering gear.

The ECU controls vehicle-velocity sensitive PS SBW AWS conversion mechatronic control systems by processing signals indicating the vehicle velocity and the rotation of the ECE or ICE.

In addition, the torque limiter is positioned between the plastic gear in the reduction gear mechanism and the pinion shaft, and protects the plastic gear from on/off road surface pressure (Figs 4.55 and 4.56) [UEKI ET AL. 2004; HAGHIGHAT-GOO AND ESFANDYARI 2005; COSC/PSYCH 2006].

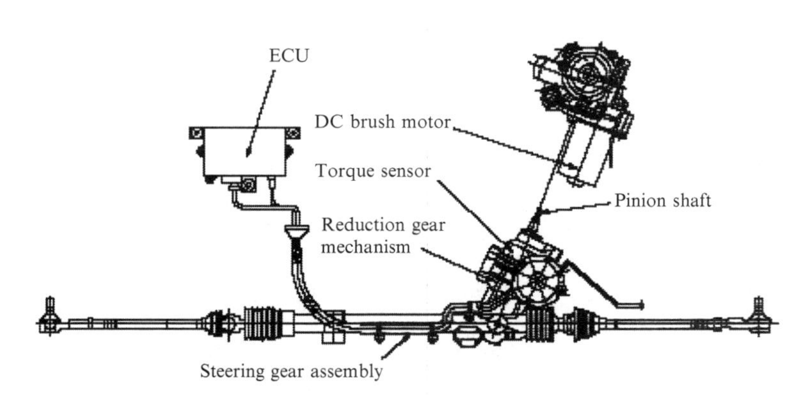

Fig. 4.55 Electric power steering (EPS) [UEKI ET AL. 2004].

Of particular interest to complex embedded mechatronic control systems such as SBW 2WS conversion, are software-implemented hazard mechatronic controls. These tasks monitor the state of the system for signs of hazards and take the necessary action.

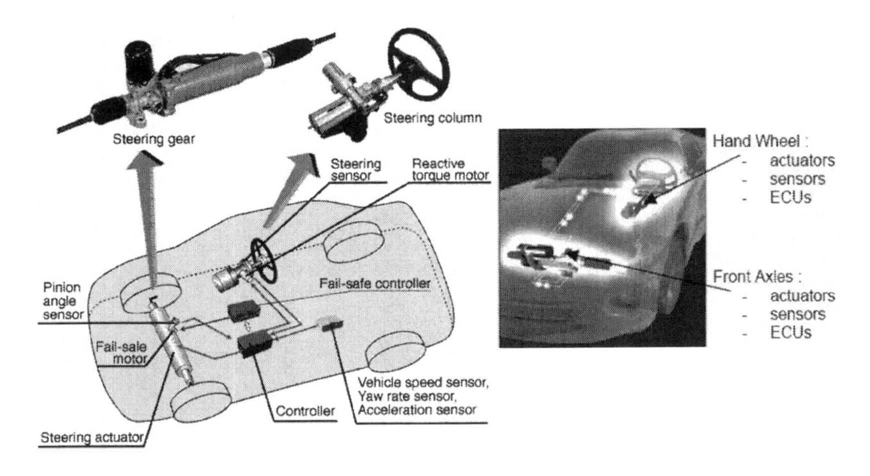

Fig. 4.56 Electric power steering (EPS) [COSC/PSYCH 2006 – Left image; HAGHIGHATGOO AND ESFANDYARI 2005 -- Right picture taken from: www.delphi.com].

Some potential hazards in mechatronic control systems, such as SBW 2WS conversion, may require fault tolerance because of inherent system limitations. This implies that some wiring, and/or ECUs (Controllers), and/or actuators, and so on may be unnecessary. Starting with the simple fault tree of Figure 4.57 [BERTRAM ET AL. 1999; AMBERKAR ET AL. 2003], but introducing the impact of hazard mechatronic controls for reducing risk, it is now possible to demonstrate improvements in the safety of the AE E-M EPS SBW 2WS conversion mechatronic control system. The same likelihood of occurrence is assumed for each event in the fault tree. By introducing hazard mechatronic controls into the tree, the likelihood that certain branches of the tree may lead to the top event can be reduced, thus reducing the risk of the hazard. For instance, if a redundant ECU (Controller) is added, it may take over for the primary ECU if it fails. The addition of the ECU (Controller) reduces the likelihood that the system may fail due to an ECU (Controller) failure (Fig. 4.58), since ECU 1 (Controller 1) and ECU 2 (Controller 2) must now both fail [BERTRAM ET AL. 1999; AMBERKAR ET AL. 2003].

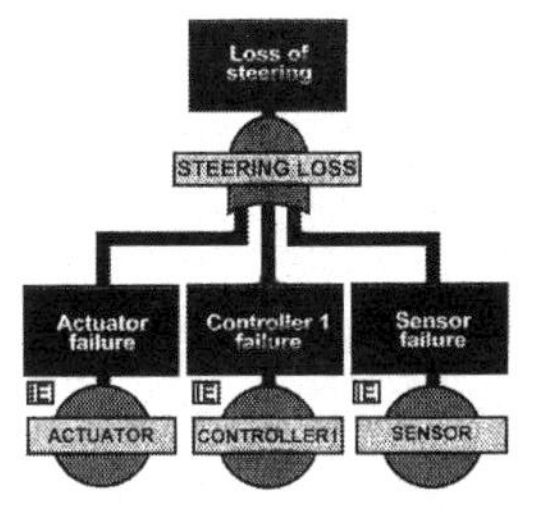

Fig. 4.57 Principle layout of the fault tree [AMBERKER ET AL. 2003].

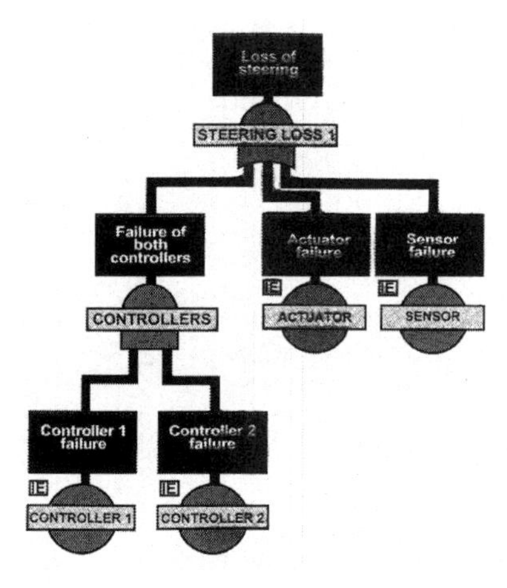

Fig. 4.58 Principle layout of the modified fault tree [AMBERKAR ET AL. 2003].

From a design perspective, it is important to know how this additional hazard mechatronic control should be added to the system so that it can take over when necessary, for example, warm standby, system voting, and so on.

Since hazard mechatronic controls may be added at a high level, as just illustrated in Figure 4.58, or at lower hyposystem or component levels, the hazard mechatronic controls are being implemented, where they are being implemented, and how many exist.

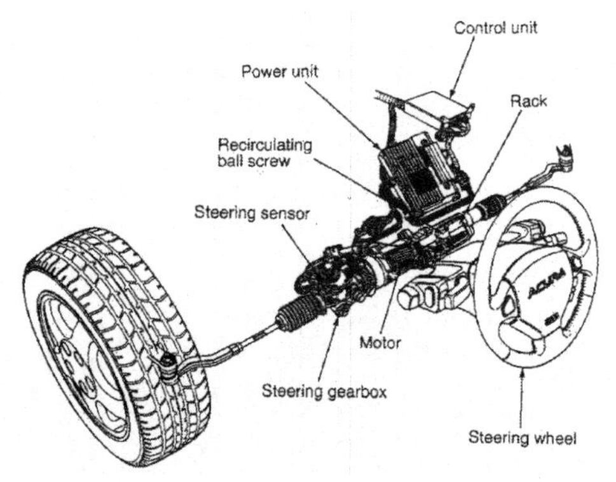

Fig. 4.59 Structure of E-M EPS SBW 2WS conversion mechatronic control system (rack assist-type ball screw drive) [ACURA 1991]

An AE E-M EPS SBW 2WS conversion mechatronic-control system, as illustrated in Figure 4.59 [ACURA 1991], is one that weakens the quantity of steering effort by basically using the output value of the mechanical energy from an E-M motor to it. This mechatronic control system contains of vehicle-velocity sensors, steering-wheel angular velocity and torque sensors, an ECU, a drive unit, and an E-M motor.

Signal outputs from each sensor are input to the ECU, where the essential steering assistance is computed and used by the drive unit to control the operation of the E-M motor. Since the motor's output value of a power is mechatronically controlled in this mechatronic control system, the adjusting range for the steering effort is great. In addition, because it is able to provide only the quantity of output power that is essential when the steering wheel is turned, great attenuation in output power needs may be efficiently fulfilled with no power losses. This shows when a contrast is made to M-F-M FPS SBW 2WS conversion mechatronic control systems that it is not essential for the M-F pump to maintain operating continuously when the steering wheel is not being turned.

In R&P steering mechanisms, the E-M EPS SBW 2WS conversion mechatronic control system uses the E-M motor's mechanical energy to the pinion gear shaft or to the rack shaft. Individual reduction gears are incorporated to amplify the torque of the motor. This system may be arranged systematically according to the drive method, as given in Table 4.4 and Figures 4.60 and 4.61.

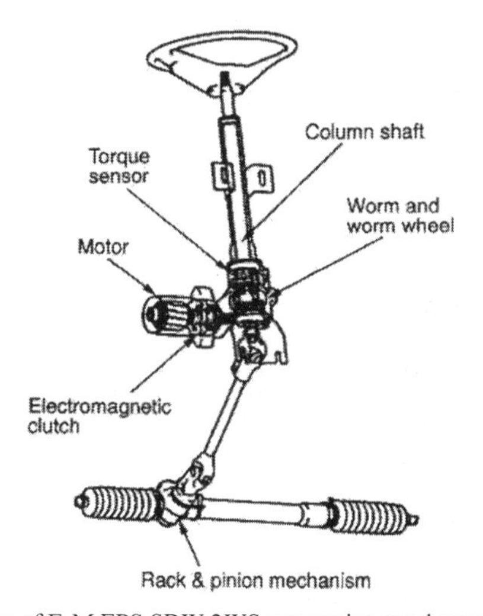

Fig. 4.60 Structure of E-M EPS SBW 2WS conversion mechatronic control system (column shaft drive) [SATO 1991].

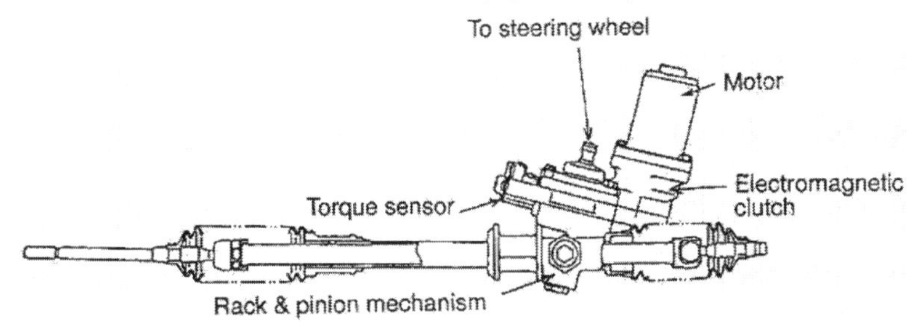

Fig. 4.61 Structure of E-M EPS SBW 2WS conversion mechatronic control system (pinion shaft drive) [SATO 1991].

The maximum amount of assistance, the softness of the steering feel, and the level of noise present during steering are, as a rule, specified by the power transmission systems in Table 4.4 [SATO 1995].

Table 4.4 Classification of EM Motor Mechanism Drive in EPS SBW 2WS Conversion Mechatronic Control

Method	Rack assist	Pinion assist	Figure
E-M motor drive mechanism	Another shaft pinion drive.	E-M motor → planetary gear train → another shaft pinion → rack shaft.	
	Ball screw drive.	E-M motor → ball screw → rack shaft.	0.10
Power transmission mechanism	Column shaft drive.	E-M motor→worm gear → column shaft → pinion shaft.	0.11
	Pinion shaft drive.	E-M motor → gear train → pinion shaft.	0.12

In most cases, it is able to acquire a greater amount of assistance from the rack assist method than from the pinion assist method which is optimal for vehicles in which the FWD load is high. Particulars of the individual sensors, controls, and the results fulfilled thereby are thus set under informal headings [SATO 1995].

Summing up, an AE E-M EPS SBW 2WS conversion mechatronic control system is designed to use an E-M motor to provide directional control to the driver. Most E-M EPS SBW 2WS conversion mechatronic control systems have variable assist that allows for more assistance as the vehicle velocity decreases and less assistance from the mechatronic control system during high-velocity situations. This functionality requires a delicate balance of power and control that has only been available to vehicle manufacturers in recent years.

The E-M EPS SBW 2WS conversion mechatronic control system is replacing the M-F-M FPS SBW 2WS conversion mechatronic control system and is destined to soon become a mainstream among vehicle manufacturers. E-M EPS SBW 2WS conversion mechatronic control systems do not require ECE or ICE power to operate. Thus, a vehicle equipped with an E-M EPS SBW 2WS conversion mechatronic control system may achieve an estimated 3% greater fuel economy than the same vehicle with conventional M-F-M FPS SBW 2WS conversion mechatronic control systems. As an added benefit, more of the ECE's or ICE's power is transmitted to its intended location -- the wheels.

Sensors -- The E-M EPS SBW 2WS conversion mechatronic control system applies a diversity of sensors to control the E-M motor. These sensors contain a torque sensor that measures the steering effort of the steering wheel; a steering wheel angular-velocity sensor that measures the angular velocity of the steering wheel; a DC CH-E/E-CH storage-battery sensor that measures the storage-battery voltage; an electrical-current sensor that measures the motor's armature current and the storage-battery current; and a vehicle-velocity sensor [SATO 1995]. Of these sensors, the torque sensor and the steering wheel angular-velocity sensor, that constitute the essential part of the SBW 2WS conversion mechatronic control system, are represented as follows.

Torque Sensor -- The pinion shaft in the R&P steering mechanism is separated into two parts -- the input shaft and pinion gear. The torque sensor includes a torsion bar that joins the two parts: a slider with a movable soft-iron core, a cam mechanism that changes the twist torque of both parts of the shaft into an axial direction displacement, and a differential transformer that changes the axial direction displacement of the slider into an electrical signal. For example, the torque sensor detects the magnitude and sense of direction of the slider displacement. The main signal is the output signal of torque and others are output signals for fault diagnosis instrumentality. The differential transformer-type torque transducer has a dual electrical structure to supply differential output signals so that a high sensing accuracy and satisfactory temperature characteristics can be acquired and detection of faults can be correctly made.
Angular Velocity Sensor -- The angular-velocity sensor is made up of a gear train that is situated around the input shaft and an AC-DC macrocommutator generator that is driven by the gear train to enhance the angular velocity. The right or left sense of the turning direction and the angular velocity of the steering wheel are measured by the right or left sense of the turning direction and the angular velocity of this AC-DC macrocommutator generator, respectively. An output signal shows the signal from the steering HW angular-velocity sensor.
Electronic Control Unit (ECU) -- The ECU is made up of an interface circuit that equals in importance the signals from the different sensors, an **analog-to-digital** (A/D) converter and **pulse-width-modulation** (PWM) unit that are all built into a 32- or 64 bit single-chip microprocessor, a **watch-dog timer** (WDT) circuit that checks the operation of this microprocessor, and a PWM drive ASIM macrocommutator matrixer that drives the previously mentioned electric power unit. The ECU allows a search for data according to a table lookup method used as the basis for the signals input from each sensor and fulfils a necessary computation employing the data to acquire the assistance force. Also, fault diagnosis for the sensors and the single-chip microprocessor is fulfilled. When a difficulty is discovered, electrical energy to the E-M motor is turned off, an indicator lamp lights up, and the difficulty circumstance memorised, then this difficulty mode appears instantly on a display [SATO 1995].
Power Unit -- The electric power unit comprises a power **metal oxide semiconductor field effect transistor** (MOSFET) DC-DC matrixer that drives the E-M motor in a forward or reverse sense of direction, a drive circuit that controls the respective power MOSFET of this power MOSFET DC-DC matrixer, an electrical current sensor, and a relay that turns the motor's armature current '*ON*' and '*OFF*'. The motor is driven based on instructions from the ECU. The electrical current at this time is monitored by the ECU, and the electrical energy supplied to the motor is turned off in the case of no uniformity. Relying on the magnitude of the motor's armature current, certain mechatronic control systems are furnished with an integrated ECU and electric power unit, while other systems include each part separately [SATO 1995].

E-M Motor Mechatronic Control Systems -- In the physical model (equivalent circuit) of the E-M motor, the relationship between the terminal armature voltage u_a, the armature self-inductance L_a, the armature resistance R_a, and the induced armature voltage constant k, the rotor angular velocity ω, the armature current i_a, and the time t, is represented by the subsequent mathematical model [SATO 1995]:

$$u_a = L_a (\mathrm{d}i_a/\mathrm{d}t) + R_a i_a + k\,\omega, \tag{4.12}$$

$$\approx R_a i_a + k\,\omega. \tag{4.13}$$

Moreover, it is recognised that the motor's armature current i_a is proportional to its assistance torque T_M. As can be comprehended from Eq. (4.12), there are two mechatronic control systems. In the motor's armature-current mechatronic control system, the reference (desired) value of a motor's armature current that is proportional to the value of the motor's assistance torque T_M , is resolved from i_a^* the signal output T_M from the torque sensor. Moreover, control is achieved so that there is no difference between the reference value of an armature current i_a^* and the actual value of an armature current i_a measured through a feedback loop from the armature-current sensor.

In a motor's armature voltage mechatronic control system, the reference value of an armature voltage component ($u_{a1}^* = R_a i_a = k_T T_M$; k_T is a proportional constant) that is equal to the value of the motor's assist torque as computed from the output signal T_M from the torque sensor, and the reference value of the motor's armature voltage component ($u_{a2}^* = k\,\omega$) that is equal to the motor's rotor angular velocity as computed from the output signal θ_s from the steering wheel angular-velocity sensor. These two voltage components are, in that case, added and output [SATO 1995].

Armature Current Mechatronic Control System -- In this mechatronic control system, the reference value of the motor's armature current that is equal to the motor's assistance torque, is adjusted so that it corresponds to the vehicle velocity response type received from the signal of the vehicle-velocity sensor.

Armature Voltage Mechatronic Control System -- In this mechatronic control system, both the E-M motor torque and the motor's rotor angular velocity can be controlled by the output signals from the torque sensor and the steering wheel's angular velocity sensor. When the vehicle is cruising at low velocity, '*normal mechatronic control*' is performed. With this mechatronic control system, the expression $R_a i_a + k\,\omega$ in Eq. (4.13) is output to the motor to obtain a good steering response rate (E-M motor response rate) and thus provide a comfortable steering performance. When the vehicle is travelling at high velocity, it is possible to perform two forms of mechatronic control systems. In the first one, '*return mechatronic control*', the value for $k\,\omega$ is made smaller so that a damping torque that is proportional to the motor's rotor angular velocity generates.

In the second one, '*damper mechatronic control*', the motor's torque generates in the opposite sense of the motor rotation with $u_a = 0$ when the steering wheel is released in turning.

Normal Mechatronic Control System --This is a mechatronic control system of drive control for steering with poor steering effort and an excellent steering response rate. As is displayed in Table 4.5 [ACURA 1991], when the steering wheel is turned to the right, MOSFET electrical valve Q1 is '*ON*' at the same time that MOSFET electrical valve Q4 is performing PWM-driver ASIM macrocomutator and armature-current flows to the MOSFET DC-DC macrocommutator, as shown in Figure 4.62 [KIFUKU AND WADA 1997], uses as a basis for the value of u_a in Eq. (4.13). If the angular velocity of the steering wheel increases, the PWM function also increases.

Table 4.5 DC-AC Commutator Motor Drive During Normal Mechatronic Control [ACURA 1991]

Steering condition	Steering to right	Straight ahead	Steering to left
Q1	ON	OFF	OFF
Q3	OFF	OFF	ON
Q4	PWM	OFF	OFF
Q2	OFF	OFF	PWM
DC-AC commutator E-M motor operation	Operates in a right sense of steering direction	Stops	Operates in a left sense of steering direction

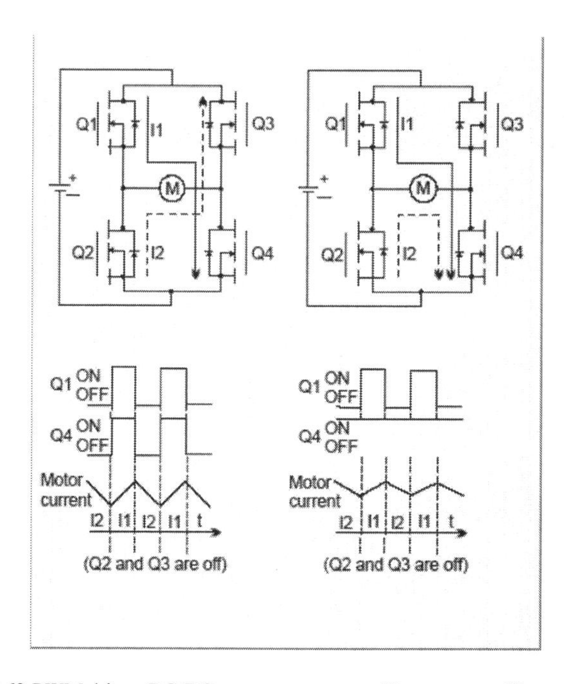

Fig. 4.62 PWM driven DC-DC macrocommutator [KIFUKU AND WADA 1997].

Return Mechatronic Control System -- This mechatronic control system of drive control is for altering the steering wheel return characteristics. When the driver is returning the steering wheel to the neutral position at low-velocity driving, the E-M motor's armature current is instantaneously and uniformly weakened to cause the motor to operate in the reverse sense of the motor's rotor rotation to the torque originating senses of motion direction.
As a result, a satisfactory returnability of the steering wheel should be gained. At high-velocity driving, the motor's armature current is slowly and uniformly decreased to suppress returnability and gain more stable steering characteristics.

Table 4.6 DC-AC Commutator E-M Motor Drive during Return Mechatronic Control [Source: ACURA 1991]

Steering condition	Return from right steering to straight ahead	Return from left steering to straight ahead
Q1	PWM	OFF
Q3	OFF	PWM
Q4	PWM	OFF
Q2	OFF	PWM

As represented in Table 4.6 [ACURA 1991], when the steering wheel returns to the basic position after being turned to the right, MOSFET electrical valve Q1 performs PWM-driver ASIM macrocommutator used as a basis for the signals from the steering wheel angular-velocity sensor and, at the same time, MOSFET electrical valve Q4 also performs PWM-driver ASIM macrocommutator used as a basis for the signals from the torque sensor.
Damper Mechatronic Control System -- This mechatronic system of drive control is for enhancing the convergence of the steering wheel when the vehicle is cruising at high velocity and for suppressing wandering of the steering wheel supplied by the wheel-tyre inputs. When the motor's armature terminals are short-circuited, it is able to generate torque in the reverse sense of direction in proportion to the angular velocity of the motor's rotor and this characteristic is used for mechatronic control.

4.3 Energy-Saving Effectiveness

4.3.1 Foreword

Since the E-M EPS SBW 2WS conversion mechatronic control system is a **power-on-demand** (POD), one that provides only the essential amount of mechanical energy at the required time, very little energy loss is encountered- when the steering HW is not being turned. As a consequence of this, the system has exceptionally high fuel efficiency [SATO 1995].

Table 4.7 Measurements Results on Mode Specific Fuel Consumption [SHIMIZU ET AL. 1991]

Mode			Highway driving	Urban road driving
Electrically powered steering (EPS)		mpg	28.88	20.51
Fluidically powered steering (FPS)		mpg	28.18	20.01
Betterment in specific fuel consumption (SFC)	SFC difference	mpg	1.14	0.80
	Bettered rate	%	2.52	2.50

Table 4.7 represents the measurement results of **specific fuel consumption** (SFC) during various driving modes to compare EPS, that is, an E-M EPS SBW 2WS conversion mechatronic control system with the FPS, or an M-F-M FPS SBW 2WS conversion mechatronic-control system [SHIMIZU ET AL. 1991].

4.3.2 Tendency in Research and Development (R&D)

The requirements for greater vehicle velocity, higher quality, and decreased energy demands in automotive vehicles are chronically enhancing. In uniformity to react to these requirements, R&D is preparing the application of mechatronic control systems with the objective of further enhancing normal actions and performance [SATO 1995]. Features that are being investigated include the insertion of **neural networks** (NN) and **fuzzy logic** (FL), that is, **neuro-fuzzy** (NF), as well as the application of power steering that reacts to the driving surroundings by altering the assistance amount in uniformity with the traffic circumstances or the on- and/or off-road surface situation to supply steering feel to adjust the sensitiveness of drivers. The most necessary of these is most likely to be active reaction power steering that stipulates feedback to the driver concerning the behaviour of the vehicle in the form of steering reaction force. Such a SBW 2WS conversion mechatronic control system supplies the driver with information concerning the operating circumstances of the vehicle, for instance, the yaw velocity and/or lateral acceleration, as steering reaction forces.

Not only would it better the relationship between the driver and the vehicle to make it possible to gain a steering feel that satisfies the sensitivities of the driver, but a normal action that automatically resituates for abnormalities in vehicle behaviour provided by secondary disturbances would be desirable.

Figure 4.63 is an active reaction SBW 2WS conversion mechatronic control system's structural and functional block diagram in which the yaw rate is fed back as a steering reaction force [SATO 1995].

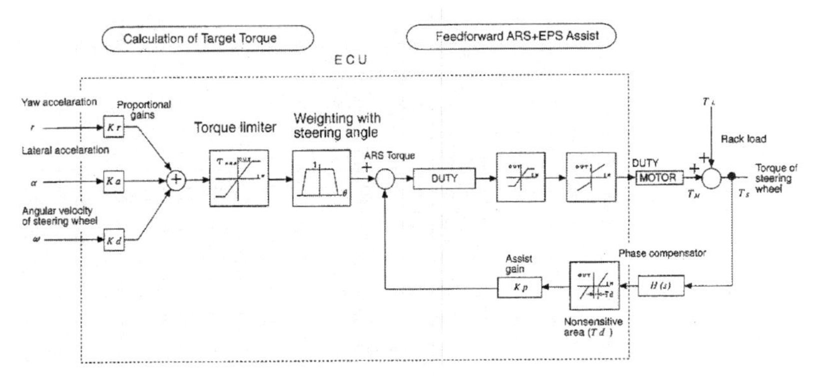

Fig. 4.63 Structural and functional block diagram of active reaction SBW 2WS conversion mechatronic control [SATO 1995]

Figure 4.64 illustrates an example of the effect of suppressing abnormalities in vehicle behaviour provided by secondary disturbances by comparing the amount of lateral removal when braking in a rut on a road surface during normal power steering [SATO 1995].

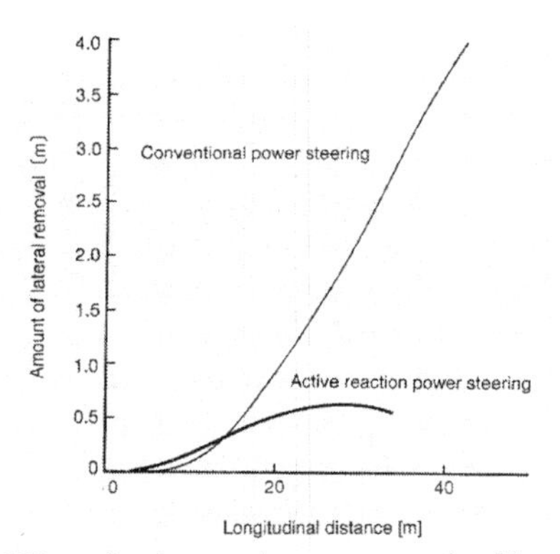

Fig. 4.64 Effect of active reaction power steering [SATO 1995].

Summing up, an SBW AWS conversion mechatronic control system means the end of the mechanical connection between the steering wheel and the road wheels, with mechatronically-controlled actuators setting the steering angles and electronically-controlled force feed-back to the driver's steering *'device'*.

The SBW AWS conversion mechatronic control system may be the enabler for fully integrated vehicle stability mechatronic control systems, for collision avoidance mechatronic control systems, and potentially one day, even for autoenormous driving.

The challenge for very advanced SBW AWS vehicle technology is defining the fault-tolerant electrical architecture with internal redundancies that enables the mechatronic control system to function when the vehicle has no mechanical connection between the steering wheel and the road wheels.

Automotive scientists and engineers are now working to develop an SBW AWS conversion mechatronic control system that delivers advanced functionality facilitated through a RBW or XBW integrated unibody, space-chassis, skateboard-chassis, or body-over-chassis mechatronic control hypersystem, such as variable steer ratio according to vehicle velocity (speed) and a choice of driver feel, but at a cost viable for the mass market. For instance, Delphi Corporation has introduced an innovative **active front steering** (AFS) SBW AWA conversion mechatronic control system that may provide variable ratio control with virtually no compromise in steering performance. This system may enable vehicle manufacturers to improve vehicle stability and to program the desired variable steering ratio while virtually eliminating trade-offs to base steering performance such as noise, lash, return-ability, and on-centre feel.

The Delphi's SBW AWS conversion mechatronic control system is designed for efficient interaction with conventional M-F-M steering and may be installed with no increase in M-F-M FPS SBW AWS conversion mechatronic control system or component size, resulting in efficiencies comparable to traditional M-F-M steering [GAUT 2003B].

Delphi Corporation has demonstrated its AFS AWA SBW conversion mechatronic control system capabilities and received very positive remarks from several vehicle manufacturers. The design addresses many of the vehicle manufacturers' concerns and, as a result, provides a transparent, smooth AWA SBW conversion mechatronic control system with virtually no compromise or trade-offs to base steering performance.

AFS helps provide the driver with simplified city driving and parking by reducing the turning required at low vehicle velocity so that a hand-over-hand parking manoeuvre may be accomplished in as little as two-thirds of a turn of the steering wheel.

Delphi's AFS smoothly transitions from a low-speed steering ratio to a high-speed steering ratio, providing a tighter, sportier feel for driving enjoyment and better control on the highway. It accomplishes this by modifying the steering kinematics, or motion, of the vehicle in a manner similar to other SBW. The system electronically influences the steering angle on the wheels, enabling it to be greater or less than the driver's steering wheel angle input. Thus, turning into a parking spot or even negotiating a hairpin turn at moderate velocities may be accomplished with significantly fewer turns of the steering wheels.

In essence, the system electronically turns the road wheels at a rate different from the rate the driver turns the steering wheel.

Although some may think this could be intrusive or controlling, those that have experienced Delphi's AFS realise it helps make driving very easy and enjoyable with a very natural, transparent feel. Unlike other SBW, AFS maintains the mechanical link and uses the existing electrical architecture. This mechanical link helps ensure system safety. If the system is switched off or inadvertently loses power, the system may default smoothly to the base steering ratio without disturbing or alarming the driver.

AFS is the newest technology in Delphi's line of RBW or XBW integrated unibody, space-chassis, skateboard-chassis, or body-over-chassis motion mechatronic control hypersystems, which use integral control technology to help enhance safety and improve vehicle performance, ride, and control. It may be integrated to provide a seamless connection between SBW AWS conversion, BBW AWB dispulsion, and ABW AWA suspension mechatronic control systems. It also may be integrated with controlled braking to provide a more effective vehicle system solution to stability control than brakes alone. It may instantaneously deliver steering control; counter steering the vehicle to bring it back on its intended course and blending in braking, if needed, through finely tuned mechatronic controls that are virtually transparent to the driver. In addition, this integration may help minimise stopping distances on split and mixed coefficient on/off road surfaces while maintaining directional stability [GAUT 2003B].

4.4 Steer-By-Wire (SBW) Four-Wheel Steering (4WS) Conversion Mechatronic Control Systems

4.4.1 Foreword

The aim of a **steer-by-wire** (SBW) **four-wheel steering** (4WS) conversion mechatronic control system is better stability during overtaking manoeuvres, reduction of vehicle oscillation around its vertical axis, reduced sensibility to lateral wind, neutral behaviour during cornering, and so on, that is, improvement of active safety [BRABEC ET AL. 2004].

The SBW 4WS conversion mechatronic control system is computer-controlled and automatically adjusts rear wheel angles according to the steering wheel's position, vehicle velocity, and other variables.

More than four decades ago, hardly any automotive scientists and engineers had arrived at the far-sighted and well-calculated conclusion that was quite revolutionary for the time. For instance, in their technical presentation at the *1962 Japanese Automotive Engineers' Society Technical Conference*, they summarised their arduous research concerning vehicle dynamics as follows [MAZDA 2002].

The basic differences in the characteristics of oversteer and understeer lies in the magnitude of time delay and response:

- A vehicle that is stable under high speed must possess understeer characteristics;
- The rear wheel tyre reflects heavily on the stability;
- A major improvement on control and stability may be anticipated by means of the automatic rear wheel steering system.

The automotive scientists' and engineers' ultimate objective was still a positive measure to generate forces for positive controls; a SBW 4WS conversion mechatronic control system.

The degree of **rear wheel steering** (RWS) was determined by the measurement of both the front wheel steering angle and vehicle velocity, by means of a central microcomputer unit.

Automotive vehicles of the future may combine a refined mechatronically-controlled SBW 4WS conversion mechatronic control system with a continually varying torque-split, DBW 4WD propulsion mechatronic control system and powerful *'crankless'* mechatronic commutator ICEs [FIJALKOWSKI 1986, 1998B, 1999D, 2000C].

The mechatronically-controlled, vehicle-velocity-sensing SBW 4WS conversion mechatronic control system steers the rear wheels in the direction, and to a degree, most suited to the corresponding vehicle velocity range. The system may be mechanically and fluidically actuated, creating greatly enhanced stability, and within certain parameters, agility.

The driver of a 4WS-equipped vehicle derives five strategic benefits, over and above the conventional vehicle chassis [MAZDA 2002]:

- ❖ Superior cornering stability;
- ❖ Improved steering responsiveness and precision;
- ❖ High-speed straight-line stability;
- ❖ Notable improvement in rapid lane-changing manoeuvres;
- ❖ Smaller turning radius and tight-space manoeuvrability at low vehicle speed range.

The most outstanding advantage of the SBW 4WS conversion mechatronic control system contributes to a notable reduction in driver fatigue over high vehicle velocity and extended travelling. This is achieved by optimally [MAZDA 2002]:

- ❖ Reducing the response delay to steering input and action;
- ❖ Eliminating the automotive vehicle's excessive reaction to steering input.

In essence, by providing the optimum solution to the phenomena researched by automotive scientists and engineers in the early 1960s -- using the method advocated by them -- the SBW 4WS mechatronic control system has emerged as a fully beneficial technology.

The SBW 4WS conversion mechatronic control system delivers improved active safety while providing optimal handling characteristics without traditional chassis tuning trade-offs.

For vehicles with very long wheelbases and ones that require to be operated in narrow places, the concept of SBW 4WS conversion mechatronic control system is desired.

In such SBW 4WS conversion mechatronic control systems, the rear wheels are turned in the opposite sense of the turning-motion (yawing) direction to the sense of turning-motion direction of the front wheels in order to make the turning radius as small as possible and to better handling ability. Such SBW 4WS conversion mechatronic control systems have been under investigation for some time.

However, the concept of the SBW 4WS conversion mechatronic control system being applied in passenger vehicles with the intention of enhancing vehicle stability and steering response at medium to high vehicle velocities is slightly new. The system for passenger vehicles has the following two objectives:

- ❖ Decreasing the turning-motion (yawing) of the vehicle by steering the rear wheels in the identical sense of turning-motion direction as the front wheels, thus bettering the vehicle stability at high values of the vehicle velocity;
- ❖ Enhancing the steering response at medium vehicle velocity, while at the same time decreasing the turning circle radius at low velocity, by steering the rear wheels in the opposite sense of the turning-motion direction to the front wheels.

4.4.2 Philosophy of SBW 4WS Conversion Mechatronic Control Systems

Reduction of Minimum Value of the Turning Radius - As illustrated in Figure 4.65, showing the origin point of the coordinate situated at the centre of the rear tread, the coordinates of the turning centre of the vehicle $P(x_0\ y_0)$ when the rear wheels are turned in the opposite sense of the turning-motion direction from the front wheels are specified in the subsequent equations [SATO 1991]:

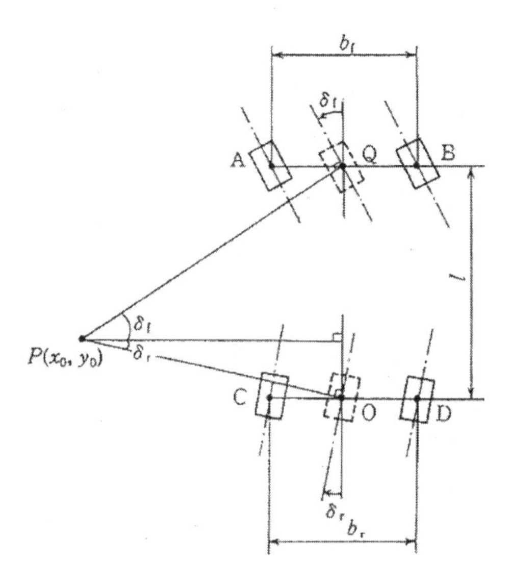

Fig. 4.65 Single-track half vehicle (bicycle) physical model of 4WS [JSAE 1991].

$$x_0 = \frac{l}{\tan\delta_f + \tan\delta_r}, \tag{4.15}$$

$$y_0 = \frac{l\tan\delta_f}{\tan\delta_f + \tan\delta_r}. \tag{4.16}$$

Furthermore, if the turning radius for the front outer wheel is R and the difference between the turning radius of the front and rear outer wheels is ΔR, then

$$R = \overline{BP} = \sqrt{\left(\frac{\delta_f}{2} + x_0\right)^2 + (l - y_0)^2} \cdot \sqrt{\left(\frac{\delta_f}{2} + \frac{l}{\tan\delta_f + \tan\delta_r}\right)^2 + \left(\frac{l\tan\delta_f}{\tan\delta_f + \tan\delta_r}\right)^2}, \tag{4.17a}$$

$$\Delta R = \overline{AP} - \overline{CP} = \sqrt{\left(-\frac{\delta_f}{2} + x_0\right)^2 + (l - y_0)^2} - \sqrt{\left(-\frac{\delta_r}{2} + x_0\right)^2 + y_0^2}$$

$$= \sqrt{\left(-\frac{\delta_f}{2} + \frac{l}{\tan\delta_f + \tan\delta_r}\right)^2 + \left(\frac{l\tan\delta_f}{\tan\delta_f + \tan\delta_r}\right)^2}$$

$$- \sqrt{\left(-\frac{\delta_r}{2} + \frac{l}{\tan\delta_f + \tan\delta_r}\right)^2 + \left(\frac{l\tan\delta_f}{\tan\delta_f + \tan\delta_r}\right)^2}, \quad (4.17b)$$

$$R = \overline{PQ} = \frac{l}{\sin\delta_f + \cos\delta_f + \tan\delta_f}, \quad (4.18)$$

$$\Delta R = \overline{PQ} - \overline{PO} = \frac{l}{\sin\delta_f + \cos\delta_f + \tan\delta_f} - \frac{l}{\sin\delta_f + \cos\delta_f + \tan\delta_f}. \quad (4.19)$$

It may be noted from Eq. (4.18) that when the rear wheel steer in the opposite sense of the turning motion direction as the front wheels, the turning radius becomes less than when the rear wheels are not steered ($\delta_r = 0$).

Furthermore, it may be noted from Eq. (4.19) that when the amount of steering for the front and rear wheels is the same ($\delta_f = \delta_r$), then it becomes able to gain no difference between the turning radius of the front and rear wheels.

Enhancement in Stability and Manoeuvrability When Driving at Medium to High Vehicle Velocities -- The steering characteristics in yaw velocity and lateral acceleration of the vehicle with a SBW 4WS conversion mechatronic control system in which the rear wheels are steered relative to the front wheel steering angle in identical sense of the turning motion direction, are presented in the Nomenclature, along with that for a vehicle yaw SBW 2WS conversion mechatronic control system [SATO 1995].

However, the single-track half-vehicle (bicycle) physical model displayed in Figure 4.65 is applied as a vehicle physical model [SATO 1991].

From Table 4.8, in a SBW 4WS conversion mechatronic control system in which the rear wheels are turned in the identical sense of the turning-motion direction as the front wheels, the stability factor K, the damping ratio ζ, and the natural oscillation of yawing are unchanged from the values for a conventional vehicle with a SBW 2WS conversion mechatronic control system, so that the intrinsic stability of the vehicle may not alter [SATO 1991].

On the other hand, because the lateral acceleration response delay will become less as the coefficients of the complex frequency (*Laplace* numerator) s and s^2 increase and the steady- state gain in the yaw angular velocity causes a fall in proportion to *(1 -- k)*, the yawing movement that occurs at the same time as the lateral movement of the automotive vehicle, will become smaller and the stability within the range of practical use may be improved.

Table 4.8 Steering Characteristics [SATO 1991]

	Automotive vehicle with SBW 2WS	Automotive vehicle with SBW 4WS in which the rear steering angle is proportional to the front
Stability factor	$K=\frac{m\left(l_r K_r - l_f K_f\right)}{2l^2 K_f K_r}$	K
Steady state gain in yaw velocity	$\frac{1}{1+KV^2}\frac{V}{l}$	$\frac{1}{1+KV^2}\frac{V}{l}$
Damping rate	$\varsigma=\frac{1}{2}\frac{\left(K_f+K_r\right)I_z+\left(K_f l_f^2+K_r l_r^2\right)m}{\sqrt{m I_z K_f K_r l^2\left(1+KV^2\right)}}$	ς
Response angular frequency	$\omega_n=\frac{2l}{V}\sqrt{\frac{K_f K_r\left(1+KV^2\right)}{I_z m}}$	ω_n
$\frac{\phi}{\delta_H}$	$\frac{1}{i_s}G_\varphi(0)\frac{T_f s+1}{\frac{1}{\omega_n^2}s^2+\frac{2\varsigma}{\omega_n}s+1}$ $G_\varphi(0)=\frac{1}{1+KV^2}\frac{V}{l}$	$\frac{1-k}{i_s}G_\varphi(0)\frac{T_f s+1}{\frac{1}{\omega_n^2}s^2+\frac{2\varsigma}{\omega_n}s+1}$ $G_\varphi(0)=\frac{1}{1+KV^2}\frac{V}{l}$ $\lambda_\varphi=\frac{k}{1-k}\frac{T_f-T_r}{T_f}$
$\frac{\alpha_y}{\delta_H}$	$\frac{1}{i_s}G_{\alpha_y}(0)=\frac{T_{\alpha_y,2}s^2+T_{\alpha_y,1}s+1}{\frac{1}{\omega_n^2}s^2+\frac{2\varsigma}{\omega_n}s+1}$ $G_{\alpha_y}(0)=\frac{1}{1+KV^2}\frac{V^2}{l}$ $T_{\alpha_y,1}=\frac{l_r}{V},\ T_{\alpha_y,2}=\frac{I_z}{2K_r l}$	$\frac{1-k}{i_s}G_{\alpha_y}(0)=\frac{\left(1+\lambda_{\alpha_y,2}\right)T_{\alpha_y,2}s^2+\left(1+\lambda_{\alpha_y,1}\right)T_{\alpha_y,1}s+1}{\frac{1}{\omega_n^2}s^2+\frac{2\varsigma}{\omega_n}s+1}$ $G_{\alpha_y}(0)=\frac{1}{1+KV^2}\frac{V^2}{l}$ $\lambda_{\alpha_y,1}=\frac{1}{1-k}\frac{T_{\alpha_y,1}+T'_{\alpha_y,1}}{T_{\alpha_y,1}}$ $\lambda_{\alpha_y,2}=\frac{1}{1-k}\frac{T_{\alpha_y,2}+T'_{\alpha_y,2}}{T_{\alpha_y,2}}$ $T_{\alpha_y,1}=\frac{l_r}{V},\ T_{\alpha_y,2}=\frac{l_f}{V}$ $T_{\alpha_y,2}=\frac{I_z}{2K_r l},\ T'_{\alpha_y,2}=\frac{I_z}{2K_f l}$

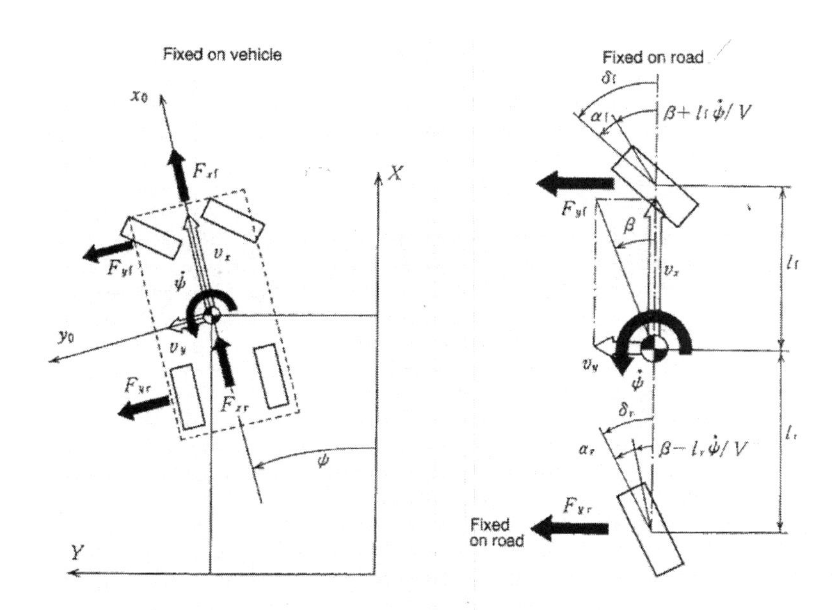

Fig. 4.66 Double-track full vehicle physical model fixed on vehicle and single-track half vehicle physical model fixed on road [SATO 1991].

If, however, the rear wheels are steered in the opposite sense of turning direction to the front wheels, $k < 0$ and so $1 -- k > 1$. This means that the steady-state gain of the yaw velocity may increase with the result that the steering response will be improved. The symbols applied in Table 4.8 and Figure 4.66 [SATO 1991] are interpreted in the Nomeclature.

4.4.3 Dynamic Analysis of SBW 4WS Conversion Mechatronic Control Systems

The focus of research may be on the development of vehicles with improved performance, increased stability, and enhanced manoeuvrability. In this section an existing technology, such as SBW 4WS conversion mechatronic control systems, will be discussed. Recent research on SBW 4WS devices has been constantly advancing the technology. It may be observed that little research has been done that examines the important vehicle dynamic aspects. This section principally may be focused on the development and analysis of a physical model, for example, as shown in Figure 4.67 [VILLEGAS AND SHORTEN 2005], before considering control aspects of 4WS.

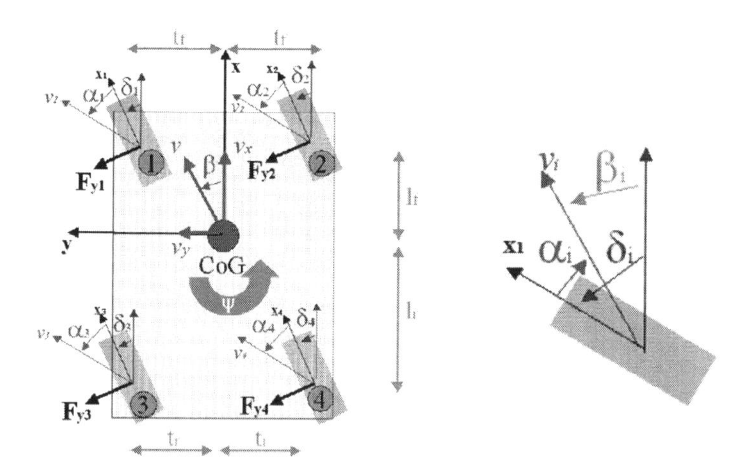

Fig. 4.67 A lateral dynamics physical model with a wheel-tyre index for every wheel ($i = 1, 2, 3, 4$) and wheel-tyre angles [VILLEGAS AND SHORTEN 2005].

Such results may be presented to make the application of the mechatronic control system on an automotive vehicle easier. This research may set an objective of making data useful for the development of a SBW 4WS device through computer simulation physical models and dynamic analyses [LAKKAD 2003].

A single-track half-vehicle (bicycle) physical model for the 4WS vehicle will be developed and vehicle-handling characteristics using *Matlab*® simulations will be discussed. Also, a mathematical model relating RWS input to FWS input will be developed. An **automatic dynamic analysis of mechanical systems** (ADAMS) computer program may be used to simulate physical model (see Fig. 4.68) [LAKKAD 2003].

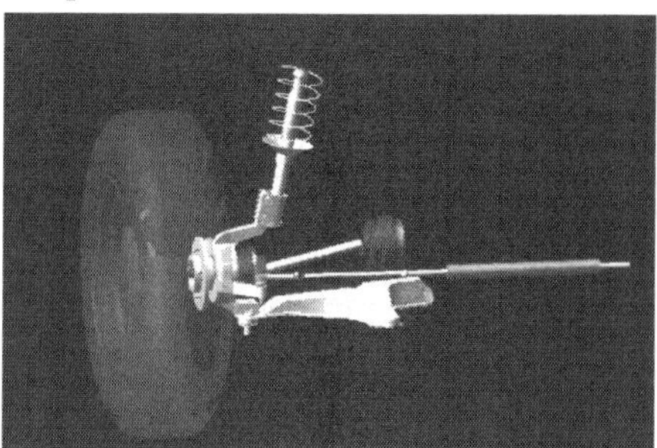

Fig. 4.68 Steering demonstration physical model in ADAMS [LAKKAD 2003].

ADAMS, in the developer's words, is the world's most commonly used mechanical system simulation software. It helps automotive scientists and engineers to produce virtual prototypes, realistically simulating the full-motion behaviour of sophisticated mechanical systems on their computers and quickly analysing multiple design variations until an optimal design is acquired.

All the components may be installed except for the ADAMS/ PRE® (a suspension/chassis dynamics tool analogous to ADAMS/ CAR®). ADAMS may interact in various ways with the following software in the ENGINEERING BUILDING: UNIGRAPHICS, ANSYS, ABAQUS, AND MATLAB.

A 3D vehicle physical model of the XUV may be developed in ADAMS. To validate the physical model, an ADAMS demonstration steering physical model results may be used. The same input shown in Figure 4.69 is given to the XUV physical model [LAKKAD 2003]. The results of both physical models may be compared for validation of the XUV physical model.

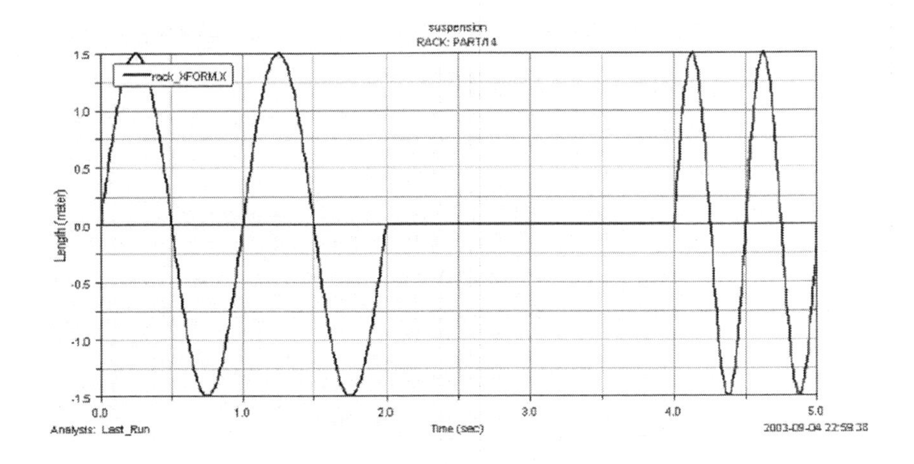

Fig. 4.69 Rack input [LAKKAD 2003].

The requirements for SBW 4WS conversion mechatronic control systems may become even more diverse.

The demand for optimal feedback leads inevitably to the employment of R&P steering systems, especially for vehicles with high front axle loads.

Because R&P steering typically shows a significantly higher efficiency from the tie rod to the steering shaft, the desired feedback may be achieved. Figure 4.70 shows the XUV ADAMS physical model [LAKKAD 2003].

A common R&P arrangement may be used for FWS. An E-M motor may drive the pinion (not shown in the physical model). As mentioned earlier, RWS input may be a function of steering input.

The ratio of RWS to FWS used here is --1. This may be used since the maximum velocity of XUV may be circa 60 km/h (40 mph) and falls into the low velocity category.

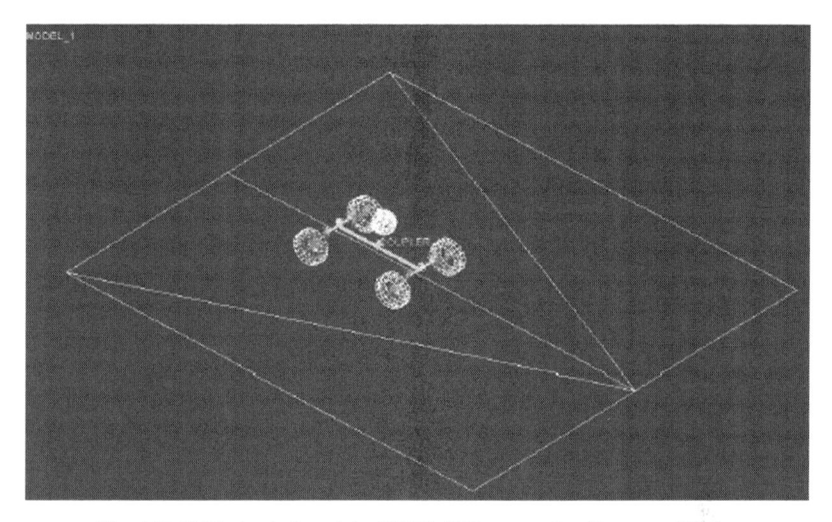

Fig. 4.70 XUV physical model of SBW 4WS conversion [LAKKAD 2003].

Referring to Figure 4.71, at low values of vehicle velocity, the rear wheels may move in the opposite sense of direction to the front wheels which gives greater manoeuvrability for the automotive vehicle [LAKKAD 2003]. As the velocity increases, the ratio K_s may also change and the rear wheel may move accordingly. The results in this section was discussed for $K_s = -1$ [LAKKAD 2003].

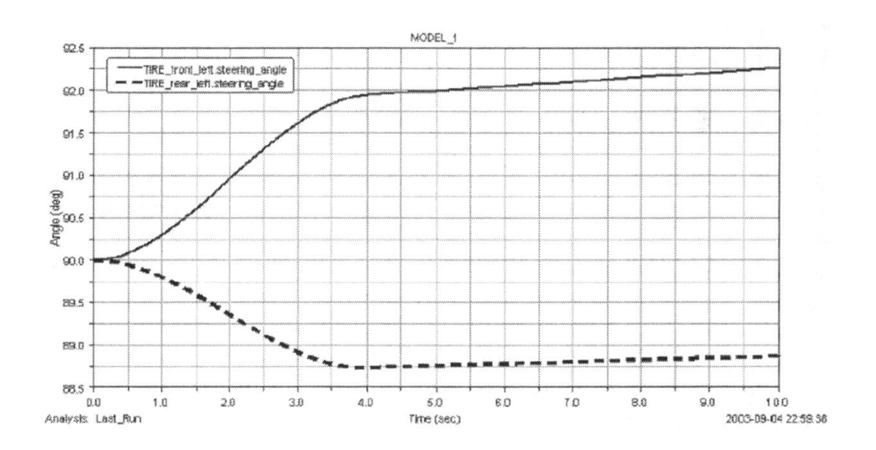

Fig. 4.71 Comparison of front left and rear left steering angles [LAKKAD 2003].

The input to the rack may be given as shown in Figure 4.69. As seen from Figure 4.71, the steering angle may follow the steering input. The angular velocity also follows the input.

4.4.4 Categories of SBW 4WS Conversion Mechatronic Control Systems

SBW 4WS conversion mechatronic control systems that are presently being developed in automotive vehicles are arranged in conformity with their normal actions and mechanisms.

The objectives and characteristics of each control system are interpreted in Tables 4.9 and 4.10 [SATO 1991].

Because the SBW 4WS conversion mechatronic control system represented in Table 4.10 has a low degree of control freedom, there is little incentive for it to be applied. Regardless of its low disbursement, however, this system causes every fundamental normal action of SBW 4WS, so that there is a possibility that it may be applied to vehicles in the not-to-distant future, especially in smart vehicles.

Table 4.9 [SATO 1991]

Classification by function	Small range of rear steering angle only controlled mechatronically. Not only a small range in medium to high values of vehicle velocity but also a large range in low values of the vehicle velocity of rear steering angle is controlled mechatronically.
Aims	Improvement of steering response and vehicle stability in medium to high values of vehicle velocity. In addition to the above, making the minimum turning radius smaller.

Table 4.10 [SATO 1991]

Classification by mechanism	Mechano-mechanical (M-M) SBW 4WS conversion mechatronic control system Mechano-fluido-mechanical (M-F-M) SBW 4WS conversion mechatronic control system Electro-mechano-fluido-mechanical (E-M-F-M) SBW 4WS conversion mechatronic control system Electro-mechanical (E-M) SBW 4WS conversion mechatronic control system
Feature	Simple mechanism High degree of control freedom (compact actuator) High degree of control freedom (mechanism is not simple) High degree of control freedom (simple mechanism)

SBW 4WS conversion mechatronic control systems may be installed on some vehicles where the rear wheels also help to turn the vehicle and they are operated by mechanical or fluidical, or even electrical systems.

❖ *Mechano-mechanical (M-M) SBW 4WS conversion*
  ➢ Two steering gears are used:
    • one for the front;
    • one for the rear.
  ➢ A steel shaft connects the two steering gearboxes;
  ➢ The shaft terminates at an eccentric shaft that is fitted with an offset pin;
  ➢ The pin engages a second offset pin that fits into a planetary gear.

❖ *Mechano-fluido-mechanical (M-F-M) SBW 4WS conversion*
  ➢ The rear wheels turn in the same direction as the front wheels;
  ➢ The rear wheels turn no more than 1½ deg;
  ➢ The system activates at speeds above 50 km/h;
  ➢ The system will not operate when the vehicle moves in reverse.

❖ *Electro-mechano-fluido-mechanical (E-M-F-M) SBW 4WS conversion*
  ➢ It is a combination of computer electronic controls and fluidics;
  ➢ A velocity sensor and steering HW angle sensor feed information to the ECU;
  ➢ The ECU commands the fluidical system to steer the rear wheels;
  ➢ The rear wheels don't play a large part in the steering process at low velocity.

4.4.5 Foreword to Each SBW 4WS Conversion Mechatronic Control System

Mechano-Mechanical (M-M) SBW 4WS Conversion Mechatronic Control System - This was the first SBW 4WS conversion mechatronic control system applied in passenger vehicles. This system is a **front-wheel steering** (FWS) angle responsive-type SBW 4WS conversion mechatronic control system in which the **rear-wheel steering** (RWS) angle is fixed precisely by the steering angle of the front wheels.

The rear axle of a vehicle equipped with a conventional wheel suspension has a tendency to oversteer in curves under the action of lateral forces. Large slip angles at the rear axle cause undesirable sideslip angles.

In order to maintain favourable directional stability, measures must be taken to prevent rear-wheel oversteer, even during sudden alterations in direction at high velocity. This may be achieved by taking up the lateral forces elastically, or through steering the rear wheels; in the latter circumstance, the steer angle does not need to be more than $\pi/180$ rad or $\pi/90$ rad [SATO 1995].

The shorter the delay time of this compensating reaction, the larger is the safety reserve towards the limit of adhesion.

Figure 4.72 shows an '*intelligent*' rear axle that is used in some vehicles [SEIFFERT AND WALZER 1991]. It is elastically connected to the vehicle body by means of an asymmetrically designed axle support bearing (shown in the fore-ground); under lateral loads, this arrangement is '*track-correcting*'.

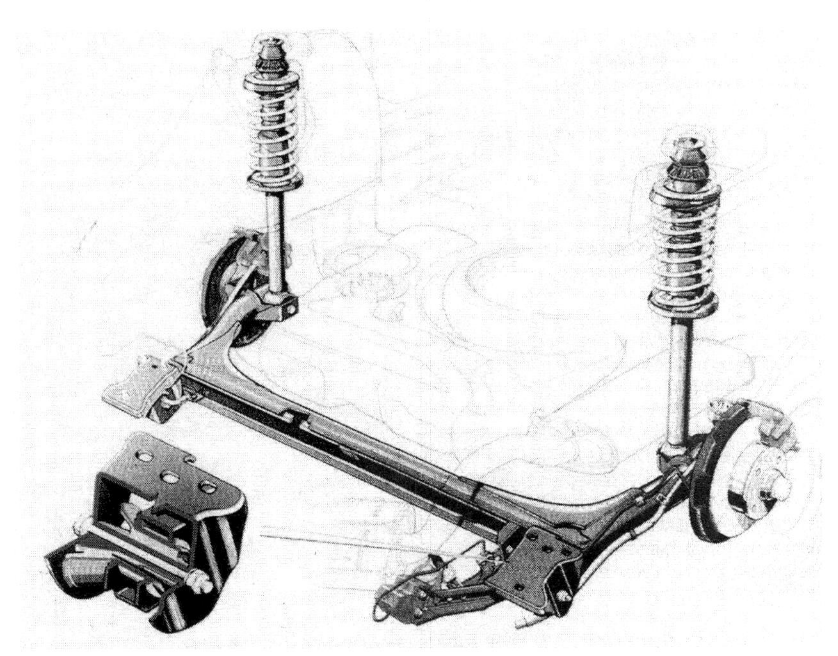

Fig. 4.72 Track-correcting rear axle of the VW *Passat*; left lower corner: the asymmetrically designed axle support bearing [SEIFFERT AND WALZER 1991].

Figure 4.73 explains this action in comparison to a conventional bearing [SEIFFERT AND WALZER 1991]. Since the bearing's kinematics compensates for the oversteering influence of lateral loads, only a small sideslip angle remains that is the angle between the longitudinal vehicle axis and the direction of the vehicle velocity at the barycentre (centre of gravity).

The advantage of this elasto-kinematic arrangement is the ability to react to any kind of lateral forces, including wind gusts, in addition to its simple construction that guarantees high reliability.

M-M SBW 4WS conversion mechatronic control systems are offered today that keep the sideslip angle small through the deliberate counter-steering action of the rear wheels. For this purpose, the rear wheels must be turned in the same sense of direction as the front wheels but at a much reduced steering angle. It is also desirable to make this action depend on the driving vehicle velocity. Given existing cost limitations, an M-M SBW 4WS conversion mechatronic control system may yield only minimal, if any, advantages over the afore-described elasto-mechanical one.

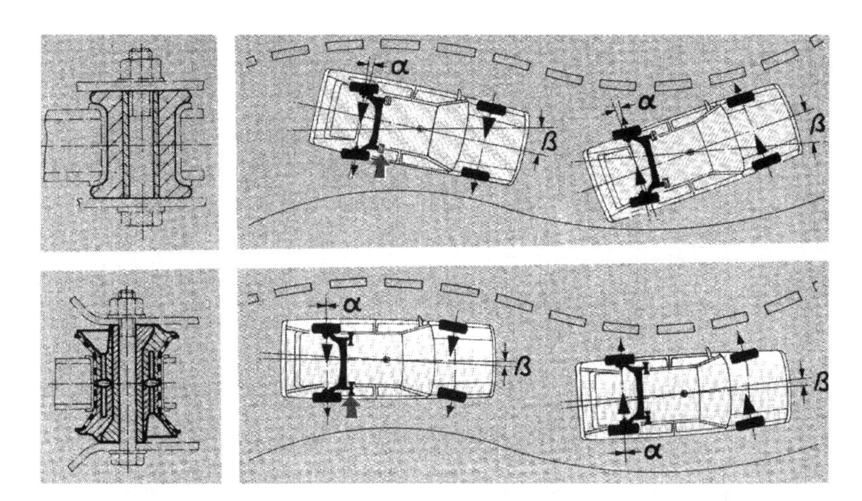

Fig. 4.73 Function of the track-correcting rear axle with the asymmetrically designed support bearing (upper picture) and comparison to the function of a rear axle with conventional support bearings (lower picture) [SEIFFERT AND WALZER 1991].

Further improvements in vehicle dynamics are possible; however, through electronically controlled M-M SBW 4WS conversion mechatronic control systems. It would be possible, for instance, to automatically correct the influence of wind forces from the side or yawing moments around the vertical axis that result from uneven brake forces. The structure and assembly of this SBW4WS conversion mechatronic control system are illustrated in Figures 4.74 and 4.75, respectively [FURUKAWA 1989].

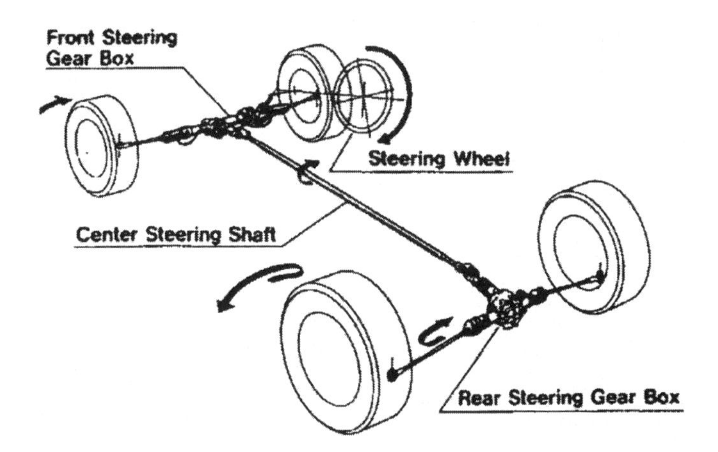

Fig. 4.74 Principle layout of a mechano-mechanical (M-M) four-wheel steering (4WS) [FURUKAWA 1989].

Fig. 4.75 Mechano-mechanical (M-M) four-wheel steering (4WS) [FURUKAWA 1989].

In R&P type steering gearboxes for the front wheels, a RWS pinion is arranged accordingly to transfer the steering angle of the front wheels to the rear wheels.

The displacement in the steering angle is transferred to the RWS gearbox towards the centre steering shaft. The RWS gearbox contains an aggregation of an eccentric shaft and a planetary gear. The result of this is that when the value of a steering angle for the front wheels is small, the rear wheels turn in the same sense of direction, but when the value of a steering angle for the front wheels is large, the rear wheels turn in the opposite sense of direction, as illustrated in Figure 4.76 [FURUKAWA 1989].

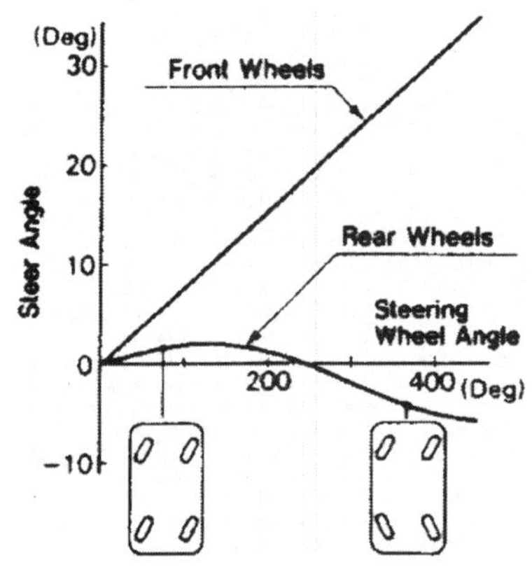

Fig. 4.76 Characteristics of front wheel steering (FWS) vehicle velocity-responsive 4WS conversion mechatronic control system [FURUKAWA 1989].

During high-velocity driving, since the front wheels are only turned by very small amounts, the rear wheels turn in the identical sense of direction, and driving stability is improved.

At very low values of vehicle velocity where the front wheels are steered through much larger angles, the rear wheels turn in the opposite sense of direction and the turning radius becomes small, thus, the working ability should be improved.

M-F-M SBW 4WS Conversion Mechatronic Control System -- *Vehicle Velocity/ Lateral Acceleration Responsive Mode* - This mechatronic control system was set up in advanced automotive vehicles. The structure of the system and the arrangement of the RWS gearbox are illustrated in Figure 4.77 [SATO 1991, 1995].

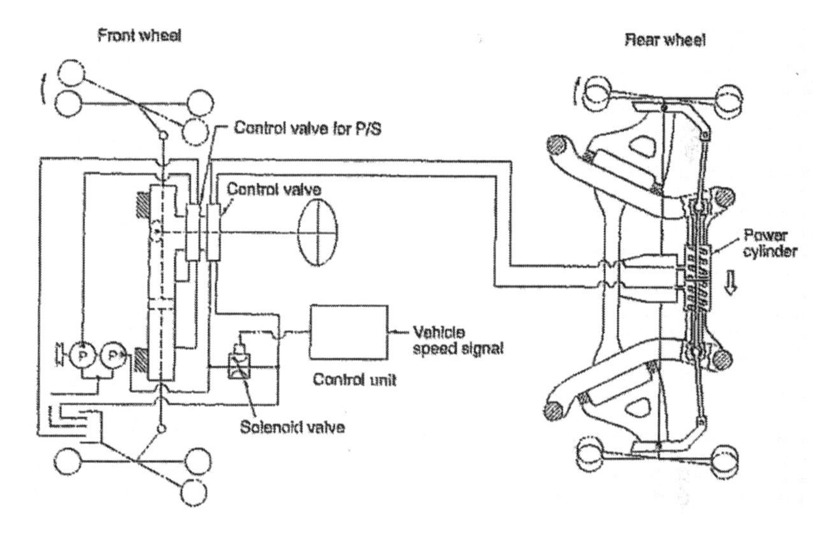

Fig. 4.77 Principle layout of a vehicle velocity/lateral acceleration-responsive SBW 4WS conversion mechatronic control system [SATO 1991, 1995 --1986 Nissan *Skyline*].

In this system, a special fluidical valve that generates oily-fluid pressure balanced with the reaction force from the front wheels in proportion to the lateral acceleration, is prepared in the front-wheel PS system, and this oily-fluid pressure is transferred to the rear-wheel steering-actuator. The latter comprises a high-rate spring that permits displacement of the output rod to the position that balances the allowed oily-fluid pressure. The rear wheels are steered in the same sense of direction as the front wheels with the aid of the displacement of this rod. Therefore, the relationship between the vehicle velocity and the rear-wheels' steering angle alters in conjunction with the lateral acceleration, as illustrated in Figure 4.78 [SATO 1991].

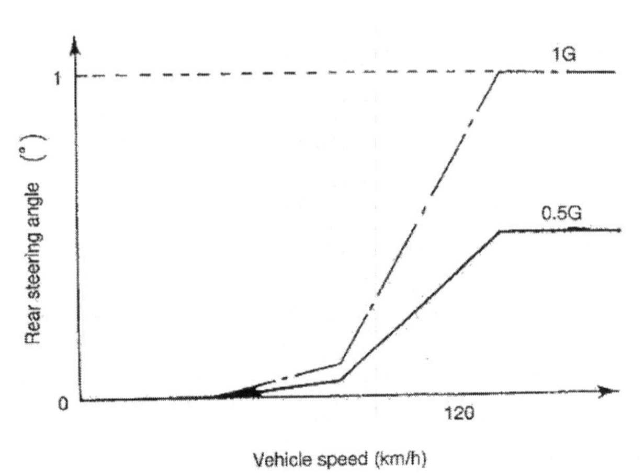

Fig. 4.78 Characteristics of rear wheel steering (RWS) angle [SATO 1991].

In this mechatronic control system, the maximum value of the steering angle of the rear wheels is kept within certain limits of a completely low value, so that it is not supposed to be a control system for improving the minimum value of the vehicle turning radius at very low values of vehicle velocity.

Vehicle Velocity/FWS Angle/Steering Hand-Wheel Angular Velocity Responsive Mode -- This mechatronic control system was introduced in advanced vehicles and was developed with the intention of further enhancing vehicle stability and manoeuvrability during medium- to high-velocity driving. The structure of the mechatronic control system and the arrangement of the rear-wheel steering-actuator (solenoid servo fluidical valve) are illustrated in Figure 4.79 [SATO 1991, 1995]. In this system, the working oily-fluid discharged from the M-F pump is directly inserted into the solenoid servo fluidical valve and controlled by coded commands from an ECU, after which it is transmitted to the power F-M cylinder.

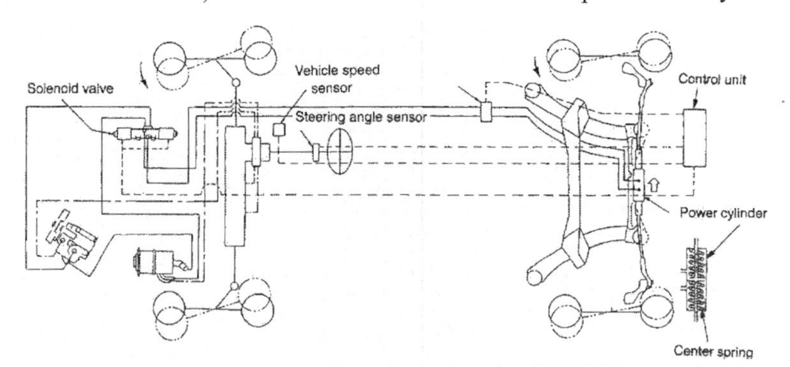

Fig. 4.79 Principle layout of a front wheel-responsive SBW 4WS conversion mechatronic control system [SATO 1991, 1995 -- 1988 Nissan *Sylvia*]

The ECU computes the front-wheel turning angle and the steering-wheel angular velocity with the aid of signals from the steering angle sensor situated in the steering wheel. The values thus computed and the vehicle velocity is provided to fix precisely the rear wheels' steering angle. The maximum value of the rear-wheels' steering angle is kept within the limits of a completely low value in this control system so that it was purposed only to be a control system for improving vehicle stability and manoeuvrability at medium to high values of vehicle velocity. On the other hand, it does not only improve the stability by turning the rear wheels in identical sense of turning-motion direction, but also improves the manoeuvrability by transitorily turning the rear wheels in the opposite sense of direction to the front wheels when the steering wheel is turned quickly at medium values of vehicle velocity. That is, the initial yawing movement (yaw angular velocity) of the vehicle is enhanced and the steering response is thus improved.

Figure 4.80 shows the steering angle pattern of the rear wheels with time expressed along the horizontal axis [SATO 1991].

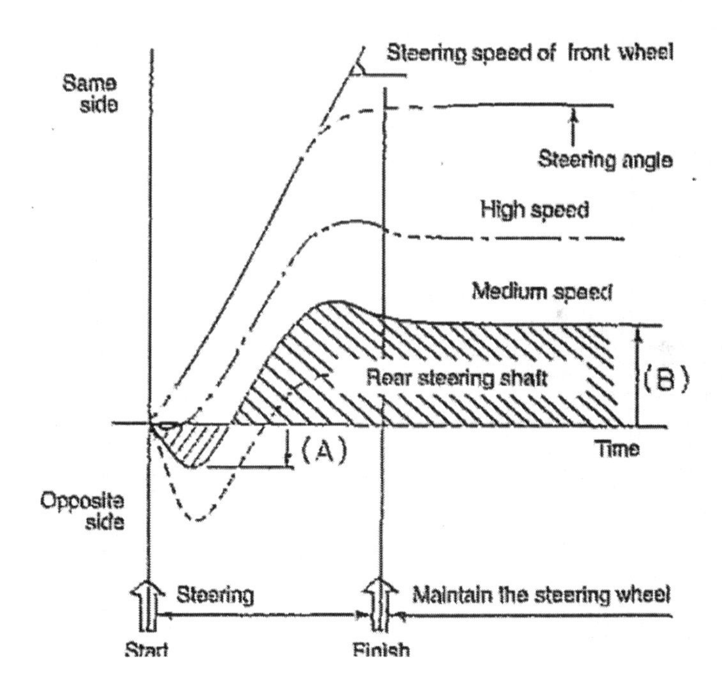

Fig. 4.80 Characteristics of front wheel steering (FWS) angle-responsive 4WS conversion mechatronic control [SATO 1991].

Another SBW 4WS conversion mechatronic control system consists of an R&P FWS system (Fig. 4.81) that is fluidically assisted by a twin-tandem M-F pump main power source, with an overall steering ratio of 14.2:1.

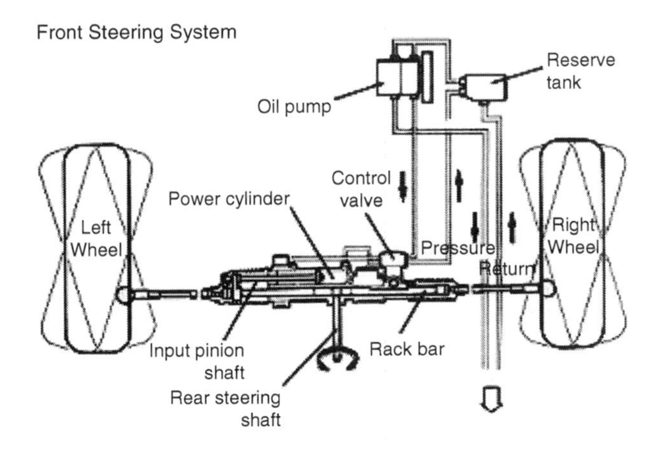

Fig. 4.81 Front wheel steering (FWS) system [MAZDA 2002].

The RWS mechanism (Figure 4.82) is also fluidically assisted by the main M-F pump and mechatronically controlled, according to the front steering angle and vehicle velocity.

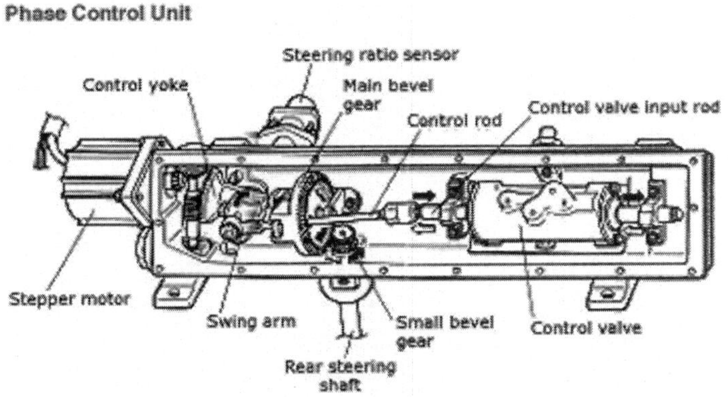

Phase Control Unit
Steering ratio sensor
Control yoke
Main bevel gear
Control rod
Control valve input rod
Stepper motor
Swing arm
Small bevel gear
Control valve
Rear steering shaft

Fig. 4.82 Phase control unit [MAZDA 2002].

The rear steering shaft extends from the rack bar of the front steering gear assembly to the rear steering-phase control unit. The rear steering system comprises the input end of the rear steering shaft, vehicle velocity sensors, a steering-phase control unit (determining sense of direction and degree), a power F-M cylinder and an output rod. A centring lock spring is incorporated that locks the rear system in a neutral (straightforward) position in the event of hydraulic failure.

Additionally, a solenoid fluidical valve that disengages fluidical (hydraulical) assist (thereby activating the centring lock spring) in the case of an electrical failure, is included.

The SBW 4WS mechatronic control system varies the phase and ratio of the RWS to the front wheels, according to the vehicle velocity. It steers the rear wheels toward the opposite phase (sense of direction) of the front wheel during values of the vehicle velocity of less than 35 km/h (22 mph) for a tighter turn and *'neutralises'* them (to a straightforward direction, as in a conventional 2WS principle) at 35 km/h (22 mph). Above that value of vehicle velocity, the system steers toward the same phase-direction as the front wheels, thereby generating an increased cornering force for stability. The maximum value of the steering angle of the rear wheels extends 5 deg to left or right, a measurement that has been determined to be optimally effective and natural to human sensitivity [SATO 1991].

The structure of the SBW 4WS mechatronic control system is illustrated in Figures 4.83 and 4.84 [MAZDA 1989, 2002].

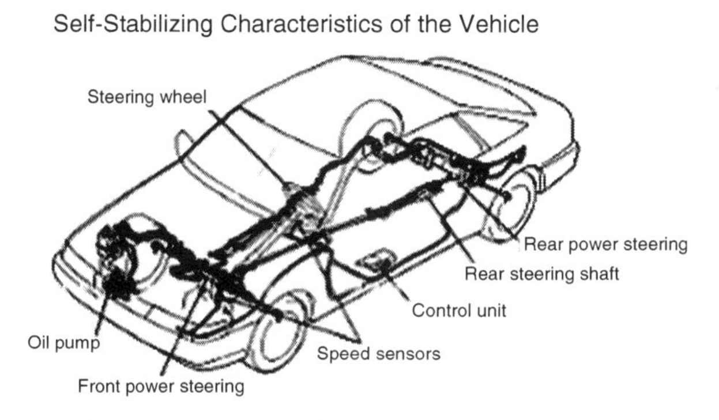

Fig. 4.83 Self-stabilising characteristics of the 4WS automotive vehicle [MAZDA 2002].

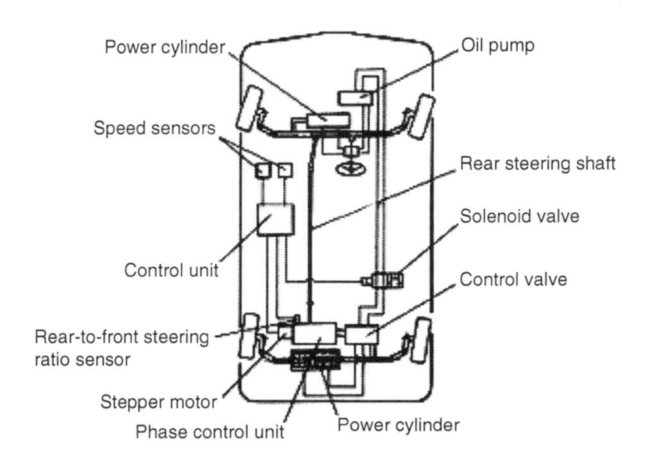

Fig. 4.84 Principle layouts of a front wheel steering (FWS) angle and vehicle velocity-responsive 4WS conversion mechatronic control system [MAZDA 1989, 2002 -- Mazda *Capella*].

This mechatronic control system may be widely separated into the power assist part and the phase control part.

The power assist part contains a linear spool fluidical valve and a power piston, and employs oily-fluid pressure as the dynamical force. The phase control part includes a bevel gear that connects with the input shaft, a control yoke, and a control rod that is linked to a fluidical valve.

A stepping E-M motor towards a worm gear, thus controlling phase-direction of the rear wheels so that they may be turned, drives the angle of the control yoke, not only in same phase-direction, but also in the opposite sense of direction from the front wheels [SATO 1995].

Primary components are as follows [MAZDA 2002]:

- ❖ Vehicle velocity sensors; interpret speedometer shelf revolutions, and send signal to the electronic computer unit (ECU); two sensors, one within the speedometer and the other at the transmission output, are used to cross-check the other for accuracy and fail-safe measures;
- ❖ Steering phase control unit; conveys to the power steering fluidical cylinder booster fluidical valve the sense of direction and stroke of rear wheel steering by the combined movement of the control yoke angle and bevel gear revolutions;
- ❖ Stepper E-M motor; performs altering the yoke angle and bevel gear phasing;
- ❖ Rear steering shaft; transmits front wheel steering angle by turning the small bevel gear in the steering phase control unit that rotates the main bevel gear in the assembly;
- ❖ Control valve; feeds hydraulical pressure to the steering actuator, according to the phase and stroke required for appropriate rear wheel steering;
- ❖ Power F-M cylinder; operates the output rod by oily-fluid pressure and steers the rear wheels; it locks the rear wheels in a *'neutral'* (straightforward) position with a centring lock spring that is activated by a solenoid fluidical valve in the case of failure to ensure a normal 2WS function for the vehicle;
- ❖ M-F pump provides hydraulical pressure to both the front and rear steering systems.

The steering phase control unit alters the sense of direction and degree of rear wheel steering. It consists of a stepper E-M motor that controls the rear steering ratio, a control yoke, a swing arm, a main bevel gear engaged to the rear steering shaft by means of a small bevel gear, and a control rod connected to the control fluidical valve.

It functions [MAZDA 2002]:

- ❖ Opposite phase-direction steering under 35 km/h (22 mph):
  - ➢ Control yoke is at an angle activated by the stepper E-M motor;
  - ➢ Front wheels are steered to the right; the small bevel gear is rotated in direction *X* by the rotation of the rear steering shaft; the small bevel gear, in turn, rotates the main bevel gear;

- ➢ Rotation of the main bevel gear causes movement of the control rod toward the control valve; input rod of the control valve is pushed to the right, according to the degree of the control rod's movement (determined by the disposition of the swing arm), which is positioned to move in an upward direction, to the right; the rear wheels are thus steered to the left, in the opposite sense of direction from the front wheels;
- ➢ As the angle of the control yoke is increased in direction as vehicle speed decreases, the rear-to-front steering ratio proportionately increases and the vehicle's steering lock tightens.

❖ Same phase (sense of direction) over 35 km/h (22 mph); the operation of this phase is the reverse of the opposite phase one, because the control yoke is angled toward *'positive'* in this vehicle speed range, as illustrated; the phasing of the swing arm, yoke rod, and bevel gear steers the rear wheels toward the right in same sense of direction as the front wheels; neutral phase, at 35 km/h (22 mph); the control yoke's angle is horizontal (neutral); thus, the input rod is not affected, even if the control rod is moved with the rotation of the bevel gear unit; as a result, the rear wheels are not steered in this mode.

The movement of the input rod of the control valve unit may be transmitted to the power F-M cylinder's spool.

The spool's displacement to the sleeve causes a pressure difference between the right and left side chambers in the power F-M cylinder is shown in Figure 4.85 [Bootz 2004].

The oily-fluid pressure difference overcomes the output shaft load and initiates sleeve movement. The sleeve-power rod assembly is moved in the direction of the input rod by a proportionate degree. The output rod transmits the steering action to the tie rod at either end of the RWS control mechanism unit, thereby steering the rear wheels.

The system automatically counteracts possible causes of failure, both electronic and fluidical. In either case, the centring lock spring housed in the steering system unit returns the output rods in the *'neutral'* straightforward position, essentially alternating the entire steering system to a conventional 2WS principle.

Specifically, if a fluidical defect should cause a reduction in pressure level (by a movement malfunction or a broken driving belt), the rear wheel steering mechanism is automatically locked in a neutral position, activating a low-level warning light. In the event of an electrical failure, such would be detected by a self-diagnostic circuit integrated within the 4WS control unit that stimulates a solenoid valve and then neutralises the oily-fluid pressure and return lines, thereby alternating the system again to that of a 2WS principle. So, the warning light referencing the SBW 4WS conversion mechatronic control system within the main instrument display is activated, indicating a system failure.

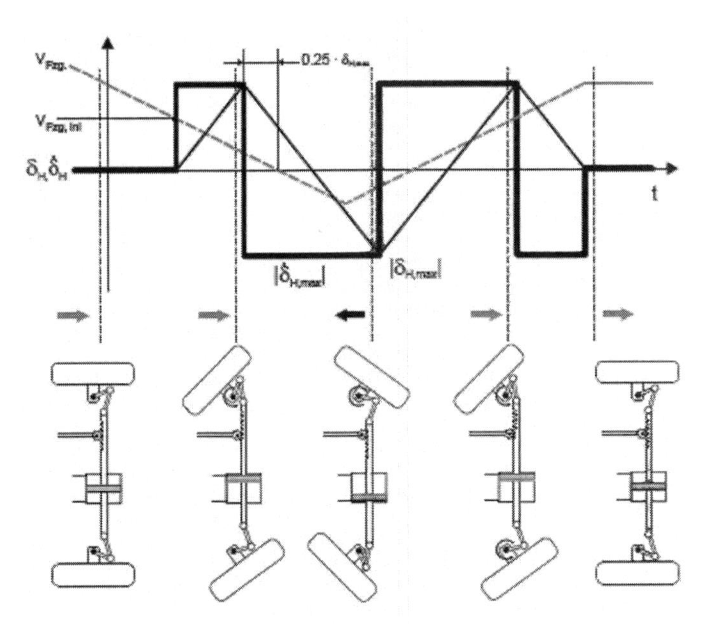

Fig. 4.85 Closed-centre (CC) M-F-M SBW 4WS conversion mechatronic control system may be investigated as an alternative to E-M SBW 4WS conversion mechatronic control-systems and conventional open-centre (OC) M-F-M SBW 4WS conversion mechatronic control system for the use in passenger vehicles [BOOTZ 2004].

M-F-M Frame Articulated SBW 4WS Conversion Mechatronic Control System - With a dual system; up to two swivel axles may be steered. While the three-axle 4WS tractor-semi-trailer (Figure 4.86) is being steered, the steering motion is transmitted to the fifth wheel plate (1) through a wedge that drops into the fifth wheel [TRIDEC 2003]. The fifth wheel plate (1) is connected to the control F-M cylinders (2) that transmit the steering movement to the steering F-M cylinders (3) of the steerable rear axles by means of oily-fluid pressure. To provide the system with additional manoeuvrability, manual steering is available as an option, with infrared, radio-controlled, or compact operation.

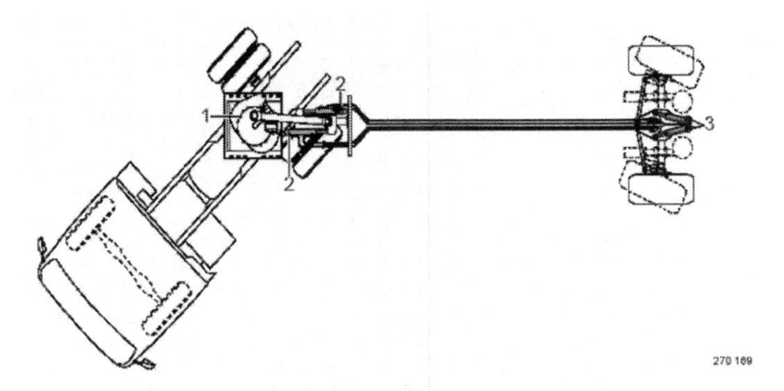

Fig. 4.86 M-F-M frame articulated SBW 4WS conversion [TRIDEC 2003].

Linear and progressive steering systems - The steering movement of the fifth wheel plate is transmitted to the steerable rear axle(s) by means of oily-fluid pressure. The control F-M cylinders on the fifth wheel plate generate the oily-fluid pressure. When the tractor makes the steering movement, the movement may be passed on to the steerable rear axle(s), depending on the arrangement of the control F-M cylinders (parallel or progressive). With a progressive system, the maximum wheel angle is reached faster as a result of the design. This means that with a progressive system, the maximum wheel angle is reached at an angle of 700 between tractor and semi-trailer.

With the linear system, the maximum angle is reached only at 900. If the system is fitted with manual steering (only possible on a single steerable axle) the semi-trailer axle may be steered irrespective of the tractor movements. This gives the semi-trailer greater manoeuvrability in restricted areas.

The oily-fluid pressure for the steering movement is not generated by the steering F-M cylinders on the fifth wheel plate but by an electrically driven E-M-F pump. Besides providing greatly improved semi-trailer manoeuvrability, the steering system has a favourable effect on maintenance costs, due to longer wheel-tyre life, lower **specific fuel consumption** (SFC), and less wear on the axle suspension [TRIDEC 2003].

Analysis of dynamic lateral response for a multi-axle AWS tractor and trailer -- The objective of this study is to present a theoretical analysis of the directional dynamics for an active multi-axle AWS tractor and full-trailer. It may be applied to on/off road vehicles.

In WU & HAI [2003], a linear yaw plane physical model is established to analyse the effect of an additional steering wheel in the front axle of the full-trailer on its performance characteristics, including the directional response simulation during lane change at high speed and a $\pi/2$ rad (90 deg) turn.

The application of additional steering axle on the full-trailer is shown to increase the directional stability and reduce lateral force during steady turning.

The control algorithm of multiple steering axles for a tractor and full-trailer front wheels is investigated using the tractor zero sideslip angle criteria. Moreover, the critical speed of the system related to the location of the steer axle for a full-trailer is examined in detail.

E-M-F-M SBW 4WS Conversion Mechatronic Control Systems -- *Vehicle Velocity/FWS Angle Responsive Mode* -- This mechatronic control system was introduced into advanced vehicles, with the objective of being a mechatronic control system for enhancing both stability when driving at high values of velocity, like the all mechanic SBW 4WS conversion mechatronic control system, and a vehicle's minimum turning radius at low velocity.

Figure 4.87 illustrates the circumstances in which the ratio of the rear wheels' steering angle in relation to the front wheels' steering angle is continuously altering in relation to vehicle velocity [SATO 1991, 1995].

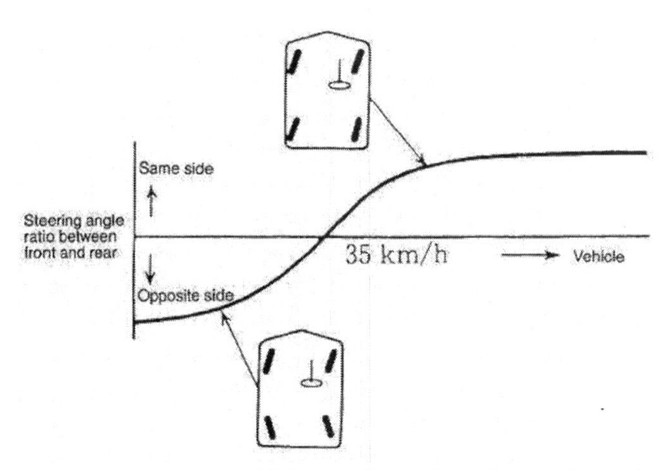

Fig. 4.87 E-M-F-M SBW 4WS conversion mechatronic-control system's vehicle velocity/FWS angle responsive SBW 4WS conversion mechatronic-control system [SATO 1991]

Vehicle Velocity/FWS Angle/Yaw Velocity Responsive Mode -- This mechatronic control system was introduced in advanced vehicles. Figures 4.88 represent the system's structure [TOYOTA 1989, SATO 1991]. The system comprises two parts: an M-F-M steering mechanism to steer the rear wheels with a substantially large steering angle in the opposite sense of direction from the front wheels to constitute a small turning radius at low velocity, and E-M-F-M steering mechanism to steer the rear wheels by a substantially smaller value to not only improve the steering response and the stability during medium- to high-velocity driving, but also to suppress the sudden movements due to exterior disturbances [SATO 1995].

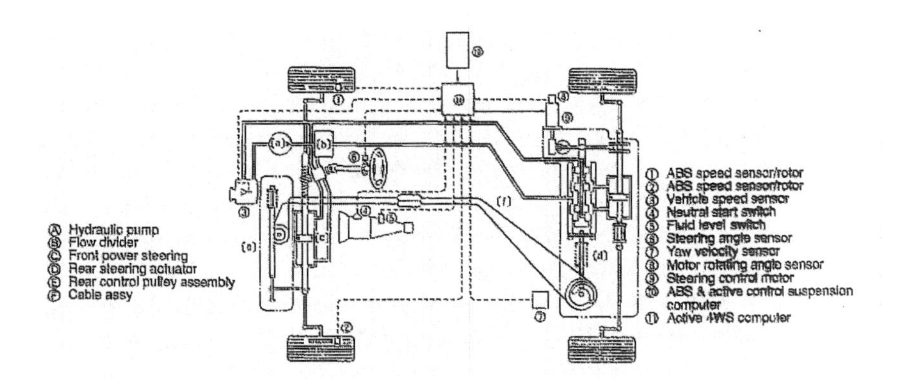

Fig. 4.88 Principle layout of an active SBW 4WS conversion mechatronic control system [TOYOTA 1989 -- 1992 Toyota *Soarer*].

M-F-M Steering Mechanism -- This steering mechanism contains steering angle transference cables which transfer data regarding the front wheels' steering angle to the rear-wheel steering actuator; a junction to conjoin the front and rear transference cables; a pulley assembly that converts the movement of the transference cable into displacement of the front-wheel steering rack into transference-cable movement; a cam that transforms the movement of the transference cable into displacement of the sleeve fluidical valve with an inferior the rear-wheel steering actuator; and a copy fluidical valve that steers the rear wheels as far as the command provided by the sleeve fluidical valve and upholds them. In addition, a dead zone is given in the pulley assembly so that the rear wheels are not steered in the opposite sense of direction to the front wheels when the vehicle is cruising at high velocity [SATO 1995].

E-M-F-M Steering Mechanism --This steering mechanism is purposed to improve the steering response and the vehicle stability and to suppress the sudden movements from outside disturbances when the vehicle is cruising at medium and high values of the vehicle velocity. For this determination, the mechatronic control system contains five types of sensors for measuring the driving/braking and steering circumstances, an ECU that computes the steering angle for the rear wheels used as a basis for the signals from these sensors, a pulse E-M motor which drives the gear mechanism used as a basis for command from the ECU, a gear mechanism which transforms the rotation of the pulse E-M motor into a spool fluidical valve displacement, and a copy fluidical valve that steers the rear wheels as far as the command provided by the sleeve fluidical valve and retains them in that position [SATO 1995].

Mechatronic Control System -- The fundamental algorithm to compute the rear wheels' steering angle is shown in Figure 4.89 [TOYOTA 1989]. When driving in the medium- to high-velocity range, the FWS angle proportional gain for the rear wheels is adjusted to null; instead of that, the yaw velocity proportional gain in the sense of turning-motion direction which suppresses the yaw rate, is adjusted.

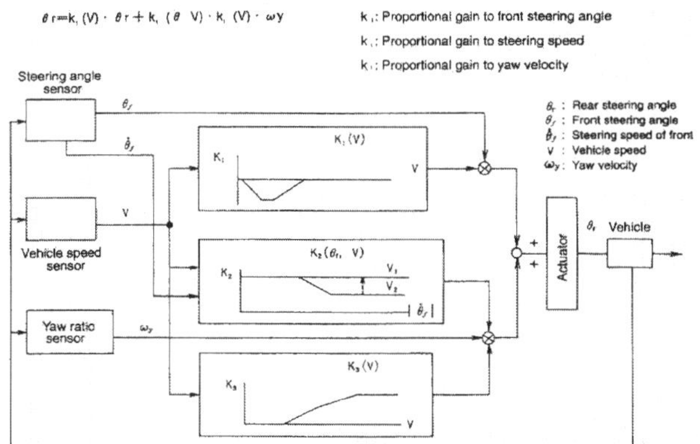

Fig. 4.89 Structural and functional block diagram of the SBW 4WS conversion mechatronic control [TOYOTA 1989].

Consequently, in the medium- to high-velocity range, the rear wheels are turned in an identical sense of direction as the front wheels, with identical result to the other control system afore mentioned.

The especially prominent peculiarity of this mechatronic control system is that a yaw rate sensor is employed to identify the vehicle's dynamics and applies its signal as a control parameter. The result of this is the aim of a normal action. This automatically resituates the disturbance of a normal vehicle's behaviour originated by exterior disturbance [SATO 1995].

4.4.6 E-M SBW 4WS Conversion Mechatronic Control Systems

All four wheels are equally designed, except for some minor differences between the front and rear wheels. The wheel modules contain actuators and sensors for the wheel angles and the wheel brakes [JOHANNESSEN 2001].

The SBW 4WS conversion mechatronic control system consists of two redundant sensors for determining the angle of the wheel, one digital revolving sensor and one resistive linear sensor. To change the angle there is one E-M motor connected to the wheel with a ball screw. By having four individually controlled wheels, it is possible to have different steering options. In the current implementation, there is normal 2WS, 4WS and parallel steering (Fig. 4.90) [JOHANNESSEN 2001].

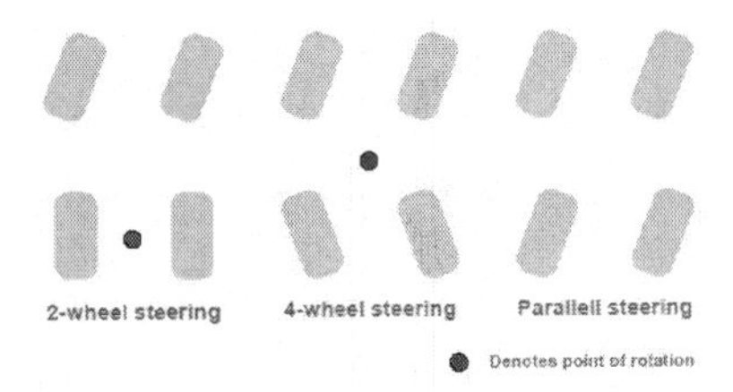

Fig. 4.90 Three types of steering [JOHANNESSEN 2001].

The steering HW has two redundant revolving digital sensors measuring the steering HW angle (Fig. 4.91) [Source: JOHANNESSEN 2001].

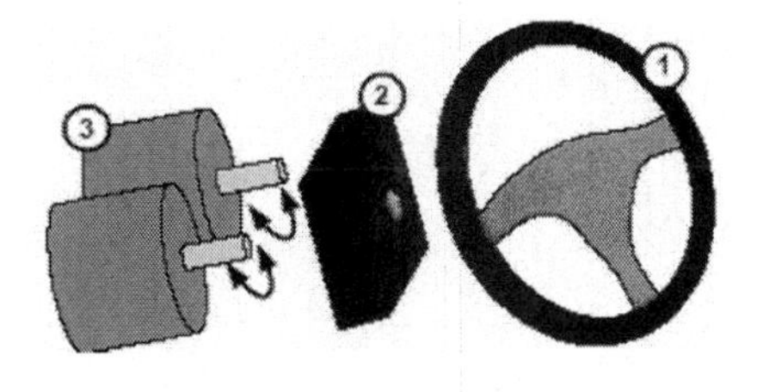

Fig. 4.91 Components of the driver's module: 1 - steering hand wheel (HW); 2 - MRF brake for the steering HW; 3 - redundant angle sensors for the steering HW [JOHANNESSEN 2001].

There is also a **magneto-rheological fluid** (MRF) or **nano-magneto-rheological fluid** (NMRF) brake connected to the steering wheel to give some feedback to the driver. By applying an electric current to this MRF or NMRF brake, it is possible to create a torque when the driver rotates the steering wheel. There have been several ideas of how to best provide the feedback to the driver. Currently, the torque is only dependent on the steering wheel angle. Another alternative is to let the difference between the steering wheel angle and the roadwheels' angles generate the torque. There should also be a vehicle velocity-dependent behaviour, restricting the turning angle at high values of vehicle velocity. The RBW or XBW integrated unibody, space-chassis, skateboard-chassis, or body-over-chassis motion mechatronic control hypersystem may be divided into four individual mechatronic control systems. There is a global hypersystem control that handles the behaviour of the vehicle as a whole and local system control that controls individual actuators. Figure 4.92 shows this relationship [JOHANNESSEN 2001].

In Figure 4.92, there is also one layer called sensor adaptation. This layer combines and interprets the result of individual sensors. It could also provide fault detection and handling of erroneous or missing sensor data. The global control coordinates the actuators in the vehicle to achieve the intentions of the driver. Normally this requires more actuators than **degrees of freedom** (DoF) or actuators that, to some extent, have the same functionality. Examples of this are the possibility to slow down a vehicle by using its ECE or ICE instead of the wheel brakes or to steer the vehicle at high velocity by braking both wheels on one side of the vehicle. One important application of global control is safety. By being able to use actuators for more than its primary purpose, there is a form of redundancy. This type of redundancy is termed intrinsic redundancy. It may be used to let the vehicle function at a degraded mode in the case of actuator failure. For the individual wheel steer angles, there is a proportional controller implemented in all wheel nodes. The characteristics of a proportional controller are slow response and a remaining error. Therefore, the intention is to upgrade the controller to a PID-controller with faster response and no remaining error.

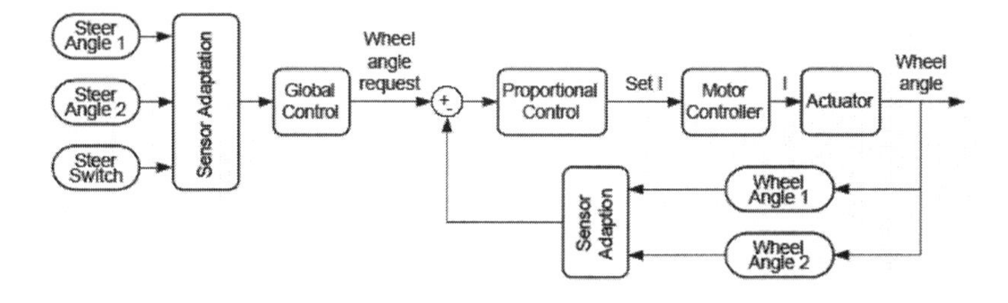

Fig. 4.92 Structural and functional block diagram of the SBW 4WS conversion mechatronic control system [JOHANNESSEN 2001].

The wheel angles depend on the angle of the steering wheel and the selected steering option. Currently there are three options implemented: two-wheel, four-wheel, and parallel steering. To close the loop, the wheel angle needs to be measured. This is done with two sensors, out of which only one sensor is used for backup and monitoring. Figure 4.92 shows how the SBW 4WS conversion mechatronic control system is set up [JOHANNESSEN 2001].

Vehicle Velocity/FWS Angle/Steering Hand-Wheel Angular-Velocity Responsive Mechatronic Control System -- This mechatronic control system was brought into practice in advanced vehicles. The mechatronic control system structure is illustrated in Figure 4.93 [HONDA 1991]. This mechatronic control system has a structure whereby the rotation of an E-M motor's rotor that is controlled by the ECU, is transformed into linear motion with the aid of a ball screw to directly drive the rear wheels. The control system is easy to set up.

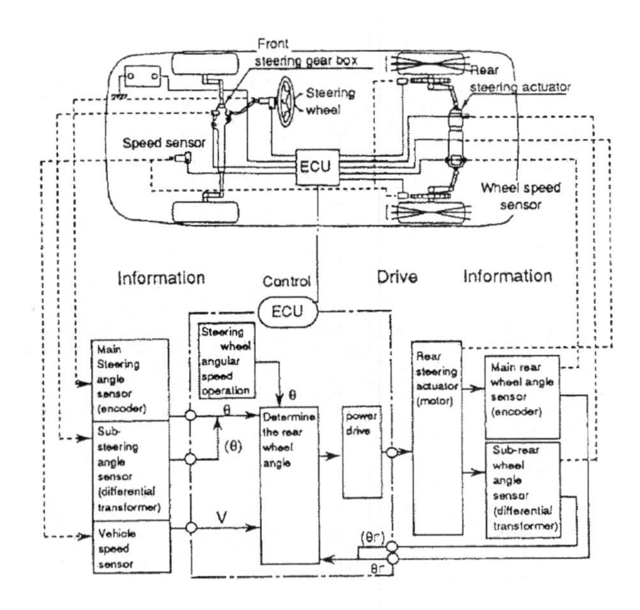

Fig. 4.93 Structural and functional block diagram of the E-M SBW 4WS conversion mechatronic control system [HONDA 1991 --1991 Honda *Prelude*].

In addition, in this system, precisely the vehicle velocity, the FWS angle, and the steering wheel angular velocity fix the turning of the rear wheels. The computing algorithm is like the mechatronic control system mentioned above. In this system, improving in the vehicle stability at high velocity, improving in the steering response when the steering wheel is turned rapidly, and improving in the vehicle's turning radius at low velocity have been gained synchronously [SATO 1995].

A recent example of an AE E-M SBW 4WS conversion mechatronic control system is one [NISSAN 1989] that has been brought into use by vehicle manufacturers.

The means by which control may be reached in this system is identical to the vehicle velocity/FWS angle/ steering wheel angular velocity responsive system afore mentioned, but by employing an E-M actuator, the limitless area in which all mechatronic control systems exist has been diminished and ease of location improved.

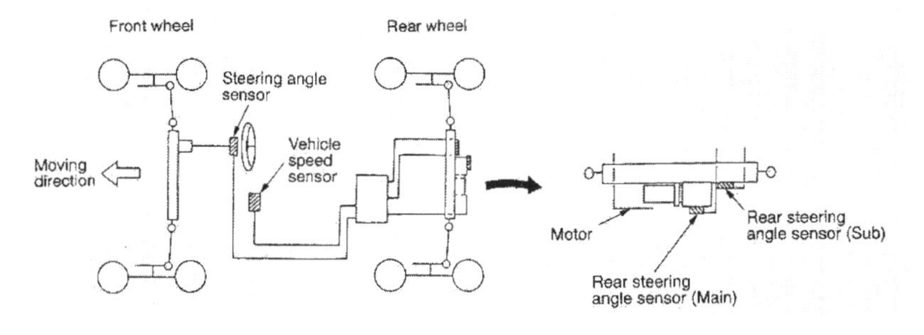

Fig. 4.94 Principle layout of the E-M SBW 4WS conversion mechatronic control system [NISSAN 1989 -- *HICAS*].

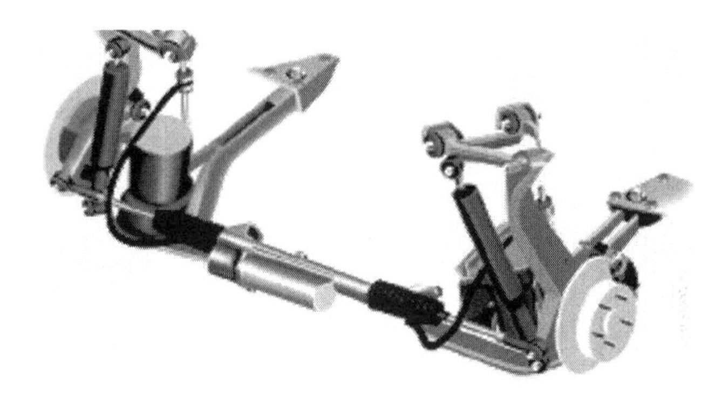

Fig. 4.95 Rear-wheel steering unit of the E-M SBW 4WS conversion mechatronic control system [HOED 2004 – EuroNCAP-sterren].

Figures 4.94 and 4.95 illustrates the AE E-M SBW 4WS conversion mechatronic control system structure and Figure 4.96 displays the steering-angle generation modes for the rear wheels [NISSAN 1989; HOED 2004]. Automotive vehicles equipped with the SBW 4WS conversion mechatronic-control systems emphasised a new trend in R&D in the form of FWS and RWS [IGA ET AL. 1988]. The world's potentates in the automotive mechatronics market worked out innovative SBW 4WS conversion mechatronic control systems. They are unique SBW 4WS conversion mechatronic control systems. They have been elaborated for the needs of large vehicles, including trucks with long platforms and elongated cabins, vans, pickups, as well as SUVs, see Figures 4.97 and 4.98 [DELPHI 2005A]. They combine transportable, fully-dimensioned vehicles with the turning diameter and manoeuvrability of much smaller vehicles.

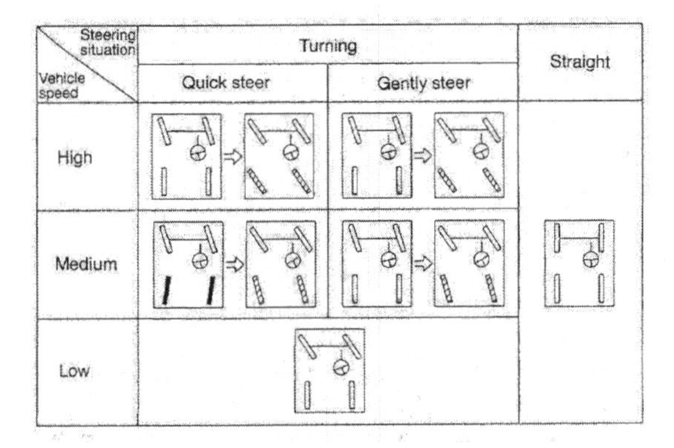

Fig. 4.96 Rear wheel steering (RWS) angle generation mode [NISSAN 1989].

Fig. 4.97 Pickup equipped with the SBW 4WS conversion mechatronic control system [SMITH 2002 -- Delphi '*QUADRASTEER*'™].

Fig. 4.98 Half-truck equipped with the SBW 4WS conversion mechatronic control system [Delphi '*QUADRASTEER*™'].

Operation of an RWS is realised by means of the electrically powered E-M actuator (Fig. 4.99) [LEVINE AND GILLIES 2001].

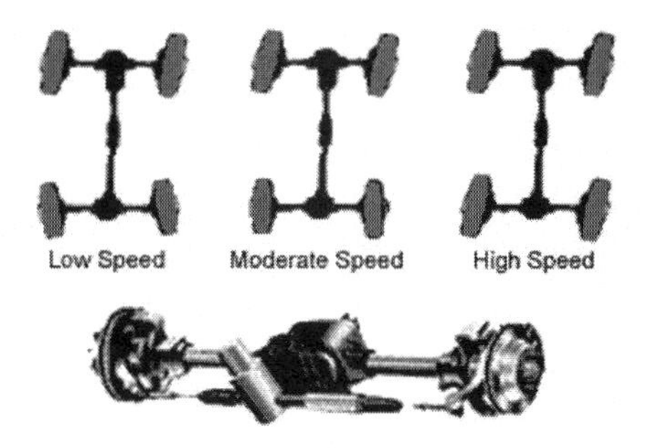

Fig. 4.99 Mechatronically-controlled rear wheel steering (RWS) actuator [LEVINE AND GILLIES 2001].

No mechanic link exists of the both FWS and RWS. From the mechanical viewpoint, RWS does not differ much from conventional FWS used in vehicles with rigid front axles. This mechatronic control system may be equipped with an auxiliary, second RWS located inside the rear bridge of the vehicle. In the case of pickups and half-trucks, it is most often rigid-bridge suspended on leaf springs. ***Operation principle of the SBW 4WS conversion mechatronic control system -*** This system functions in three basic phases: *'Negative', 'Neutral'* and *'Positive',* as is shown in Figure 4.100.

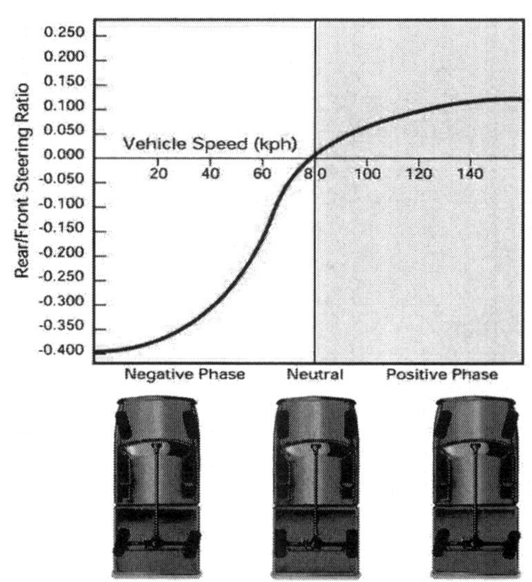

Fig. 4.100 Alterations of values of the RWS angle as a function of the vehicle velocity (speed) and three basic phases of the SBW 4WS conversion mechatronic-control system [Delphi '*QUADRASTEER*'].

Negative Phase: at low values of vehicle velocity, wheels turn in the opposite sense of direction in relation to the front wheels. By mechatronic control of the rear wheels' lateral inclination angle, the diameter of turning at small values of vehicle velocity has been limited, which enhance manoeuvrability. In a case of the large delivery vehicle, the diameter of turning with a normal value 13.5 m reaches a value of 10 m when using of this mechatronic control system. Thus, it decreases to a level nearly that of compact vehicles [DELPHI 2005A]. Alterations in the sense of steering direction of the rolling rear wheels also cause the vehicle to react much better on the motion of the steering wheel, it needs less rotations of the steering wheel during manoeuvres with low velocity. This enhances the **comfort-and-handling stability** (C&HS) and causes a sensation of steering similar to that of compact vehicles but not large delivery vehicles.
Neutral Phase: at medium values of vehicle velocity, wheels remain in a neutral position (ride forward), as in a vehicle with a normal SBW 2WS conversion mechatronic control system.
Positive Phase: at high values of vehicle velocity, the rear wheels turn in the same sense of steering direction as the front wheels. Due to this, the vehicle reacts less rapidly (for instance, when changing a motion lane and quick reversion on a previous motion lane), what enhances its C&HS. The driver than gains a safety feeling and better vehicle mechatronic control. It is especially seen during driving on wet on/off-road surfaces or during gusts of wind. Both at low and high velocities, the SBW 4WS conversion mechatronic control system considerably increases the towability of the vehicle. At low velocities, when the rear wheels are turning in the opposite sense of direction to the front wheels, a trailer is bending the towing a vehicle's path, as it has a place in the conventional SBW 2WS conversion mechatronic control system. Driving on the streets of a town and reversing as well as parking and deparking become decidedly easier. However, towing a trailer at higher velocity due to the SBW 4WS conversion mechatronic control system means the driving at a lower lateral force which generally acts on the rear part of the vehicle. Thus, it enhances the trailer's C&HS, decreases its inclinations, and the driver can correct the driving path after meeting gusts of wind or those caused by passing trucks as well bumps in the road. For instance, the unique Delphi *QUADRASTEER*™ SBW 4WS conversion mechatronic control system sets a 21% reduction in large-platform SUV turning radius.

Fig. 4.101 Rear-wheel steering (RWS) unit of the *QUADRASTEER*™ ^{E}-M SBW 4WS conversion mechatronic control system [JUST-AUTO.COM/Lotus Engineering].

In other words, if the SUV currently makes a U-turn of circa 9 m (27 feet), this option may cut 1.5 m (5 feet) off that radius. Additionally, the system is mechatronically based, with three buttons on a dashboard-mounted switch to control its three modes [SENEFSKY 2003]. The computer-controlled *QUADRASTEER*™ SBW 4WS conversion mechatronic control system is **electro-mechanical** (E-M) and the **rear-wheel steering** (RWS) steering unit is mounted on a heavy-duty, axle housing (see Figs 4.101 and 4.102) with a 17 cm (9-3/4 inch) ring gear. An extra-thick skid plate protects the unit from damage [SMITH 2002; LAW 2003; LEWIS AND WRIGHT 2004].

Fig. 4.102 Rear-wheel steering unit of the *QUADRASTEER*™ v.2.0 E-M SBW 4WS conversion mechatronic control system [SMITH 2002; LAW 2003].

It is possible that whatever other vehicles come with the *QUADRASTEER*™ SBW 4WS conversion mechatronic control system, it may be the enhanced version that Delphi Corporation recently announced. Delphi Corporation's engineers have taken this electronic steering system to the next level by *'closing loop'* and setting the sensory inputs (vehicle velocity, steering HW position, vehicle yaw rate, and lateral acceleration) to be continuously fed back to the controller so that the SBW 4WS conversion mechatronic control system may actively respond to the vehicle's actions. In addition, Delphi has integrated this closed-loop version of the system with its brake-based *TraXXar* stability control program (Fig. 4.103) [LAW 2003].

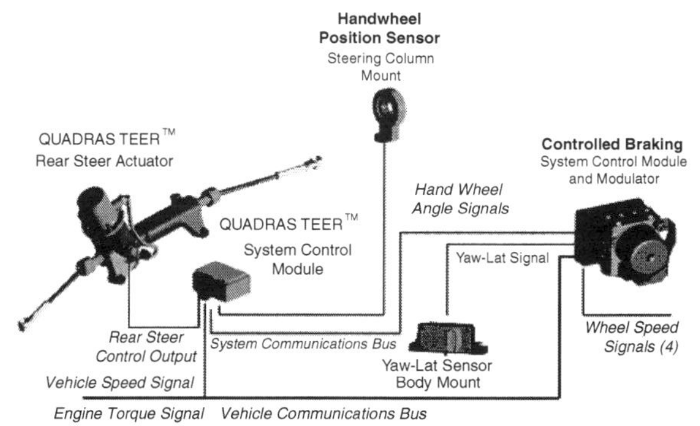

Fig. 4.103 *QUADRASTEER*™ integrated with *TraXXar* system components and interfaces [LAW 2003].

The mechatronic integration of these two vehicle systems helps take safety to a new level. Until now, brake-based systems have been the standard for achieving vehicle stability control. However, these systems react only during emergency driving situations, applying the brakes to help maintain control. Loss of directional control is a primary cause of rollovers, by bringing rear steering into the equation; the next *QUADRASTEER*™ constantly helps reduce loss of directional control before the driver even realises that the vehicle might become unstable.

A 2WS mode sets the platform to function conventionally. The 4WS mode sets the rear wheels to function (out-of-phase steering) until the platform reaches circa 72 km/h (45 mph), when the rear wheels (in-phase-steer) turn in the same sense of direction as the front wheels.

A 4WS tow mode functions in the same manner the 4WS mode does, with the exception of the rear-wheel angle that is set tighter to improve control of the rear-mounted trailer.

An internal on-board computer uses three main components:

- A front-wheel position sensor;
- A steerable solid hypoid rear-axle assembly;
- An E-M motor driven E-M actuator.

Sensors positioned all over the platform continually generate ongoing functioning data to the controller which automatically adjusts the amount and the direction angle in which the rear wheels should move and, further, whether they should turn in the same sense of direction as the front wheels or move in the opposite sense of direction. Algorithms are used to decide how much steering input should go into the rear wheels. For instance, Delphi Corporation has identified three principal parameters for the system, depending on vehicle velocity. As expected, in the *'Negative Phase'*, when the vehicle is moving at slow velocity, the rear wheels turn in the opposite sense of direction to the front wheels, making parking chores easier. On the pavement, at moderate values of vehicle velocity, in the *'Neutral Phase'*, the rear wheels now move in the same straight sense of direction as the front wheels, because neither affect the platform's manoeuvrability. In the *'Positive Phase'*, with the vehicle moving at higher rates of velocity, the rear wheels turn in the same sense of direction for easier and crisper lane changing. What comes into play here is the reduction of vehicle yaw or sway, the rotational motion needed to accomplish a manoeuvre. Stable platform response during evasive manoeuvres or under adverse on/off road surface conditions is a definite plus. When pulling a trailer, the appreciation is immediate and stability of both vehicles is notably enhanced. Positive rear steering lowers the articulation angle between vehicle and trailer, reducing the lateral forces pressed on the rear of the towing vehicle. SUV may now reduce yaw velocity gain and increase yaw damping of the trailer and towing vehicle. Increased trailer stability, combined with reduced sway and lower corrective steering manoeuvres, tackles external annoyances such as wind gusts, curvy on-ramps and clover-leafs, and lane-change manoeuvres without the caterpillar sway-motion. Trailering with this system also improves the SUV true vehicle path, as well as making it easier to manoeuvre large trailers into position [SENEFSKY 2003].

Summing up, *QUADRASTEER™* SBW 4WS conversion mechatronic control system combines conventional **front-wheel steering** (FWS) with an electrically powered **rear-wheel steering** (RWS) system composed of a steering wheel position and vehicle-velocity sensors and a central **electronic control unit** (ECU). The system works at three different values of vehicle velocity [DELPHI 2005B]:

- At low values of vehicle velocity, below 64 km/h (40 mph), in the negative phase, the rear wheels turn in the opposite sense of direction to the front wheels:
  - Reduces turn radius up to 20%;
  - Improves manoeuvrability;
  - Eases parking;
  - Simplifies trailer positioning.
- At moderate values of vehicle velocity, around 64 km/h (40 mph), the rear wheels remain straight, or neutral.
- At higher values of vehicle velocity, above 64 km/h (40 mph), the rear wheels are in the positive phase, turning in the same direction as the front wheels:
  - Helps improve handling;
  - Virtually eliminates trailer sway;
  - Enables more stability during lane changing.

Benefits of the *QUADRASTEER™* SBW 4WS conversion mechatronic control system are as follows [DELPHI 2005B]:

- Dramatically enhances low-velocity manoeuvrability, higher-velocity stability, and trailering capability for full-size vehicles (trucks, vans, and SUVs);
- May be integrated with active safety chassis systems;
- Instantly returns vehicle to two-wheel steering if the four-wheel system is damaged;
- May be fully automatic or driver selectable;
- The driver may adjust the RWS for different driving conditions;
- Designed for comfort, convenience, and safety:
  - No loss of front leg room;
  - Does not reduce ground clearance.

Conventional vehicle suspensions are traditionally tuned with a bias for either handling or driving comfort. Now, automotive vehicle manufacturers may optimise their vehicle's driving and handling characteristics using SBW 4WS conversion mechatronic control systems. They may serve as a primary mechanism for enhancing the vehicle's handling performance by using its highly tuneable software.

Using dynamic control algorithms, the SBW 4WS conversion mechatronic control system may serve as a primary mechanism for enhancing a vehicle's handling performance by using its highly tuneable software.

Using dynamic control algorithms, the SBW 4WS conversion mechatronic control system provides the ability to specifically *'dial in'* desired handling characteristics [GAUT 2003C].

For instance, if the desired vehicle character is for a smoother ride through a softer suspension, the SBW 4WS conversion mechatronic control may be used to help regain the desired handling by using an algorithm that dynamically adjusts the rear-wheel angle according to a vehicle behaviour physical model.

The result is optimised handling performance and driving comfort. The real beauty of the SBW 4WS conversion mechatronic control system is in the way it helps balance drive and handling performance with improved vehicle dynamics, or active safety. This combination delivers value to the customer who wants to be safe but also desires superior ride and handling characteristics for every-day driving comfort and enjoyment [GAUT 2003C].

The SBW 4WS conversion mechatronic control system separates the yaw and lateral dynamics of the vehicle. This gives chassis design and tuning experts a new degree of freedom to control vehicle motion.

When combined with the latest in advanced algorithms, the SBW 4WS conversion mechatronic control system allows customers to achieve a superior handling performance while also increasing dynamic safety through active rear steering.

The SBW 4WS conversion mechatronic control system helps minimise over-steering and under-steering at all values of vehicle velocity, and on virtually all on/off road surfaces, even during normal driving, without slowing the vehicle.

Emergency lane changes become more predictable, more manageable, and less stressful when RWS is added to the equation.

The SBW 4WS conversion mechatronic control system may be integrated with controlled BBW 4WB dispulsion mechatronic control system to provide a more effective vehicle system solution to stability control than brakes alone. Together, these mechatronic control systems help deliver instantaneous RWS control to bring a vehicle back on its intended course and blended braking as needed. This approach minimises any slowing of the vehicle, making the correction less intrusive to the driver.

In addition, by allowing steering to maintain directional control and braking to slow the vehicle, this integration may help reduce vehicle stopping distances on split and mixed coefficient on/off road surfaces, such as snow and ice, in a stable, controlled manner.

The SBW 4WS conversion mechatronic control system complements and expands the impact of brake-based stability control systems on vehicle dynamics by improving handling and yaw stability. Bringing steering into the equation allows our customers to deliver the ultimate in active safety combined with a comfortable ride and superior handling.

Conventional 4WS benefits are also maintained when applied to passenger vehicles. The system may reduce the vehicle's turning circle for added manoeuvrability during city driving or parking.

Figure 4.104 shows a comparison of the driving path after the vehicle moves in the case of both SBW 2WS and SBW 4WS conversion mechatronic control systems [EGUCHI 1989].

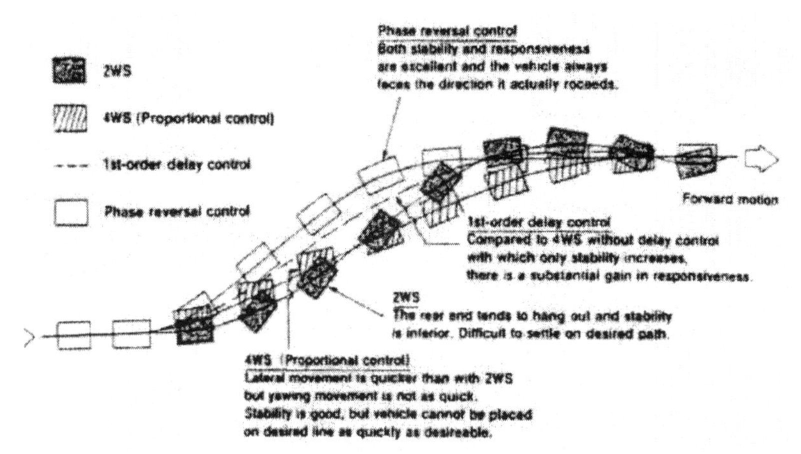

Fig. 4.104 Comparison of the driving path after the vehicle moves in the case of both SBW 2WS and SBW 4WS conversion mechatronic control systems [EGUCHI 1989].

The SBW 4WS conversion mechatronic control system may also be configured with unique algorithms for improved handling and safety while towing a caravan or utility trailer.

The SBW 4WS conversion mechatronic control system represents a high-value enhancement for vehicle manufacturers using proven, reliable technology. It features a low-cost, light mass and modular actuator flexible enough to function with different suspension configurations.

Very advanced SBW 4WS conversion mechatronic control systems may be equipped with on-board instruments for more up-to-date driver information, including transmitters, receivers and microcomputers needed for the **guidance and information system** (GIS).

Figure 4.105 shows the programmable display where advice for the driver can be visually displayed [SEIFFERT AND WALZER 1991].

Fig. 4.105 Driver information received by a programmable screen [SEIFFERT AND WALZER 1991 -- VW *Futura*].

By means of sensors, the distance of a vehicle from obstacles in the direction of travel is continuously measured. If an obstacle is approached too fast, the driver receives an optical warning. The vehicle may also be equipped with a very advanced SBW 4WS conversion mechatronic control system.

A special mechatronic control system is provided that is capable of parking and/or deparking the vehicle automatically in parking spaces of only 30 cm longer than the vehicle.

The automatic parking and/or deparking processes are depicted in Figure 4.106 [SEIFFERT AND WALZER 1991].

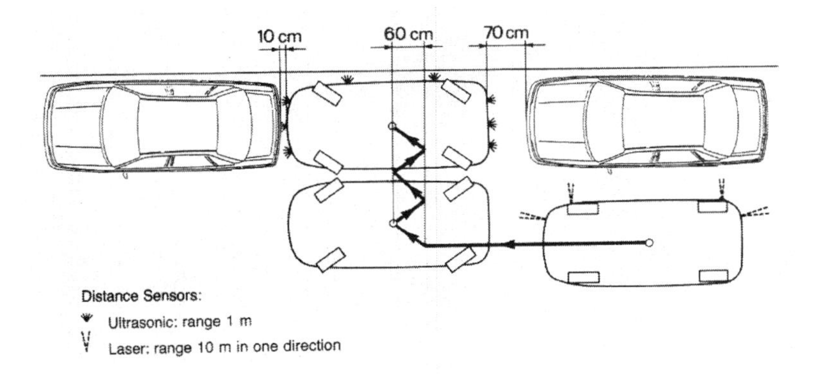

Fig. 4.106 Automatic parking and/or deparking in a narrow parking space by means of a very advanced SBW 4WS conversion mechatronic control system: the heavy lines indicate the path of the vehicle's centre [SEIFFERT AND WALZER 1991 -- VW *Futura*].

For this purpose, the rear wheels of the vehicle are steerable up to $2\pi/9$ rad. The parking and/or deparking procedure is initiated by the driver and is then carried out automatically. **Laser** (L) and **ultrasonic** (US) distance sensors survey all sides of the parking space. The L sensors determine the size of the available parking slot while the US sensors observe the parking slot for any obstacle. A microcomputer or transputer determines the actions required and issues commands to the servo-mechanisms for steering, accelerator and brake foot-pedals. Microcomputer-controlled actuators initiate movements of the accelerator, and brake foot-pedals and steering. L and US beams scan the parking gap.

4.4.7 Tendency in Research and Development (R&D)

SBW 4WS conversion mechatronic control systems that have been set up in current vehicles chiefly accept the programme control technique in which the RWS angle is programmable, computed used as a basis for the earlier scheduling relationship with the vehicle velocity, the FWS angle, and steering-wheel angular velocity.

The objective of the SBW 4WS conversion mechatronic control system is to improve the insufficient handling performances of the automotive vehicle under special steering circumstances.

Until quite recently, however, the SBW 4WS conversion mechatronic control system, applied as a foundation for the new concept was earnestly researched in relation to improving the handling characteristics in any area for practical use. Such technology is like that of a **control configured vehicle** (CCV) that is employed in aeroplanes and usually called **active control technology** (ACT). As examples, a fundamental introduction to the subsequent two characteristic concepts regarding active SBW 4WS conversion mechatronic control systems will be presented below.

Vehicle's Slip Angle Mechatronic Control System **-** As a rule, the moving direction of the barycentre, that is, the centre of gravity of the vehicle is not identical to the direction in which the vehicle while the latter is turning, and the angle between these two directions is called the sideslip angle of the vehicle.

When the value of the sideslip angle is large, the driver drives with an inclined line of vision. This can be regarded as one reason for augmented driving difficulties. From this point of view, the mechatronic control concept has been suggested for consideration by which the value of the side-slip angle of the vehicle, while in motion, is always held to be nearly equal to null.

Also, the reason for augmenting of a phase delay in lateral acceleration in conformity with the vehicle velocity is that the steady-state gain in the sideslip angle is reduced together with the vehicle velocity and becomes negative at high values of vehicle velocity. Along with ignoring such phenomena, the concept that holds the sideslip angle as nearly equal to null, may also be advisable. The sideslip angle β may is usually equated as

$$\beta = \frac{B_f(\tau_f s+1)\delta_f + B_r(\tau_r s+1)\delta_r}{POL(s)},$$
$$B_f : \frac{2l_r l K_r - l_f m V^2}{2l K_r V} A, \quad B_r : \frac{2l_f l K_f - l_r m V^2}{2l K_f V} A, \tag{4.20}$$

$$k = \frac{\delta_r}{\delta_f} = \frac{B_f \tau_f s - B_f}{B_r \tau_r s + B_r} = -\frac{\left(\frac{I_z}{2K_r l}\right) V s - \left(\frac{l_f m}{2K_f l}\right) V^2 + l_{f+}}{\left(\frac{I_z}{2K_f l}\right) V^2 s - \left(\frac{l_r m}{2K_r l}\right) V^2 - l_{f+}}. \tag{4.21}$$

Because of the complex frequency (*Laplace* numerator) $s = 0$ in Eq. (4.20), the steering angle ratio between the front and rear wheels required for obtaining a sideslip angle of zero may be gained from Eq. (4.21) as a function of the vehicle's velocity. This is illustrated in Figure 4.107 [SATO 1991].

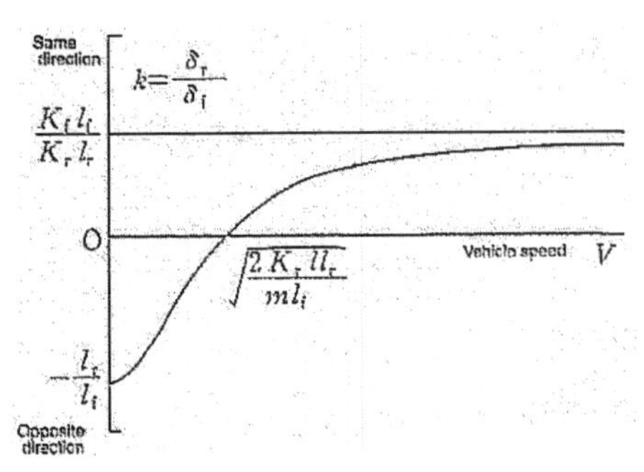

Fig. 4.107 Steering angle ratio for making the vehicle sideslip and equal to null [SATO 1991].

If the front- and rear-wheel control principle that contains the phase advance function for the FWS angle is used, the vehicle's characteristics can be adjusted more independently. An example of the simulation analysis when the front and rear wheels are steered using the control principle mentioned above is illustrated in Figure 4.108 [SATO 1991].

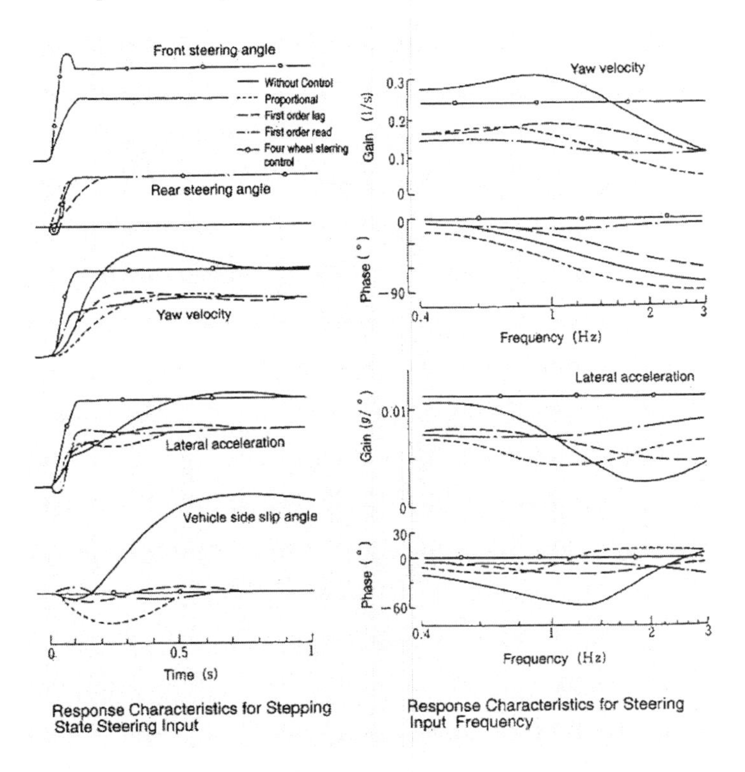

Fig. 4.108 Affection in steering response characteristics by using the delay/advanced SBW 4WS conversion mechatronic control means [SATO 1991].

Mathematical Model Following SBW 4WS Conversion Mechatronic-Control System - In this system, the advisable vehicle behaviour in conformity with the driver's steering operations has been earlier determined as a reference mathematical model and the mechatronic control is fulfilled so that the real behaviour of the vehicle makes the mathematical model very suitable.

Usually, the lateral acceleration and the yaw velocity are provided as the parameters for making known the lateral movement of the vehicle. If the unrelated mathematical models are arranged for both of these characteristics, and constitute the real behaviour of the vehicle, it is essential to actively control mechatronically both of the front and rear wheels [SATO 1995].

Research into SBW 4WS conversion mechatronic control systems such as this is presently being fulfilled, but as an example, a system for controlling the rear wheels' steering angle that applies a range mathematical model for M^* specified by the linear Eq. (4.22) for lateral acceleration and the yaw angular velocity, may be presented as

$$M^* = c\,\alpha_y + (1 - c)\,V\varphi\,. \tag{4.22}$$

If the weighted constant $c = 0$ in Eq. (4.22), this system comes to be a yaw angular velocity mathematical model following the SBW 4WS conversion mechatronic control system, and if $c = 1$, then it is a lateral acceleration mathematical model following the SBW 4WS conversion mechatronic control system.

Distributed SBW 4WS Conversion Mechatronic Control - Ongoing efforts aim to develop a common standard for communication protocols within the FBW and RBW or XBW industry [JOHANSSON ET AL. 2003].

- Four major standards:
  - FlexRay -- Currently under development by BMW, Daimler-Chrysler, Motorola, Philips, et al.;
  - SAFEbus -- Developed by Honeywell and in operation in *Boeing* 777 aeroplanes;
  - TTCAN -- Developed by Bosch and extends the existing CAN protocol; sample chips are available;
  - TTP/C -- Developed by the Technical University of Vienna and TTTech; Commercial chips are available.
- Instead of starting with the protocol specification, automotive scientists and engineers have investigated the requirements derived from the application.

As an example, the principle layouts of the two proof-of-concept prototype vehicles are shown in Figure 4.109 [JOHANSSON ET AL 2003]:

- *FAR* - Scale 1:5 experimental SBW 4WS conversion mechatronic control system;
- *SIRIUS* -- Scale 1:1 experimental SBW 4WS conversion mechatronic control system.

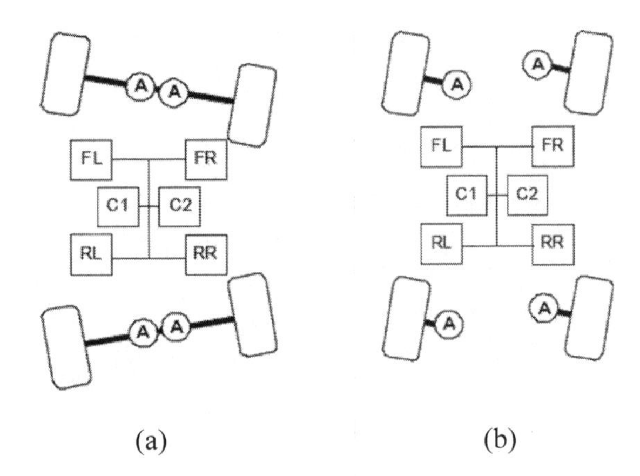

Fig. 4.109 Layouts of the SBW 4WS conversion mechatronic control systems:
(a) -- *FAR – TTTCAN* network – Independent front/rear axles steering;
(b) -- *SIRIUS – TTP/C* network – Independent four-wheel steering (4WS)
[JOHANSSON ET AL. 2003].

<u>*Inherent redundancy:*</u> In the case of a wheel node failure, it is still possible to steer the vehicle, for example *SIRIUS*, provided that [JOHANSSON ET AL. 2003]:

- The faulty wheel is locked in a fixed position (fail-safe mode);
- The global steering algorithm is modified (mode change).

An example SBW 4WS communication bandwidth is displayed in Table 4.11 [JOHANSSON ET AL 2003].

Table 4.11 SBW 4WS Communication Bandwidth [JOHANSSON ET AL.2003].

Central node (C1 and C2)		
100 Hz: steering wheel angle (14 bits) 10 Hz: steering mode (2 bits) Sensor redundantly allocated to C1 and C2.	14 bits x 100 Hz 2 bits x 10 Hz x 2	2840
Wheel node (FL,FR,RL and RR)		
100 Hz: wheel speed and steer angle (12 bits) Steer angle sensor duplicated 100 Hz: Four command words (14 bits) and one status word (16 bits) from all four nodes	3 x 12 bits x 100 Hz ((4 x 14 bits) + 16 bits) x 4 x 100 Hz	32400
	Resulting bandwidth	35.240 bits/s

SBW 4WS Communication System is represented by the following features [JOHANSSON ET AL. 2003]:

- The communication system is vital to assure system safety;
- The RBW or XBW vehicle dynamics mechatronic control system should not host any other function than vehicle dynamics control;
- Inherent redundancy should be used by the application for cost-effective fault-tolerance;
- The communication system must guarantee a sufficient data rate, as well as constant and limited time delays;
- The communication system should be a **commercial off-the-shelf** (COTS) with basic required functional and limited autonomous behaviour;
- The communication protocol should not guarantee application consensus; it is best accomplished at the application level.

4.5 Tri-Mode Hybrid SBW AWS Conversion Mechatronic Control Systems for Future Automotive Vehicles

4.5.1 Foreword

The steering of an automotive vehicle is achieved, not only by means of **hand wheel** (HW) but also by varying actual values of the angular velocity and sense of rotation of all the **electro-mechanical/mechano-electrical** (E-M/M-E) **steered, motorised and/or generatorised wheels** (SM&GW) for AWD × AWS vehicles. The high-tech improved tri-mode hybrid SBW AWS conversion made significant progress during the 1990s. One evolutionary factor behind this has been the increasing requirements for active safety as well as **ride comfort and road handling** (RC&RH) of vehicles.

A major contribution to this progress is the introduction and fast-growing application rate of electrically powered and mechatronically controlled R&P steering gears.

The automotive world is moving toward an **all-electric vehicle** (AEV) in every respect other than actual conventional SBW AWS conversion, DBW AWD propulsion, BBW AWB dispulsion, and ABW AWA suspension. Enthusiasm for AEV, a feature of the early 1990s, has declined.

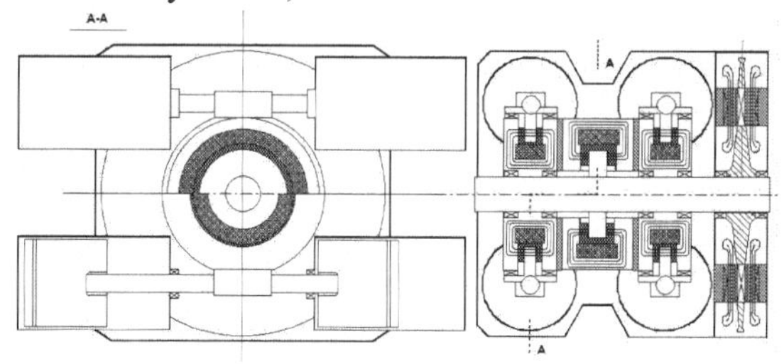

Fig. 4.110 Principle layout of the Fijalkowski engine with the two-in-one silent, electronically-commutated, magnetoelectrically-excited, brushless and high-temperature superconductor AC-DC/DC-AC macrocommutator composite-flywheel onboard generator/starter motor [FIJALKOWSKI 1999D, 2000C].

Present wisdom appears to be that **wheeled vehicles** (WV) are within reach of the crankless, **magneto-rheological fluid** (MRF) or **nano-magneto-rheological fluid** (NMRF) mechatronic commutator, constant volume, nearly ideal 2-, 4- or even 5-stroke thermodynamic cycle, twin-opposed-piston-type **internal combustion engine** (ICE) termed the Fijalkowski engine [FIJALKOWSKI 1986, 1998B, 1999D, 2000C], shown in Figure 4.110. This may be achieved with the aid of very advanced technical features such as fully **inlet/outlet** (I/O) **variable valve timing**

(VVT), highly capable **engine management units** (EMU), and **electrically-heated catalytic converters** (ECC) designed for greater conversion efficiency and improved durability.

The automotive industry would much prefer to move towards this trend and be well on the way to automotive mechatronic control as well as E-M linear and rotary actuating of every AEAV's automotive mechatronic control system.

Completely integrated AEAV like the 6 × 6.6 or 8 × 8.8 and 4 × 4.4 **all electric combat vehicles** (AECV), shown in Figure 4.111, encourage thoughts of what automotive scientists and engineers call a **ride-by-wire** (RBW) or **x-by-wire** (XBW) unified chassis mechatronic control hypersystem [FIJALKOWSKI 1999E].

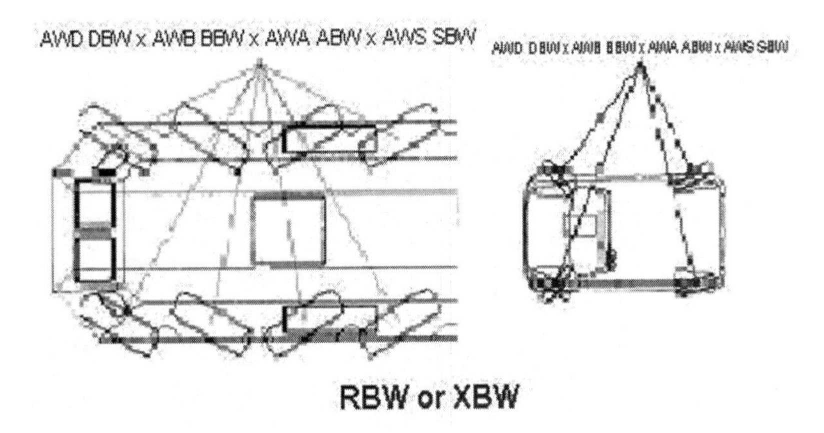

Fig. 4.111 Principle layout of a high-performance all-round energy efficient AECV and AEIV with a family of SBW AWS conversion, DBW AWD propulsion, BBW AWB dispulsion, and ABW AWA suspension [FIJALKOWSKI 1999E]

Particular explanations of these SBW 2WS conversion mechatronic control systems are as follows. The aim of the AECVs is to test new concepts of tri-mode hybrid SBW AWS conversion mechatronic control: by conventional steering or skid steering or by a mixture of both. This embraces linear and rotary E-M actuating of SBW AWS conversion, DBW AWD propulsion, BBW AWB dispulsion, and ABW AWA suspension as well as **throttle-by-wire** (TBW) traction and/or cruise controls. The RBW mechatronic control system could even take over full control of an AECV in response to signals from very advanced collision-evasion (avoidance) sensors.

As the conventional uncontrolled fluidically powered **rack-and-pinion** (R&P) steering gear can only provide constant power assistance, whereas the AECV ideally requires a variable support, the introduction of an electrically-powered and mechatronically controlled R&P steering gear is introduced mainly to adapt the level of power assistance to off-road vehicle velocity.

To achieve these objectives, a further increase in the flexibility and controllability of the R&P steering gear is required.

Only the expanded use of the SBW AWS conversion ECU combined with more sophisticated smart EM actuators can make these objectives achievable.

The tri-mode hybrid SBW AWS conversion control system allows steering to respond to **human- and/or telerobotic driver** (H&TD) input and external disturbances. AECVs thus equipped would therefore respond to crosswind or split braking in a mechatronically manner.

The series hybrid DBW AWD propulsion system allows the omission of the conventional **mechano-mechanical** (M-M) **front-wheel drive** (FWD); **middle-wheel drive** (MWD), and **rear-wheel drive** (RWD) M-M differentials. Therefore, a superimposed off-road vehicle-motion control is necessary to coordinate all EM FWD, MWD, and RWD units. All front, middle, and rear **steered, motorised, and/or generatorised wheels** (SM&GW) are controlled individually. Obviously, the distribution of actual values of the wheel torque may influence the driving/braking behaviour very much. If all actual values of the wheel torque are equal, the DBW AWD propulsion system behaves like a conventional one with M-M differentials lock.

In practice, each distribution of actual values of the wheel torque is possible within the technical limits. This means, for instance, positive maximum values of the wheel torque at the right-hand front and rear SM&GWs and negative maximum values of the wheel torque at the left-hand at the same time.

Such inner front and rear maximum values of the wheel torque distribution leads to actual values of the yaw torque around the vertical axis and cause a turn of the off-road vehicle.

Consequently, the influence of the DBW AWD propulsion control system on the driving/braking behaviour of the AECV may be compared with a skid steering or torque steering mode of the full-time tri-mode hybrid SBW AWS conversion mechatronic control system. Thus, the skid steering (torque-steering) can be used to control the yaw rotation of the AECV simultaneously with the conventional steering HW steered **front-wheel steering** (FWS), **middle-wheel steering** (MWS), and/or **rear-wheel steering** (RWS) gears. If a yaw angular velocity sensor is applied, it may be useful to combine this sensor with an SBW AWS conversion ECU based on the **application specific integrated circuit** (ASIC) **artificial intelligence** (AI) **neuro-fuzzy** (NF) PID microcontroller. Besides, AECVs ought to be equipped with day/night **infra-red** (IR) cameras.

At the heart of the SBW AWS conversion mechatronic control system is an optical torque sensor. Its simple contactless mechanical design makes for an extremely reliable torque sensor. It also allows all possible sensor-failure modes to be detected through software and measurement of steering angle and velocity are also available from the sensor *'for free'*.

The SBW AWS conversion mechatronic control embeds a multi-layer hierarchy of feedback loops within its AI NF **microprocessor control unit** (MCU).

A perfect full-time tri-mode hybrid SBW 4WS conversion mechatronic control system should use not only a feedback control signal for vehicle stability, but also a feedforward control signal for better steering response.

It may be vehicle-speed controlled by an AI NF ECU, and **electronic control bit** (ECB) for each of the SM&GWs stations. Tighter turning and reduced scrub may be the benefits.

It may allow the omission of the conventional front- and rear-axle as well as inter-axle mechanical differentials. Therefore a superimposed on- and/or off-road-motion control would be necessary to co-ordinate both **front-wheel-drive** (FWD) and **rear-wheel-drive** (RWD) units. Both front and rear independent-sprung SM&GW can be controlled individually.

The tri-mode hybrid SBW AWS conversion mechatronic control system study comes from the necessity to develop optimised steering and drive architecture for future **automotive vehicles** (AV), especially for AECVs. As future projects of wheeled AECVs are 6 × 6.6 or 8 × 8.8, and 4 × 4.4 (Fig. 4.104), possibilities of a steering gear are numerous: conventional steering of 1, 2, 3, or 4 axle-free (allows the SM&GW to rotate freely on an axle), skid steering or combination of wheel angle and skid-steering.

Naturally, the choice of a steering concept would be the best considering power consumption, manoeuvrability, and extremely high mobility. As E-M DBW AWD propulsion mechatronic control systems are more and more used in a lot of fields, the eventuality of having an E-M drive in one of the future concepts has to be considered and in the same way, an optimised steering for that kind of transmission has to be considered too.

Automotive scientists and engineers studied various kinds of tri-mode hybrid SBW AWS conversion mechatronic control systems to find the best compromise amongst integration, power, manoeuvrability, and extremely high mobility. This section presents a short overview of the tri-mode hybrid SBW AWS conversion mechatronic control system concept.

The improvement of high-tech SBW AWS conversion mechatronic control has made significant progress during the 1990s. An evolutionary factor behind this has been the increasing requirements for an active safety and the riding comfort of the vehicle. A major contribution to this progress is the introduction of electrically powered and mechatronically controlled R&P steering gears. As the standard fluidically powered and uncontrolled R&P steering gear can only provide constant assistance, whereas the AECV ideally requires variable support; the introduction of electrically powered and mechatronically controlled R&P gear is introduced, mainly to adapt the level of power assistance to the vehicle velocity.

To achieve these objectives a further increase in the flexibility and controllability of the R&P steering gear is required. Only the expanded use of ECU combined with more complex smart electromechanical SBW AWS conversion actuators can make these objectives achievable.

Currently, a great deal of attention has been focused on the research and development of tri-mode hybrid SBW AWS conversion systems for AECVs in order to enhance their handling properties. At high velocities, AECV response may be enhanced by steering the middle and/or rear SM&GWs in the same sense of direction as those in the front, while at low velocities, AECV manoeuvrability may be enhanced by turning the front, middle and/or rear SM&GWs in the opposite sense of direction.

It is well known that the addition of tri-mode hybrid SBW 4WS conversion systems improved the responsiveness of AECVs by reducing transient response time, and also reduced undesirable AECV motion such as fishtailing, making an automotive vehicle easier to SBW AWS conversion mechatronic control during a potential accident situation. However, it is also well known that the addition of a tri-mode hybrid SBW 4WS conversion mechatronic control system did not appreciably extend the overall stability of an AECV.

The emergence of full-time SBW AWS × DBW AWD × BBW AWB × AAW ABW presents an opportunity for automotive scientists and engineers to enhance AECV handling properties at high values of vehicle velocity as well as increasing low values of vehicle-velocity manoeuvrability. It is known too that significant improvements in AECV manoeuvrability cannot be expected in such situation as parallel parking.

4.5.2 Philosophy of Tri-mode Hybrid SBW AWS Conversion Mechatronic Control

The path of an AECV in a manoeuvre is dependent on the forces and torques that act on it. If aerodynamic drag is not taken into account, the only forces and torques that act on the vehicle are gravity and those generated by wheel-tyres at the eight, six, or four contact patches. The lateral force caused by a wheel-tyre is dependent on the values of slip angle and normal loads that are imposed on it. In a conventional SBW 2WS conversion mechatronic control system, driver input to HW generates a slip angle on the front SM&GWs that in turn produces lateral forces on the front wheel-tyres.

These forces acting at the front of the vehicle cause the AECV to yaw and develop a side-slip angle. This generates slip angles at the rear SM&GWs. The lateral forces acting on the rear SM&GWs are built until they balance with those acting on the front. When the front and rear wheel-tyre forces become balanced, the AECV reaches a steady-state condition [FIJALKOWSKI 1999E]. For instance, at high velocities, the main advantage of SBW 4WS is that, by turning the rear SM&GWs at or nearly the same time, and in the same sense of direction as the front SM&GWs, the AECV may generate rear SM&GW slip angles without the need for a vehicle sideslip angle. This eliminates the time lag between steering input and the generation of the rear SM&GW's lateral force, and reduces the time required for the AECV to reach steady-state conditions.
Furthermore, by properly SBW AWS conversion mechatronic controlling the amount of RWS, zero vehicle sideslip may be maintained.

A tri-mode hybrid SBW AWS conversion mechatronic control system may also be used to enhance the low values of the vehicle-velocity manoeuvrability of an AECV by turning the rear SM&GWs in a sense of direction opposite to the front SM&GWs (counter phase steering), thereby reducing its radius of turn [FIJALKOWSKI 1999E].

It is well known that SBW AWS conversion and ABW AWA suspension mechatronic control systems can have substantial influence on vehicle handling. ABW AWA suspension kinematics determines the position of the vehicle roll axis that influences the distribution of lateral mass transfer between the front and rear axles, thereby affecting vehicle stability. It is also been known that the roll axis rolls, causing significant variations in the distribution of lateral mass transfer between front and rear SM&GWs. Because of these considerations, a fixed-roll axis approximation is not a justified assumption in the analysis of longitudinal and lateral dynamics of automotive vehicles with ABW AWA suspension mechatronic control systems such as the McPherson strut double wish-bone system.

4.5.3 EM SBW AWS Conversion Actuators

Electro-mechanical (E-M) SBW AWS conversion actuators are complete E-M dynamic automotive hypersystems comprising a brushless tubular linear **alternating current** (AC) and/or **direct current** (DC) electronic macrocommutator E-M SBW AWS conversion-actuator motor with an **application-specific integrated matrixer** (ASIM) low-, medium-, or high-power electronic commutator termed micro- meso- and/or macrocommutator, respectively, and an **electronic control unit** (ECU) based on an **artificial intelligence** (AI) **application-specific integrated circuit** (ASIC) **neuro-fuzzy** (NF) that is a **neural network** (NN) learning and **fuzzy logic** (FL) programmable microcomputer- or transputer-based PID controller termed an ASIC NF PID microcontroller.

Application to the front- and/or rear-wheel steering mechatronic control of an AECV shows some very interesting characteristics of the modular automotive dynamic hypersystem. The modular concept and the ability of integration of such systems are emphasised.

The core of the proof-of-concept, full-time tri-mode SBW 4WS conversion mechatronic control system is not only independently suspended front and rear SM&GWs with anti-lock and anti-spin E-M drum, ring or disc brakes, but also two smart reciprocating E-M SBW AWS conversion actuators for each of the front-, middle-, and rear-wheel R&P steering gears, and a single smart rotary E-M SBW AWS conversion actuator that is attached to the steering HW column shaft. Reciprocating E-M SBW AWS conversion actuators and a rotary E-M SBW AWS conversion actuator complete E-M dynamic automotive mechatronic control hypersystems comprising tubular linear DC-AC macrocommutator E-M conversion-motors, and a rotary DC-AC macrocommutator E-M steering-actuator motor with an application of the ASIM inverter DC-AC macrocommutators and an AI automotive mechatronic control system ECU based on the ASIC NF PID microcontroller. As a base a conventional R&P steering gear, shown in Figure 4.112, is used.

The brushless tubular linear DC-AC macrocommutator conversion-actuator motor is placed concentrically to the rack, and drives so that produces the necessary rack force that is needed in the order of 12 kN for a full-time tri-mode hybrid AWB SBW conversion mechatronic control system. Hardware of the all-electric steering gear is also shown in Figure 4.112 [FIJALKOWSKI 1999E].

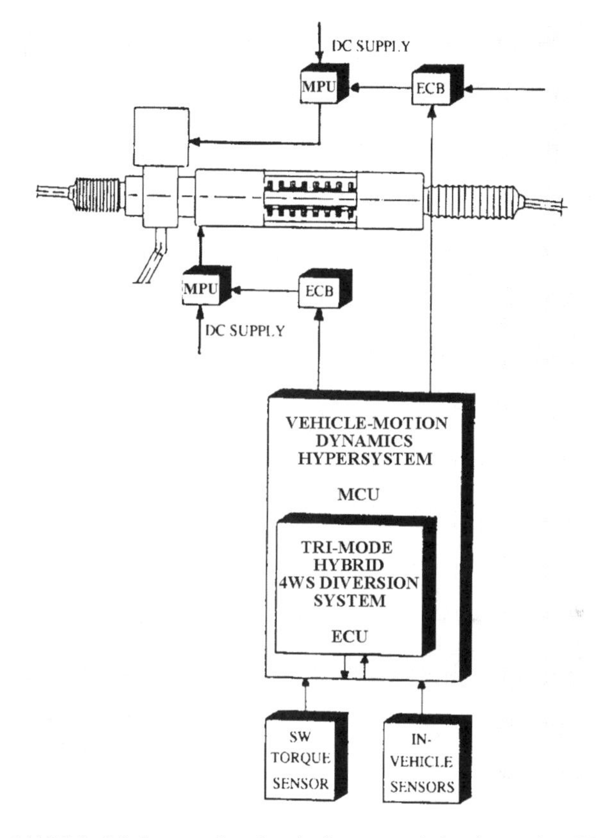

Fig. 4.112 Principle layout of an electrically-powered electric steering (EPES) [FIJALKOWSKI 1999E].

The advantages of **all-electrical** (AE) steering gear being compared to current **all-fluidical** (AF) ones are [FIJALKOWSKI 1999E]:

- More easily controllable and adaptable boost characteristics;
- Programmable features like self-centring and returnability;
- Lower noise-pollution level;
- Improved situation for front-end packaging;
- Modular assembly to aid easy final AECV assembly;
- Elimination of components like **mechano-fluidical** (M-F) pumps, hoses, reservoir, belts, etc.;
- Potential fuel economy of up 0.5 l/100 km.

The steering diverts the steering HW rotary motion into a turning motion of the SM&GWs of the full-time SBW AWS ×DBW AWD × BBW AWB × ABW AWA AECVs [FIJALKOWSKI 1999E].

A perfect tri-mode hybrid SBW AWS conversion mechatronic control system should not only use a feedback control signal for AECV stability, but also a feed-forward control signal for improved steering response. It is easy to assume without analysis that low values of vehicle velocity manoeuvrability is improved due to a smaller turning radius.

Really, contrary steering of the rear SM&GWs at low vehicle velocity may be counter-productive for certain manoeuvres. For instance, at high vehicle velocity, stability is improved by a tri-mode hybrid SBW 4WS conversion mechatronic control system. It is usually accepted that maintaining zero vehicle sideslip is desirable, but the cause why this stabilises the AECV is not known.

Well-known analyses of SBW 4WS stability at high values of vehicle velocity have focused on fixed (steer angle input, including constant angle) or free (driver steering torque normalised to front kingpin axis, including zero torque) control response by taking into account lateral acceleration and yaw rate as being dependent on the steering HW angle. It must be possible to drive the AECV accurately, that is, without any unusual steering corrections. Play in the **mechano-mechanical** (M-M) parts is impressible.

The entire M-M transmission devices must be able to cope with all loads and stresses occurring during operation. Unusual driving manoeuvres, such as driving over obstacles, accident-like occurrences, and so on, must not lead to any cracks or breakages. This inherent steering behaviour is a consequence of different requirements on the slip angles of the SM&GWs that arise when, with increasing centrifugal force, the ratio of lateral force-to-wheel load develops differently at the front and rear axles.

Normally, neutral cornering behaviour is required. Even though, it may allow the optimum use of lateral forces (that is, maximum values of the cornering velocity). It may also reduce the subjective impression of the stability limit of the AECV. In addition, the breakaway of the AECV may be incalculable, since it may break away both at the front and the rear. For this reason, the goal of most AECV manufacturers is to achieve a light understeering in case breaking away of the AECV may lead to a calculable straight-ahead course.

4.5.4 SBW 4WS Conversion Mechatronic Control

The tri-mode hybrid SBW 4WS conversion mechatronic control system (Figs 4.113a and b) allows steering to respond to both **human and/or telerobotic driver** (H&TD) inputs and external disturbances. AECVs thus equipped would therefore respond to crosswinds or split braking in a mechatronically-controlled manner.

There is also less sensitivity to vehicle load and wheel-tyre performance – unlike open loop and predictive-type strategies that can only be optimised for specific operating conditions.

The tri-mode hybrid SBW 4WS conversion mechatronic control system comprises a brushless tubular linear DC-AC macrocommutator conversion-actuator E-M motor with an ASIM inverter DC-AC macrocommutator and an AI automotive mechatronic control system's ECU based on the ASIC NF PID microcontroller.

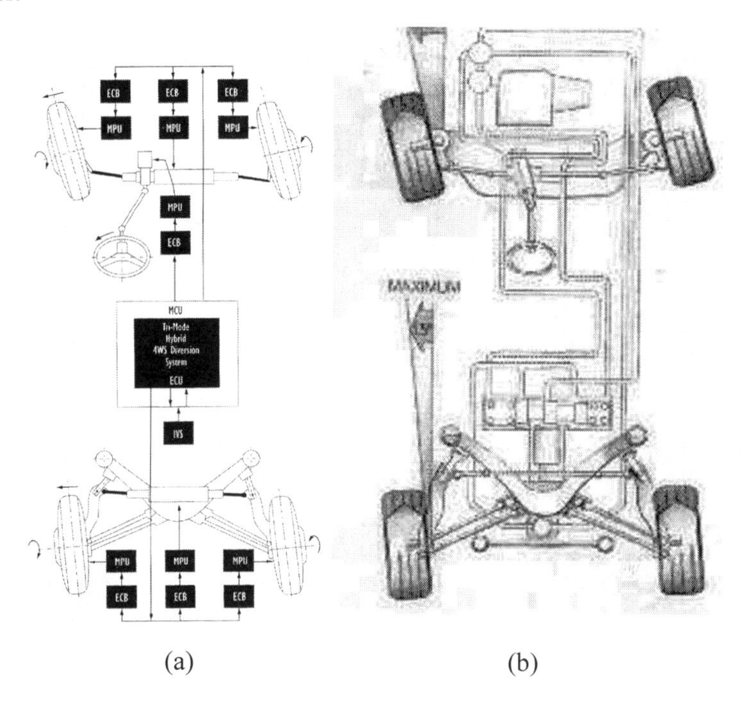

Fig. 4.113 Principle layout of a full-time tri-mode hybrid SBW 4WS conversion mechatronic control system with front and rear electromechanical (E-M) rack-and pinion (R&P) steering gears, two right-side and two left-side steered, motorised and/or generatorised wheels (SM&GW), steering hand wheel (HW) and its hardware [FIJALKOWSKI 1987A, 2000D (a); AMAHOSER 2003 (b)].

The tri-mode hybrid SBW 4WS conversion mechatronic control system is vehicle-velocity controlled by an AI automotive mechatronic control system based on the ASIC NF PID microcontroller. Tighter turning and reduced scrub are the benefits.

Assuming an AECV as a rigid body, the mathematical model of only lateral motion dynamics may be derived from the system-hypomatrix differential equation for forces and torques.

The solution may then be given for yaw angular velocity $\dot{\psi}$ and the sideslip angle β by system hypomatrices with constant coefficients a_{ij}.

For an AECV with tri-mode SBW 4WS conversion mechatronic control system, this first order system-hypomatrix differential equation for the vehicle-motion dynamics, one-lane mathematical-model has the following form [ACKERMANN 1991; LANGHEIM AND FETZ 1992]:

$$\begin{Vmatrix} \dot{\beta} \\ \ddot{\psi} \end{Vmatrix} = \begin{Vmatrix} a_{11} & a_{12} & a_{13} & a_{14} \\ a_{21} & a_{22} & a_{23} & a_{24} \end{Vmatrix} \begin{Vmatrix} \beta \\ \dot{\psi} \\ \delta_f \\ \delta_r \end{Vmatrix} . \tag{4.23}$$

The series hybrid DBW 4WD propulsion mechatronic control system described in FIJALKOWSKI AND KROSNICKI [1993] allows the omission of the conventional front- and rear-axles as well as inter-axle M-M differentials. Therefore a superimposed vehicle-motion control is necessary to co-ordinate both front- and rear-axle units. Both front and rear SM&GWs are controlled individually.

Obviously, the distribution of wheel torques may greatly influence the driving behaviour. If both front and rear wheel torques are equal, the DBW 4WD propulsion mechatronic control system behaves like a conventional one with M-M differentials lock. In practice, each distribution of wheel torques is positive wheel torques within technical limits. This means, for instance, maximum positive wheel torques at the right-hand front and rear SM&GWs and maximum negative wheel torques at the left-hand side at the same time. Such inner-front- and/or inner-rear-wheel torques distribution leads to yaw torques around the vertical axis that is a turn of the AECV. Consequently, the influence of the DBW 4WD propulsion mechatronic control system on the driving behaviour of the AECV may be compared with a torque-steering mode of a tri-mode hybrid SBW 4WS conversion mechatronic control system.

For a smart, full-time SBW AWS ×DBW AWD × BBW AWB × ABW AWA **all-electric intelligent vehicle** (AEIV) *Poly-Supercar* [FIJALKOWSKI 1995, 1997A] with not only a conventional dual-mode SBW 4WS conversion mechatronic control system, but also an unconventional tri-mode hybrid SBW 4WS system with front- and/or rear-axle torque-steering, a similar first order system-hypo matrix differential equation for forces and torques may be derived. The solution may then be given for the yaw angular velocity $\dot{\psi}$ and sideslip angle β by system-hypo-matrices with constant coefficients a_{im}. This first order system-hypomatrix differential equation for the AEIV-motion dynamics, one-lane mathematical model is derived in the form [FIJALKOWSKI AND CROSSKICK 1993]:

$$\begin{Vmatrix} \dot{\beta} \\ \ddot{\psi} \end{Vmatrix} = \begin{Vmatrix} a_{11} & a_{12} & a_{13} & a_{14} & a_{15} & a_{16} \\ a_{21} & a_{22} & a_{23} & a_{24} & a_{25} & a_{26} \end{Vmatrix} \begin{Vmatrix} \beta \\ \dot{\psi} \\ \delta_f \\ \delta_r \\ T_{fz} \\ T_{rz} \end{Vmatrix} . \tag{4.24}$$

Thus, the torque steering may be used to control the yaw motion of the AEIV *Polycar* simultaneously with the conventional front- and rear-axle steering.

If a yaw angular velocity sensor is applied, it may be useful to combine an open-loop control with superimposed closed-loop control of the yaw angular velocity [FIJALKOWSKI AND KROSNICKI 1993].

One possible control of the yaw angular velocity is the application of an AI automotive mechatronic control system's ECU based on the conventional ASIC PID microcontroller. Also, an AI automotive mechatronic control system's ECU based on an unconventional ASIC NF PID microcontroller can already give satisfactory results with little knowledge of the lateral-motion dynamics of the tri-mode hybrid SBW 4WS conversion mechatronic control system that has to be controlled. NF lateral AEIV-motion control may be easier adapted.

Conventional dual-mode hybrid 4WS conversion mechatronic control systems give full-time SBW AWS ×DBW AWD × AWD BBW × ABW AWA AEIVs with a real wheelbase l the driving behaviour of one with another virtual wheelbase l'. The virtual wheelbase l' is smaller than the real wheelbase l if the rear SM&GWs are steered in the opposite sense of direction.

Especially at low values of vehicle velocity, this leads to higher manoeuvrability of the AEIV. If the rear SM&GWs are steered into the same sense of direction, the virtual wheelbase l'' becomes greater than the original real one. This behaviour may also be considered to be, at torque steering (skid-steering), an unconventional tri-mode hybrid SBW 4WS conversion mechatronic control system. This means that short-wheelbase AEIV that are equipped with such unconventional SBW 4WS conversion mechatronic control systems can give proof of having a stable driving behaviour like long-wheelbase AEIVs.

Thus, using the torque and/or angular velocity controls of the SBW 4WS conversion mechatronic control system's inner SM&GWs can increase this effect, especially at recuperative braking with the inner SM&GWs acting as the AC-DC macrocommutator wheel-hub generators, because the front gravitational forces into the AEIV become greater than the respective rear forces. At the same creep, this leads to greater horizontal (longitudinal and lateral) forces.

SBW 4WS Conversion Mechatronic Control System Configuration - Figure 4.113 shows the tri-mode hybrid SBW 4WS conversion control system configuration. The rear SM&GWs are steered according to the front-wheel steer angle by installing the steering mechanism to steer the rear SM&GWs at the rear ABW 4WA suspension and electrically connecting it with the front steering mechanism by electrical connections (cabling) only. The relation of steer angles between the front and rear SM&GWs are mechatronically controlled according to the front- or backward movement of the AEIV, and the actual value of the vehicle velocity.

SBW 4WS Conversion Mechatronic Control -- <u>*Tri-mode SBW 4WS Function*</u> -- The driver can select tri-modes with a SBW 4WS conversion mechatronic control map for the phase and steering angle of the rear SM&GWs to the steer angle of the front SM&GWs by selecting the '*Tri-Mode Selector*'.

Cancelling SBW 4WS Function for Backward Movement – To provide the driver with a natural feeling for backward movement of the AEIV and to allow easy movement of the AV ride in tandem parking lot, the SBW 4WS conversion

mechatronic control system is designed so that the driver may cancel the SBW 4WS only for backward movement by selecting '*Forward/ Backward Selector*'.
Vehicle Velocity-Sensitive SBW 4WS Function – The phase of the steering angle of the front to the rear SM&GWs changes according to the actual value of vehicle velocity. At low velocity, controllability is improved because the rear SM&GWs are steered in the opposite sense of direction (reverse phase) to the front SM&GWs. At high velocity, travelling stability is improved because the rear SM&GWs are steered in the same sense of direction (same phase) as the front SM&GWs. When the front-wheel steering angle is constant, the rear SM&GWs are steered up to $\pi/36$ rad (5 deg) to decrease the minimum radius by 0.5 m (see Fig. 4.113 b). Because the control map is tuned exclusively for the predictive and adaptive ABW 4WA suspension mechatronic control system, it is different from the coil suspension. A same-phase 4WS ABW suspension mechatronic control system lets all four wheels respond in unison to steering input. By limiting the RWS angle to 1.5 deg, the simultaneous wheel-tyre movement is subtle, yet extremely effective in its influence on yaw, transient response, and lateral acceleration [AMAHOSER 2003].
Brushless Tubular Linear DC-AC Macrocommutator Conversion Actuator E-M Motor - The tri-mode hybrid SBW AWS conversion mechatronic control system uses conversion E-M actuators for actuation. The conversion E-M actuator shown in Figure 4.112 requires poly-phase AC electric power that is inverted from DC electric power at the inverter ASIM macrocommutator of the tubular linear DC-AC macrocommutator conversion-actuator motor at the frequency determined by a mover-position sensor. Since the EM conversion actuators are considered for full-time SBW AWS × DBW AWD × BBW AWB × ABW AWA AEIV *Polycars* all have linear outputs, and the EM conversion actuator achieves gear reduction through an R&P steering gear. A brushless tubular linear DC-AC macrocommutator conversion-actuator motor may be developed from a flat linear electrical machine by rolling it about its longitudinal axis.

Figure 4.112 shows a tubular linear DC-AC macrocommutator conversion-actuator motor, and it is observed that **interior permanent magnets** (IPM) are in the form of cylindrical shells, whereas the AC armature winding is formed from simple circular coils. A tubular linear DC-AC macrocommutator conversion-actuator E-M motor that is supplied from a variable frequency-changer (inverter), the ASIM DC-AC macrocommutator may be controlled so that it behaves like a brushed rotary DC-AC mechanocommutator conversion-actuator E-M motor. To do this, the DC-AC macrocommutator-supplied armature current is phase-locked to the signals received from mover-position sensors. A conversion-actuator E-M motor controlled in this way is often termed an '*autosynchronous electric machine*'.

The mover-position signals are received from *Hall Effect* cells on the stator teeth and, after decoding, are used to control the electric valves (switches) of the inverter ASIM DC-AC macrocommutator. It performs all the necessary decoding and inverter functions as well as providing a chopped current limit condition.

Electrically Powered Electric Steering - The use of the AE HW column-, pinion- or rack-assist torque/force actuation rather than a fluidical one, gives a wide range of benefits: fuel economy, cheaper and quicker installation, in-vehicle tuning, mass and volume, functionality, more environmentally friendly, and maintenance free [FIJALKOWSKI 1999E].

Conventional FPSs employ an ECE- or ICE-driven M-F pump that must run with some output pressure all the time and must be sized to provide maximum assistance with the ECE or ICE idling. Hence, steering power is continually taken from the ECE or ICE. On the other hand, EPS uses power only when it is required and ultimately leads to a significant fuel saving. There is no oily-fluid to fill or pipework to fit, therefore no end-of-line venting and leakage checks.

Automotive vehicle manufacturers have estimated that the EPS unit could be fitted in only 15 -- 20% of the time that it takes to fit a FPS assembly. A wide range of steering characteristics can be tuned extremely rapidly in the AECV through software. The EPS unit is 3 -- 5 kg lighter in mass and smaller in volume than the comparable FPS assembly.

Its single unit integrated design may be compared with the M-F pump, pulley, belt, pipework, oily-fluid, reservoir, and fixtures required for FPS assembly. A simple FPS includes velocity sensitivity, yaw damping, self-centring and optional *'light'* or *'sport'* steering feel settings.

In contrast, FPS functionality is severely limited and static. FPS carries the risk of fluid spillage and contamination. This risk does not exist with EPS. In fact, the fuel savings offered also mean that it has a less adverse impact on the environment, helping to save natural resources and reduce tailpipe emissions.

The EPS may be a sealed-for-life maintenance-free single unit. There are no fluids to top up or drive belts to tighten as with FPS. It is available as EFPS. Both may be modular full-time, tri-mode hybrid SBW AWS conversion control systems, and can offer distinct performance advantages over conventional F-M steering packages. Because the tri-mode hybrid SBW AWS conversion control system operates independently of the off-road vehicle's ICE, placement within its compartment is extremely flexible.

In an emerging EPFS, a rotary, electronically-commutable, magnetoelectrically excited, brushless DC-AC mechanocommutator steering-pump E-M motor drives an FPS, M-F pump in a conventional R&P steering gear.

The necessary rack-assisted force yields 8.5 kN. This tri-mode hybrid SBW AWS conversion control system offers brushless steering-pump motor technology, eliminating brush wear and bettering its efficiencies.

It also offers HW angular velocity sensitive steering for improved feel at various values of the hand SW angular velocity. An M-F pump's rotor angular velocity is matched to the steering HW angular velocity to minimise energy consumption.

In an emerging EPES, a high-output, tubular linear, electronically commutated, magnetoelectrically excited or reluctance, brushless DC-AC commutator steering-actuator motor drives the rack of the R&P steering gear through a ball screw mechanism. The necessary rack -assisted force yields about 10 kN.

In the novel EPS, shown in Figure 4.112, a conventional R&P steering gear may be used as a base. The brushless tubular linear DC-AC macrocommutator steering actuator motor is placed concentrically to the rack, and drives it so that it produces the necessary rack-assisted force (12 kN) for a full-time, tri-mode hybrid SBW AWS conversion control system [FIJALKOWSKI 1999E]. Its steering-actuator motor's low inertia provides improved dynamical performance, including improved returnability and improved yaw-stability. Safety is another benefit. Because the steering-actuator motor operates independently of the ECE or ICE, the off-road vehicle may be steered easily should ECE or ICE power fail. The steering-actuator motor must also be carefully designed to give low levels of ripple and cogging torque and hence a smooth progressive feel at the steering HW. This SBW AWS conversion control system uses an AI NF **microprocessor controller unit** (MCU) as well as an overseer designed for full, safe operation.

The torque sensor must be an ultra-reliable, fault tolerant design with performance and cost compatible with SBW AWS conversion mechatronic control system requirements.

At the core of the SBW AWS conversion control system is an optical torque sensor. Its simple contactless mechanical design makes for an extremely reliable torque sensor.

Its design also allows all possible sensor failure modes to be detected through software, and measurement of steering angle and velocity are also available from the sensor '*for free*'.

The SBW AWS conversion control system embeds a multi-layer hierarchy of feedback loops within its AI NF MCU.

At a core of the tri-mode hybrid SBW AWS conversion control system is a steering-actuator-motor control and management. This is a fast closed-loop designed to ensure that the steering-actuator-motor output rapidly tracks the net torque demand. A stabilising controller ensures good stability margins when the SBW AWS conversion control system is fitted to the AECV and protects against parametrical changes in mechanical components throughout the AECV's life.

The assistance control and steering feel enhancement loops add the software '*hooks*' that make in-vehicle tuning of steering characteristics possible. Assist gain characteristics are adjusted in the software, so that the possibilities are virtually infinite. Other control functions are added to integrate the tri-mode hybrid SBW AWS conversion control system with an on-board RBW automotive unified chassis mechatronic control hypersystem on the AECV.

The OEM may optimise the steering feel in minutes with a simple laptop computer. Damping may be adjusted to allow more precise steering and optimum feel at various values of the steering HW angular velocity.

As an example, the hardware of an AE EM SBW 4WS gear is shown in Figure 4.112. The steering diverts the HW rotary motion into a turn motion of the SM&GWs of the off-road vehicles [FIJALKOWSKI 1999E].

A perfect full-time tri-mode hybrid SBW 4WS conversion control system should use not only a feedback control signal for off-road vehicle stability, but also a feedforward control signal for better steering response.

The full-time, tri-mode hybrid SBW 4WS conversion control system may control the vehicle's speed with an AI NF ECU, and ECB for each of the SM&GW stations. Tighter turning and reduced scrub may be the benefits.

The hybrid DBW 4WD propulsion control system mentioned may allow the omission of the conventional front and rear-axle as well as inter-axle **mechano-mechanical** (M-M) differentials. Therefore, a super-imposed off-road motion control may be necessary to coordinate both FWD and RWD units. Both front and rear independently sprung SM&GWs may be controlled individually. Obviously, the distribution of wheel torques may influence the driving behaviour very much. If both front and rear wheel torques are equal, the DBW 4WD propulsion control system may behave like a conventional one with an M-M differentials lock.

In practice, each distribution of wheel torques can be possible within the technical limits. This means, for instance, maximum positive wheel torques at the right-hand front and rear SM&GWs and maximum negative wheel torques at the left-hand ones at the same time. Such inner-front and/or inner-rear-wheel torques distribution may lead to yaw torques around the vertical axis and, a turn of a smart off-road vehicle.

Consequently, the influence of the DBW 4WD propulsion mechatronic control system on the driving behaviour of an off-road vehicle may be compared with a torque steering mode called the skid-steering mode of a tri-mode hybrid SBW 4WS conversion mechatronic control system. Thus, the skid steering (torque steering) may be used to control the yaw motion of the off-road vehicle simultaneously with the conventional FWS, MWS and RWS gears. If a yaw angular velocity sensor is applied, it may be useful to combine an open loop control with superimposed closed loop control of the yaw angular velocity. One possibility to control the yaw angular velocity is the application of a conventional PID controller. Also, an AI NF MCU may already give satisfactory results with little knowledge of lateral motion dynamics of the tri-mode hybrid SBW 4WS conversion control system that has to be controlled.

NF reasoning-based lateral off-road-vehicle motion control may easily be adapted. Thus, using torque and/or velocity controls of a DBW 4WD propulsion control system's inner SM&GWs may increase this effect, especially at recuperative braking with the inner SM&GWs acting as the AC-DC macrocommutator wheel-hub generators, because the front gravitational forces on the off-road vehicle become greater than the respective rear ones. At the same creep this leads to greater horizontal (longitudinal and lateral) forces.

Figure 4.112 shows the tri-mode hybrid SBW 4WS conversion control system configuration. The rear independent-sprung SM&GWs may be steered according to the front wheel steer angle by installing a steering mechanism to steer the rear SM&GWs at the rear suspension and electrically connecting it with the front steering mechanism by electrical connections (cabling) only.

The relation of steer angles between the front and rear SM&GWs may be mechatronically controlled according to the front or backward movement of an off-road vehicle, and the vehicle velocity. An active RWS gear that operates from normal to critical conditions and extends a safety margin on all situations may be used. Its aims are the following: to improve vehicle stability when low values of cornering force divided by slip angle of wheel tyres are used or during loaded conditions; to reduce the driven burden and to give the driver advanced steering feel, and to assure enhancement of vehicle stability from usual to critical condition.

For instance, the active RWS gear's goal is to realise the above-mentioned aims, so the rear steering angle is limited within $\pi/36$ rad (5 deg).

The steering angle sensor of an EMI housing is manufactured via mass construction, where two precision bearings are incorporated. These bearings allow for data measuring via a hollow shaft. The sensor functions according to the principle of allowing for precise measurement of angles of rotation. Applications include measuring the wheelbase, registering the angle of steering, or measuring torque.

The tri-mode hybrid SBW AWS conversion mechatronic control system concept comes from the combination of the conventional steering of **wheeled vehicle** (WV) and the skid steering of **tracked vehicle** (TV).

Conventional steering of WV is largely sufficient for mobility on the road but the angle of the **steered, motorized and/or generatorised wheel** (SM&GW) lowers the volume available inside the hull. Skid steering is particularly efficient for manoeuvrability, especially on all ground surfaces (dry tarmac, grass, snow, and sand) but needs high power.

Tri-mode hybrid steering is the combination of a percentage of the SM&GW angle and a percentage of skids; this combination provides the advantages of both steering. Advantages of tri-mode hybrid SBW AWS conversion for DBW AWD propulsion are as follows [FIJALKOWSKI 1999E]:

- ❖ Extremely high manoeuvrability by additional skid steering to the conventional steering, especially in heavy terrain;
- ❖ Optimal extremely high mobility and control provided by independent SM&GWs with the brushless AC-AC/AC-AC or DC-AC/AC-DC macrocommutator magnetoelectrically excited and reluctance wheel-hub motors /generators;
- ❖ Lower power in comparison with pure skid steering for the same performance; greater volume under the vehicle's body since the SM&GWs angle is lowered.

Tri-mode hybrid SBW AWS conversion mechatronic control system for a WV implies integration of the laws [FIJALKOWSKI 1999E]. The first law concerns the angular velocity ratio between two SM&GWs of the same axle-tree according to the angle of steering HW. The second law concerns the angle of the SM&GW according to the angle of the steering wheel. The third law concerns the angular velocity ratio modulation according the vehicle velocity.

Indeed, for high values of vehicle velocity (above 14 m/s), conventional steering is sufficient and skid steering useless. It is the reason why in every law of tri-mode hybrid SBW AWS conversion there is no more skid above 4 m/s.

When using an M-M drive, tri-mode hybrid steering laws are applied to the vehicle by a special M-M differential.

One of the advantages of tri-mode hybrid steering with independent SM&GWs with brushless AC-AC/AC-AC or DC-AC/AC-DC macrocommutator magneto-electric and reluctance wheel-hub motors/generators is that tri-mode hybrid steering laws are easily taken into account to command automotive mechatronics. For instance, the angular velocity ratio reaches a maximum value of 2.5. The latter was established not only by trials but also through calculations. It is the best compromise solution for 4 × 4.4 AV to reach a high performance of manoeuvrability without requiring too much power. Until 4 m/s, additional skid steering is easy for tri-mode hybrid SBW AWS conversion mechatronic control by the driver. Above that value of vehicle velocity, an angular velocity ratio of 2.5 could lead the vehicle to unsteady steering. This is the reason why above 4 m/s, the maximum value of the angular velocity ratio obtainable is modulated. For instance, at 8 m/s, the maximum value of the angular velocity ratio is 60% below, but over 14 m/s there is no more skid steering.

Explicit steering is realised on a vehicle by altering the heading of the wheels to initiate an alteration in vehicle orientation.

Skid steering, alternatively, realises a rotation motion by the velocity difference on each side of the vehicle.

An AECV is an example of a normally (explicit) steered vehicle or a skid-steered vehicle. Despite the fact that some dynamic features of skid-steered vehicles are analogous to tracked vehicles (two tracks covering the wheels on each side, similar to a main battle tank), most of the literature is derived from tracked vehicles [WONG 1993; SCHILLER ET AL. 1993].

These categories of vehicles are often used in robotic mobile systems: they are more beneficial because of the lack of a steering mechanical system while maintaining high mobility and good manoeuvrability. Still, skid steering continues to be a convoluted phenomenon dependant on the properties of the soil, the wheel-tyre constitution, the wheel-tyre/road friction and the wheel skidding. On a one-axis explicit steered vehicle, the **instantaneous centre of rotation** (ICR) is always positioned on the axis traversing the rear wheels [SHAMAH 1999].

For a skid-steering vehicle, the ICR is located on the rear wheel's axis if and only if no slippage between the vehicle and the soil subsists. If slippage arises, the ICR may move forward and sometimes out of the vehicle's base, thereby losing motion stability [CARACCIOLO ET AL. 1999].

In order to maintain control of the vehicular platform, the control design must afford a method to remain the ICR between the vehicle's bases. One method to do this is to include an operational unholonomic kinematics constraint in the motion controller [D'ANDREA-NOVEL ET AL. 1995].

This restrains the values of the vehicle velocity (angular and lateral) so that the ICR continues in the vehicle's base, thus presenting stability for the moving vehicular platform, supposing an accurate control scheme is appropriate. The interest of tri-mode hybrid steering lies in the possibility to considerably reduce the SM&GW angle without lowering the manoeuvrability.
An option of SBW AWS conversion control system is an **ultrasonic parking assistance** (USPA) that protects drivers and pedestrians from obstructed-view collisions. The USPA helps the driver to avoid accidents, protects off-road vehicles from bumper damage and makes parallel parking easier and safer. The USPA includes **ultrasonic** (US) sonars that emit ultrasonic pulses that detect objects in the path of the off-road vehicle, and then immediately analyses the pulses to determine the distance from the object. If the off-road vehicle is close to an object, the SBW AWS conversion control system's USPA emits a series of beeps, which increase in frequency, as the driver gets closer, if an AECV is within approximately 0.25 m of an object, the tone becomes continuous to warn the driver of an impending collision. The SBW AWS conversion control system's USPA emits a higher-pitched tone if the object is in front of the off-road vehicle, so that drivers can distinguish whether the object is in the front or the rear of the off-road vehicle. The USPA uses ultrasonic sonars (distance measurement) in four, six, or eight channel configurations. It is adaptable to many control/logic inputs, and offers automotive vehicle manufacturers additional options such as stereo mute capability and visual display. The USPA is automatically activated when an AECV or AEIV is put into reverse, but may also be turned *ON/OFF* manually.

The proposed on-board AI NF MCU has a concentric configuration in which all AI NF RBW ECUs are centred in one place [FIJALKOWSKI 1999E]. This tends to connect the electrical connections (cabling) to AECV's **in-vehicle sensors** (IVS), as well as linear and rotary actuator E-M motors. It stands to reason that the hybrid (concentrical/dispersal) configuration, shown in Figure 4.114, ought to be used Exclusive AI NF ECB placed at each SM&GW station may be combined together with a bi-sensual-direction **fibre-optic** (FO) **in-vehicle data link** (IVDL) for information exchange. Thus, the on-board AI NF MCU may be placed in the centre and can do the computations for the whole off-road-vehicle-motion control. It may also exchange data between SBW 4WS conversion, DBW 4WD propulsion, BBW 4WB dispulsion, and ABW 4WA suspension controls AI NF ECUs that may be translated to the AI NF ECBs at the SM&GWs. In this way, it may be possible to fit together linear and rotary actuator E-M motors with unified ones and their AI NF ECBs which seems to be a promising AECV motion mechatronic control for the future. The distinction between randomness and imprecision may perhaps is made clear by consideration of crisp or fuzzy sets.

The universal set U may be an on-board AI NF MCU for RBW automotive vehicle dynamics mechatronic control and for SBW AWS, DBW AWD, BBW AWB or ABW AWA controls or all, for instance, heterogeneous integrated RBW automotive unified chassis-motion mechatronic control, the intersection is the crisp or fuzzy set of all the AI NF ECUs for all individual AWA SBW, DBW AWD, BBW AWB and ABW AWA controls.

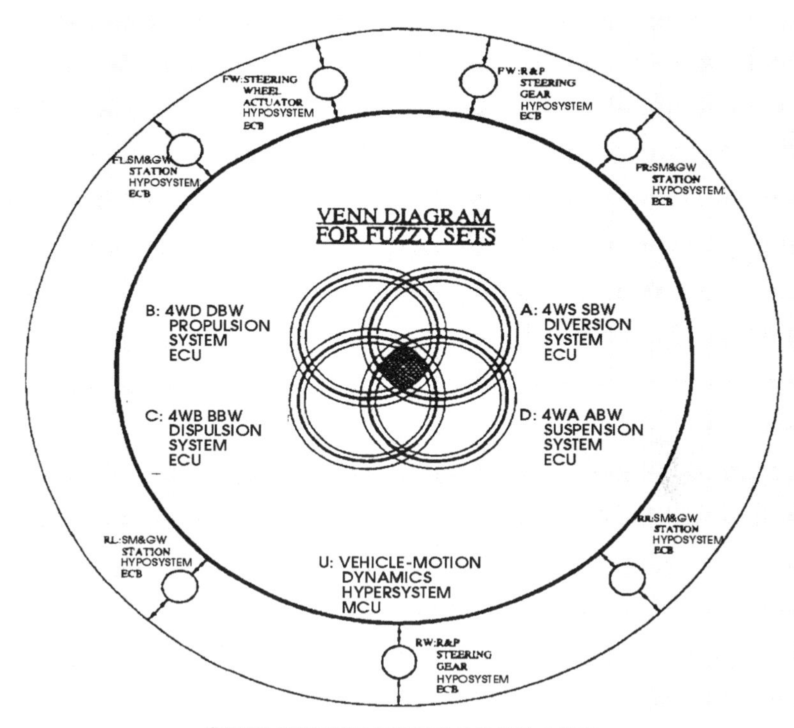

Fig. 4.114 Concept of heterogeneous RBW integrated unibody or body-over-chassis motion mechatronic control between SBW 4WS conversion, DBW 4WD propulsion, BBW 4WB dispulsion, and ABW 4WA suspension controls [FIJALKOWSKI 1999E].

The *Venn* diagram shown in Figure 4.114 may adequately explain the primary purpose of the integrated RBW or XBW integrated unibody, space-chassis, skateboard-chassis, or body-over-chassis motion mechatronic control between individual dynamics automotive controls. *'SBW AWS + DBW AWD + BBW AWB + ABW AWA'* in Fig. 4.113 represent simple aggregation of individual SBW 4WS conversion, DBW 4WD propulsion, BBW 4WB dispulsion, and ABW 4WA suspension controls, while *'SBW AWS × DBW AWD × BBW AWB × ABW AWA'* represent heterogeneous integrated RBW automotive unified chassis mechatronic control between SBW 4WS conversion, DBW 4WD propulsion, BBW 4WB dispulsion, and ABW 4WA suspension controls that shows *'SBW AWS × DBW AWD × BBW AWB × ABW AWA'* improves dynamic performance more than *'SBW AWS + DBW AWD + BBW AWB + ABW AWA'* does. This heterogeneous integrated RBW unified chassis-motion mechatronic control between dedicated automotive functional controls is sophisticated because it consists of many components.

4.5.5 Conclusion

The well-proven technology of a full-time, tri-mode hybrid SBW AWS conversion mechatronic control system has been adapted to DBW 4WD propulsion and BBW 4WB dispulsion as well as ABW 4WA suspension mechatronic control systems. Besides regenerative steering, it offers best efficiency, high steering precision, and a good stabilization effect for precise straight-ahead driving.

Steering is achieved not only by means of steering wheel but also by varying the values of the angular velocity and senses of rotation of all the SM&GWs for vehicles. The AECVs as well as AEIVs are highly mobile [FIJALKOWSKI 1999E]. For instance, the SBW AWS conversion and DBW AWD propulsion mechatronic control system's mechanism for vehicles are contained within the hubs of the SM&GWs (very little ground clearance being needed) giving a very low centre of gravity (barycentre).

It was testified that the rolling resistances of driving SM&GWs are lower than in the case of the free rolling wheels.

Tri-mode hybrid SBW AWS conversion mechatronic control is a great opportunity for vehicles to increase the volume under the vehicle's body by lowering the SM&GW angle without decreasing manoeuvrability or requiring high power.

Furthermore, the additional SM&GWs angle to skid steering allows better manoeuvrability than pure skid steering. All things considered, tri-mode hybrid steering has to be designed considering the whole vehicle because of the consequences on the SM&GWs angle, on size of the vehicle's body, on command automotive electronics, and on transmission.

Mechatronically controlled tri-mode hybrid SBW AWS conversion control system is an efficient means to influence an off-road vehicle's yaw and roll dynamics. The physical limits in terms of maximum value of the force between an SM&GW's wheel-tyre and the off-road surface may be further exploited to provide additional safety margins by improving the yaw disturbance attenuation and diminution of rollover risk, respectively.

For instance, the core of the mechatronically controlled tri-mode hybrid **six-wheel steered** (6WS) SBW conversion mechatronic control system for 6 × 6.6 AECV may not only be three smart reciprocating E-M R&P FWS, MWS, and RWS actuators, and a single smart rotary EM steering hand wheel actuator that may be attached to the steering wheel column shaft, but also six independently suspended variable-angular-velocity SM&GWs with the EMBs.

There are already mechatronic tools that provide the possibility to set an auxiliary steering angle to the one directly transmitted by the steering wheel.

In the axleless AECV of the future, heterogeneous integrated RBW or XBW integrated unibody, space-chassis, skateboard-chassis or body-over-chassis motion mechatronic control (SBW AWS conversion, DBW AWD propulsion, BBW AWB dispulsion, ABW AWA suspension as well as traction and/or cruise controls); and in-vehicle route guidance navigation and collision-avoidance controls might not only do much of the work, they might also monitor themselves, respond adaptability to changing demands and emergencies, and have the intelligence to keep the heterogeneous RBW or XBW integrated unibody or body-over-chassis motion mechatronic control system operating smoothly and efficiently.

Thanks to on-line RBW or XBW integrated unibody or body-over-chassis motion mechatronic control of normal reactions on SM&GWs of the AEAV, it is possible to decrease the so-termed norm coefficient form stability, and consequently, increase the carrying capacity.

Practically it means that the AEAV of higher capacity may be installed on the same AEAV that increase its efficiency. Active safety system for AECVs and AEIVs can be easily mounted on serially manufactured AEAVs. Because of the above advantages, the novice idea makes a serious product offer for automotive manufacturers of AECVs and AEIVs. Such is the hope of the author who is working towards this goal using several AI NF techniques.

4.6 SBW 4WS Conversion Mechatronic Control System for Automotive Vehicle Lane Keeping

4.6.1 Foreword

As a recent advance in automotive vehicle automation, research into lane-keeping mechatronic control is being carried out extensively in the field of automated manned and/or unmanned vehicles as well as steering and driving assistance mechatronic control systems. From the viewpoint of automotive vehicle dynamics and mechatronic control, however, it is impossible to achieve the desirable lane-keeping performance in both lateral and yaw senses of direction by using the only FWS SBW angle as the control input. Thus additional control input is necessary to overcome this problem. From a viewpoint of mechatronic control theory, SBW 4WS seems to be an attractive alternative in enhancing lane-keeping mechatronic control performance.

As a recent advance in automatic steering, research into lane-keeping control is being carried out extensively in the field of automated vehicles as well as driving assistance systems.

Since the mid 1950s, many control algorithms based on not only classical control theories but also on modern control theories, have been proposed for the design of lane-keeping control systems [HEDRICK 1994; TSUGAWA 1998; FUJIOKA ET AL. 1999].

At the beginning of our research, a lane-keeping controller based on a conventional front-wheel steering vehicle (2WS) was designed to regulate lateral deviation at the centre of gravity with the application of optimal control theory. As a result, the lateral deviation is well regulated with satisfactory dynamics, while yaw dynamics tends to have undesirable damping behaviour [MOURI ET AL. 1997].

From the viewpoint of vehicle dynamics and control, however, it is impossible to achieve the desirable lane-keeping performance in both lateral and yaw senses of direction by using the only front-wheel steering angle as control input. Thus, additional control input is necessary to overcome this problem.

From a viewpoint of control theory, a **four-wheel-steering** (4WS) system seems to be an attractive alternative in enhancing lane-keeping control performance.

In the field of active safety, 4WS has been extensively studied for a long time and regarded as an effective tool to enhance vehicle-handling performance.

In the last decade, various types of four-wheel-steering vehicles have been developed and appeared on the market. Many control algorithms for a four-wheel-steering system have been developed for various desirable control objectives such as reduction of phase-lag of acceleration and yaw rate, reduction of side slip angle, and so on [FURUKAWA ET AL. 1989].

Automotive scientists' and engineers' main interest, here, is what type of control law of 4WS is suitable for the automatic lane-keeping control system.

This section employs 4WS that applies the mathematical model matching control technique to realise the desired steering response [NAGAI 1989].

Two representatives desired steering responses are presented in this section. Computer simulation is carried out to verify the effectiveness of each control law on lane keeping control performance [RAKSINCHAROENSAK ET AL. 2003].

4.6.2 Automotive Vehicle Physical and Mathematical Models

This section for the most part concentrates on the routine lane-keeping performance of automotive vehicles where the vehicle's-body sideslip angle may be so diminutive that the vehicle velocity may be almost constant.

The results of roll, heave, and pitch motions as well as the dynamic characteristics of the wheel-tyre and steering system may also be neglected.

The sideslip angle of right and left wheel-tyres may be in the region where the linear wheel-tyre physical model is valid.

Derived from these statements, this section uses an equivalent bicycle physical model to exemplify the vehicle dynamics as shown in Figure 4.115 [RAKSINCHAROENSAK ET AL. 2003].

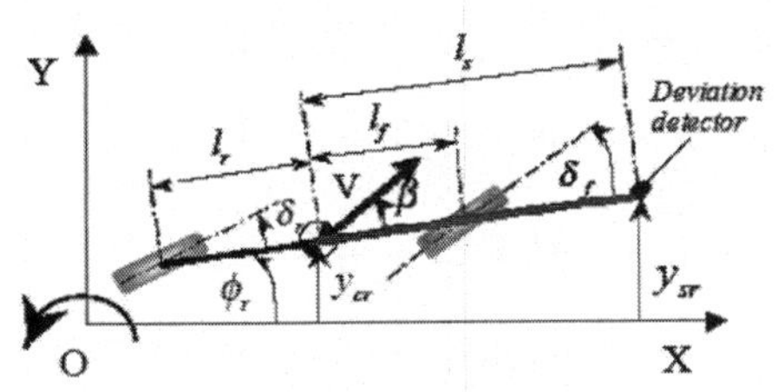

Fig. 4.115 Equivalent bicycle physical model [RAKSINCHAROENSAK ET AL. 2003].

The equivalent bicycle mathematical model corresponding to the *Euler--Lagrange* equations of motions may be described as

$$m\ddot{y}_c = mV(\dot{\beta} + \dot{\phi}) = 2F_f + 2F_r \quad , \tag{4.25}$$

$$I\ddot{\phi} = 2l_f F_f - 2l_r F_r \quad , \tag{4.26}$$

$$F_f = C_f\left(\delta_f - \frac{l_f}{V}\dot{\phi} + \phi_r - \frac{\dot{y}_{cr}}{V}\right) \quad , \tag{4.27}$$

$$F_r = C_r\left(\delta_r + \frac{l_r}{V}\dot{\phi} + \phi_r - \frac{\dot{y}_{cr}}{V}\right) \quad , \tag{4.28}$$

where m - the automotive vehicle mass;
I - the yaw moment of inertia;
Y_c - the lateral displacement at centre of gravity (CG);
Φ - the yaw angle;
V - the vehicle velocity;
β - the vehicle's body sideslip angle;
l_f and l_r - the distances from the front and rear axles to the centre of gravity, respectively;
F_f and F_r - the front and rear cornering forces, respectively;
C_f and C_r - the front and rear cornering stiffnesses, respectively;
δ_f and δ_r - the front and rear steering angles, respectively.
(Note: subscript r denotes the variable relative to reference line.)

In the case of a curved trajectory with a constant radius of curvature ρ, the relative variables with respect to the desired course are expressed as

$$\dot{\phi}_r = \dot{\phi} - \rho V \quad , \tag{4.29}$$

$$\ddot{y}_{cr} = \ddot{y}_c - \rho V^2 \quad . \tag{4.30}$$

4.6.3 SBW 4WS Conversion Mechatronic Control System Design

The objective of the SBW 4WS conversion mechatronic control system is to regulate the lateral deviation at the **centre of gravity** (CoG) and yaw deviation of the vehicle to be zero.

The mechatronic control system, as shown in Figure 4.116, comprises the mathematical model matching controller that causes the vehicle to ensue the desired vehicle response by making use of the active front and rear steering angle and the lane-keeping controller designed by **linear quadratic control** (LQC) theory [RAKSINCHAROENSAK ET AL. 2003].

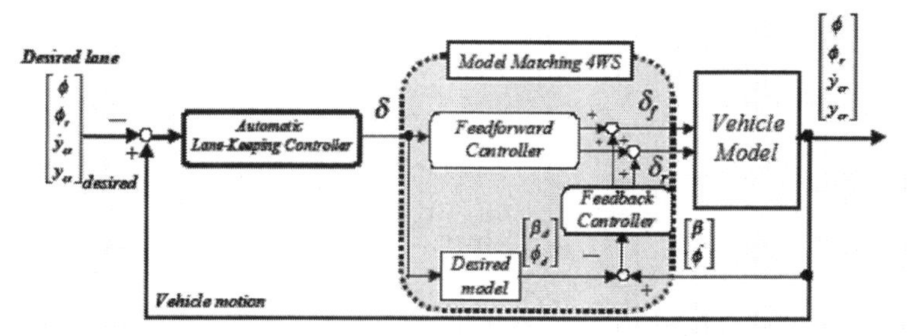

Fig. 4.116 Description of lane-keeping mechatronic control system by a four-wheel-steering (4WS) automotive vehicle [RAKSINCHAROENSAK ET AL. 2003].

Mathematical Model Matching Controller - The function of mathematical model matching control is to create the outputs of a 4WS vehicle to ensue the output of the desired mathematical model that has an ideal steering response. The mathematical model matching controller comprises a feedforward controller that relies on the steering wheel angle input from the automatic lane-keeping system, and feedback controller that may adjust the state deviations of the side slip angle and yaw rate.

Feedforward Controller -- By deriving the inverse mathematical model of the *Euler--Lagrange* equations of the 2 DoF motions, the control law of front and rear steering angles may be determined as the following transfer functions:

$$\frac{\delta_f(s)}{\delta(s)} = \frac{1}{2lC_f}\left[\left(ml_rVs + 2lC_f\right)\frac{\beta(s)}{\delta(s)} + \left(Is + ml_rV + \frac{2l_f lC_f}{V}\right)\frac{\dot{\phi}(s)}{\delta(s)}\right] \quad , \tag{4.31}$$

$$\frac{\delta_r(s)}{\delta(s)} = \frac{1}{2lC_r}\left[\left(ml_fVs + 2lC_r\right)\frac{\beta(s)}{\delta(s)} + \left(Is - ml_fV + \frac{2l_r lC_r}{V}\right)\frac{\dot{\phi}(s)}{\delta(s)}\right] \quad . \tag{4.32}$$

From the above transfer functions, the feedforward control law of 4WS is determined in relation to the steering responses of the desired mathematical model that may be determined in many ways.

This section may examine the subsequent two kinds of steering responses; those are [RAKSINCHAROENSAK ET AL. 2003]:

- Zero-sideslip-angle response that may be referred as *'4WS-A'* (Fig. 4.118);
- Steering response that has no phase-lag in lateral acceleration that may be referred as *'4WS-B'* (Fig. 4.119).

First, for a 4WS-A automotive vehicle, the desired steering response may be described as

$$\frac{\beta(s)}{\delta(s)} = 0 \quad , \tag{4.33}$$

$$\frac{\dot{\phi}(s)}{\delta(s)} = \frac{k_r}{1+\tau_r s} \quad , \tag{4.34}$$

where δ indicates the steering wheel angle input.

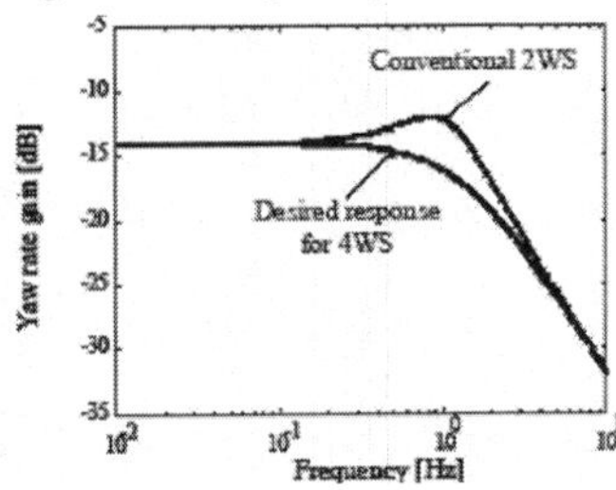

Fig. 4.117 Approximated first-order lag yaw rate response [RAKSINCHAROENSAK ET AL. 2003].

As shown in Figure 4.117 [RAKSINCHAROENSAK ET AL. 2003], k_r and τ_r are the steady-state gain and the time constant of yaw rate response respectively, which are established from conventional front-wheel steering vehicle as

$$k_r = \frac{1}{1 - \frac{m(C_f l_f - C_r l_r)V^2}{2l^2 C_f C_r}} \cdot \frac{V}{l} \cdot \frac{1}{i_{st}} \quad , \tag{4.35}$$

$$\tau_r = \frac{k_r I}{2 l_f C_f} \quad , \tag{4.36}$$

where l indicates the vehicle wheelbase, is the steering gear ratio.

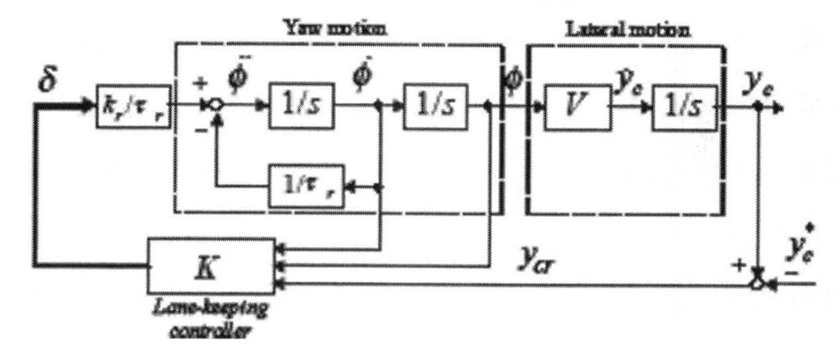

Fig. 4.118 Structural and functional block diagram of the lane-keeping system by a 4WS-A automotive vehicle (4WS-A = sideslip zeroing 4WS) [RAKSINCHAROENSAK ET AL. 2003].

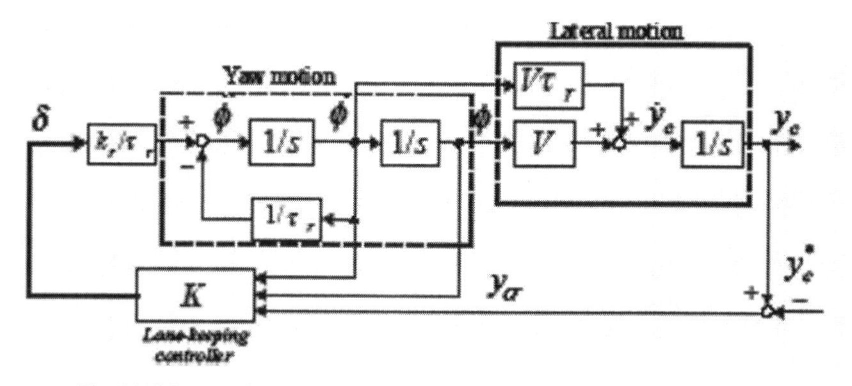

Fig. 4.119 Structural and functional block diagram of the lane-keeping system by a 4WS-B automotive vehicle (4WS-B - lateral acceleration phase-lag zeroing 4WS) [RAKSINCHAROENSAK ET AL. 2003].

To realise the desired responses of 4WS-A, substituting Eqs (4.33), and (4.34) into Eqs (4.31) and (4.32), the active front and rear steering law may be obtained as

$$\frac{\delta_f(s)}{\delta(s)} = \frac{k_r}{2lC_f}\left[\frac{I}{\tau_r} + \left(ml_r V + \frac{2ll_f C_f}{V} - \frac{I}{\tau_r}\right)\frac{1}{1+\tau_r s}\right] , \tag{4.37}$$

$$\frac{\delta_r(s)}{\delta(s)} = \frac{k_r}{2lC_r}\left[-\frac{I}{\tau_r} + \left(ml_f V - \frac{2ll_r C_r}{V} + \frac{I}{\tau_r}\right)\frac{1}{1+\tau_r s}\right] . \tag{4.38}$$

Next, for a 4WS-B automotive vehicle, the desired steering response may be expressed as

$$\frac{\ddot{y}_c(s)}{\delta(s)} = k_r V , \tag{4.39}$$

$$\frac{\dot{\varphi}(s)}{\delta(s)} = \frac{k_r}{1+\tau_r s} . \tag{4.40}$$

Note that the yaw rate response is set to be identical to the 4WS-A automotive vehicle. According to this desired steering response, the side slip angle response becomes a first-order lag response as

$$\frac{\beta(s)}{\delta(s)} = \frac{1}{V}\left(\frac{\ddot{y}_c(s)}{\delta(s)}\right) - \frac{\dot{\varphi}(s)}{\delta(s)} = \frac{k_r \tau_r}{1+\tau_r s} . \tag{4.41}$$

To realise the desired responses of 4WS-B, substituting Eqs (4.34), and (4.39) into Eqs (4.31) and (4.32), the steering law for active front and rear steering angle may be obtained as follows:

$$\frac{\delta_f(s)}{\delta(s)} = \frac{k_r}{2lC_f}\left[\frac{I}{\tau_r} + ml_r V + \left(\frac{2ll_f C_f}{V} + 2lC_f\tau_r - \frac{I}{\tau_r}\right)\frac{1}{1+\tau_r s}\right] , \tag{4.42}$$

$$\frac{\delta_r(s)}{\delta(s)} = \frac{k_r}{2lC_r}\left[-\frac{I}{\tau_r} + ml_f V + \left(-\frac{2ll_r C_r}{V} + 2lC_r\tau_r + \frac{I}{\tau_r}\right)\frac{1}{1+\tau_r s}\right] . \tag{4.43}$$

Feedback Controller -- In addition to a feedforward controller, a feedback controller is used to compensate the influence of external disturbances or dynamic uncertainties. With the application of optimal control theory, the feedback controller may be designed in the following manner. When the controlled variables are defined to be the errors of sideslip angle and yaw rate, the equation concerning the error variables is derived as

$$\begin{bmatrix} \Delta\dot{\beta} \\ \Delta\ddot{\varphi} \end{bmatrix} = \begin{bmatrix} A_{11} & A_{12} \\ A_{21} & A_{22} \end{bmatrix}\begin{bmatrix} \Delta\beta \\ \Delta\dot{\varphi} \end{bmatrix} + \begin{bmatrix} B_{11} & B_{12} \\ B_{21} & B_{22} \end{bmatrix}\begin{bmatrix} \delta_{fb} \\ \delta_{rb} \end{bmatrix} , \tag{4.44}$$

where all elements in the matrix are determined from equations of 2 DOF motions as follows:

$$A_{11} = -\frac{2(C_f + C_r)}{mV},\ A_{12} = -1 - \frac{2(l_f C_f - l_r C_r)}{mV^2},$$
$$A_{21} = -\frac{2(l_f C_f - l_r C_r)}{I},\ A_{22} = -\frac{2(l_f^2 C_f + l_r^2 C_r)}{IV},$$
$$B_{11} = \frac{2C_f}{mV}, B_{12} = \frac{2C_r}{mV},\ B_{21} = \frac{2l_f C_f}{I},\ B_{22} = -\frac{2l_r C_r}{I}$$

This equation means that the state feedback controller may compensate the stability of the error:

$$\begin{bmatrix} \delta_{fb} \\ \delta_{rb} \end{bmatrix} = -\begin{bmatrix} K_{11} & K_{12} \\ K_{21} & K_{22} \end{bmatrix} \begin{bmatrix} \Delta\beta \\ \Delta\dot{\phi} \end{bmatrix} \quad . \tag{4.45}$$

By using LQR control design theory, the gain matrix of the feedback controller may be determined to minimise the following performance index:

$$J_{MMC} = \int_0^\infty \left[\left(\frac{\Delta\beta}{\beta_{max}} \right)^2 + \left(\frac{\Delta\dot{\phi}}{\dot{\phi}_{max}} \right)^2 + \left(\frac{\delta_{fb}}{\delta_{f\,max}} \right)^2 + \left(\frac{\delta_{rb}}{\delta_{r\,max}} \right)^2 \right] dt \quad , \tag{4.46}$$

where the denominators of each term indicate the allowable values of state errors and control inputs.

Lane-keeping Controller Design - To study how 4WS enhances lane-keeping performance, this section may subordinate the 4WS vehicle to a simple lane-keeping control function. In view of the fact that the steering responses of the vehicle are determined in a vehicle-fixed coordinate system, they must be transformed into the earth-fixed coordinate system in lane-keeping control design. In the case of 4WS-A, the vehicle's body sideslip angle is zero; hence the lateral velocity may be converted in proportional to the heading angle as:

$$\dot{y}_{cr} = V(\beta + \phi_r) = V\phi_r \quad . \tag{4.47}$$

From the yaw rate response in Eq. (4.34), yaw motion may be described as

$$\ddot{\phi} = -\frac{1}{\tau_r}\dot{\phi} + \frac{k_r}{\tau_r}\delta \quad . \tag{4.48}$$

From Eqs (4.46) and (4.47), the state-space equation of 4WS-A automotive vehicle for lane-keeping control may be achieved as

4WS-A Automotive Vehicle:

$$\begin{bmatrix} \ddot{\phi} \\ \dot{\phi}_r \\ \dot{y}_{cr} \end{bmatrix} = \begin{bmatrix} -1/\tau_r & 0 & 0 \\ 1 & 0 & 0 \\ 0 & V & 0 \end{bmatrix} \begin{bmatrix} \dot{\phi} \\ \phi_r \\ y_{cr} \end{bmatrix} + \begin{bmatrix} k_r/\tau_r \\ 0 \\ 0 \end{bmatrix} \delta \quad . \tag{4.49}$$

On the other hand, vehicle's-body sideslip angle of 4WS-B is not zero but it is proportional to the yaw rate as:

$$\dot{y}_{cr} = V(\beta + \phi) = V(\tau_r \dot{\phi} + \phi) = V\tau_r \dot{\phi} + V\phi \ . \tag{4.50}$$

From Eqs (4.23) and (4.25), the state-space equation of 4WS-B in lane-keeping control may be achieved as

4WS-B Automotive Vehicle:

$$\begin{bmatrix} \ddot{\phi} \\ \dot{\phi}_r \\ \dot{y}_{cr} \end{bmatrix} = \begin{bmatrix} -1/\tau_r & 0 & 0 \\ 1 & 0 & 0 \\ V\tau_r & V & 0 \end{bmatrix} \begin{bmatrix} \dot{\phi} \\ \phi_r \\ y_{cr} \end{bmatrix} + \begin{bmatrix} k_r/\tau_r \\ 0 \\ 0 \end{bmatrix} \delta \quad . \tag{4.51}$$

The control input of the lane-keeping controller is determined by using LQC theory. All state variables of the vehicle are fed back to determine the steering wheel angle as the following expression:

$$\delta = -K_{\dot{\phi}} \dot{\phi} - K_{\phi} \phi_r - K_{yc} y_{cr} \quad , \tag{4.52}$$

where the state feedback, gains, $K_{\Phi'}$, K_{Φ}, K_{yc} , are determined to minimise the performance index that regards only lateral deviation from the desired lane as follows:

$$J = \int_0^{\infty} \left(q_{yc} y_{cr}^2 + r\delta^2 \right) dt \quad , \tag{4.52}$$

where q_{yc} indicates the weighting coefficient of lateral deviation that refers to the level of the control achievement and r indicates that of steering wheel angle that refers to the lane-keeping controller effort.

Road Curvature Estimation using a Kalman Filter - In the case of a curved roadway, road curvature is treated as a disturbance that reflects in steady-state error. To deal with this problem, we propose a control algorithm, with the application of a *Kalman* filter (Fig. 4.120), for estimating road curvature by detecting only the vehicle's lateral deviation, without requiring feedforward of the road curvature [RAKSINCHAROENSAK ET AL. 2003].

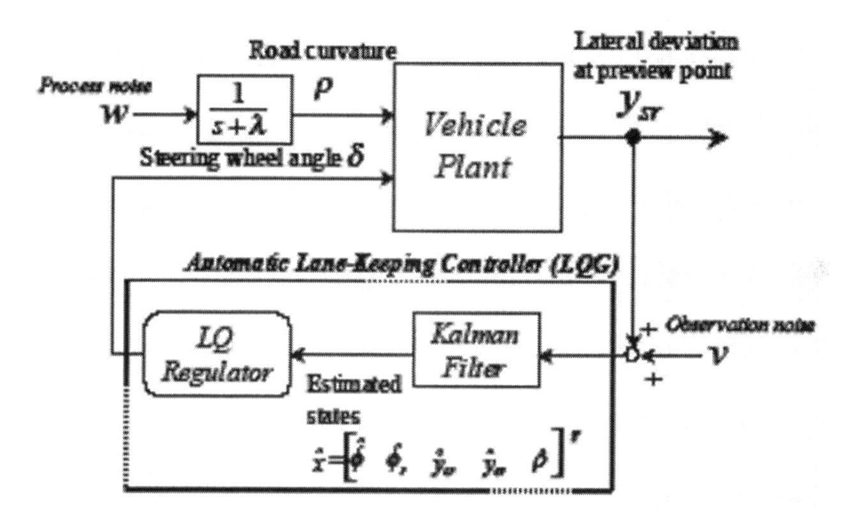

Fig. 4.120 Description of lane-keeping system coupled with curvature estimation by a *Kalman* filter [RAKSINCHAROENSAK ET AL. 2003].

The change in road curvature is approximated as a first-order system disturbed by *Gaussian* white noise that may be expressed as

$$\frac{d}{dt}\rho = -\lambda\rho + w \quad , \tag{4.53}$$

where w indicates the process noise of the control plant, and λ indicates the dynamic characteristics of curvature variation.

The lateral deviation at the sensor near the front bumper, employed for estimating vehicle states together with curvature, may be expressed as

$$, \quad y_{sr} = y_{cr} + l_s\phi_r - \frac{l_s^2}{2}\rho + v \tag{4.54}$$

where l_s indicates the distance from CG to front bumper and v indicates observation noise.

Determining the covariances of process noise and observation noise, the *Kalman* filter may be designed [RAKSINCHAROENSAK ET AL. 2003].

To construct the lane-keeping controller with a curvature estimation, the curvature may be included as a state variable so that the state-space equation of each type of vehicles may be rewritten as

4WS-A Automotive Vehicle:

$$\begin{bmatrix} \ddot{\phi} \\ \dot{\phi}_r \\ \dot{y}_{cr} \\ \dot{\rho} \end{bmatrix} = \begin{bmatrix} -1/\tau_r & 0 & 0 & 0 \\ 1 & 0 & 0 & -V \\ 0 & V & 0 & 0 \\ 0 & 0 & 0 & -\lambda \end{bmatrix} \begin{bmatrix} \dot{\phi} \\ \phi_r \\ y_{cr} \\ \rho \end{bmatrix} + \begin{bmatrix} k_r/\tau_r \\ 0 \\ 0 \\ 0 \end{bmatrix} \delta + \begin{bmatrix} 0 \\ 0 \\ 0 \\ 1 \end{bmatrix} w \quad ; \tag{4.55}$$

4WS-B Automotive Vehicle:

$$\begin{bmatrix} \ddot{\phi} \\ \dot{\phi}_r \\ \dot{y}_{cr} \\ \dot{\rho} \end{bmatrix} = \begin{bmatrix} -1/\tau_r & 0 & 0 & 0 \\ 1 & 0 & 0 & -V \\ V\tau_r & V & 0 & 0 \\ 0 & 0 & 0 & -\lambda \end{bmatrix} \begin{bmatrix} \dot{\phi} \\ \phi_r \\ y_{cr} \\ \rho \end{bmatrix} + \begin{bmatrix} k_r/\tau_r \\ 0 \\ 0 \\ 0 \end{bmatrix} \delta + \begin{bmatrix} 0 \\ 0 \\ 0 \\ 1 \end{bmatrix} w \quad . \tag{4.56}$$

The steering wheel angle may be calculated as

$$\delta = -K_{\dot{\phi}}\dot{\phi} - K_{\phi}\phi_r - K_{ycr}y_{cr} - K_{\rho}\rho \quad , \tag{4.57}$$

where feedback gains are determined to minimise the same performance index shown in Eq. (4.52).

4.6.4 4WS Automotive Vehicle Lane-Keeping Simulation

Lane-Keeping Control on a Straight Roadway - The lateral desired course, step input with magnitude of 0.2 m, is applied to the vehicle running at a constant vehicle velocity (speed) of 100 km/h, at 1 s after simulation started. Specification of the vehicle used in the simulation is given in Table 4.12 [RAKSINCHAROENSAK ET AL. 2003].

Table 4.12 Specification of the vehicle used in the simulation.

Model parameter	Notation	Unit	Value
Vehicle mass	m	kg	1980
Yaw moment of inertia	I	kgm^2	3758
Distance from center of gravity to front axle	l_f	m	1.358
Distance from center of gravity to rear axle	l_r	m	1.472
Distance from center of gravity to lateral sensor	l_s	m	2.57
Equivalent front cornering stiffness (1 wheel)	C_f	N/rad	42000
Equivalent rear cornering stiffness (1 wheel)	C_r	N/rad	74000

The simulation result is shown in Figure 4.121 [RAKSINCHAROENSAK ET AL. 2003] which compares the step desired lane-keeping responses in the case of 4WS-A, 4WS-B, and 2WS automotive vehicles. Clearly, both 4WS-A and 4WS-B vehicles provide the improved damping behaviour of both -- lateral and yaw motions compared to a 2WS automotive vehicle. Especially, in the case of a 2WS automotive vehicle, the yaw angle has a dominant oscillatory response.

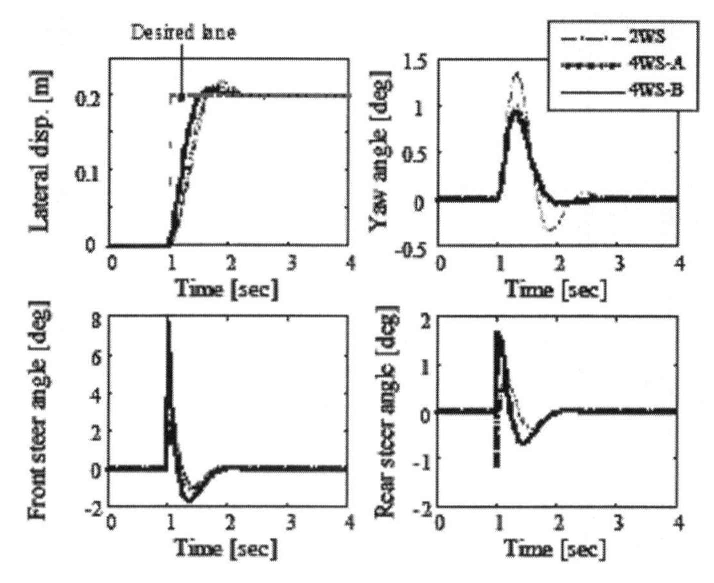

Fig. 4.121 Lane-keeping responses on straight lane
(Weights of lane-keeping controller: $q_{yc} = 102$, $r = 1$)
[RAKSINCHAROENSAK ET AL. 2003].

Moreover, the 4WS-B vehicle also provides better response and damping behaviour in the lane-keeping response when comparing with the 4WS-A vehicle. The lane-keeping performances of the three types of automotive vehicles are evaluated by using '*trade-off curve*'.

The trade-off curves, shown in Figure 4.116 [RAKSINCHAROENSAK ET AL. 2003], plot the square of lateral error along the vertical axis and the square of the steering wheel angle along the horizontal axis.

The simulation result shows that under the same steering angle control input, the 4WS-B vehicle shows the best performance in lane keeping with the smallest deviation. On the other hand, as may be noticed from the graph, lane-keeping control performance by a 4WS-A vehicle is not significantly improved from the 2WS vehicle.

Moreover, to have a better understanding of the improved lane-keeping response of the yaw direction, Figure 4.122 plots the square of the lateral deviation along the horizontal axis and the square of the yaw deviation along the vertical axis [RAKSINCHAROENSAK ET AL. 2003].

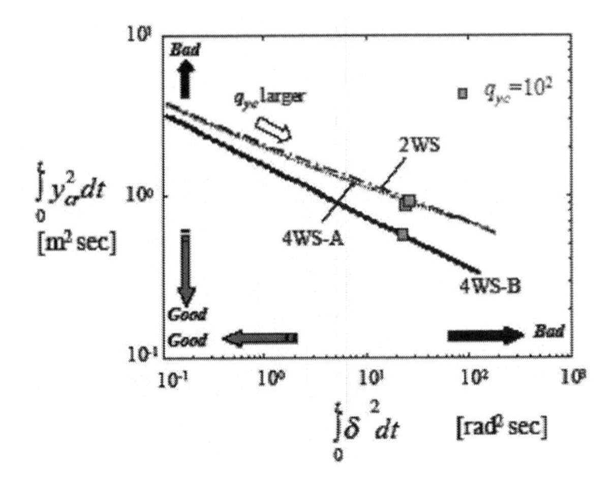

Fig. 4.122 Trade-off curves of lane-keeping system on a straight lane (Weights of lane-keeping controller: q_{yc} = 101 to *103*, r =1, simulation stop time t = 5 s) [RAKSINCHAROENSAK ET AL. 2003].

Clearly from the graph, the 4WS-B automotive vehicle shows the best performance in lane-keeping control with the smallest lateral and yaw deviations.

Lane-keeping Control on a Curved Roadway - The simulation under the condition that vehicle runs in a straight-line at a constant speed of 100 km/h for 1 s and enters the curved trajectory with a constant radius of 400 m, is conducted. Figure 4.123 shows the curved lane-keeping responses of three types of vehicles [RAKSINCHAROENSAK ET AL. 2003].

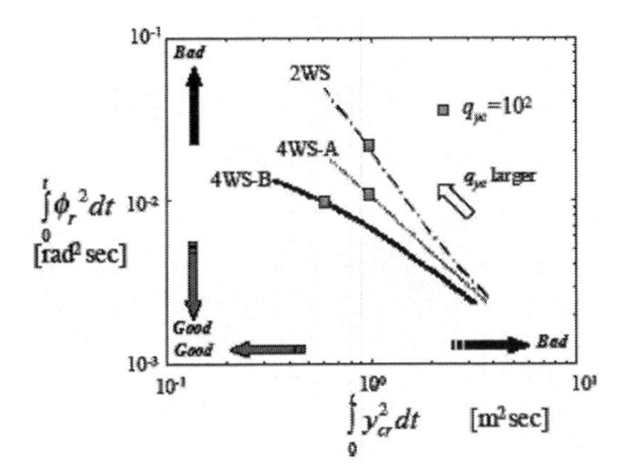

Fig. 4.123 Lane-keeping performance curves on a straight lane (Weights of lane-keeping controller: q_{yc} = 101 to 103, r =1, simulation stop time t = 5 s) [RAKSINCHAROENSAK ET AL. 2003].

After the curvature estimation is conducted, the lateral deviation is regulated to be zero in a steady state. The 4WS-B vehicle provides the most satisfactory response of both yaw and lateral directions in a transient state, whereas the 4WS-A and 2WS vehicles still have an oscillatory response. The 4WS-B vehicle also provides the smallest lateral deviation in tracking the curved lane with a satisfactory response.

However, the 4WS-B vehicle has its drawback concerning the heading error due to its own desired model. On the other hand, the 4WS-A vehicle, regulating the sideslip angle during cornering, is superior in the tracking performance of the yaw direction. To evaluate the performance of lane-keeping system, Figure 4.124 [RAKSINCHAROENSAK ET AL. 2003] shows the trade-off curve that plots the square of the steering wheel angular velocity along the horizontal axis and the square of the lateral deviation along the vertical axis.

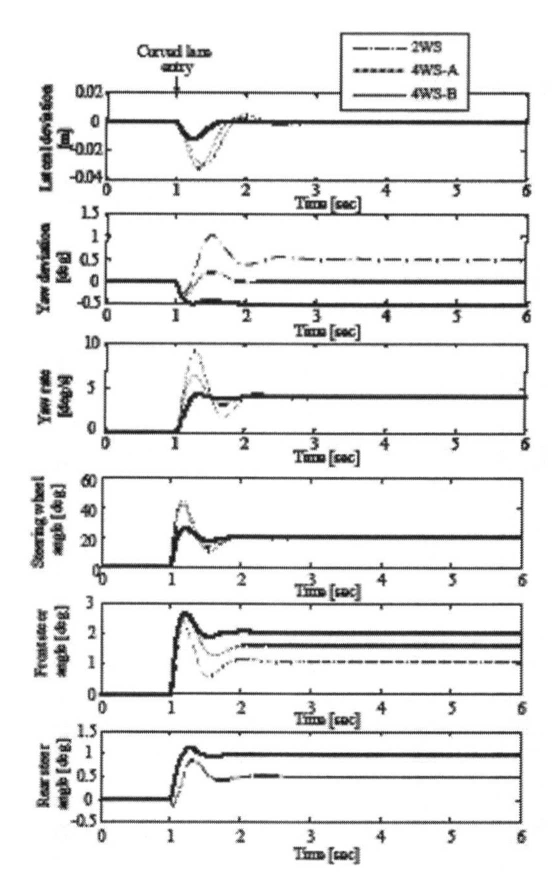

Fig. 4.124 Lane-keeping responses on a curved lane (Weights of lane-keeping controller: q_{yc} = 102, r = 1) [RAKSINCHAROENSAK ET AL. 2003].

As may be noticed from the graph, the 4WS-B vehicle provides the best performance of the lane-keeping system with less steering effort and less deviation compared with other vehicles. Figure 4.125 [RAKSINCHAROENSAK ET AL. 2003] shows the lane-keeping performance in lateral and yaw directions.

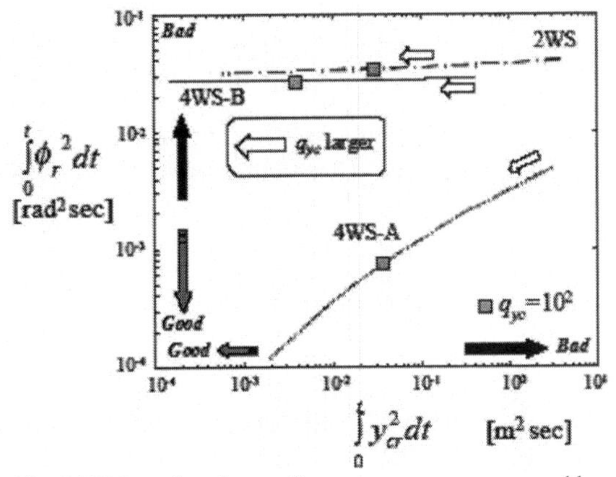

Fig. 4.125 Lane-keeping performance curves on a curved lane (Weights of lane-keeping controller: q_{yc} = 101 to 104, r =1, simulation stop time t = 5 s) [RAKSINCHAROENSAK ET AL. 2003].

The 4WS-A vehicle provides a superior lane-keeping performance in the yaw direction, whereas the 4WS-B vehicle has a limit in controlling yaw deviation since it always generates a vehicle-body sideslip angle with respect to the steering wheel angle.

4.6.5 Conclusion

A theoretical analysis of the 4WS system used for an automatic lane-keeping control system was conducted. The comparative study on the lane-keeping control performance between two types of steering responses of the 4WS system was clarified.

- The major conclusions to be drawn from the simulation study are summarised as follows [RAKSINCHAROENSAK ET AL. 2003]: By applying an active 4WS system to the lane-keeping task, the lateral and yaw dynamics during a lane-keeping manoeuvre may be effectively controlled to have the desirable characteristics, compared to the 2WS vehicle;
- It is clarified that an active 4WS system that makes the phase-lag of the lateral acceleration become zero (4WS-B) provides a superior lane-keeping performance in the lateral direction to the one that makes the vehicle-body sideslip angle become zero (4WS-A); however, 4WS-B system has a drawback that the heading error of the vehicle cannot be regulated during tracking of the curved roadway.

4.7 Model-Based Design with Production Code Generation for SBW AWS Conversion Mechatronic Control System Development

4.7.1 Foreword

Production code generation (PCG) technology has become a key component in the evolution of software development because it is practical: one may add details to designs directly on a physical model and then automatically generate the final code. PCG has done well in its early acceptance; however, further growth requires a supporting software engineering framework that integrates processes, methods, and tools.

Model-based design (MBD) provides such a framework. Using a SBW 2WS conversion mechatronic control system example, this section describes the application of PCG and MBD to the development of embedded automotive systems [LANGEWALTER AND ERKKINEN 2005].

4.7.2 Model-Based Design with Production Code Generation

Scientists and engineers use MBD in nearly every industry that requires the development of embedded control systems. It is particularly well entrenched in development processes for embedded applications such as large-scale automotive **electronic control units** (ECU). The high-volume nature of mass-produced ECUs demands low-cost, fixed-point microcontroller units and DSPs, so the final executable code must be extremely compact, fast, and traceable. Signal processing and communications applications also use this approach, with emphasis on modelling and prototyping rather than PCG.

MBD may provide control and signal processing systems for scientists, engineers, and software developers with a common environment for graphical specification and analysis. Physical models serve to specify systems data, interfaces, feedback control logic, discrete and state logic, and real-time behaviour.

In MBD, a structural and functional block diagram or a state diagram physical model may serve as the system design, software requirements, software design, or even (in some developer's minds) as a source code.

MBD may emphasise process iterations, early testing, and code re-use throughout the development process. PCG is an automatic translation from detailed design. To satisfy growing demand, MBD must address the necessities of safety-critical systems such as SBW 2WS conversion mechatronic control systems, which often require additional process rigour.

Through simulation and prototyping, this approach supports safety-related system development by providing extensive **verification and validation** (V&V) prior to a final build. The benefit of early V&V is clear: system integration and testing result in fewer bugs and less reworking. Identifying a bug on the desktop may be highly preferable to encountering it during a winter test drive.

The following sections discuss the development activities, key V&V methods, and integration tasks in the development of a SBW AWS conversion mechatronic control system. These activities and methods are detailed in ERKKINEN [2003].

4.7.3 Behavioural Modelling

In MBD, one uses physical models to specify requirements and design for all aspects of each system (e.g., the SBW 2WS conversion mechatronic control system).

A typical system includes [LANGENWALTER AND ERKINNEN 2005]:

- Input (for example, steering HW sensors);
- Controller or signal processing physical model;
- Plant or environment model (E-M motor, R&P, wheels, and communications);
- Output (change of direction).

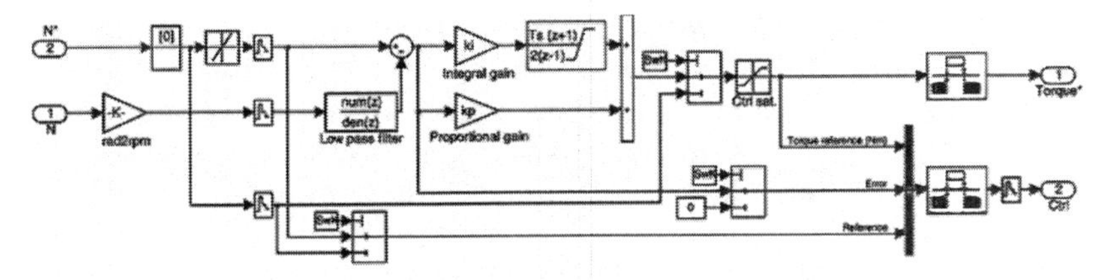

Fig. 4.126 Feedback physical model of a PI controller for SBW 2WS [LANGENWALTER AND ERKINNEN 2005].

Automotive scientists and engineers, as well as software developers, may create a system physical model to represent the desired behaviour using mechatronic control system block diagrams for feedback control, state machines for discrete events and conditional logic, and signal processing blocks for filters (Fig. 4.126) [LANGENWALTER AND ERKINNEN 2005].

4.7.4 Simulation and Analysis

Automotive scientists and engineers, as well as software developers, execute the physical model and then analyse it to ensure that the requirements are satisfied, using methods such as time- or event-based simulation and frequency domain analysis.

For instance, a SBW 2WS conversion mechatronic control system must respond to a sensor failure, attenuate a high-frequency response below 3 db, and not lag the commanded rate by more than 1.5 m/s [LANGENWALTER AND ERKINNEN 2005].

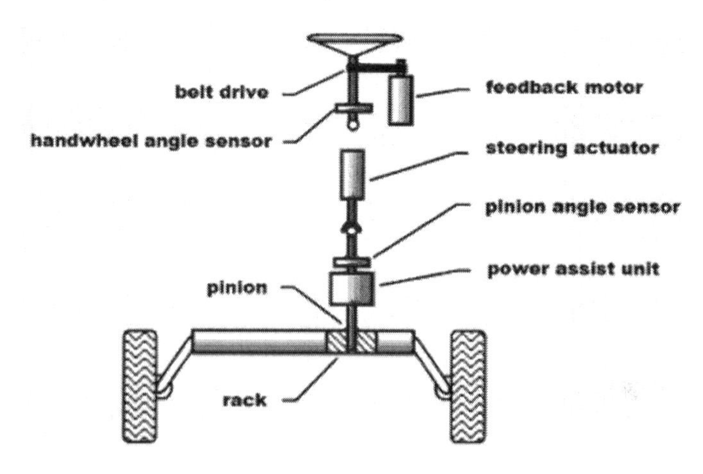

Fig. 4.127 SBW 2WS conversion mechatronic control system [LANGENWALTER AND ERKINNEN 2006].

Automotive scientists and engineers as well as software developers model and simulate the SBW AWS conversion mechatronic control system (shown in Figs 4.127 and 4.128) to determine whether the requirements are valid.

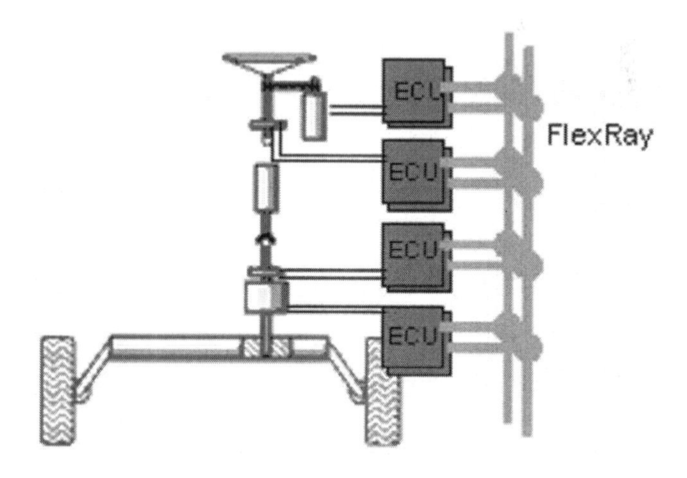

Fig. 4.128 SBW 2WS conversion mechatronic control system with fault-tolerant redundant bus system (*FlexRay*™) [LANGENWALTER AND ERKINNEN 2005].

Simulation, a core validation activity, ensures that a SBW 2WS conversion mechatronic control system can be developed to satisfy the requirements.

4.7.5 Rapid Prototyping

Because plant physical models may be inaccurate and the production processor may not provide sufficient processing power to get a working result, modelling alone does not provide a total solution.

Rapid prototyping overcomes these shortcomings because it replaces the plant physical model with the physical plant. In the SBW 2WS example, the plant might be an automotive vehicle, so an actual vehicle is used.

However, because the new ECU system is not yet built, a real-time or embedded platform runs the controller software and interacts with the plant [LANGENWALTER AND ERKINNEN 2005].

There are two forms of rapid prototyping: functional and on-target. Functional prototyping uses a powerful real-time computer such as a multiprocessor floating-point PowerPC or DSP. The goal is to determine whether the system controls the physical vehicle as well as it controlled the modelled vehicle. If so, plant model inaccuracies are considered insignificant and the control strategy is validated.

On-target rapid prototyping executes the software in the same or a similar production MCU or DSP, rather than a high-end PowerPC core or other dedicated high-end rapid prototyping hardware. The aim is to download the code into the actual production target for quick testing with the physical plant. If it performs well, the controller is not only deemed valid, but also feasible: it may be accepted into production.

4.7.6 Detailed Software Design

Software design activities include fixed-point data specification, real-time tasking, data typing, built-in-test, and diagnostics [LANGENWALTER AND ERKINNEN 2005].

With MBD, the same physical model used for algorithm specification and validation is refined and constrained by the software scientists and engineers as part of the PCG process.

Automated scaling and data type override help convert floating-point physical models to a fixed point and provide mechanisms to assess a fixed-point design by simulating it in the floating point.

4.7.7 Physical Model Testing

Testing the physical model on a desktop is preferable to deploying it on hardware for build and integration. Source-code-based testing has existed for many years, and recent methods have enabled physical-model testing and structural coverage analysis. Using simulation and coverage, developers can fully stress the controller to verify its design integrity.

Failure mode effect analysis (FMEA) ensures the safe operation of the **steer-by-wire** (SBW) under fault conditions (Fig. 4.129) [LANGENWALTER AND ERKINNEN 2005].

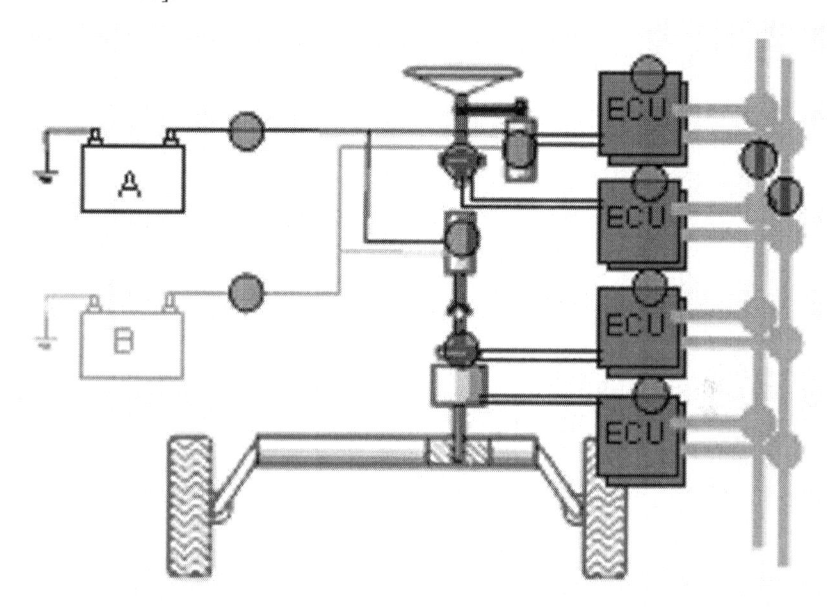

Fig. 4.129 Possible faults on the SBW 2WS conversion mechatronic control system [LANGENWALTER AND ERKINNEN 2005].

Numerical overflow and dead code are examples of poor design integrity. Stress testing of the physical model using minimum and maximum numerical values helps ensure that overflow conditions may not occur. Simulation facilitates this method of stress testing, but a dead code is not easy to find because detection requires structural coverage analysis. Dead code differs from a deactivated code in that the latter is known to the developer and is deactivated for a reason. Actual dead code indicates that details were omitted during requirements, design, or test case development. Model coverage assesses the cumulative results of a test suite to determine which blocks were not executed or which states were not reached. Certain types of coverage are well established in source code languages (such as *C* and C^{++}). However, these languages do not possess constructions such as blocks or states, so we need a new theory and tools to obtain model coverage. The FAA considers **modified condition/decision coverage** (MC/DC) the most stringent coverage level necessary to satisfy safety-critical system requirements. MBD enables AC/DC that is, in many cases, required for RBW or XBW integrated unibody, space-chassis, skateboard-chassis, or body-over-chassis motion mechatronic control hypersystem designs Figure 4.130 shows the result of coverage analysis for the power management design shown in Figure 4.131 [LANGENWALTER AND ERKINNEN 2005].

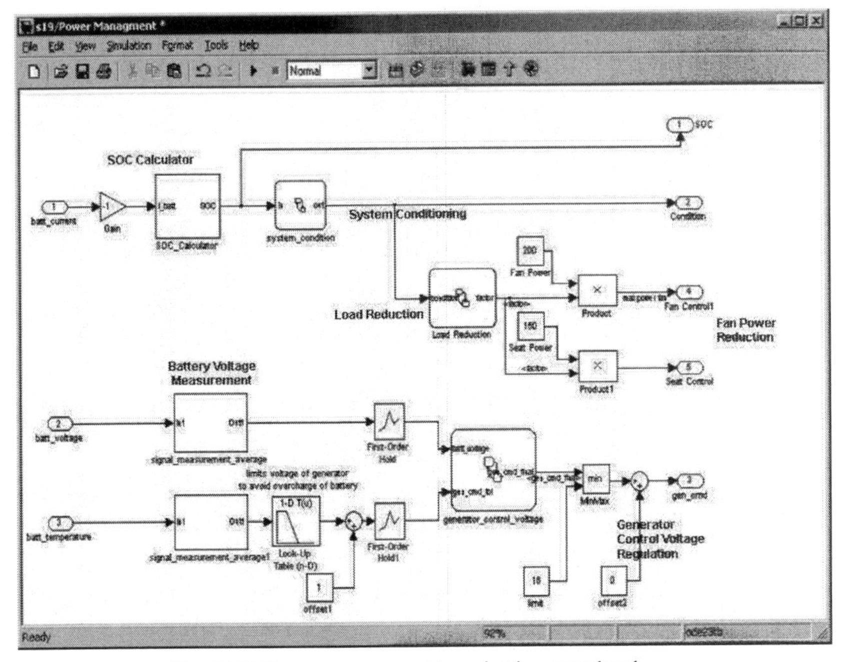

Fig. 4.130 Power management to maintain power level for RBW or XBW integrated unibody or body-over-chassis motion mechatronic control system [LANGENWALTER AND ERKINNEN 2005].

Model Hierarchy/Complexity: **Test 1**

		D1	C1	MCDC
1. req_test	35	84%	70%	50%
2. . . . Logic	25	78%	75%	50%
3. . . . . . . SF: Logic	24	78%	75%	50%
4. . . . . SF: Altitude	11	100%	83%	67%
5. . . . . SF: Active	4	100%	NA	NA
6. . . . . SF: GS	13	61%	67%	33%
7. . . . . SF: Active	6	50%	NA	NA
8. . . . . SF: Coupled	3	33%	NA	NA

Fig. 4.131 Coverage for the power management design shown in Figure 4.130 [LANGENWALTER AND ERKINNEN 2005].

4.7.8 Distributed Architecture Design

Modern embedded systems contain several distributed ECUs which communicate in real time with each other over a fault-tolerant communication system such as FlexRay™. BMW's latest **dynamic stability control** (DSC) contains ABS as one of 15 subfunctionalities. Adding blocks of DECOMSYS network components such as hosts, tasks, signals, and so forth, to the individual subsystems enables the embedded functions to be connected and mapped onto the architecture of ECUs. Using the blocks in this way makes it easier to simulate the temporal behaviour of task activations of a time-triggered operating system such as OSEK-time/OS. Scientists and engineers design and simulate clusters, hosts, tasks, and connections within the model environment [LANGENWALTER AND ERKINNEN 2005]. The distributed network design solution from Vector (*DaVinci*) then integrates the code generated from physical models with a legacy code from other sub-systems and different suppliers and maps the resulting code onto the ECU or system architectures for verification.

4.7.9 Production Code Generation

After the model has been verified and validated, it is time to generate the code. As with a standard C compiler, this process is straightforward. Various optimisation settings and user configuration options exist. The key is to keep the code efficient, accurate, and integrated with the legacy code or other tools. In safety-related software, it is also important for the code to be traceable to the diagram so that it can be reviewed and verified [LANGENWALTER AND ERKINNEN 2005].

4.7.10 In-the-Loop Testing

Once the controller is built, software scientists and engineers may perform a series of open- and closed-loop tests with the real-time plant model in the loop.

Some tests may involve only the software or processor and are known as *'software-in-the-loop'* or *'processor-in-the-loop'* testing, respectively. Another test, termed *'hardware-in-the-loop'*, may use actually built ECU hardware.

In either case, automotive scientists and engineers test the physical controller with the plant model. Through a series of tests, perhaps the same test used during requirements validation, the controller must be proved to be acceptable to the customer [LANGENWALTER AND ERKINNEN 2005].

4.7.11 Integration Components

Most software standards require traceability of requirements, perhaps originating in other requirements tools, throughout development.

Also, **software configuration management** (SCM) is needed to store, version, and retrieve the various development artefacts. Documentation by means of report generators ensures that management, customers, and suppliers may see the physical model. MBD supports all of these components. Figure 4.132 shows the SCM interface [LANGENWALTER AND ERKINNEN 2005].

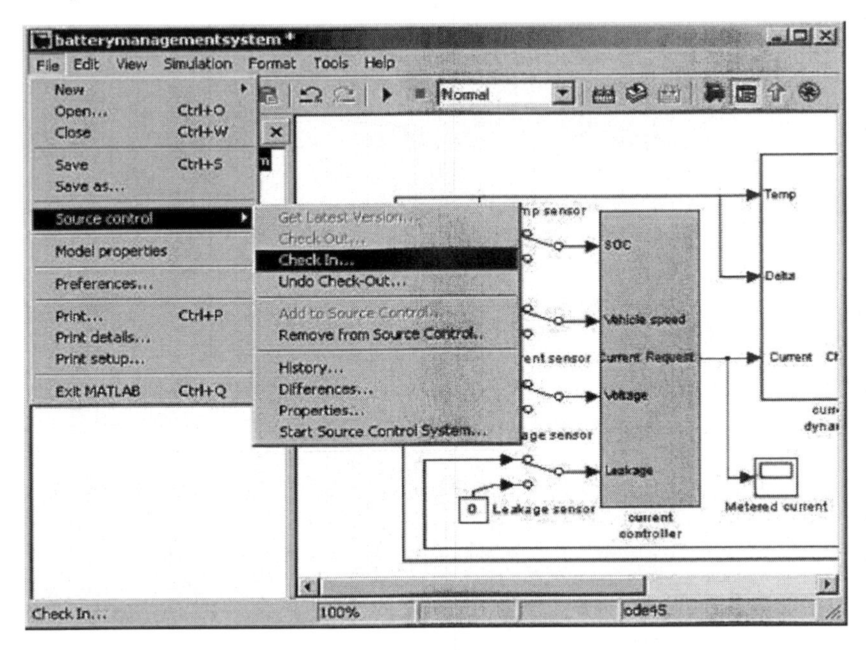

Fig. 4.132 Secondary energy-source control interface [LANGENWALTER AND ERKINNEN 2005].

4.7.12 Additional Resources

LANGENWALTER AND ERKINNEN [2005] introduced a full software engineering framework that focuses on MBD with **production control generation** (PCG). The specific methods and tools described here demonstrate the feasibility of this approach to embedded system development. To learn more or to exchange ideas regarding additional methods and use cases, please contact the authors.

For a detailed comparison of traditional software development processes and MBD, see HOSAGRAHARA AND SMITH [2004].

4.8 SBW AWS Conversion Mechatronic Control System Using Fault-Silent Units

4.8.1 Foreword

RBW or XBW integrated unibody, space-chassis, skateboard-chassis, or body-over-chassis motion mechatronic control hypersystems have great potential in increasing the automotive vehicle's safety and comfort in comparison with conventional M-M, M-F-M and E-M-F-M mechatronic control systems.

For safety critical electronic systems in automotive applications, the **time-triggered architecture** (TTA) offers a cost effective way of implementing a fully fault-tolerant real-time system.

The philosophy behind a TTA system is to use non-fault-tolerant components, combining them to an SBW AWS conversion mechatronic control system that behaves altogether fault-tolerant. The software development for SBW AWS conversion mechatronic control systems creates great challenges to be met in the next few years.

Because of the high demands in system safety, all processes have to be guarantee hard real-time constraints. This means that all system tasks have to guarantee all temporal requirements in 100% of all cases. This is also valid in situations system error. It is not acceptable that, because of a system error a task fails

A precondition is a deterministic system behaviour based on a **time triggered protocol** (TTP) and the introduction of redundant components for processes and communication.

SBW AWS conversion mechatronic control systems must not only be reliable but also meet real-time requirements.

In WILWERT ET AL. [2003] the authors present an integrated approach for evaluating both the temporal performance and the behavioural reliability of SBW AWS conversion mechatronic control systems taking into account the delay variation introduced by network transmission errors.

The considered temporal performance is the **quality of service** (QoS) perceived by the driver, that is, vehicle stability.

Tests in vehicles and simulations have been realised to estimate the maximum tolerable response time of the SBW AWS conversion mechatronic control system, and to evaluate the impact of this delay on the QoS.

The authors WILWERT ET AL. [2003] quantify the worst case response time of the SBW AWS conversion mechatronic control system for a generic architecture based on the **time division multiple access** (TDMA) protocol but independent of the communication network (could actually be TTP/C or FlexRay™), and apply these generic results to a case study.

Table 4.13 Safety Integrity Levels [IEC].

Safety Integrity Level (SIL)	Probability of dangerous failure per hour (P_{FAIL})
4	$\geq 10^{-9}$ to $< 10^{-8}$
3	$\geq 10^{-8}$ to $< 10^{-7}$
2	$\geq 10^{-7}$ to $< 10^{-6}$
1	$\geq 10^{-6}$ to $< 10^{-5}$

WILWERT ET AL. [2003] further define the notion of *'behavioural reliability'* as the probability that *'the worst case response time is less than a threshold'*. In the latter case study, this behavioural reliability is evaluated and linked to the **safety integrity levels** (SIL) in IEC61508-1 standard, as shown in Table 4.13. Based on this behavioural reliability concept, the final objective of their work was to propose a new dependability analysis method for SBW AWS conversion mechatronic control systems by taking into account dynamic performance, fault-tolerance mechanisms, and static redundancy of the mechatronic control system.

WILWERT ET AL. [2003] first define the pure delay introduced by an SBW AWS conversion mechatronic control system between the steering wheel requests to the reception of the driver's request by the front axle's actuators.

The total response time of the SBW AWS conversion mechatronic control system's steering wheel request and front axle response is divided into pure delay and mechatronic delay (Fig. 4.133) [WILWERT ET AL. 2003].

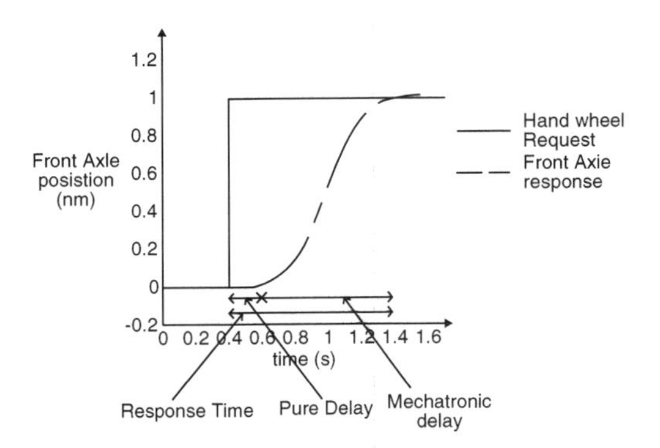

Fig. 4.133 An SBW AWS conversion mechatronic control system's response time [WILWERT ET AL. 2003].

Mechatronic delay is the time necessary for the actuator to reach the front axle position and pure delay is directly related to the SBW operation (processing time in the ECUs, network delay, and so on) as shown in Figure 4.134 [WILWERT ET AL. 2003].

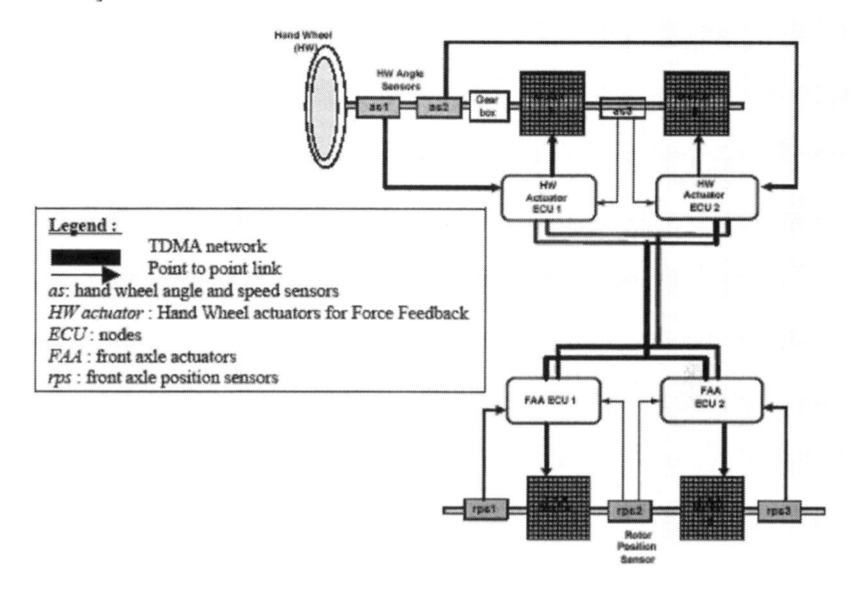

Fig. 4.134 Steer-by-wire (SBW) architecture [WILWERT ET AL. 2003].

As they consider that the mechatronic delay is constant, WILWERT ET AL. [2003] only analysed the impact of the pure delay on the QoS.

To illustrate the authors' proposition, a case study is considered showing how to apply their method to SBW architecture (Fig. 4.134). The chosen SBW architecture seems a realistic one for an automotive vehicle. In fact, a lot of SBW architectures are presented with central ECU between the steering wheel and the **front axle** (FA) [WILWERT ET AL. 2003], but in terms of cost, it seems not to be very realistic for mass manufacture.

That's why WILWERT ET AL. [2003] have chosen architecture with only four ECUs (Fig. 4.134). This is constructed with three steering **hand wheel angle** (HWA) sensors connected to two HW ECUs, a TTA communication network, two FA ECUs connected to three sensors. HW ECUs are also connected to two actuators for the **force feedback** (FF) and FA ECUs are connected to actuators to turn the wheels.

Treatment of redundant ECUs is made in parallel. For example, ECU HWA1 and ECU HWA2 (Fig. 4.135) receive the data from the sensors synchronously, so they are treated together at the same time.

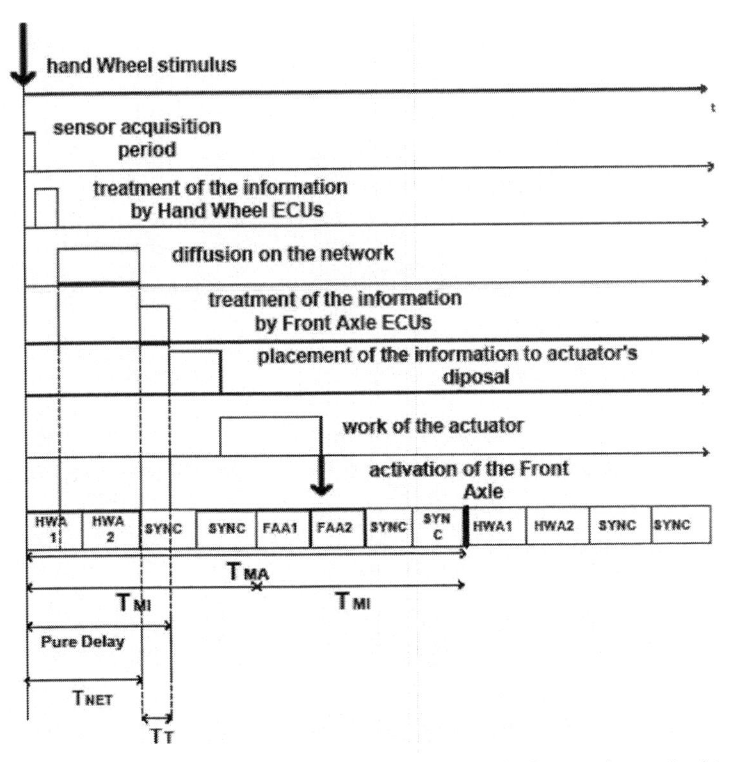

Fig. 4.135 Temporal characteristics of the function *'turning the wheels according to the driver's will'* [WILWERT ET AL. 2003].

Figure 4.135 shows that, during the temporal window, HWS1, ECU HWA1 and ECU HWA2 are treating data in parallel [WILWERT ET AL. 2003].

4.8.2 Time-Triggered Architectures for SBW AWS Conversion Mechatronic Control Systems

Because of the introduction of the innovative RBW or XBW automotive very advanced technologies, the trend for more electronic components may increase over the next few years.

These kinds of systems are designed to replace M-M, M-F-M and E-M-F-M components in the fields of propulsion (driving), dispulsion (braking), suspension (absorbing) and conversion (steering) with pure electronic solutions.

An SBW AWS conversion mechatronic control system (Fig. 4.136), requires absolutely no steering column, instead the steering movements of the driver are collected by the mechatronic control system with sensors and directly translated with a network of processors, sensors, and servomotors into the corresponding steering angle of the wheels [MÜLLER-GLASER AND KÜHL 2005].

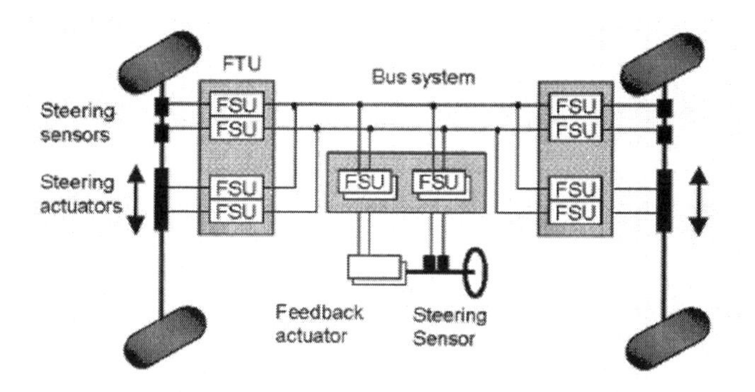

Fig. 4.136 A SBW 4WS conversion mechatronic control system using fail-silent units (FSU) [MÜLLER-GLASER AND KÜHL 2005].

MÜLLER AND PLANKENSTEINER [2002], introduce the necessary theoretical background needed for a design of a fault-tolerant system out of different kinds of non-tolerant components. Afterwards the practical application is discussed and a possible implementation of a SBW 4WS conversion mechatronic control system is given as an example. The careful design of the system is very important, because a mistake in the assumption of the component or system behaviour may have severe effects on the safety properties. Therefore design rules for practical implementation may be given.

4.8.3 Structure of Possible Four-Wheel-Steered (4WS) Steer-By-Wire (SBW) Conversion Architecture

A possible implementation of a SBW 4WS conversion mechatronic control system based on the above-mentioned principles is depicted in Figure 4.137 [MÜLLER AND PLANKENSTEINER 2002].

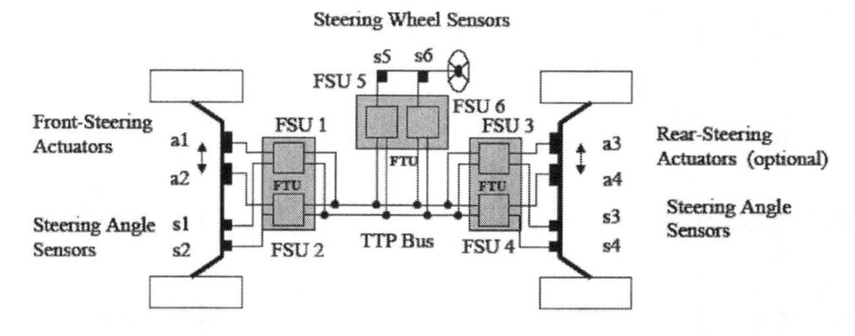

Fig. 4.137 Possible implementation of a SBW 4WS conversion mechatronic control system, using fault-silent units (FSU) [MÜLLER AND PLANKENSTEINER 2002].

The automotive vehicle has **front-wheel steering** (FWS) and optional **rear-wheel steering** (RWS). All actuators and sensors are assumed to be fail-silent and thus always two **fail-silent units** (FSU) build up one **fault-tolerant unit** (FTU).

In TTP networks, the communication channel is also replaced for safety-critical data.

The desired value of the steering angle is measured by the sensors s5 and s6, and the connected node broadcasts the value on the bus.

As the sensors are assumed to be fail-silent, the agreement protocol that may be carried out may be as simple as *"Take the first valid value"*.

This *'agreed value'* is transmitted to the application in the nodes that calculates the necessary output for the steering actuators. A control loop using the steering angle sensors is implemented.

The implementation in the above-sketched example is based on a minimum number of components. Still it allows for a very high level of safety and availability.

In the case of a FSU, it is much more expensive than a **fail-consistent unit** (FCU) and the FTUs may consist of three nodes to obtain the same level of safety [MÜLLER AND PLANKENSTEINER 2002].

Another example of a SBW 4WS conversion mechatronic control system's prototype is shown in Figure 4.138 [JEFFE 2005].

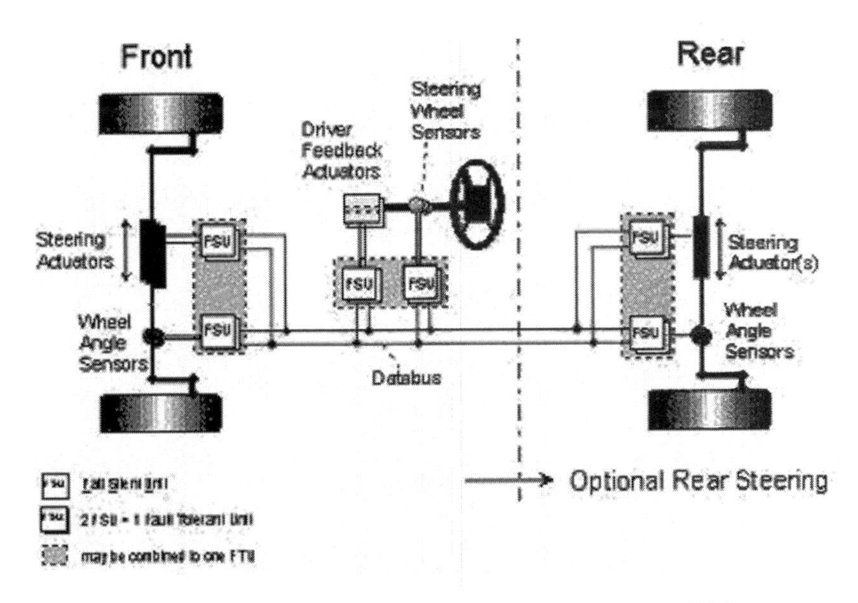

Fig. 4.138 SBW 4WS conversion mechatronic control system [JEFFE 2005].

Two other examples of 2WS and SBW 4WS conversion mechatronic control system's prototypes are shown in Figures 4.139 and 4.140 [HANSEN 2005].

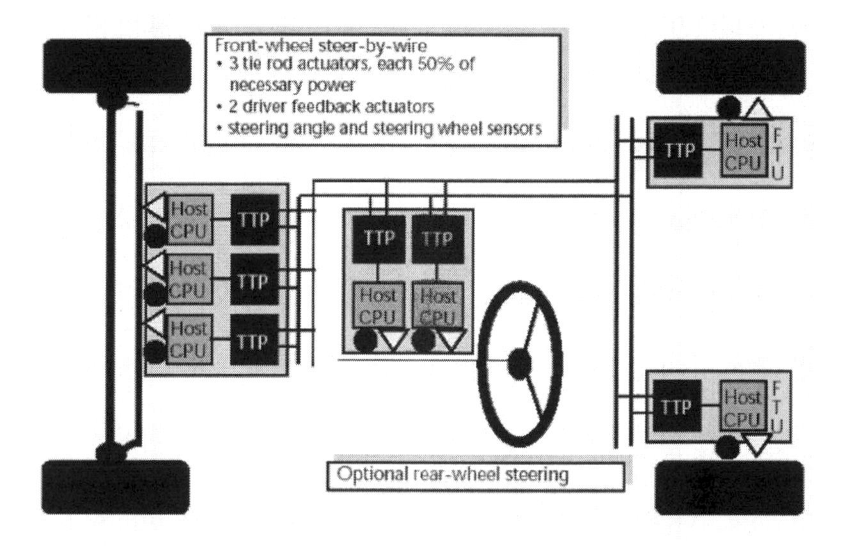

Fig. 4.139 SBW 4WS conversion mechatronic control system [HANSEN 2005].

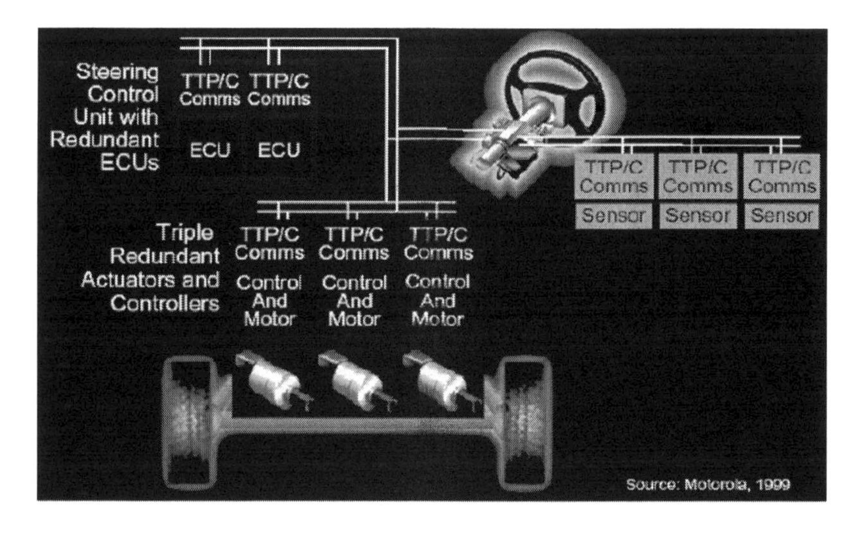

Fig. 4.140 SBW 2WS conversion mechatronic control system [Motorola 1999; HANSEN 2005].

As another example ones may use a very simple SBW 2WS mechatronic control system that consist of two mechatronic actuation modules interconnected with an electronic communication link (e.g. TTP or FlexRay™).

One of the actuation modules may be the driver command/feedback module, attached to the steering wheel/joystick. The other may be the actual steering E-M actuator that turns the front wheels of the vehicle. A structural and functional block diagram of the SBW 2WS mechatronic control system is shown in Figure 4.141 [ROOS AND WIKANDER 2005]. This marginal example is limited to the mechatronic steering rack module in Figure 4.141.

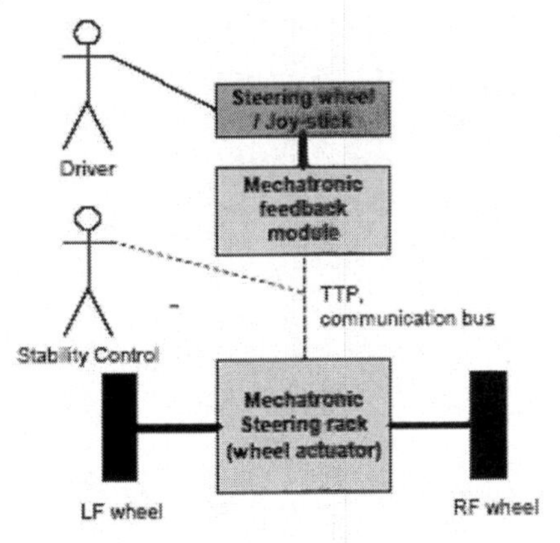

Fig. 141 SBW 2WS conversion mechatronic control system [ROOS AND WIKANDER 2005].

The mechatronic steering rack is here assumed to be a module (Fig. 4.142), with a linear tubular E-M motor as the E-M actuator [ROOS AND WIKANDER 2005].

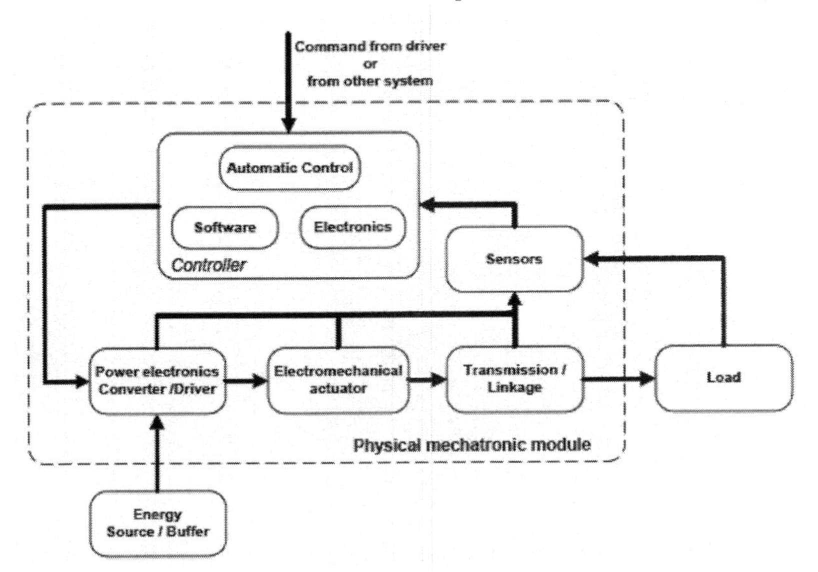

Fig. 4.142 Structural and functional block diagram of a mechatronic actuation module [ROOS AND WIKANDER 2005].

The main objective is to optimise the mechatronic steering rack module with respect to mass and/or energy efficiency given predefined yaw torque and velocity requirements on the front wheels of the vehicle.

4.8.4 Conclusion

It may be shown, that TTP is capable of providing fault isolation of single component failures. By using the appropriate replica strategy, a fully operational mode may be guaranteed even in the case of the failure of non-fault-tolerant component. Handling faults and redundancy on the protocol level relieves the application software designer from taking care of faulty components [MÜLLER AND PLANKENSTEINER 2002; JEFFE 2005]. The choice of components exhibiting different failure modes determines the architecture of a FTU. The assist the handling component is a FSU. It operates as *"The right message at the right time, or nothing at all"*.

The TTA is a systematic approach for the developer with a well-defined internal structure helping to reduce the complexity of the involved processes. Clear interfaces facilitate the re-use of components and software. This makes the TTA a powerful approach towards the realisation of RBW or XBW applications.

4.9 Discussion and Conclusions

Steering is realised by swivelling the vehicle's front wheels about an approximately vertical axis, controlled by a linkage from the driver's steering control input device – conventionally, a steering wheel – although other attitudes are realisable. Although RWS automotive vehicles exist for special purposes (fork-lift cars being the prevalent example), they are intrinsically unstable, necessitate particular driver skills, and may be safely functioned only at low values of vehicle velocity. It may be beneficial to steer all four wheels with the front wheel contribution prevalent, but FWS is adequate for all normal road-going functions.

As a general rule, technical evolution in SBW AWS conversion mechatronic control systems has been focused on three major subjects:

- ❖ Electric power assisted steering (EPAS);
- ❖ Steer-by-wire (SBW);
- ❖ All-wheel steering (AWS).

The irresistible technical tendency of the last few years has been for power-assisted steering to become more or less standard except in the smallest and cheapest automotive vehicles in the A/B segments. This has resulted in a major engineering effort to advance EPAS systems that experience much less from parasitic power losses than from conventional fluidical (hydraulical and/or pneumatical) systems and, as a result, necessitate far less of a fuel economy penalty.

At one time it was experienced that the restrictions of 14/12 V_{DC} E&INs would indicate that EPAS would be appropriate only to smaller vehicles and that wider use would have to predict the acceptance of 42/36 V_{DC} E&INs. On the other hand, the broad negative response of the 42 V_{DC} concepts has led to the development of more compact and powerful 14/12 V_{DC} EPS (as in many other areas of automotive vehicle engineering) and systems may now be undoubtedly appropriate for D-segment vehicles, as some recent functions have shown. It is expected that the majority of the European automotive vehicle park may be EPS-operational by just about 2020. The EPS system comprises the steering column, integrated E-M motor and controller, and R&P gear. The system may be used in an upcoming SUV.

The EPS design offers the advantages of modularity and reduced assembly costs for vehicle manufacturers; improved fuel economy and steering feel for drivers; and no need for oily-fluid that eliminates leaks and helps the environment.

The move to improve safety and functionality in automotive vehicles has lead to an increase in mechatronic control systems and the introduction of RBW or XBW integrated unibody, space-chassis, skateboard-chassis, or body-over -chassis motion mechatronic control hypersystems.

An SBW AWS conversion mechatronic control system offers improvements in safety with the removal of the steering column, thus improving the crash performance of the vehicle.

In addition active safety functionality may be applied providing corrective steering to improve stability control.

Other benefits include variable gain steering, tuneable steering feel, steering lead, and the opportunity to create an innovative **human-machine** (HM) interface for vehicle steering.

In order to achieve a robust implementation of an SBW AWS conversion mechatronic control system, certain key factors must be considered, including system diagnostics and redundancy, sensor and actuator performance, and control integration.

An E-M SBW AWS conversion mechatronic control system replaces the steering column with a fault-tolerant controller and E-M motors that connect to the steering rack to control direction. This type of mechatronic control system is safer than mechanical steering and improves *'road feel'* and fuel economy.

E-M SBW AWS conversion mechatronic control systems also give vehicle manufacturers more flexibility in vehicle design. Some analysts claim that by 2020 all innovative vehicles may have the E-M SBW AWS conversion mechatronic control system.

Delphi's *QUADRASTEER*™ SBW 4WS conversion mechatronic control system is mechatronically controlled, whereas the systems were mechanically controlled. Earlier systems also turned the rear wheels at a smaller angle.

The *QUADRASTEER*™ SBW 4WS conversion mechatronic control system has been designed to turn at a large angle, up to 12 deg, specifically for full size automotive vehicles.

The Honda 4WS was archaic and had only a mechanical system that had no appeal, with a short wheelbase vehicle that already had a small turning radius. The 1988 Honda became the first automaker to introduce 4WS in its compact *Prelude* sedan.

About the same time GM was showcasing a much more advanced version of 4WS in its *Blazer XT-1* concept vehicle. For whatever reason, the feature never took off, probably because the *Prelude*'s purely mechanical set-up didn't provide much benefit in the already nimble vehicle and the *XT-1*'s system was as complex and expensive as a NASA *X*-plane.

If you read the *4-Wheel & Off-Road* article in the July 2003 release, they describe the testing of it with 88.5 cm (35 inch) wheel tyres and a 15 cm (6 inch) lift without problems. The *QUADRASTEER*™ trucks have a higher capacity rear axle and increased GVWR.

A larger, 25 cm (9.75 inch) rear ring gear can be used to increase the rear axle's gross mass rating. The *ZX3 Ride Control* suspension provides selectable settings for a *'smooth or control/ trailering'* setting from the driver's seat.

With a fail-safe neutral, the *QUADRASTEER*™ unit may return to a stationary straight ahead position if a wire is cut or unplugged, returning the truck to FWS only. If at any time the two microprocessors *'disagree'* over the steering information they have received, *QS4* automatically shuts down and reverts back to traditional 2WS. The RWS has no mechanical linkage to the front and it may return to neutral 2WS if a part of the system fails.

QUADRASTEER™ 4WS is a unique, SBW 4WS conversion mechatronic control system designed for full-size vehicles, including long-bed and extended cab pickup trucks, vans, and sport utility vehicles.

QUADRASTEER™ gives full-size vehicles the turning radius and manoeuvrability of smaller vehicles, dramatically improving stability, handling and control at higher speeds for a more comfortable and safer driving experience.
Major applications can be as follows:

- Electric power steering;
- Two forms: assist E-M-F pump and direct E-M motors;
- The assist E-M-F pump uses an E-M motor to drive a conventional fluidical unit;
- The direct system uses E-M motors with the steering rack;
- In both cases, action can be controlled independent of the ICE or ECE.

Summing up, a unique SBW 4WS conversion mechatronic control system is the structure where all four wheels can steer or can operate as steering wheels.

The advantage is that in lower velocity motion, the E-M motors are controlled out of phase, which enables smaller vehicle turning radius with higher safety and more efficient manoeuvring. The vehicle even can rotate with zero radii and move laterally.

SBW AWS conversion technology that can make the vehicle more flexible in low-speed manoeuvres and parking; and DBW AWD propulsion and BBW AWB dispulsion technology that improves the efficiency of the powertrain by replacing the M-M differentials by mechatronic control.

The development of an in-vehicle RBW or XBW integrated unibody, space-chassis, skateboard-chassis, or body-over-chassis motion technology not only improves the transmission efficiency and enables SBW AWS conversion, it also enhances the mobility and flexibility for low-velocity driving and parking.

Glossary

AC-DC commutator - The commutator is a mechanical AC-DC rectifier; for a rotary DC-AC commutator generator, the commutator mechanically switches the armature windings so that the resultant induced source AC armature voltages always act with the same sense of voltage polarisation; this requires a reversal of the armature winding connection every π rad; the induced source AC armature voltages are mechanically rectified to induced source DC armature voltage via commutator segments that contact the carbon brushes.

AC-DC macrocommutator - The macrocommutator is an ASIM AC-DC rectifier; for a rotary DC-AC commutator generator, the macrocommutator electronically switches the armature windings so that the resultant induced source AC armature voltages always act with the same sense of voltage polarisation; this requires a reversal of the armature winding connection every π rad; the induced source AC armature voltages are electronically rectified to induced source DC armature voltage via inputs of the ASIM that contact via bipolar electrical valves an output of the ASIM.

Actuator - The component of an open-loop or closed-loop mechatronic control system that connects the electronic control unit (ECU) with the process; the actuator consists of a commutator and a final-control element; positioning electrical signals are converted to mechanical output.

Algorithm - A set of software instructions causing a computer to go through a prescribed routine; because embedded computer ECE or ICE controls have become so common, the algorithm has become essentially synonymous with control law for automotive scientists and engineers.

***Alternating injection suppression* -** An adaptation of the number of active ICE's cylinders by one after another every two ICE's crankshaft rotations that modulate the ICE torque.

Analog input - Sensors usually generate electrical signals that are directly proportional to the mechanism being sensed; the signal is, therefore, an analog or may vary from a minimum limit to a maximum limit.

Analog signal - A signal in which the information of interest is communicated in the form of a continuous signal; the magnitude of this signal is proportional (or analogous) to the actual quantity of interest.

Analog-to-digital (A/D) converter - An electronic device that produces a digital result that is proportional to the analog input voltage.

ASIC - Application-specific integrated circuit, an IC designed for a custom requirement, frequently a gate array, single-chip microprocessor, or programmable logic device.

ASIM - Application-specific integrated matrixer, an IM designed for a custom requirement, frequently a gate array or single-chip macrocommutator.

Bus - Topology of a communications network where all nodes are reached by links that allow transmission in both sense of direction.

Capacity - Energy storage capability of the CH-E/E-CH storage battery, ultracapacitor, ultrainductor, or ultraflywheel.

Central processing unit (CPU) - The portion of a computer system or microcontroller that controls the interpretation and execution of instructions and includes arithmetic capability.

CH-E/E-CH storage battery - Self-contained CH-E/E-CH cell/cells or system that converts chemical energy to electrical energy in a reversible process.

Closed-loop mechatronic control - A process by which a variable is continuously measured, compared with a reference variable, and changes as a result of this comparison in such a manner that the deviation from the reference variable is reduced; the purpose of closed-loop mechatronic control is to bring the value of the output variable as close as possible to the value specified by the reference variable in spite of disturbances; in contrast to open-loop mechatronic control, a closed-loop mechatronic control system acts to offset the effect of all disturbances.

Convergence - Yaw stability of the automotive vehicle when the human-driver inputs a rapid steering hand wheel movement and releases the steering wheel.

Coupling - A means or a device transferring power between systems.

D controller - A controller with derivative characteristics.

DC-AC commutator - The commutator is a mechanical DC-AC inverter; for a rotary DC-AC commutator motors or actuators, the commutator mechanically switches the armature windings so that the resultant force always acts in the same sense of rotary direction; this requires a reversal of the armature winding connection every π rad; the DC supply to the armature is via carbon brushes that contact the commutator segments.

DC-AC macrocommutator - The macrocommutator is an ASIM DC-AC inverter; for a rotary DC-AC commutator motors or actuators, the macrocommutator electronically switches the armature windings so that the resultant force always acts in the same sense of rotary direction; this requires a reversal of the armature winding connection every π rad; the DC supply to the armature is via the input of the ASIM that contacts via bipolar electrical valve outputs of the ASIM.

Defuzzification - The process of translating output grades to analog output values.

Depth of discharge (DoD) - Percentage of capacity [Ah] that has been removed from the CH-E/E-CH storage battery, ultracapacitor, ultrainductor, or ultraflywheel.

Device - A machine equipped with an ECE or ICE but not self-propelled; devices include, but are not limited to, chain, M-F pumps and M-P compressors, etc.

Digital signal - A signal in which the information of interest is communicated in the form of a number; the magnitude of this number is proportional to (within the limitations of the resolution of the number) the actual quantity of interest.

Digital signal processor (DSP) - A monolithic integrated circuit (IC) optimised for digital signal-processing applications; portions of a device are similar to a conventional microprocessor; the architecture is highly optimised for the rapid, repeated additions and multiplications required for digital signal processing; digital signal processors may be implemented as programmable devices or may be realised as dedicated high-speed logic.

Driver - A solid state device used to transfer electrical energy to the next stage that may be another driver, an electrical load (power driver), a wire or cable (line driver), a display (display driver), etc.

Final-control element - The second or last stage of an actuator to control mechanical output.

Fuzzification - The process of translating analog input variables to input memberships or labels.

Fuzzy logic (FL) - Software design based upon a reasoning model rather than fixed mathematical algorithms; a FL design allows the automotive system engineer to participate in the software design because the fuzzy language is linguistic and built upon easy-to-comprehend fundamentals.

Inference engine - The internal software program that produces output values through fuzzy rules for given input values; the inference process involves three steps: fuzzification, rule evaluation, and defuzzification.

Input memberships - The input signal or sensor range is divided into degrees of membership, i.e., low, medium, high or cold, cool, comfortable, warm, hot; each of these membership levels is assigned a numerical value or grade.

Kickback Steering - torque and angle arising from inverse input through the wheel tyres from an uneven concrete road or other surface.

Microcontroller unit (MCU) - A semiconductor device that has a CPU, memory, and I/O capability on the same chip.

Open-loop mechatronic control - A process within a mechatronic control system in which one or more input variables act on output variables based on the inherent characteristics of the mechatronic control system; an open loop is a series of elements that act on one another as links in a chain; in an open loop, only disturbances that are measured by the control unit can be addressed; the open loop has no effect on other disturbances.

Output memberships - The output signal is divided into grades such as off, slow, medium, fast, and full-on; numerical values are assigned to each grade; grades can be either singleton (one value) or *Mandani* (a range of values per grade).

PI controller - A controller with the proportional and integral characteristics.

PID controller - A controller with the proportional, integral, and derivative characteristics.

Protocol - The rules governing the exchange of information (data) between networked elements.

Pulse-width modulation (PWM) - The precise and timely creation of negative and positive waveform edges to achieve a waveform with a specific frequency and duty cycle.

Robust - Able to survive and operate properly in a severe environment.

Rule evaluation - Output values are computed per the input memberships and their relationship to the output memberships; the number of rules is usually set by the total number of input memberships and the total number of output memberships; the rules consist of *IF inputvarA* is *x*, *AND inputvarB* is *y*, *THEN outvar* is *z*.

Semicustom MCU - An microcontroller unit (MCU) that incorporates normal MCU elements plus application-specified peripheral devices such as higher-power port outputs, special timer units, etc.; mixed semiconductor technologies, such as high-density CMOS (HCMOS) and bipolar analog, are available in a semicustom MCU; generally, HCMOS is limited to 10 V_{DC}, whereas a bipolar-analog is suitable to 60 V_{DC}.

Slip threshold switch - A switch for escalation of indispensable slip threshold on sand and loose gravel that realise maximum traction on these on-off road surfaces.

Specific energy (energy density) - Energy storage capability per unit mass of the CH-E/E-CH storage battery, ultracapacitor, ultrainductor, or ultraflywheel [Wh/kg].

Specific power (power density) - Power delivery capability per unit mass of the CH-E/E-CH storage battery, ultracapacitor, ultrainductor, or ultraflywheel [W/kg].

State of charge (SoC) - The CH-E/E-CH storage battery, ultracapacitor, ultrainductor, or ultraflywheel level of charge can be stated as either DoD or SOC.

References and Bibliography

1. ABE M (1991): A study on effects of roll moment distribution control in active suspension on improvement of limit performance of vehicle handling. *JSAE Review*, Vol. 12, July 1991.
2. ABE M ET AL. (1996): A Direct Yaw Moment Control for Improving Limit Performance of Vehicle Handling-Comparison and Cooperation with 4WS. *Vehicle System Dynamics Supplement*, Vol. 25, 1996.
3. ACKERMANN J (1990): Robust car steering by yaw rate control. *Proceedings of the 29th Conference on Decision and Control,* Honolulu, Hawaii, December 1990, pp. 2033-2034.
4. ACKERMANN J (1991): Entkopplung von zwei Teilaugaben der Lenkung. *Allradlenksysteme bei Personnenwagen*, Essen, Germany, December 3-4, 1991.
5. ACKERMANN J (1993): *Robust Control – Systems with Uncertain Physical Parameters*. Springer-Verlag, London, 1993.
6. ACKERMANN J AND W SIENEL (1993): Robust Yaw and Damping of Cars with Front and Rear Wheel Steering. *IEEE Transactions on Control Systems Technology*, Vol. 1, No. 1, 1993, pp. 15-20.
7. ACKERMANN J (1994): Robust decoupling, ideal steering dynamics and yaw stabilization of 4WS cars. *Automatica*, Vol. 30, 1994, pp. 1761–1768.
8. ACKERMANN J, J GULDNER, W. SIENEL, R STEIHAUSER AND V UTKIN (1995): Linear and Nonlinear Controller Design for Robust Automotive Steering. *IEEE Transactions on Control Systems Technology*, Vol. 3, No. 1, 1995, pp. 132-143.
9. ACKERMANN J (1996): Yaw disturbance attenuation by robust decoupling of car steering. *Proceedings of the IFAC World Congress*, San Francisco, CA, 1996.
10. ACKERMANN J (1997): Yaw disturbance attenuation by robust decoupling of car steering. *Control Engineering Practice*, Vol. 5, No. 8, 1997.
11. ACKERMANN J AND T BÜNTE (1997): Actuator rate limits in robust car steering control. *Proceedings of the 36th Conference on Decision and Control*, San Diego, CA, USA, December 1997, pp. 4726-4731.
12. ACKERMANN J (1998): Active steering for better safety, handling and comfort. *Proceedings of the Conference on Advances in Vehicle Control and Safety*, (Amiens, France), 1998.
13. ACKERMANN J, T BÜNTE AND D ODENTHAL (1999): Advantages of active steering for vehicle dynamics control. *Proceedings of the 32nd International Symposium on Automotive Technology and Automation - ISATA*, Vienna, Austria, 1999.
14. ACURA (1991): *Service Manual*. Acura NIX, 1991; (In Japanese).

15. ADAMS (2005): Mechanical Dynamics Inc, 2005. Available online at http:// www. adams.com/ .
16. ADLER U – ED. (1988): *Automotive Electric/Electronic Systems*. Bosch – VDI -Verlag, Düsseldorf 1988.
17. AGA M, H KUSUNOKI, Y SATOH, R SAITOH AND M ITO (1990): Design of a 2-Degrees-of-Freedom Control System for Active Front-and-Rear -Wheel Steering. *SAE Technical Paper Series*, 1990, Paper No. 901746, pp. 69-76.
18. AGOGINO AM AND M-L TSENG (1990): *Research on Integrating Neural Networks with Influence Diagrams*. Final Report to the Institute for Scientific Computing Research, Lawrence Livermore National Laboratory, 1990, 26 p.
19. AHRING E AND M. MITSCHKE (1995): Comparison of All-Wheel Steerings in the System Driver-Vehicle. *Vehicle System Dynamics*, Vol. 24, 1995, pp. 283-298.
20. ALBERTI V AND E BABBEL (1996): Improved Driving Stability by Active Braking of the Individual Wheels. *Proceedings of AVEC 1996*, Aachen, Germany, 1996.
21. ALEXANDER D (2004): Chassis Integration Keeps the Rubber on the Road. *Automotive Engineering International*, Vol. 112, No. 5, 2004, SAE International, pp. 54-58.
22. ALLEYNE A (1997A): A Comparison of Alternative Intervention Strategies for Unintended Roadway Departure (URD) Control. *Vehicle System Dynamics*, Vol. 27, 1997, pp. 157-186.
23. ALLEYNE A (1997B): A Comparison of Alternative Obstacle Avoidance Strategies for Vehicle Control. *Vehicle System Dynamics*, Vol. 27, 1997, pp. 371-392.
24. AMAHOSER (2003): Four-Wheel Steering. *Amahoser's Stilth Site – All-Wheeel Steering*, 2003, Available online at: http://linkline.com/personal/ amahoser/4WS.htm .
25. AMBERKAR S, Y ESCHTRUTH, Y DING AND F BOLOURCHI (1999): Failure Mode Management for an Electric Power Steering System. *Proc. ISATA*, 1999, Paper No. 99AE002.
26. AMBERKAR S, BJ CZERNY, JG D'AMBROSIO, JD DEMERLY AND BT MURRAY (2001): A Comprehensive Hazard Analysis Technique for Safety-Critical Automotive Systems. *SAE Technical Paper Series*, 2001, Paper No. 2001-01-0674, pp. 1-13.
27. AMBERKAR S, BJ CZERNY, JD AMBROSIO, B MURRAY AND J WYSOCKI (2003): A System-safety process by "by-wire" automotive systems. *Delphi Automotive Systems and HRL Laboratories*, 2003. E-mail contact: brian.t.murray@ delphiauto.com .
28. APOSTOLOPOULOS D (1997): *Analytical Configuration of Wheeled Robotic Locomotion*, Ph.D. Thesis, Carnegie Mellon University, Robotics Institute, 1997.

29. ARMIN V, F SYLVIA, R JOSEF AND T ANSGAR (2004): Vehicle Dynamics Management – Benefits of Integrated Control of Active Brake, Active Steering and Active Suspension Systems. *Proceedings of the FISITA 2004*, 2004, Paper No. F2004F185.
30. ASGARI J AND D HROVAT (1997): On-Demand Four Wheel-Drive Transfer Case Modeling. *SAE Technical Paper Series,* 1997, Paper No. 970969.
31. BANDORE RT (1968): *A Primer on Vehicle Directional Control.* General Motors Technical Centre, Warren, Michigan Engineering Publication A-2739, September 19, 1968.
32. BAKKER E, L NYBORG AND H PACEJKA (1987): Tyre Modelling for Use in Vehicle Dynamics Studies. *SAE Technical Paper Series*, 1987, Paper No. 870421.
33. BAKKER E, H PACEJKA AND L LIDNER (1989): A New Tire Model with an Application in Vehicle Dynamics Studies. *SAE Technical Paper Series*, 1989, Paper No. 890087.
34. BEDNER, JR., EJ AND HH. CHEN (2004): A Supervisory Control to Manage Brakes and Four-Wheel-Steer Systems. *SAE Technical Paper Series*, 2004, Paper No. 2004-01-1059.
35. BEKKER MG (1960A): *Off-the-Road Locomotion*, The University of Michigan Press, Ann Arbor, MI, 1960.
36. BEKKER MG (1960B): Mobility of Cross Country Vehicles -- Thrust for Propulsion - Floatation and Motion Resistance -- Track and Wheel Evaluation - Optimum Performance and Future Trends. Series of articles in: *Machine Design*, Penton Publishing Co., Cleveland, Ohio, December 24, 1959; January 7, 1960; January 21, 1960; February 4, 1960.
37. BEKKER MG (1964): Mechanics of Locomotion and Lunar Surface Vehicle Concepts. *Transactions of the Society of Automotive Engineers*, Vol. 72, 1964, pp. 549-569.
38. BEKKER MG (1965): A proposed system of physical and geometrical terrain values for the determination of vehicle performance and soil trafficability. *Interservice Vehicle Mobility Symposium*, Vol. 2, Hoboken, N.J., Stevens Institute of Technology, 1965.
39. BEKKER MG (1969): *Introduction to Terrain-Vehicle Systems.* The University of Michigan Press, Ann Arbor, MI, 1969.
40. BEKKER MG (1982): *Optimization of Terrain-Vehicle Systems in Agroforestry.* Part I. -- Methods and Procedures, Prepared for National Research Council of Canada under Contract OSX81-00161, Santa Barbara, CA, March 1982.
41. BEKKER MG (1983): *Parametric Evaluation of Pneumatic Tires in On and Off the Road Locomotion.* Lectures delivered at CEMOTER, IMTCNR, Italy, 1983.
42. BERNARD JE, MJ VANDERPLOEG AND JE SHANNAN (1988): Liner Analysis of a Vehicle with Four Wheel Steering. *SAE Technical Paper Series*, 1988, Paper No. 880643, pp. 4680--4687.

43. Bertram T, P Dominke and B Mueller (1999): The Safety Related Aspect of CAR-TRONIC. *Proceedings of the SAE International Congress*, 1999, Paper No. 1999- 01-0488.
44. Birch TB (1999): *Automotive Chassis Systems*. Delamar Publishers, NY, USA, 165 p.
45. Boezeman AH and KF Drenth (1998): *Road-Train Stability Optimisation Using ADAMS*. The Delft University of Technology, 1998.
46. Bodie M and A Hac (2000): Closed loop yaw control of vehicles using magneto-rheological dampers. *Proceedings of the SAE 2000 World Congress*, Detroit, Michigan, March 2000. *SAE Technical Paper Series*, Paper No. 2000-01-0107.
47. Bootz A (2004): *Konzept eines Energiesparenden Elektrohydraulischen Closed-Center-Lenksystems für PKW mit Hoher Lenkleistung*. PhD Dissertation, Technische Universität Darmstadt. Darmstadt, Germany, 2004.
48. BOSCH (1986): *Automotive Handbook*. VDI Verlag, 1986.
49. Brabec P, M Malý and R Voženílek (2004): Controls System of Vehicle Model with Four Wheel Steering (4WS). *Mezdunarodni naučni simpozijum Motorna Vozila i Motori - International Scientific Meeting Motor Vehicles & Engines*, Kragujevac, October 4-6, 2004, Paper No. YU04017, pp. 1-7.
50. Brienza DM and CE Brubaker (1999): A steering linkage for short wheelbase vehicles: Design and evaluation in a wheelchair power base - A technical note. *Journal of Rehabilitation Research and Development*, Vol. 36, No. 1, January 1999.
51. Canada's de Wit C, H Olsen, KJ Astor and P Lischinsky (1995): A New Model for Control of Systems with Friction. *IEEE Transactions on Automatic Control*, Vol. 40, No. 3, March 1995, pp.419-425.
52. Canudas de Wit C and Ptsiotras (1999): Dynamic Tire Friction Models for Vehicle Traction Control. *Proceedings of the 38th Conference on Decision & Control*, Phoenix, Arizona, USA, December 1999.
53. Caracciolo L, A De Luca and S Iannitti (1999): Trajectory Tracking Control of a Four-Wheeled Differentially Driven Mobile Robot, *Proc. IEEE. Conference on Robotics and Automation*, 1999, pp. 2632-2638.
54. Chen C and M Tomizuka (2000): Lateral Control of Commercial Heavy Vehicles. *Vehicle System Dynamics*, Vol.33, 2000, pp.391-420.
55. Chen X (2006): *Trend and Challenges in Vehicle Electrification*. Visual Presentation, Department of Electrical and Computer Engineering, University of Windsor, Canada, 2006.
56. Citroën (2005): Citroën C5 "by Wire". *Babez.de – Cars and More,* July 7, 2005, http://www.babez.de/citroen/c5.php .
57. Cho D and JK Hedrick (1989): Powertrain Modeling for Control. *Trans. of the ASME, Journal of Dynamic Systems Measurement and Control*, Vol. 111, 1989, pp. 568-576.
58. Cho YH and J Kim (1995): Design of Optimal Four-Wheel Steering System. *Vehicle System Dynamics*, Vol. 24, 1995, pp. 661-682.

59. CLIMATRONIC (1990): Nur für den internen Gerbuch in der V.A.G Organisation Volkswagen AG, Wolfsburg, July 1990.
60. COLLETTI JB (1990): Control of Future Steering and Suspension Systems. *Proc 22nd ISATA: International Symposium on Automotive Technology & Automation,* Florence, Italy, Paper No. 90308.
61. COOPER N, D CROLLA, M LEVESLEY AND W MANNING (2004): Integration of active suspension and active driveline to improve vehicle dynamics. *Motorsports Engineering Conference and Exhibition*, Dearborn, Michigan, 2004. SAE Technical Paper Series, Paper No. 2004- 01-3544.
62. COSC/PSYCH (2006): Cosc6326/Psych6750X Introduction – 8. Vehicular Applications Drive or Fly by Wire, 2006, ss. 1-33.
63. CROLLA DA (1989): Intelligent Suspensions. *Agricultural Engineer*, Winter 1989, pp. 111-115.
64. DAIMLERCHRYSLER (2004): Drive by wire. *DaimlerChrysler's Special Reports,* 1998-2004.
65. D'ANDREA-NOVEL B, G CAMPION AND G BASTIN (1995): Control of Nonholonomic Wheeled Mobile Robots by State Feedback Linearization. *International. Journal of Robotics Research*, Vol. 14, No. 6, 1995, pp. 543 -449, 1995.
66. DEFALCO F. (1989): Materiale rotabile e trazione, *ATTI Relazioni Generali, 1-mo Convegno Internazionale "LA FERROVIE NEI TRASPORTI DEGLI ANNI 2000",* Bologna, Italy, pp. 39-50 [425].
67. DELPHI (2005A): *Delphi Tomorrow's Technology*. Delphi Corporate Brochure, DA-02--001, September 30, 2005. Available online at: http:// delphi.com/pdf/ corp_english.pdf .
68. DELPHI (2005B): *Delphi Quadrasteer™ Four Wheel Steering.* Delphi Corporation, 2005. Available online at http://www.delphi.com/pdf/ppd/ch-steer/str_quadra.pdf .
69. DJH (2003): *Off-Road/All Terrain*. DJH Engineering Center, Inc. International, 2003. Available online at http://djhec.com/pages/offroad_all_terrain/ index.htm .
70. DONGES E (1978): Ein regelungstechnischesn Zwei-Ebenen-Modell des menschlichen Lenkverhaltens im Kraftfahrzeug. *Zeitschrift für Verkehrssucherheit*, Vol. 3, 1978.
71. DONISELLI C, G. MASTINU, AND M. GOBBI (1996): Aerodynamic Effects on Ride Comfort and Road Holding of Automobiles. Vehicle System Dynamics Supplement, Vol. 25, 1996.
72. DUDZINSKI PA (1989): Design Characteristics of Steering Systems for Mobile Wheeled Earthmoving Equipment. *Journal of Terramechanics*, Vol. 26, No. 1, 1989, pp. 25-82.
73. EGUCHI T, Y SAKITA, K KAWAGOE, S KANEKO, K MORI AND T MATSUMOTO (1989): Development of "Super Hicas", a New Rear Wheel Steering System with Phase Reversal Control. *SAE Technical Paper Series*, 1989, Paper No. 891978, pp.1495-1504.

74. EICKER C, M NYENHUIS AND U DIERKES (2002): Mechatronical Design and Realization of an Electro-Hydraulical "Steer-by-Wire" System. *3rd International Fluid Power Conference Aachen*, IFK, Aachen, Germany, 2002.
75. ERKKINEN T (2003): High-Integrity Production Code Generation. *2003 AIAA GN&C Conference*, 2003, Paper No. 2003-5488. Available online at: http://www.mathworks.com/products/featured/12945_Production_Code_Generation_AIAA_GNC_2003.pdf .
76. FIJALKOWSKI B (1985): On the new concept hybrid and bi-modal vehicles for the 1980s and 1990s. *Proc. DRIVE ELECTRIC Italy '85*, Sorrento, Italy, October 1985, pp. 4.04.1-4.04.8.
77. FIJALKOWSKI B (1986): Future hybrid electromechanical very advanced propulsion systems for civilian wheeled and tracked vehicles with extremely high mobility. *Proc. EVS-8: The 8th International Electric Vehicle Symposium*, Washington, DC, USA, October, 1986, pp. 426-442.
78. FIJALKOWSKI B (1987A): *Modele matematyczne wybranych lotniczych i motoryzacyjnych mechano-elektro-termicznych dyskretnych nadsystemow dynamicznych* (*Mathematical models of selected aerospace and automotive mechano-electro-thermal discrete dynamic hypersystems*), Monografia 53, Politechnika Krakowska im. Tadeusza Kosciuszki, Krakow, 1987, 274 p. (In Polish).
79. FIJALKOWSKI B (1987B): Choice of Hybrid Propulsion Systems - Wheeled City and Urban Vehicle - Tracked All-Terrain Vehicle (Part I). *Electric Vehicle Developments*, Vol. 6, No. 4, 1987, pp. 113-117.
80. FIJALKOWSKI B (1988): Choice of Hybrid Propulsion Systems - Wheeled City and Urban Vehicle - Tracked All-Terrain Vehicle (Part II). *Electric Vehicle Developments*, Vol 7, No. 1, 1988, pp. 31-34.
81. FIJALKOWSKI B (1989): The Novel Means of Very Advanced Propulsion for 2000s Thermo-Electric High-Speed Trains. *Proceedings of the 1st International Congress "THE ROLE OF RAILWAYS IN THE YEAR 2000"*, University of Bologna, Bologna, Italy, April 12-14, 1989 pp. 425-431.
82. FIJALKOWSKI B (1990A): Very Advanced Propulsion Spheres for High Speed Tracked Vehicles. *Proc. of 10th International Conference of the ISTVS*, Vol. III, Kobe, Japan, August 20-24, 1990.
83. FIJALKOWSKI B (1990B): Artificial Intelligence Very Advanced Propulsion Spheres. *Proc. ED&PE '90: International Conference on Electric Drives and Power Electronics*, Vol 2. The High Tatras, Czecho-Slovakia, November 5-7, 1990, pp. 23-28.
84. FIJALKOWSKI B (1991): Mechatronically fuzzy-logic controlled full-time 4WA - 4WC - 4WD - 4WS intelligent motor vehicle. *Proc. ISATA 91: International Dedicated Conference and Exhibition on MECHATRONICS, USE OF ELECTRONICS FOR PRODUCT DESIGN, TESTING, ENGINEERING AND RELIABILITY in conjunction with the 24th International Symposium on Automotive Technology and Automation (ISATA)*, Florence, Italy, May 20-24, 1991, Paper No. 911238, pp. 101-108.

85. FIJALKOWSKI B. (1993): Small All-Weather & All-Terrain Surveillance Unmanned Autonomous Vehicle for Law Enforcement Applications. *SPIE Proc., OE/Aerospace Sensing Symposium - Surveillance Technology II,* Orlando, Florida, USA, Vol. 1693, 1993, Paper No. 1693-30, pp.20-24.
86. FIJALKOWSKI B (1995): The concept of a high performance all-round energy efficient mechatronically-controlled tri-mode supercar. Special Issue on Automotive Electronics: Part 2, Guest Editor: B Fijalkowski. *Journal of Circuits, Systems and Computers*, Vol. 5, No. 1, 1995, pp. 93-107.
87. FIJALKOWSKI B (1996A): Mathematical model of the steer-, autodrive- and autoabsorbable wheels for smart three-mode 4WS × 4WA × 4WD × 4WB supercars. *Proc. ISATA 96: International Dedicated Conference on ELECTRIC, HYBRID & ALTERNATIVE FUEL VEHICLES in conjunction with 29th International Symposium on Automotive Technology and Automation (ISATA)*, Florence, Italy, June 3-6, 1996, Paper No. 96EL031, pp. 159-166.
88. FIJALKOWSKI B (1996B): Emerging and future intelligent aviation and automotive applications of MIMO ASIM macrocommutators and ASIC microcontrollers, pp. 397-405. *Proceedings of the NATO Advanced Research Workshop on Future Trends in Microelectronics - Reflections on the Road to Nanotechnology* (Luryi S., J. Xu and A. Zaslavsky -- Eds), Ille de Bendor, France, July 17-21, 1995, NATO ASI Series, Series E: Applied Sciences - Vol. 323. Kluwer Academic Publishers, Dordrecht, Boston, London: 1996, Published in cooperation with NATO Scientific Affairs Division, 421 pp.
89. FIJALKOWSKI B (1996C): Novel AC-DC/DC-AC, DC-AC/AC-DC and AC-AC macrocommutators for intelligent main battle tank propulsion and dispulsion. *Proc. IEEE Workshop on Power Electronics in Transportation,* Dearborn, MI, USA, October 24-25, 1996, pp. 191-196.
90. FIJALKOWSKI B (1997A): Intelligent automotive systems: Development in full-time chassis motion spheres for intelligent vehicles, pp. 125-142. Chapter 5 in the book: *Advanced Vehicle and Infrastructure Systems: Computer Applications, Control, and Automation* (Christopher O. Nwagboso, Ed.), John Wiley & Sons, New York, Weinheim, Brisbane, Singapore, Toronto, 1997, 502 p.
91. FIJALKOWSKI B (1997B): Ultrainductors, ultracapacitors and ultraflywheels for all-electric and hybrid vehicles. *Proc. International Dedicated Conference on ELECTRIC, HYBRID AND ALTERNATIVE FUEL VEHICLES IN THE AUTOMOTIVE INDUSTRY in conjunction with the 30th ISATA: International Symposium on Automotive Technology and Automation,* Florence, Italy, June 16-19, 1997, Paper No. 97EL051, pp. 225-232.
92. FIJALKOWSKI B (1997C): The concept of a mechatronic engine management neuro-fuzzy microcontroller for the Fijalkowski engine. *Proc. ROVA '97 INTERNATIONAL: 3rd International Conference on Road Vehicle Automation,* Salamanca, Spain, September 22-24, 1997.

93. FIJALKOWSKI B (1997D): Neuro-fuzzy AWD × AWB × AWA × AWS all-terrain vehicles. *Proceedings of the 7th European ISTVS Conference,* Ferrara, Italy, October 8-10, 1997, pp. 433-440.
94. *FIJALKOWSKI B (1998A): Mechatron*ics electromechanic-activated inlet/ outlet fluidical valve control - The anciliary benefits, *Proc. International Dedicated Conference on MECHATRONICS in conjunction with the 31st ISATA: International Symposium on Automotive Technology and Automation,* Duesseldorf, Germany, June 2-5, 1998, Paper No. 98AE017, pp. 271 -278.
95. FIJALKOWSKI B (1998B): Prime mover for hybrid electric propulsion systems. *Proc. IEEE Workshop on Power Electronics in Transportation,* Dearborn, MI, USA, October 23-24, 1998, pp. 102-112.
96. FIJALKOWSKI B (1999A): A family of driving, braking, steering, absorbing, rolling and throttling controls –"X-By-Wire Automotive Control". *Proc. Programme Track on Auto-motive Electronics and New Products, in conjunction with the 32nd ISATA: International Symposium on Automotive Technology and Automation -- Advances in Automotive and Transportation Technology and Practice for the 21st Century,* Vienna, Austria, June 14-18, 1999, Paper No. 99AE016, pp. 287-294.
97. FIJALKOWSKI B (1999B): A novel concept of the air independent propulsion and/or dispulsion for attack submarines. U*EES'99: The 4th International Conference on UNCONVENTIONAL ELECTROMECHANICAL AND ELECTRICAL SYSTEMS,* St. Petersburg, Russia, June 21-24, 1999, pp. UES: 1-6.
98. FIJALKOWSKI B (1999C): A two-in-one automotive AC-DC/DC-AC & AC-AC commutator flywheel generator/motor. *Proc. Programme Track on Automotive Electronics and New Products, in conjunction with the 32nd ISATA: International Symposium on Automotive Technology and Automation -- Advances in Automotive and Transportation Technology and Practice for the 21st Century,* Vienna, Austria, June 14-18, 1999, Paper No. 99AE015, pp. 279-286.
99. FIJALKOWSKI B (1999D): A novel automotive MRF or ERF reciprocating prime mover. *Proc. Barcelona 1999 European Automotive Congress - Vehicle Systems Technology for the Next Century,* Barcelona, Spain, June 30 – July 2, 1999, Paper No. STA99P410, p. 356.
100. FIJALKOWSKI B (1999E): Mechatronically controlled tri-mode hybrid SBW AWS diversion control system for off-road vehicles. *Proc. of the 13th International Conference of the ISTVS (International Society for Terrain-Vehicle Systems),* Vol. II, Munich, Germany, September 14-17, 1999, pp. 471-478. FIJALKOWSKI B (2000A): DBW 4WD propulsion & BBW 4WB dispulsion control system for intelligent automotive vehicles. *Proc. Telematics Automotive 2000 - Conference and Exhibition,* Birmingham, England, April 11-13, 2000.

101. FIJALKOWSKI B (2000B): Technology potential of ultrainductor, ultracapacitor and ultraflywheel storage. *Proc. of the 39th Power Source Conference*, Cherry Hill, New Jersey, June 12-15, 2000.
102. FIJALKOWSKI B. (2000C): Hybrid DBW propulsion control systems with the MRF commutator prime mover. *Proc ISATA 2000 – Electric, Hybrid, Fuel Cell and Alternative Fuel Vehicles*, Dublin, Ireland, September 25-27, 2000, pp. 363-370.
103. FIJALKOWSKI B (2000D): Advanced Chassis Engineering. *World Market Series Business Briefing* – Global Automotive Manufacturing & Technology, World Markets Research Centre, August, 2000, pp. 109-116.
104. FIJALKOWSKI B (2003A): Novel mobility and steerability enhancing concept of all-electric intelligent articulated tracked vehicles. *Intelligent Vehicles Symposium, 2000 – Proceedings IEEE*, June 9-11, 2003, pp. 225-230.
105. FIJALKOWSKI B (2003B): Novel Mobility and Steerability Enhancing Concept of Articulated Tracked Vehicles. *Proceedings of the 9th European Conference of the ISTVS,* Harper Adams, Newport, UK, September 8-11, 2003, pp. 1-13.
106. FIJALKOWSKI B AND J KROSNICKI (1990A): Electromechanical wheel-hub motors applied to four-wheel-drive automobile. *Papers: The 6th Symposium on Motor Vehicle and Motors '90,* October 1990; also in *Motorna Vozila-Motors Saopstenja* (XVI-92/93), Yugoslavia, 1990, Vol. 30, pp. 217-225.
107. FIJALKOWSKI B AND J KROSNICKI (1990B): Novel experimental proof-of -concept electro-mechanical 'single-shaft' automotive very advanced propulsion spheres. *Proc. EVS-10: The 10th International Electric Vehicle Symposium,* Hong Kong, December, 1990.
108. FIJALKOWSKI B AND J KROSNICKI (1991): All-wheel Driven Track - A Key Component of Future Tracked Electric/Hybrid Vehicles, *Proc 24th ISATA: International Symposium on Automotive Technology & Automation, Dedicated Conference on Electric/Hybrid Vehicles*, Florence, Italy, May 20-24, 1991, pp. 87-94.
109. FIJALKOWSKI B. AND J KROSNICKI (1992A): Mechatronically Controlled Intelligent Mobile Battle Tank. *Proc. 25th ISATA, Dedicated Conference on Mechatronics*, Florence, Italy, June 1-5, 1992, Paper No. 920816, pp. 343-354.
110. FIJALKOWSKI B AND J KROSNICKI (1992B): Very Advanced Propulsion Spheres for Loop-wheeled Electric/Hybrid Vehicles, *Proc. 25th ISATA, Dedicated Conference on Zero Emission Vehicles - The Electric/Hybrid and Alternative Fuel Challenge*, Florence, Italy, Paper 920218, pp. 419-426,
111. FIJALKOWSKI B. AND J KROSNICKI (1992C): All-Electric Articulated Triad Martian Roving Vehicle. *Proc EVS-11: The 11th International Electric Vehicle Symposium*, Florence, Italy, September 27-30, 1992, Vol 2, pp 13.05.1-13.05.12.

112. FIJALKOWSKI B AND J KROSNICKI (1993): Smart electro-mechanical conversion actuators for intelligent road vehicles: Application to front- and/or rear-wheel rack-and-pinion steering gears, pp. 16-25. *Proceedings of the 1st International Conference on Road Vehicle Automation* held at Vehicle Systems Research Centre, School of Engineering, Bolton, UK, May 24-26, 1993 (Christopher O Nwagboso, Ed.), PENTECH PRESS Publishers: London, 1993, 309 pp.
113. FIJALKOWSKI B AND J KROSNICKI (1994): Concepts of electronically-controlled electro-mechanical/mechano-electrical steer-, autodrive- and autoabsorbable wheels for environmentally-friendly tri-mode supercars. Special Issue on Automotive Electronics: Part 1, Guest Editor: B Fijalkowski. *Journal of Circuits, Systems and Computers*, Vol. 4, No. 4, 1995, pp. 501-516.
114. FIJALKOWSKI B AND K TROVATO. (1994): A concept for a mechatronically controlled full-time 4WD × 4WB × 4WA × 4WS intelligent vehicle for drivers with special needs. *Proc. for the Dedicated Conferences on Mechatronics & Supercomputing Applications in the Transportation Industries in conjunction with the 27th ISATA,* Aachen, Germany1994.
115. FILIPI F AND R SIVORI (1990): Four Wheel Steering Vehicle (4WS) with Continuous *k* Ratio. *Proc. 22nd ISATA: International Symposium on Automotive Technology & Automation,* Florence, Italy, 1990, Paper No. 90063, pp. 865-872.
116. FLECK R (2003): Aktivlenkung - ein wichtiger erster Schritt zum Steer-by-Wire. *3. PKW-Lenksysteme*, Hausder Technik, Essen, Germany, 2003.
117. FRITZ A (2001): Lateral and Longitudinal Control of a Vehicle Convoy. *Vehicle System Dynamics,* Supplement 35, 2001, pp. 149-164.
118. FUJIOKA T ET AL. (1999): Overview of Control for Automatic Driving System. *Journal of the Robotics Society of Japan*, Vol. 17, No.3, 1999, pp. 18-23 (In Japanese).
119. FUJITA K, K OHASHI, K FUKATANI, S KAMEI, Y KAGAWA AND H MORI (1998): Development of Active Rear Steer System Applying H_∞-μ synthesis. *SAE Technical Paper Series*, 1998, Paper No. 981115, pp. 1694-1701.
120. FUKADA Y (1998): Estimation of vehicle slip-angle with combination method of model observer and direct integration. *Proceedings of the International Symposium on Advanced Vehicle Control (AVEC)*, Nagoya, Japan, 1998.
121. FUKAO T, S MIYASAKA, K MORI, N ADACHI AND K OSUKA (2001): Active steering systems based on model reference adaptive non-linear control. *IEEE Intelligent Transportation System Conference*, Oakland, CA, USA, August 2001, p. 502.
122. FUKUI K, K MIKI, Y HAYASHI AND J HASEGAWA (1988): Analysis of Driver and a "Four Wheel Steering Vehicle" System Using a Driving Simulator. *SAE Technical Paper Series*, Paper No. 880641, pp. 4657-4667.

123. FUKUNADA Y, N IRIE, J KUROKI AND F SUGASAWA (1987): Improved Handling and Stability Using Four-Wheel-Steering. *The 11th International Conference on Experimental Technical Safety Vehicles*, Washington, D.C., May 1987.
124. FURUKAWA Y, N YUHARA, S SANO, H TAKEDA AND Y MATSUSHITA (1989): A Review of Four-Wheel Steering Studies from the Viewpoint of Vehicle Dynamics and Control. *Vehicle System Dynamics*, Vol. 18, 1989, pp. 151 -186.
125. FURUKAWA Y AND S SANO (1989): Effects of Nonlinear Rear Steer Control On Steering Response During Higher Lateral Acceleration Cornering. *Vehicle System Dynamics*, Vol. 18, 1989, pp. 248-262.
126. FURUKAWA Y, M ABE (1996): On-Board-Tire-model Reference Control for Cooperation of 4WS and Direct Yaw Moment Control for Improving Active Safety of Vehicle Handling. *Proceedings of AVEC 1996*, Aachen, Germany, 1996.
127. GÁSPÁR P, I SZÁSZI, T BARTHA, I VARGA, J BOKOR, L PALKOVICS AND L GIANONE (2000): Visual Lane and Obstruction Detection System for Commercial Vehicles. *Proceedings of the 4th IFAC Symposium on Fault Detection Supervision and Safety for Technical Processes*, Budapest, 2000.
128. GAUT S (2003A): Delphi's Electric Power Steering to be available on seven European vehicles by 2006. *Business News*, Welcome IAA (Frankfurt), Press Conference: September 10, 2003, p. 9.
129. GAUT S (2003B): Delphi unveils new Active Front Steering system. *Technology News*, Welcome IAA (Frankfurt), Press Conference: September 10, 2003, p. 41.
130. GAUT S (2003C): Delphi's Quadrasteer provides high value solution for handling enhancement on passenger vehicles. *Technology News*, Welcome IAA (Frankfurt), Press Conference: September 10, 2003, p. 53.
131. GENTA G (1997): *Motor Vehicle Dynamics.* Modeling and Simulation, World Scientific 1997, pp.213-334.
132. GIANONE L, L PALKOVICS AND J BOKOR (1995): Design of an active 4WS system with physical uncertainties. *Control Engineering Practice*, Vol. 3, No. 8, 1995, pp. 1075–1083.
133. GILLESPIE TD (1992): *Fundamentals of Vehicle Dynamics.* Society of Automotive Engineers, Warrendale, 1992.
134. GERDES JC, P YIH AND *K SATYAN (2002): Safety Performance and Robustness of Heavy Vehicle AVCS.* California PATH PrograM REPORT FOR MOU 390, JANUARY 2002.
135. *GODDARD PL (1998): Automotive Embedded Computing: The Cu*rrent Non-Fault-Tolerant Baseline for Embedded Systems. *Proc. 1998 Workshop on Embedded Fault-Tolerant Systems*, May 1998, pp. 76-80.

136. GORDON T, M HOWELL AND F BRANDAO (2003): Integrated control methodologies for road vehicles. *Vehicle system dynamics*, Vol. 40, No. 1–3, 2003, pp. 157–190.
137. GÜVENÇ BA AND L GÜVENÇ (2002): Robust Steer-By-Wire Control based on the Model Regulator. *Proceedings of IEEE International Conference on Control Applications*, Glasgow, Scotland UK, September 18-20, 2002, p. 435.
138. GÜVENÇ L (2005): *The EU FP6 Funded Automotive Controls and Mechatronics Research Center at İstanbul Technical University*. İstanbul Technical University (İ.T.Ü.) MEKAR: MECHATRONICS RESEARCH, May 18, 2005, pp. 1-44.
139. HAC A (1998): Evaluation of Two Concepts in Vehicle Stability Enhancement Systems. *Proceedings of the 31st ISATA: International Symposium on Automotive Technology and Automation*, Duesseldorf, Germany, June 2-5, 1998.
140. HAC A AND M BODIE (2002): Improvements in vehicle handling through integrated control of chassis systems. *International Journal of Vehicle Autonomous Systems*, Vol. 1, No. 1, 2002.
141. HAGHIGHATGOO H AND A ESFANDYARI (2005): Datorer gör bilarna säkrare. *Delta 3*, 2005. Available online at http://www.mdh.se/7-datorer_sakrara bilar.wbk ; http://auto.howstuff works.com/steering5.htm .
142. HAGIWARA T AND S HIROSE (1999): Development of dual mode x-screw: A novel load-sensitive linear actuator with a wide transmission range. *Proceedings of the 1999 IEEE International Conference on Robotics and Automation*, Detroit, MI, USA, May 1999, pp. 537-542.
143. HAMILTON A. (2002): Transit & Talk: HY-WIRE CAR. Inventor: General Motors, *TIME Invention,* 2002, page 5, Web Site: http://www.Monster-design.co.kr/reference/ 2002 _time_invention.pdf.
144. HAMMETT RC AND PS BABCOCK (2003): Achieving 10-9 Dependability with Drive-by-Wire Systems. *2003 SAE World Congress,* Detroit, March 3-6, 2003, Paper No. SAE 2003-01-1290.
145. HANSEN FO (2005): Introduction to TTR and FlexRay real-time protocols. *Presentation for LUNA,* Odense, Ingeniørhøjskolen I Århus Denmark, May 31, 2005, ss. 1-39. Available online at http://www.robocluster.dk/luna/ .
146. HARADA M AND H HARADA (1999): Analysis of lateral stability with integrated control of suspension and steering systems. *JSAE Review*, Vol. 20, 1999, pp. 465–470.
147. HARADA H ET AL. (2000): Control Effects of Active Rear-Wheel-Steering on Driver Vehicle Systems. *Proceedings of the AVEC'96*, Aachen, 1996.
148. HARTER W, W PFEIFFER, P DOMINKE, G RUCK AND P BLESSING (2000): Future Electrical Steering Systems Realizations with Safety Requirements. *SAE Technical Paper Series*, 2000, Paper No. 2000-01-0822.

149. HASKARA I, C HATIPOGLU, U OZGUNER (1997): Combined Decentralized Longitudinal and Lateral Controller Design for Truck Convoys. *Proc. of the 1997 IEEE Conference on Intelligent Transportation Systems*, 1997, pp. 123–128.
150. HATIPOGLU C, K REDMILL AND U OZGUNER (1998): Automated Lane Change: Theory and Practice. Presented at Advances in Automotive Control 1998, *Proceedings of the 2nd IFAC Workshop*, Mohican State Park, Loudonville, Ohio, USA, 1998.
151. HAYAMA R, K NISHIZAKI, S NAKANO AND K KATOU (2000): The vehicle stability control responsability improvements using Steer-By-Wire. *Proceedings of IEEE Intelligent Vehicle Symposium 2000*, Dearborn, MI, USA, October 3-5, 2000, p. 596.
152. HAZELDEN RJ (1992): Application of an optical torque sensor to a vehicle power steering system. *IEEE Colloqium on Automotive Sensors*, Solihull, UK, 1992, p. 9/1.
153. HEBDEN RG, C EDWARDS AND SK SPURGEON (2004): Automotive Steering Control in a Split-μ Maneuver Using an Observer-Based Sliding Mode Controller. *Vehicle System Dynamics*, Vol. 41, No.3, 2004, pp.181-202.
154. HEDRICK JK ET AL. (1994): Control Issues in Automated Highway Systems. *IEEE Control Systems*, Vol. 12,1994, pp.21-32.
155. HEITZER H-D (2003): Entwicklung eines fehlertoleranten Steer-by-Wire Lenksystems. *3. PKW-Lenksysteme*, Hausder Technik, Essen, Germany, 2003.
156. HEITZER H-D, AND A SEEWALD (2004): Development of a fault tolerant steer-by-wire steering system. *Convergence 2004*, Paper No. 2004-21-0046.
157. HERNER A (2001): *Elektronika w samochodzie*. WKŁ, 2001 (In Polish).
158. HIRANO Y, Y SATO, E OHNO AND K. TAKANAMI (1992): Integrated control system of 4WS and 4WD by H_∞ control. *Proceeding of the International Symposium on Advanced Control*, Yokohama, Japan, September 1992.
159. HIRANO Y (1994): Non-linear robust control for an integrated system of 4WS and 4WD. *Proceedings of the International Symposium on Advanced Vehicle Control AVEC'94*, Tsukuba, Japan, 1994.
160. HIRANO Y AND E ONO (1994): Nonlinear Robust Control and Integrated System of 4WS and 4WD. *Proceedings of AVEC'94*, 1994.
161. HIRANO Y AND K FUKATANI (1996): Development of robust active rear steering control. *Proceedings of the International Symposium on Advanced Vehicle Control AVEC'96*, 1996, pp. 359–376.
162. HITACHI (2004): Power Steering System. *Automotive Systems,* 2004. Available online at http://www.hitachi.co.jp/Div/apd/en/products/dcs/ dcs_005.html .
163. HOED E VAN (2004): Actief sturen met Delphi. *Auto & Motor TECHNIEK*, May 20, 2004. De Internetsite voor de Automotive Professional. Available online at http://www.amt.nl (In Dutch).

164. HONDA (1987): Honda Prelude Si 4WS: It Will Never Steer You Wrong. *Car and Driver*, Vol. 33, No. 2, August 1987, pp. 40-45.
165. HONDA (1991): *Service Manual*. Honda Today, 1991; (In Japanese).
166. HORIUCHI S, N YUHARA AND A TAKEI (1996): Two Degree of Freedom H-infinity Controller Synthesis for Active Four Wheel Steering Vehicles. *Vehicle System Dynamics,* Supplement 25, 1996, pp. 275-292.
167. HOSAGRAHARA S AND P SMITH (2004): Measuring Productivity and Quality in Model-Based Design. *Aerospace and Defense Digest* – August, 2004. Available online at http://www.mathworks.com/company/newsletters/aero_digest/aug04/measuringprod.html .
168. HOSAKA A, M TANIGUCHI, S UEKI, K KURAMI, A HATORI AND K YAMADA (1989): Steering control of an autonomous vehicle using a fuzzy logic controller', *Proc IMechE,* C391/005, 1989, pp.291-296.
169. HROVAT D (1997): Survey of advanced suspension developments and related optimal control applications. *Automatica*, Vol. 33, No. 10, 1997, pp. 1781–1817.
170. HUH K, C SEO, J KIM AND D HONG (1999): *Proceedings of American Control Conference*, San Diego, CA, June 1999, p. 729.
171. HUH K AND J KIM (2001): Active steering control based on the estimated tire forces. *Journal of Dynamic Systems, Measurement and Control*, Vol. 123, 2001, pp. 505-511.
172. HUNT KJ, A TA JOHANSEN, J KALKKUHL, H FRITZ AND TH GOTTSCHE (2000): Speed Control Design for an Experimental Vehicle Using a Generalized Gain Scheduling Approach. *IEEE Transactions on Control Systems Technology*, Vol. 8, May 2000, pp. 381-395.
173. HUNTER (2005): What Everyone Should Know About Wheel Alignment. *Undercar Information*, Hunter Engineering Company, 2005. Available online at http://www.hunter. com/pub/undercar/2470T/index.htm .
174. IGA S ET AL. (1988): Motor driven power steering for the maximum steering sensation in every driving situation. *SAE Technical Paper Series*, 1988, Paper 880705.
175. INAGAKI S ET AL. (1994): Analysis on Vehicle Stability in Critical Cornering Using Phase-Plane Method. *Proceedings of AVEC'94*, 1994.
176. INOUE H, KAWAI, S INAGAKI, H TANAKA AND H KAWAKAMI (1991): Development of active hydropneumatic suspension and active four wheel steering. *Proceedings of the JSAE*, Vol. 911, 1991, Paper No. 1991-5911060.
177. INOUE H (1991): Allradlenksystem im Toyota Soarer. *driverT-Tagung "Allradlenksysteme bei Personnenwagen"*, Essen, Germany, December 3-4, 1991.
178. INOUE H AND F SUGASAWA (2002): Comparison of feedforward and feedback control for 4WS. *Proceedings of the International Symposium on Advanced Vehicle Control AVEC' 02*, Hiroshima, Japan, 2002, pp. 258–263.

179. IRIE N, Y SHIBAHATA, H ITO, AND T UNO (1986): HICAS - Improvement of vehicle stability and controllability by rear suspension steering characteristics. *Proc. FISITA 21st Congress*, 1986, Paper No. 865114, pp. 2.81-2.88.
180. IRIE N AND J KUROKI (1990): 4WS Technology and the Prospects for Improvement of Vehicle Dynamics. *SAE Technical Paper Series*, Paper No. 901167, pp. 1334~1342.
181. ISERMANN R, R SCHWARTZ AND S STOLTZ (2002): Fault-Tolerant Drive-By-Wire Systems. *IEEE Control Systems Magazine*, Vol. 22, No. 5, October 2002, p. 64.
182. ISHIDA S, J TANAKA, S KONDO AND M SHINGYOJI (2003): Development of a Driver Assistance System. *SAE Technical Paper Series*, 2003, Paper No. 2003-01-0279.
183. ITO H, H ARAKAWA, K SUMI AND H YAMAGUCHI (1991): Controller for Experimental Vehicle Using Multi-Processor System. *SAE Technical Paper Series,* 1991, Paper No. 910086.
184. ITO K, T FUJISHIRO, T KAWABE, K KANAI AND Y OCHI (1986): A new way of controlling a four wheel steering vehicle. *Transaction of the SICE*, Vol. 23, 1986, pp. 828-834.
185. ITO M ET AL. (1987): Four Wheel Steering System Synthesized by Model Matching Control. *Proceedings of IEE-I Mech 6th International Conference on Automotive Electronics*, London, 1987.
186. ITOH H AND A OIDA (1990): Dynamic Analysis of Turning Performance of 4WD-4WS Tractor on Paved Road. *Journal of Terramechanics*, Vol. 27, No. 2, 1990, pp. 125-143.
187. ITOH ET AL. (1994): Meaurement of Forces Acting on 4WD-4WS Tractor Tires During Steady-State Circular Turning on a Paved Road. *Journal of Terramechanics*, Vol. 31, No. 5, 1994, pp. 285-312.
188. ITOH ET AL. (1995): Measurement of Forces Acting on 4WD-4WS Tractor Tires during Steady-State Circular Turning in a Rice Field. *Journal of Terramechanics*, Vol.32, No. 5, 1995, pp. 263-283.
189. JB (2004): *Drive By Wire Pinifarina Autosicura.* White Paper, Date: June 8, 2004, Revision: 1.3, Pi Technology, 11/10/04, pp. 1-9.
190. JEFFERSON C AND R BARNARD (1991): Emission Reduction in Road Vehicles by Kinetic Energy Recuperation. *Proc 24th ISATA: International Symposium on Automotive Technology & Automation*, Dedicated Conference on Mechatronics, Florence, Italy, 1991, Paper No. 910825, pp. 161-168.
191. JEONG W, J JANG AND CH HAN (1994): Modeling and Analysis of four wheel steering vehicle. *International ADAMS Conference*, 1994, Paper 5.
192. JIANG L, Y WANG AND M NAGAI (2000): A Theoretical Study on Front Steering Angle Compensation Control for Commercial Vehicles. *Seoul 2000 FISITA World Automotive Congress,* June 12-15, 2000, Seoul, Korea, Paper No. F2000G343.

193. JOHANNESSON P (2001): *SIRIUS 2001: A University Drive-by-Wire Project.* Technical Report No. 01-14, Department of Computer Engineering, Chalmers University of Technology, Göteborg, Sweden, 2001.
194. JOHANSSON R, P JOHANNESSEN, K FORSBERG, H SIVENCRONA AND J TOIN (2003): On Communication Requirements for Control-by-Wire Applications, *ISSC 21 Presentation*, August 7, 2003, Chalmers University of Technology, Volvo Car Corporation, 2003.
195. JORDAN M (1999): Drive-by-Wire will End the Era of the Handbrake Turn. *Electronic Engineering*. Vol. 71, No. 875, 1999, pp. 28-30.
196. JORDAN TC AND MT SHAW (1989): Electrorheology. *IEEE Trans. on Electrical Insulation,* Vol. 24, No. 5, 1989, pp.849-878.
197. JOZÍF M (2000): QUADRASTEER. *Automobil*, No. 10, 2000.
198. JUNKER HK (1991): Electronically Enhanced Steering Systems. *Proc. 24th ISATA: International Symposium on Automotive Technology & Automation, Dedicated Conference on Mechatronics,* Florence, Italy, 1991, Paper No. 910848, pp. 285-391.
199. JURGEN RK -- ED. (1995): *Automotive Electronics Handbook,* McGraw -Hill, Inc., New York, London, Tokyo 1995.
200. KARR J (1988): Mazda 626 4WS: It Won't Steer You Wrong. *Motor Trend*, Vol. 40, No. 9, September 1988, pp. 58-62.
201. KASSELMANN J AND T KERANEN (1969): Adaptive steering. *Bendix Technical Journal*, Vol. 2, 1969, pp. 26-35.
202. KATAJIMA K (2000): *H_∞ Control for Integrated Side-Slip, Roll and Yaw Controls for Ground Vehicles*. Master Thesis, University of Michigan, Ann Arbor, MI, 2000.
203. KATAJIMA K AND H PENG (2000): H_∞ Control for Integrated Side-Slip, Roll and Yaw Controls for Ground Vehicles. *Proceedings of AVEC 2000 – 5th International Symposium on Advanced Vehicle Control*, Ann Arbor, Michigan, August 22-24, 2000.
204. KATSUYAMA E AND N FUKUSHIMA (2000): Improvement of turning behavior using yaw moment feedback control. *Proceedings of the JSAE*, 2000, No. 5, 2000, Paper No. 20005171.
205. KAWAI T, Y SHIBAHATA, Y SHIMIZU, F KOHNO AND S SANO (2001A): Variable Gear Ratio Steering System. *2. PKW-Lenksysteme-Vorbereitung auf die Technik von Morgen*, Essen, Germany, 2001.
206. KAWAI T ET AL. (2001B): Improvement in driver-vehicle system performance by VGS. *2. PKW-Lenksysteme-Vorbereitung auf die Technik von Morgen*, Essen, Germany, 2001.
207. KELLY A AND A STENTZ (1998): Rough Terrain Autonomous Mobility -- Part 1: A Theoretical Analysis of Requirements. *Autonomous Robots*, No. 5, May 1998, pp. 129-161.
208. KIFUKU T AND S WADA (1997): An Electric Power-Steering System. Technical Report, Automotive Electronics Edition, *Mitsubishi Electric ADVANCE*, Vol. 78, March 1997, pp. 20-23.

209. KIM H (1996): An On-Line Learning Control of Unsupervised Neural Network for a Vehicle Four Wheel Steering System. *SAE Technical Paper Series*, 1996. Paper No. 960938, pp.1191~1457.
210. KIM J-H AND J-B SONG (2002): Control logic for an electric power steering system using assist motor. *Mechatronics*, Vol. 12, 2002, pp. 447-459.
211. KIMBROUGH S (1990): A Brake Control Strategy for Emergency Stops that Involves Steering, Part 1: Theory, Part 2: Implementation Issues and Simulation Results. *ASME Winter Annual Meeting*, AMD, Vol. 108, 1990.
212. KITAJIMA KAND P HUEI (2000): H_∞ control for integrated side-slip, roll and yaw controls for ground vehicles. *Proceedings of the International Symposium on Advanced Vehicle Control AVEC'00*, Ann Arbor, Michigan, August 2000.
213. KLEINE S AND JL VAN NIEKERK (1998): Modeling and Control of a Steer-By-Wire Vehicle. *Vehicle System Dynamics Supplement*, Vol. 29, 1998, pp. 114-142.
214. KLECZKOWSKI A (1980): Teoria skrętu tylnych kół. (Theory of Rear Wheels Turning), *AUTO - Technika Motoryzacyjna*, II-VIII, No 6, 1980 (In Polish).
215. KOHNO T, S TAKEUCHI, M MOMIYAMA, H NIMURA, E ONO AND S ASAI (2000): Development of Electric Power Steering (PS) System with H_Infinity Control. *SAE Technical Paper Series*, 2000, Paper No. 2000-01-0813.
216. KOJO T, M SUZUMURA, K FUKUI, T SUGAWARA, M MATSUDA AND JKAWAMURO (2002): Development of front steering control system. *Proceedings of the International Symposium on Advanced Vehicle Control AVEC'02*, (Hiroshima), 2002, pp. 33–38.
217. KONOPIŃSKI M (1987): *Elektronika w technice motoryzacyjnej*, WKŁ, 1987 (In Polish).
218. KOUMBOULIS FN AND MG SKARPETIS (2000): Robust Triangular Decoupling with Application in 4WS Cars. *IEEE Transactions on Automatic Control*, Vol. 45, No. 2, February 2000, pp. 344-352.
219. KRISHNASWAMI V AND G TOZZONI (1995): Vehicle steering system state estimation using sliding mode observers. *Proceedings of the 24th Conference on Decision and Control*, New Orleans, LA, December 1995, p. 3391.
220. KURISHIGE M, K FUKUSUMI, N INOUE, T KIFUKU AND S OTAGAKI (2001): A New Electric Current Control Strategy for EPS Motors. *SAE Technical Paper Series*, 2001, Paper No. 2001-01-0484.
221. KUSHIRO I, S KAWAKAMI AND R SAITOU (1992): Crosswind feedforward control by active front steering. *Proc. International Symposium on Advanced Vehicle Control*, Yokohama, Japan, 1992.
222. KWAK B (2001): *Robust Controller Design for Vehicle Stability Control System*. Ph.D. Dissertation, KAIST, 2001.

223. LAKEHAL-AYAT M, S DIOP, E FENAUX, F LAMNABHI-LAGARRIGUE, F ZARKA (2000): On global chassis control: combined braking and cornering, and yaw rate control. *Proceedings of AVEC 2000*, Ann Arbor, Michigan, USA, 2000.
224. LAKKAD S (2003): *Modeling and Simulation of Steering Systems for Autonomous Vehicles*. M.Sc. Thessis, College of Engineering, The Florida State University, Defended on September 12, 2003. Degree Awarded 2004, pp. i-xii; 1-69.
225. LANDREAU T (1989): Simulation of Dynamic Behavior of a Four Wheel Steering Vehicle by Means of a Vehicle and Driver Model *SAE Technical Paper Series*, Paper No. 890078, pp. 57-62.
226. LANGENWALTER J AND T ERKKINEN (2005): Model-Based Design with Production Code Generation for Steer-by-Wire System Development. *Newsletters - Automotive Digest*, April 2005. Available online at http://www.Mathworks.com/company/newsletters/News-letters-AutomotiveDigest/mbd.html
227. LANGHEIM J AND J FETZ (1992): Driving Behaviour of a Vehicle with Two Induction Motors for the Rear Wheels. *Proceedings of the 11th International Electric Vehicle Symposium*, Florence, Italy, September 27-30, 1992, Vol. I, Session 1-9, Paper No. 5.10, pp. 1-11.
228. LAW A (2003): *Quadraster v.2.0*. PickupTruck.com. August 5, 2003. Available online at: http://www.pickuptruck.com/html/stories/qs420/page1.html
229. LEBLANC DJ, P BENHOVENS, CF LIN, T PILUTTI, R ERVIN, AG ULSOY, C MACADAM AND G JOHNSON (1995): A Warning and Intervention System for Preventing Road Departure Accidents. The Dynamics of Vehicles on Roads and Tracks, *Proceedings of the 14th IAVSD Symposium*, Ann Arbor, MI, Vol. 25, 1995, pp. 383-396.
230. LEBLANC DJ. E JOHNSON, PJT VENHOVENS, G GERBER, R DESONIA, RD ERVIN, C-F LIN, AG ULSOY AND TE PILUTTI (1996): Road-Departure Prevention System. *IEEE Control Systems Magazine*, Vol. December 1996, pp. 61-71.
231. LEE AY (1990): Vehicle Stability Augmentation Systems Designs for Four Wheel Steering Vehicles. *ASME Journal of Dynamical Systems, Measurements and Control*, Vol. 112, No. 3, September 1990, pp. 489-495.
232. LEE AY AND AE BRYSON JR (1989): Neighboring Extremals of Dynamic Optimization Problems with Parameter Variations. *Optimal Control Applications and Methods*, Vol. 10, 1989, pp. 39-52.
233. LEE AY (1995): Performance of Four-Wheel-Steering Vehicles in Lane Change Maneuvers. *SAE Technical Paper Series*, 1995, Paper No. 950316. Available online at: http://www.itsdocs.fhwa.dot.gov/%5CJPODOCS%5CREPTS_TE/3RB01!.PDF.

234. LEE AY (1995): Emulating the lateral dynamics of a range of vehicles using a four-wheel-steering vehicle. *Proceedings of the International Congress and Exposition, (Detroit, Michigan), 1995, SAE Technical Paper Series,* 1995, Paper No. 950304.
235. LENDARIS GG, LJ SCHULTZ AND TT SHANNON (2000): Adaptive Critic Design for Intelligent Steering and Speed Control of a 2-Axle Vehicle. *Proceedings of IJCNN'2000*, Como, Italy, July 2000.
236. LEONI G AND D HEFFERMAN (2002): Expanding automotive electronic system. *IEEE Spectrum*, March 2002.
237. LEVINE M AND J GILLIES (2001): GMC Unveils Three Technologies That Will Change Your Pickup Truck Forever. *PickupTruck.com – Three Technologies*, August 7, 2001. Available online at http://www.pickup truck.com/html/stories/gmctech/three1.html.
238. LEWIS A AND CH WRIGHT (2004): Global Engineering: Myth or reality? *LOTUS pro-Active,* Issue 3, July/August 2004, Lotus Engineering, p. 10.
239. LIANG W, R RUHL AND J MEDANIC (2003): Simulation of Intelligent Convoy with Autonomous Articulated Commercial Vehicles. *SAE Technical Paper Series*, 2003, Paper No. 2003-01-3419.
240. LIANG W, J MEDANIC AND R RUHL (2004): Safety Concerns in Automatic Control of Heavy-Duty Articulated Vehicles. *SAE Technical Paper Series*, 2004, Paper No. 2004-01-2717.
241. LIN C, G ULSOY AND D LEBLANC (2000): Vehicle Dynamics and External Disturbance Estimation for Vehicle Path Prediction. *IEEE Transactions on Control Systems Technology*, Vol. 8, No. 3, 2000, pp. 508-518.
242. LIN Y (1992): Improving Vehicle Handling Performance by a Closed-Loop 4WS Driving Controller. *SAE Technical Paper Series*, 1992, Paper No. 921604, pp. 1447-1457.
243. LIU A AND S CHANG (1995): Force feedback in a stationary driving simulator. *Proceedings of the IEEE International Conference on Systems, Man and Cybernetics*, Vol. 2, 1995, Vancouver, BC, pp. 1711-1716.
244. LUGNER P AND M PLOCHL (1995): Additional 4WS and Driver Interaction. *Vehicle System Dynamics*, Vol. 24, 1995, pp.639~658.
245. LUKIC SM AND A EMADI (2003): Effects of Electrical Loads on 42V Automotive Power Systems. *SAE Technical Paper Series*, Paper No. 03FTT-28, 2003.
246. LYNCH DP (2000): *Velocity Scheduled Driver Assisted Control of a Four-Wheel Steer Vehicle.* M.Sc. Thesis, University of Illinois at Urbana-Champaign, Urbana, Illinois, 2000, 80 p.
247. MA W AND H PENG (1998): Worst-Case Vehicle Evaluation Methodology-Examples on Truck Rollover/Jackknifing and Active Yaw Control Systems. *Proceedings of AVEC'98*, 1998.
248. MACADAM CC (1981): Application of an Optimal Preview Control for Simulation of Closed-Loop Automobile Driving. *IEEE Transactions on Systems, Man and Cybernetics*, Vol. SMC-11, 1981, pp. 393-399.

249. MACADAM CC (1988): *Development of Driver-Vehicle Steering Interaction Models for Dynamic Analysis.* US Army Tank-Automotive Command RD&E Center Technical Report No. 13437, December 1988.
250. MACK J (1996): ABS-TCS-VDC – Where Will the Technology Lead Us? *SAE PT-57* 1996.
251. MAMMAR S (2000): Two-Degree-of-Freedom H_∞ Optimization and Scheduling for Robust Vehicle Lateral Control. *Vehicle System Dynamics,* Vol. 34, 2000, pp. 401-422.
252. MAMMAR S AND D KOENIG (2002): Vehicle Handling Improvement by Active Steering. *Vehicle System Dynamics*, Vol. 38, No. 3, 2002, pp. 211-242.
253. MATSUMOTO N, H KURAOKA, N OHKA, M OHBA AND T TABE (1987): Expert antiskid system. *Proc SPIE,* Vol 857, *IECON'87: Automated Design and Manufacturing*, Cambridge, Mass, USA, 1987, pp. 810-816.
254. MATSUMOTO N AND M TOMIZUKA (1992): Vehicle Lateral Velocity and Yaw Rate Control with Two Independent Control Inputs. *ASME Journal of Dynamic Systems, Measurement, and Control*, Vol. 114, 1962, pp. 606-613.
255. MAZDA (1989): *Introduction to Capella (New car)*, Mazda Motor Company, 89-5-NM4005, 1989.
256. MAZDA (2002): *4WS System – The Mazda Speed Sensitive Computerized 4-Wheel Steering System*. Mazda Motor Company, 2002. Available online at http://mazda.tetra. com.net/ 4ws.htm.
257. MILLIKEN WF, DELL'AMICO AND RS RICE (1976): The static directional stability and control of the automobile. *SAE Transactions*, 1976, Paper No. 760712, pp. 2216-2277.
258. MILLS V, J WAGNER AND D DAWSON (2001): Nonlinear Modeling and Analysis of Steering Systems for Hybrid Vehicles. *Proceedings of the ASME IMECE, Design Engineering Division,* New York, NY, November 2001.
259. MILLSAP S AND E LAW (1996): Handling Enhancement Due to an Automotive Variable Ratio Electric Power Steering System Using Model Reference Robust Tracking Control. *SAE Technical Paper Series*, 1996, Paper No. 960931.
260. MITSCHKE M (1990): *Dynamik der Kraftfahrzeuge*. Vol. C, Springer-Verlag, Berlin, 1990.
261. MOGHBELL H (1992): Electronically Assisted Steering. *SAE Automotive Suspension and Steering Systems*, Vol. 22, 1992, pp. 11-20.
262. MOKHIAMAR O AND M ABE (2002): Combined lateral force and yaw moment control to maximize stability as well as vehicle responsiveness during evasive maneuvering for active vehicle handling safety. *Vehicle System Dynamics Supplement*, Vol. 37, 2002, pp. 246–256.
263. MOURI H ET AL. (1997): Automatic Path Tracking Control Using Linear Quadratic Control Theory. *IEEE Conference on Intelligent Transportation Systems*, 1997.

264. MURRAY CHJ (2002): Delphi debuts drive-by-wire steering for pickups, SUVs, *Delphi News*, March 7, 2002, *EETimes.com*, March 9, 2000. Available online at http://www. eetimes.com/story/OEG20000309S0042?
265. MÜLLER A AND M PLANKENSTEINER (2002): Faul-Tolerant Components versus Fault-Tolerant Systems – Redundancy creates verifiable Fault-Tolerant from Non-Tolerant Components. *TTTech Computertechnik AG*, 2005, 7p. Available online at: http:/tttech. com/TTTech_2002-04-AUTO-REG-Fault-Tolerant Components.pdf .
266. MÜLLER-GLASER KD AND M KÜHL (2005): TTA Time Triggered Architectures for X-by-Wire Systems. *FZI Forschungszentrum Informatik*, 2005. Available online at http:// www.fzi.de/esm/kcms-file.pdf .
267. MYERS TT AND TJ ROSENTHAL (1993): Vehicle Stability Considerations with Automatic and Four Wheel Steering Systems, *SAE Technical Paper Series*, 1993, Paper No. 931979, pp.2191-2201.
268. NAGAI M AND M MITSCHKE (1985): Adaptive Behavior of Driver-Car Systems in Critical Situations: Analysis by Adaptive Model. *JSAE Review*, 1985.
269. NAGAI M (1989): Active Four-Wheel-Steering System by Model Following Control. *Proceedings of 11th IAVSD Symposium*, 1989, pp.428-439.
270. NAGAI M, E UEDA AND A MORAN (1995): Nonlinear Design Approach to Four-Wheel-Steering Systems Using Neural Networks. *Vehicle System Dynamics*, 1995, pp. 329-342.
271. NAGAI M, S YAMANAKA AND Y HIRANO (1996): Integrated control law of active rear wheel steering and direct yaw moment control. *Proceedings of the International Symposium on Advanced Control (AVEC)*, Aachen, Germany, 1996.
272. NAGAI M, Y HIRANO AND S YAMANAKA (1997): Integrated Control of Active Rear Steering and Direct Yaw Moment Control. *Vehicle System Dynamics,* Vol. 27, 1997, pp. 357-370.
273. NAGAI M, Y HIRANO AND S YAMANAKA (1998): Integrated Robust Control of Active Rear Wheel Steering and Direct Yaw Moment Control. *Vehicle System Dynamics Supplement*, Vol. 28, 1998, pp. 416-421.
274. NAKAYAMA T AND E SUDA (1994): The present and future of electric power steering. *International Journal of Vehicle Design*, Vol. 158, Nos. 3/4/5, 1994, pp. 243-254.
275. NALECZ AG AND AC BINDEMANN (1989): Handling Properties of Four Wheel Steering Vehicles. *SAE Technical Paper Series*, 1989, Paper No. 890080, pp. 63-82.
276. NEWTON K, W STEEDS AND TK GARRET (1987): *The Motor Vehicle*, Butterworths, London, 1987.
277. NICE K (2002): How Car Steering Works. *How Stuff Works*, April, 2002. Available online at http://www.howstuffworks.com/steering.htm .
278. NIEKERK M VAN (1992): *Electric Vehicles as Energy Processing and Conversion Systems.* Dissertation, Rand Afrikaans University 1992.

279. NIETHAMMER M (2000): *Reglerentwurf für ein fahrzeug mit lenkbarer vorder- und hinterachse*. Praktikum Bereich, DaimlerChrysler, Esslingen, Germany, 2000.
280. NISSAN (1991A): *Instruction to Laurel (New car)*. Nissan Motor Company, F005705, 1989.
281. NISSAN (1991B): *Guidebook*. Nissan Ciema, 1991; (In Japanese).
282. NWAGBOSO CHO ED. (1993): *Road Vehicle Automation*. PENTECH PRESS Publishers, London, 1993, 309 pp.
283. NWAGBOSO CHO ED. (1997): *Advanced Vehicle and Infrastructure Systems: Computer Applications, Control, and Automation*. John Wiley, New York, 1997, 502 p.
284. NYENHUIS M, C RUSTEMEIER AND U DIERKES (2002): Verteilter strukturierter Entwurf mechatronischer Systeme am Beispiel einer Steer-by-Wire-Lenkung. Fahrzeugschwingungen - Global Chassis Control, Hausder Technik, Essen, Germany, 2002.
285. OHTA T, T MIMURO AND J LEE (2001): *Robust Lateral Control System with Steering Torque Assist*. Advanced Electrical/Electronics Department, Car Research & Development Office, Mitsubishi Motors Corporation Cartoronics R&D Center, Association of Traffic and Safety Sciences, Vol. 26, No. 2, 2001.
286. ONO E, S HOSOE, H TUAN AND S DOI (1998): Bifurcation in Vehicle Dynamics and Robust Front Wheel Steering Control. *IEEE Transactions on Control Systems Technology*, Vol. 6, No. 3, 1998, pp. 412-420.
287. ÖTTGEN O AND T BERTRAM (2001): Beeinflussung des eigen lenkverhaltens eines PKW durch eine aktive wankmomenten verteilung, VDI-Berichte, No. 1631, 2001.
288. PACEJKA HB (1986): Lateral stability of road vehicles. *ICTS, Proceedings of the Third Course on Advanced Vehicle Dynamics,* 1986 Amalfi, pp. 75-120.
289. PACEJKA H, E BAKKER AND L NYBORG (1987): Tyre modelling for use in vehicle dynamics studies. *SAE Technical Paper Series*, 1987, Paper No. 870421.
290. PALKOVICS L (1992): Effect of the Controller Parameters on the Steerability of the Four Wheel Steered Car. *Vehicle System Dynamics*, Vol. 21, 1992, pp. 109-128.
291. PALKOVICS L AND M EL-GINDY (1995): Design of an Active Unilateral Brake Control System for Five-Axle Tractor-Semitrailer Based on Sensitivity Analysis. *Vehicle System Dynamics*, Vol. 24, 1995, pp. 725-758.
292. PALKOVICS L, G KOVÁCS, L GIANONE, J BOKOR, P SZÉLL AND Á SEMSEY (1999): Vision System for Avoidance of Lane-departure. *Vehicle System Dynamics*, Vol. 33, 1999, pp. 282-292.
293. PARK K ET AL. (2002): A Study for Improving Vehicle Dynamic Property using Hardware-in-the-Loop Simulation. *Proceedings of AVEC 2002*, Hiroshima, Japan, 2002.

294. PASCALI L, P GABRIELLI AND G CAVIASSO (2003): Improving Vehicle Handling and Comfort Performance Using 4WS. *SAE Technical Paper Series*, 2003, Paper No. 2003-01-0961.
295. PATWARDHAN S, H-S TAN AND J GULDNER (1997): A General Framework for Automatic Steering Control System Analysis. *Proceedings of the American Control Conference*, Albuquerque, NM, 1997.
296. PENG H AND M TOMIZUKA (1990): *Lateral Control of Front-Wheel-Steering Rubber-Tire Vehicles*. California PATH Report UCS-ITS-PRR-90-5.
297. PENG H AND M TOMIZUKA (1993): Preview Control for Vehicle Lateral Guidance in High-way Automation. *ASME Journal of Dynamic Systems, Measurement and Control*, Vol. 115, 1993, pp. 679-686.
298. PENG H, W ZHANG, M TOMIZUKA AND S SHLADOVER (1994): A Reusability Study of Vehicle Lateral Control System. *Vehicle System Dynamics*, Vol. 28, 1994, pp. 259-278.
299. PENG H AND J-S HU (1996): Traction/Braking Force Distribution for Optimal Longitudinal Motion during Curve Following. *Vehicle System Dynamics*, Vol. 26, No. 4, October, 1996, pp. 301-320.
300. PETERSEN UN, A RUKGAUER AND WO SCHIEHLEN (1996): Lateral Control of a Convoy Vehicle System. *Vehicle System Dynamics*, Vol. 25, 1996, pp. 519-532.
301. PILUTTI T, G ULSOY AND D HROVAT (1995): Vehicle Steering Intervention Through Differential Braking. *Proceedings of the American Control Conference*, Seattle, WA, 1995.
302. PIONEER (2005): Installation Manual. PIONEER -- DEH – P7000R-W.
303. POST J AND E LAW (1996): Modeling, Characterization and Simulation of Automobile Power Steering Systems for Prediction of On-Center Handling. *SAE Technical Paper Series*, 1996, Paper No. 960178.
304. POTTINGER MG, W PELZ, G FALCIOLA (1998): Effectiveness of the Slip Circle, “Combinator”, Model for Combined Tire Cornering and Braking Forces When Applied to a Range of Tires. *SAE Technical Paper Series*, 1998, Paper No. 982747.
305. QIU H, Q ZHANG, JF REID AND D WU (1999): Modeling and Simulation of an Electro-hydraulic steering system. *ASAE Meeting Presentation UILU-ENG-99-7019*, 1999, Paper No. 993076.
306. RAKSINCHAROENSAK P, H MOURI AND M NAGAI (2003): Vehicle Lane-Keeping Control by Four-Wheel-Steering System. Tokyo University of Agriculture and Technology & Vehicle Research Laboratory, Nissan Motor Co., Ltd. E-mail: pong@cc.tuat.ac.jp .
307. RAMANATA P AND M AHMADIAN (1998): *Optimal Vehicle path Generator using optimization methods.* Thesis, Department of Mechanical Engineering, Virginia Polytechnic Institute, April 1998.

308. REICHELT W (1991): Correlation Analysis of Open/Closed Loop Data for Objective Assessment of Handling Characteristics of Cars. *SAE Journal ofPassenger Cars*, Vol. 100, Section 6, 1991, Paper No. 910238, pp. 375-384.
309. REMOND B (2004): Dual steering for wheeled vehicle. *Proc. 4th International AECV Conference*, 2004, Paper 21, p. 6.
310. RILL G (1994): Simulation von Kraftfahrzeugen. Vieweg, 1994.
311. ROLLINS ET AL. (1998): Nomad: A Demonstration of the Transforming Chassis. *Proceedings of ICRA 98*, Leuven, Belgium, May 1998.
312. ROOS F AND J WIKANDER (2005): *Mechatronic design and optimisation methodology* – A problem Formulation focused on automotive mechatronics modules. KTH, Department of Machine Design, Mechatronics Lab, Stockholm, Sweden, 2005, pp. 1-6.
313. ROOS G (1990): *Vierwielbesturing: bepaling van een optimale strategie voor de besturing van de achterwielen van personenwagens (Four wheel steering: determination of an optimal strategy for steering of the rear wheels of a passenger car)*. Eindhoven: TU Eindhoven, 1990 (In Dutch).
314. RYU J AND H KIM (1999): Virtual environment for developing electronic power steering & steer-by-wire systems. *Proceedings of the International American Conference on Intelligent Robots and Systems*, IEEE, 1999, p. 1374.
315. RYU J, E ROSSETTER AND JC GERDES (2002): Vehicle sideslips and roll parameter estimation using GPS. *Proceedings of the International Symposium on Advanced Vehicle Control (AVEC),* Tokyo, Japan, 2002.
316. SAE (2004): SKF by-wire technology for Bertone concept car. *Automotive Engineering International – Online*, Focus on Electronics, SAE International, 2004.
317. SAKAI S, H SADO AND Y HORI (1999): Motion Control in an Electric Vehicle with 4 Independently Driven In-Wheel Motors. *IEEE Trans. on Mechatronics*, Vol. 4, No. 1, 1999.3, pp. 9-16.
318. SANO S, Y FURUKAWA AND S SHIRASHI (1986): Four Wheel Steering System with Rear Wheel Steer Angle Controlled as a Function of Steering Wheel Angle. *SAE Technical Paper Series*, 1986, Paper No. 860625.
319. SATO M (1991): Chapter 8 – Steering in the book: *Handbook of Society of Automotive Engineers of Japan.* Vol. Design, Society of Automotive Engineers of Japan, Inc., Tokyo, 1991.
320. SATO M (1995): Steering Control, pp.18.1-18.33. Chapter 18 in the book: *Automotive Electronics Handbook* (Ronald K. Jurgen, Ed.), McGraw-Hill, Inc., New York / London / Tokyo, 1995.
321. SCHMIDT CH (2004): Das Brenstoffzellen-Auto auf dem Weg zur Groß-serienreife. Fuel Cell Powersystem & Propulsion, GM OPEL Presentation, 2004, Slide 30.

322. SCHULTZ LJ, TT SHANNON AND GG LENDARIS (2001): Using DHP Adaptive Critic Methods to Tune a Fuzzy Automobile Steering Controller. *IFSA/NAFIPS Conference*, Vancouver, BC, July, 2001 (manuscript date: March 30, 2001).
323. SCHURING D, W PELZ AND M POTTINGER (1996): A Model for Combined Tire Cornering and Braking Forces. *SAE Technical Paper Series*, 1996, Paper No. 960180.
324. SEEWALD A AND D CHEW (2004): Integrated Vehicle Control Systems (IVCS), Integration of Steering and Braking. *Proceedings of the 2004 SAE Automotive Dynamics, Stability and Controls Conference*, Detroit, May 4-6, 2004.
325. SEGAWA M, K NISHIZAKI AND S NAKANO (2002): A study of vehicle stability control by steer by wire system. *Proceedings of the International Symposium on Advanced Vehicle Control (AVEC)*, Ann Arbor, MI, 2002.
326. SEGEL L (1957): Theoretical prediction and experimental substantiation of the response of the automobile to steering control. *Proceedings of the Automobile Division of ImechE*, Vol. 7, 1957, pp. 310-330.
327. SEGEL L (1966): On the Lateral Stability and Control of the Automobile as Influenced by the Dynamics of the Steering System. *Journal of Engineering for Industry*, August 1966, pp. 283-295.
328. SEGEL L (1982): Basic Linear Theory of Handling Stability of Automobiles. *ICTS, Proceedings of the First Course on Advanced Dynamics*, 1982 Amalfi, pp. 19-87.
329. SEIFFERT U AND P WALZER (1991): *Automotive Technology of the Future*. Society of Automotive Engineers, Inc., Warrendale, PA, 1991, 251 pp.
330. SENEFSKY B (2003): All-Wheel Steering – Control Under Pressure. *Truckin'*. World's Leading Truck Publication, 2003. Available online at http://www. todayssuv.com/tech/ 0308suv_wheelsteer/.
331. SETLUR P, J WAGNER, D DAWSON AND L POWERS (2003): A Hardware-in -the-Loop and Virtual Reality Test Environment for Steer-by-Wire System Evaluations. *Proceedings of American Control Conference*, Denver, CO, June 2003.
332. SHAMAH B (1999): *Experimental Comparison of Explicit Vs. Skid Steering for a Wheeled Mobile Robot*. M.S. Thesis, Technical Report CMU-RI-TR-99-06, Robotics Institute Carnegie Mellon University, Pittsburgh, PA, March 1999.
333. SHAMAH B, D APOSTOLOPOULUS AND M WAGNER (1999): Effect of Tire Design and Steering Mode on Robotic Mobility in Barren Terrain. *Proceedins of the International Conference on Field and Service Robots*, Wiliam "Red" Whittaker Field Robotics Center, The Robotics Institute Carnegie Mellon University, August 1999, pp. 287-292.
334. SHAMAH B, MD. WAGNER, S MOOREHEAD, J TEZA AND D WETTERGREEN (2001): Steering and Control of a Passively Articulated Robot. *Sensor Fusion and Decentralized Control in Robotic Systems IV*, Vol. 4571, Field

Robotics Center, The Robotics Institute, Carnegie Mellon University, Pittsburgh, PA, October 2001.

335. SHAO JQ AND E PLUMER (1994): *On-line adaptive control of rear wheel steering.* Los Alamos National Laboratory (LANL) Controlled Report, LA-CP-94-173, July 1994. Available online at http://www.isr.umd.edu/~jshao/resume/resume.html.
336. SHAO JQ, R TOKAR, G WRIGHT AND RD JONES (1993): *On-line adaptive four wheel steering system using CNLS neural net controller: preliminary results.* LANL Unclassified Report, LAUR-93-3382, September 1993. Available online at http://www.isr. umd.edu/ ~jshao/resume/resume.html .
337. SHIBAHATA Y, N IRIE, H ITOH K NAKAMURA (1986): The development of an experimental four-wheel-steering vehicle. *SAE Technical Paper Series*, 1986, Paper No. 860623, 1986.
338. SHIBAHATA Y, K SHIMADA AND T TOMARI (1992): The improvement of vehicle maneuverability by direct yaw moment control. *AVEC'92*, 1992, Paper No. 923081.
339. SHIBAHATA Y, K SHIMADA AND T TOMARI (1993): Improvement of Vehicle Maneuverability by Direct Yaw Moment Control. *Vehicle System Dynamics*, Vol. 22, 1993, pp. 465-481.
340. SHIBAHATA Y (2005): Progress and future direction of Chassis control technology. *Annual Reviews in Control*, Vol. 29, 2005, pp. 151–158.
341. SHILLER Z, W SERATE AND M HUA (1993): Trajectory planning of tracked vehicles. *1993 IEEE Int. Conf. On Robotics and Automation*, Atlanta, GA, 1993, pp. 796-801.
342. SHILLER Z AND S SUNDAR (1998): Emergency Lane-Change Maneuvers of Autonomous Vehicles. *ASME Journal of Dynamic Systems, Measurement, and Control*, Vol. 120, 1998, pp. 37-44.
343. SHIMADA K AND Y SHIBATA (1994): Comparison of Three Active Chassis Control Methods for Stabilizing Yaw Moments. *SAE Technical Paper Series*, 1994, Paper No. 940870.
344. SHIMIZU Y ET AL. (1991): Development of electric power steering. *SAE Technical Paper Series*, 1991, Paper 910014.
345. SHIMIZU Y ET AL. (1991): Control of electric power steering (EPS). *Proceedings of the 68th JSM Spring Annual Meeting*, Vol. C, 1991, pp. 1605.
346. SHIMOMURA H, T HARAGUCHI, Y SATOH AND R SAITOH (1991): Simulation analysis on the influence of vehicle specifications upon steering characteristics. *International Journal of Vehicle Design*, Vol. 42, No. 2, 1991.
347. SHINO M, P RAKSINCHAROENSAK AND M NAGAI (2002): Vehicle Handling and Stability Control by Integrated Control of Direct Yaw Moment and Active Steering. *Proceedings of AVEC 2002*, Hiroshima, Japan, 2002.
348. SHIOTSUKA T, A NAGAMATSU AND K YOSHIDA (1993): Adaptive Control of 4WS System by Using Neural Network, *Vehicle System Dynamics*, 1993, pp. 411-424.

349. SHLADOVER SE (1995): Review of the State of Development of Advanced Vehicle Control Systems (AVCS). *Vehicle System Dynamics*, Vol. 24, 1995, pp. 551-595.
350. SMITH DE AND JM STARKEY (1994): Effects of Model Complexity on the Performance of Automated Vehicle Steering Controllers: Controller Development and Evaluation. *Vehicle System Dynamics*, 1994, pp. 627-645.
351. SMITH DE AND JM STARKEY (1995): Effects of Model Complexity on the Performance of Automated Vehicle Steering Controllers: Model Development, Validation and Comparison. *Vehicle System Dynamics*, Vol. 24, 1995, pp. 163-181.
352. SMITH DE AND RE BENTON (1996): Automated Emergency Four-Wheel-Steered Vehicle Using Continuous Gain Equations. Vehicle System Dynamics, Vol. 26, 1996, pp. 127-142.
353. SMITH BW (2002): Right on Truck – GM Quadrasteer makes perfect sense for Chery Silverado 4 × 4. *TRUCKWORLD ONLINE*, December 11, 2002. Available online at http://truckworld.tenmagazines.com/404.asp?404; http://www.truckworld.com/copy-right.html
354. SRIDHAR J AND H HATWAL (1992): A Comparative Study of Four Wheel Steering Models Using the Inverse Solution. *Vehicle System Dynamics*, Vol. 22, 1992, pp. 1-17.
355. SZOSLAND A (2000): Fuzzy Logic Approach to Four-Wheel Steering of Motor Vehicle. *International Journal. of Vehicle Design*, Vol. 24, No. 4, 2000, pp.350~359.
356. SZOSTAK HT (1998): *Analytical Modeling of Driver Response in Crash Avoidance Maneuvering* -- Vol. II: An Interactive Tire Model for Driver/Vehicle Simulation. DOT HS 807 271, 1988.
357. TAGAWA T, H OGATA, K MORITA, M NAGAI AND H MORI (1996A): Robust Active Steering System Taking Account of Nonlinear Dynamics. *Vehicle System Dynamics Supplement*, Vol. 25, 1996, pp. 668-681.
358. TAGAWA T ET AL. (1996B): A Robust Active Front Wheel Steering System Considering the Limits of Vehicle Lateral Force. *Proceedings of ACEC'96*, 1996.
359. TAHERI S AND EH LAW (1990): Investigation of a Combined Slip Control Braking and Closed-Loop Four Wheel Steering System for an Automobile During Combined Hard Braking and Severe Steering. *American Control Conference Proceedings*, San Diego, CA, 1990.
360. TAI M, J WANG, R WHITE AND M TOMIZUKA (2001): *Robust Lateral Control of Heavy Duty Vehicles*. California PATH Research Report UCB-ITS-PRR-2001-35.
361. TAI M AND M TOMIZUKA (2003): *Robust Lateral Control of Heavy Duty Vehicles*: Final Report. California PATH Research Report UCB-ITS-PRR-2003-24.

362. TAKAHASHI T ET AL. (2000): The Modeling of Tire Force Characteristics of Passenger and Commercial Vehicles on Various Road Surfaces. *Proceedings of AVEC*, 2000, pp. 785-792.
363. TAKIGUCHI T, N YASUDA, S FURUTANI, H KANAZAWA AND H INOUE (1986): Improvement of Vehicle Dynamics by Vehicle-Speed-Sensing Four-Wheel Steering System. *SAE Technical Paper Series*, 1986, Paper No. 860624.
364. THIESEN LP (2003): Entwicklung von Brennstoffzellen-Fahrzeugen bei Opel/GM, *GM Presentation*, 2003, ss. 1-20.
365. TOMIZUKA M AND KJ HEDRICK (1995): Advanced Control Methods for Automotive Applications. *Vehicle System Dynamics*, Vol. 24, 1995, pp. 449-468.
366. TOYOTA (1989): *Toyota Technical Review*, Vol. 41, No. 1, 1989.
367. TRÄCHTLER A (2004): Integrated vehicle dynamics control using active brake, steering and suspension systems. *International Journal of Vehicle Design,* Vol. 36, No. 1, 2004.
368. TRENCSÉNYI B, P KOLESZÁR, I WAHL AND L PALKOVICS (2003): Enhanced Vehicle Stability of a Truck Using ESP COMBINED with a Steer-by-wire System. *Proceedings of the 18th International Symposium*, Dynamics of Vehicle Vehicles on Roads and Tracks, Kanagawa Institute of Technology, 2003.
369. TREVETT NR (2002): *X-by-wire, new technologies for 42 V bus automobile of the future*. Master Thesis, April 2002.
370. TRIDEC (2003): *DRIVER AND MAINTENANCE OPERATIONS – HF steering system*. TRIDEC® Transport Industry Development Centre BV, The Netherlands, 2003.
371. TRW (2003): *TRW Drives Steer-by-Wire*. Frankfurt, September 17, 2001. Downloaded from Asia Pacific Motor e-news http://www.apmotornews.com .
372. TSUGAWA S (1998): Control Algorithms for Automated Driving Systems. *Journal of JSAE*, Vol. 52, No. 2, 1998, pp. 28-33 (In Japanese).
373. UEKI N, J KUBO, T TAKAYAMA, I KANARI AND M UCHIYAMA (2004): Vehicle Dynamics Electric Control Systems for Safe Driving. *Hitachi Review*, Vol. 53, No. 4, 2004, pp. 222-226.
374. UNYELIOGLU KA, C HATIPOGLU AND U OZGUNER (1997): Design and Stability Analysis of a Lane Following Controller. *IEEE Transactions on Control Systems Technology*, Vol. 5, 1997, pp. 127-134.
375. UTKIN V, J GULDNER AND J ACKERMANN (1994): A sliding mode control approach to automatic car steering. *Proceedings of the American Control Conference*, Baltimore, Maryland, June 1994, pp. 1969-1973.
376. VELENIS E, P TSIOTRAS, C CANUDAS DE WIT (2002): Extension of the LuGre Dynamic Tire Friction Model to 2D Motion. *Proceedings, 10th Mediterranean Conference on Control and Automation (MED2002),* Lisbon, Portugal, July 9-12, 2002.

377. VILAPLANA M (2004): *Multivariable control of yaw rate and sideslip in vehicles equipped with 4-wheel steer-by-wire: Controller designed at the Hamilton Institute*. Report for Daimler-Chrysler, 2004.
378. VILAPLANA M, D LEITH, W LEITHEAD AND J KALKKUHL (2004): Control of sideslip and yaw rate in cars equipped with 4-wheel steer-by-wire. *Proceedings of the 2004 SAE Automotive Dynamics, Stability and Controls Conference, SAE Technical Paper Series*, 2004, Paper No. 2004-01-2076, pp. 143 –153, 2004.
379. VILLEGAS C AND R SHORTEN (2005): Complex Embedded Automotive Control Systems (CEMACS), *Public State of the Art of Integrated Chassis Control*, Deliverable D2, STREP project 004175 CEMACS, DaimlerChrysler, SINTEF, Glasgow University, Hamilton Institute, Lund University, February 2005, pp. 1-31.
380. VLK F (2000A): *Podvozky motorových vozidel*, Nakladatelství a vydavatelství VLK, Brno, 2000 (In Czech).
381. VLK F (2000B): *Koncepce motorových vozidel*, Nakladatelství a vydavatelství VLK, Brno, 2000 (In Czech).
382. VLK F (2001): *Dynamika motorových vozidel*. Nakladatelství a zasilatelství VLK, Brno, 2001 (In Czech).
383. WADA S ET AL. (1992): Electric power steering system. *Mitsubishi Denki Giho*, Vol. 66, No. 9, 1992.
384. WANG D AND F QI (2001): Trajectory Planning for a Four-Wheel-Steering Vehicle. *Proceedings of the 2001 IEEE International Conference on Robotics & Automation*, Seoul, Korea, May 21-26, 2001, pp. 3320-3325.
385. WANG J-Y AND M TOMIZUKA (1999): Robust H_∞ Lateral Control of Heavy-Duty Vehicles in Automated Highway System. *Proceedings of the American Control Conference*. San Diego, California, June 1999.
386. WANG L AND J ACKERMANN (1998): Robustly stabilizing PID controllers for car steering systems. *Proceedings of the American Control Conference*, Philadelphia, Pennsylvania, USA, June 1998, pp. 41-42.
387. WANG YQ (1995): *Intelligent Vehicle Control for Improving Handling Stability and Active Safety*. Ph.D. Thesis, Department of Mechanical Systems Engineering, Tokyo University of Agriculture and Technology, 1995.
388. WATANABE K, J YAMAKWA, T INAGAKI AND M KITANO (2000): Turning characteristics of articulated tracked vehicles. *Proc. 8th European Conference of ISTVS*, Umeå, Sweden, June 18-22, 2000, pp. 128-135.
389. WHITEHEAD JC (1988): Four Wheel Steering: Maneuverability and High Speed Stabilization. *SAE Technical Paper Series*, 1988, Paper No. 880642, pp. 4668-4679.
390. WILL AB AND SH ZAK (1998): Modeling and Control of an Automated Vehicle. *Vehicle System Dynamics*, Vol. 27, 1998, pp. 131-155.
391. WILLIAMS D AND W. HADDAD (1995): Nonlinear control of roll moment distribution to influence vehicle yaw characteristics. *IEEE Transactions on Control Systems Technology*, Vol. 3, March 1995.

392. WILWERT C, T CLÉMENT, Y-Q SONG AND F SIMONOT-LION (2003): Evaluating Quality of Service and Behavioral Reliability of Steer-by-Wire Systems. *LORIA-TRIO & PSA Peugeot Citroën*, 2003.
393. WILWERT C, F SIMONOT-LION, Y-Q SONG AND F SIMONOT (2005): Safety evaluation of in-car real-time applications distributed on TDMA-based networks. *3rd Nancy-Saarbrücken Workshop on Logic, Proofs and Programs*, Nancy, France, October 13-14, 2005.
394. WONG JY (1986): *Terramechanics and Its Applications*. Vol. I and II, Lectures delivered at CEMOTER, IMTCNR, Italy, 1986.
395. WONG JY (1993): *Theory of Ground Vehicles*, 2nd ed., Wiley, New York, 1993.
396. WONG T (2001): Hydraulic Power Steering System Design and Optimization Simulation. *SAE Technical Paper Series*, 2001, Paper No. 2001-01-0479.
397. WU DH AND J HAI (2003): Analysis of dynamic lateral response for a multi-axle-steering tractor and trailer. *International Journal of Heavy Vehicle System, Vol. 10, No.4, 2003, pp.281-294.* Available online at http://www.Inderscience.com/search/index.php?action=record&rec_id=3694&prevQuery=&ps=10&m=or.
398. XIA X AND EH LAW (1992): Nonlinear Analysis of Closed Loop Driver/Automobile Performance with Four Wheel Steering Control. *SAE Technical Paper Series*, 1992. Paper No. 920055, pp. 77-92.
399. YAMAMOTO M, H HARADA AND Y MATSUO (1989): A Study on Active Controlled Chassis System for Vehicle Dynamics. Vehicle System Dynamics, Vol.18, 1989, pp. 603-615.
400. YAMAMOTO, M ET AL. (1989): Improvement of Steering and Disturbance Response by Active controlled Rear Wheel Steer (Study on Active Control for Vehicle Dynamics: 1st Report), *Proceedings of the JSAE Annual Conference,* Vol. 892, 1989, Paper No. 892128.
401. YAMAMOTO M (1991): Active Control Strategies for Improved Handling and Stability. *JSAE Vehicle Dynamics and Control Symposium*, Tokyo, Japan, 1991.
402. YAMAMOTO M, M IWAMURA AND A MOHAI (1998): Time optimal motion planning of skid-steer mobile robots in the presence of obstacles. *IEEE Conference ob Robotics and Automation*, Victoria, BC, Canada, 1998.
403. YAMAKADA M AND Y KADOMUKAI (1994): A jerk sensor and its application to vehicle motion control system. *Proceedings for the Dedicated Conferences on Mechatronics & Supercomputing Applications in the Transportation Industries in conjunction with the 27th ISATA,* Aachen, Germany, 1994.
404. YIH P AND JCH GERDES (2003): Modification of Vehicle Handling Characteristics via Steer-by-Wire. *American Control Conference*, Denver, CO, 2003.

405. YOKOYA Y, R KIZU, H KAWAGUCHI, K OHASHI AND K OHNO (1990): Integrated control system between active control suspension and four-wheel steering for the 1989 CELICA. *SAE Technical Paper Series*, 1990, Paper No. 901748, pp. 87-102.
406. YOSHIMURA T, H UCHIDA, H NASU, J HINO AND R UENO (1997): Traction force control of an electric vehicle in 2WS-4WD mode using fuzzy reasoning. *International Journal of Vehicle Design*, Vol. 18, No. 5, 1997, pp. 442-454.
407. YU SH AND JJ MOSKWA (1994): A global approach to Vehicle Control: Coordination of Four Wheel Steering and Wheel Torques. *ASME Journal of Dynamic System, Measurements and Control*, Vol. 116, 1994, pp. 659-667.
408. YUHARA N ET AL. (1992): Improvements of vehicle handling quality through active control of steering reaction torque. *Proc. International Symposium on Advanced Vehicle Control*, 1992, Paper No. 923073.
409. YUNHUA L, Z ZHIHUA AND W ZHALIN (2006): *Design and Control of the 4WD System of Automobile for Transporting Bridge*. Institute of Mechatronic Control, Beijing University of Aeronautics and Astronautics, Beijing, P.R. China, 2006, Available online at http://fluid.power.net/techbriefs/hanghzau/4_7.pdf .
410. ZANTEN A VAN, R ERHARDT, G PFAFF, G KOST, U HARTMANN, AND T EHRET (1996): Control aspects of the Bosch-VDC. *Proceedings of the International Symposium on Advanced Vehicle Control AVEC'96*, Aachen, 1996.
411. ZANTEN A VAN (2002): Evolution of electronic control systems for improving the vehicle dynamic behavior. *Proceedings of the International Symposium on Advanced Vehicle Control (AVEC)*, Tokyo, Japan, 2002.
412. ZAPLETAL F (1997): *Zvyšování aktivní bezpečnosti vozidel využitím systému 4WS a AFWS*, Autoreferát disertační práce, Brno, 1997 (In Czech).
413. ZAREMBA AT, MK LIUBAKKA AND RM STUNTZ (1998): Control and steering feel issues in the design of an electric power steering system. *Proceedings of the American Control Conference*, Philadelphia, Pennsylvania, USA, Vol. 1, 1998, p. 36.
414. ZETTERSTROM S (2002): Electromechanical steering, suspension, drive and brake modules. *IEEE 56th Vehicular Technology Conference*, Vol. 3, 2002, p. 1852.
415. ZOEST P VAN (1988): *BEoordeling van lane-change-manoeuvres van vrachtwagens (Evaluating lane-change manoeuvres by trucks). Delft*: TU Delft, 1988 (In Dutch).
416. ZUURBIER J, P BREMMER (2002): State Estimation for Integrated Vehicle Dynamics Control. *Proceedings of AVEC 2002*, Hiroshima, Japan, 2002.

PART 5

5 ABW AWA Suspension Mechatronic Control Systems

5.1 Introduction

In this part of the book, we discuss vertical motion in the *z*-axis of an automotive vehicle. By vertical motion of an automotive vehicle, we mean how the vehicle responds to on/off road surface input.

The response of the vehicle to an on/off road surface input is predominantly influenced by the **absorb-by-wire** (ABW) **all-wheel-absorbed** (AWA) suspension mechatronic control system. On the other hand, the (DBW) **all-wheel-driven** (AWD) propulsion, **brake-by-wire** (BBW) **all-wheel-braked** (AWB) dispulsion and **steer-by-wire** (SBW) **all-wheel-steered** (AWS) conversion mechatronic control systems may also be used to influence the absorbing (damping) capabilities of the vehicle; it is therefore not surprising that research on controlling the lateral motions of a vehicle has recently concentrated on integrating these systems into a **ride-by-wire** (RBW) or **x-by-wire** (XBW) integrated unibody, space-chassis, skateboard-chassis, or body-over-chassis motion mechatronic control hyper-system.

The ABW AWA suspension mechatronic control system is neither revolutionary nor an innovative concept. From the time of invention of the wheel , designers have endeavoured to soften the bumps and jolts of the on/off road surface by setting up a spring of some kind between the axle of the wheel and the vehicle body itself.

The *'springs or compliances'* store energy, and precautions must be taken to avoid bouncing by inserting *'shock absorbers or dampers'* that convert the spring's mechanical energy into thermal energy (heat) that can be dispelled. Vehicular suspension components have in sequence controlled the design of vehicle body structures -- and even of wheels and tyres [AMT 2005].

A vehicular (or automotive vehicle) suspension mechatronic control system is the system that links the wheels of the vehicle to the vehicle's body in a technique that isolates the vehicle's body from jerks originating from driving on uneven on/off road surfaces.

The body of a vehicle is referred to as a *'sprung mass'*. The automotive vehicle wheels and components related to them are jointly referred to as *'unsprung masses*.

Normally, automotive scientists and engineers attempt to reduce the total unsprung mass compliance to enhance both ride comfort and ride handling by trimming down the mass of each wheel and of the vehicular suspension components associated with it.

The vehicular suspension mechatronic control system suspends the vehicle's body a short distance above the on/off road surface and upholds the vehicle body at a relatively constant height to avoid it from pitching, swaying, or rolling. In order to sustain effective acceleration, braking, and cornering -- the components of good handling--the vehicular suspension mechatronic control system must also insure all wheel-tyres firmly in contact with the on/off road surface. The vehicular suspension mechatronic control system thus has an effect on a vehicle's ride comfort, performance, and safety.

Of course, it is very problematical to design a vehicular suspension mechatronic control system that affords both a smooth ride and good handling characteristics. And like most other mechanical devices in the vehicle, the conventional vehicular suspension mechatronic control system too has a history of its own.

In the beginning, it was argued that shock absorbers (dampers) were unnecessary if automotive vehicle springs were accurately designed and ordinary values of the vehicle velocity of 16 – 24 km/h (10-15 mph) were maintained. However, most vehicular suspensions of the early period of automotive history were unproblematic and unsophisticated.

Primitive automotive vehicles had two rigid axles, one linking the two front wheels and the other linking the rear wheels. When one wheel reaches a bump, the wheel at the opposite end of the axle would also be in motion.

Shock absorbers (dampers) of the 1910's were derived from one of the four principles, particularly, friction, fluidical (hydraulical and/or pneumatic), and mechanical spring.

One *'modern-looking'* **fluido-mechanical** (F-M) shock built in 1906 was termed the *'Graygood'*. It was composed of a piston functioning in an F-M cylinder with a 10 cm (4 inch) stroke.

Conversely, in automotive vehicles where cost was of little concern, innovative designs, counting independent vehicular suspensions and telescoping F-M shock absorbers, were also accessible. For instance, the 1907 *Pilain* chassis had a fully independent rear suspension with universal joints situated at each end of the equal-length *'jack'* shafts running to the wheels.

During 1925-35, changes in road and traffic circumstances influenced the development path of vehicular suspension mechatronic control systems. For instance, Cadillac enhanced the ride quality of its vehicles by minimising the unsprung mass.

John Warren Watson, an **original equipment manufacturer** (OEM) of *'stabilisers'*, said: *"Although springs are almost entirely responsible for the way a car rides, the control mechanism determines what kind of springs can be used."* He recommended that the two be designed in a harmonic systems approach.

The A.E. Forsyth Co. estimated that a variable spring was unrivalled. Its dual-chamber air spring could be inflated or deflated to enhance the ride. Formed like a wheel-tyre, the *Forsyth* spring embraced an inflatable inner tube inside a vulcanised body.

In 1932, Packard set up the first **human driver** (HD) adjustable F-M shock absorber system. Termed the Delco-Remy unit, it contained a sleeve fluidical valve connected to a push-pull cable that controlled the amount of oily fluid accessing the F-M shock absorber's bypass. Consequently, not only was there an evident difference in ride quality but the sense of the vehicle was also magnified.

September 1934 saw the **Leaf Spring Institute** (LSI) demonstrate an experimental suspension with dual lower leaf springs for the front and rear that operated as both the control arm and springing medium.

The LSI claimed the design could recover as much as 18 kg (40 lb), when evaluated to coil spring designs. Studebaker implemented a variant of this design for its 1935 models, providing the vehicles a satisfying ride.

The 1947 British Invicta used an innovative track-correcting vehicular suspension. The front wheels were set up so that they moved up and down in an almost vertical position. The wheel was linked through a linkage to a wishbone and torsion bar. The steering arm did not, nevertheless, slide with the wheel. It was set up on a spline that set beside it vertically while the wheel assembly moved. This allowed the **rack and pinion** (R&P) to be set up firmly on the frame.

The **rear-wheel absorbed** (RWA) suspension had a wheel assembly that slid on guides against the torsion bar and F-M shock absorber (damper). As the wishbone brackets moved, the guides for the wheel swung outward. The propeller shafts used a sliding joint to change length to accommodate the movement.

Independent **front-wheel absorbed** (FWA) suspensions entirely replaced the rigid-axle type after World War II, and numerous independent RWA suspensions came into use, first on European vehicles.

The independent vehicular suspension mechatronic control systems had a number of advantages, permitting the wheels to move independently of each other so that when one wheel hits a bump in the road, only that wheel is affected. Though some automotive vehicles continue to have non-independent RWA suspension, most modern automotive vehicles have independent 4WA suspension that enables all four wheels to move separately. Independent vehicular suspension mechatronic control systems enabled a reduction of unsprung mass of the automotive vehicle, softer springs became permissible, and front-wheel vibration problems were minimised.

In view of the fact that the mass of a vehicle is rarely distributed uniformly between the front and the rear, the independent vehicular suspension mechatronic control systems allowed separate suspension mechanisms to meet the different requirements.

Classically, the mass - and consequently the vehicular suspension requirement - is normally greater at the front of the vehicle, where the **external combustion engine** (ECE) or **internal combustion engine** (ICE) is positioned, than at the rear.

Although a few vehicles have non-independent RWA suspensions (in which both rear wheels are connected to a single rigid axle that prevents them from moving independently of each other), every modern vehicle has an independent FWA suspension.

The FWA suspension mechatronic control system that is coupled to the **front-wheel steered** (FWS) mechatronic control system has a more significant function in controlling the vehicle's direction. The two mechatronic control systems must maintain the wheels in a line so that the vehicle moves in the intentional direction and the wheel-tyres continue in effect perpendicular to the on/off road surface and do not wear peculiarly.

Pivot axes that are small structures on the front and rear control arms guarantee a supplementary level ride. Front pivot axes have an anti-dive arrangement that neutralises the inclination of the vehicle's nose to dip as mass moves forward during braking. Rear pivot axes have an anti-squat arrangement to compensate for the inclination of the nose to rise as the vehicle's mass moves rearward during acceleration. There are numerous different FWA and RWA suspension mechatronic control system designs, incorporating double *A*-arm and *MacPherson* strut types (both of which have coil springs), torsion-bar springs, and leaf springs. The double *A*-arm system is more costly than other systems and is more normally used in SUVs and racing cars that normally have fully independent vehicular suspension in the front or rear. The double *A*-arm system is more compact, lowering the vehicle hood, and creating greater visibility and better aerodynamics.

The *MacPherson* strut system, a coil spring that surrounds an F-M shock absorber that functions as an upper control arm, is less costly than the double *A*-arm and is set up in most recent **front-wheel-drive** (FWD) passenger automotive vehicles. It is more complicated than the double *A*-arm suspension and necessitates a higher hood. Leaf springs are normally used in non-independent RWA suspensions and in SUVs.

Spring elements used for vehicular suspension, in escalating order of their capability of storing elastic energy per unit of mass, are leaf springs, coil springs, torsion bars, rubber-in-shear devices, and air springs. An essential factor in spring assortment is the correlation between load and deflection identified as the spring rate, described as the load in kilograms (pounds) divided by the deflection of the spring in centimetres (inches), respectively.

A soft spring has a low rate and deflects a greater distance under a specified load. A coil or a leaf spring retains a substantially constant rate within its operating range of load and may deflect 10 times as much if a force 10 times as great is applied to it. The torsion bar, a long spring-steel element with one end held firmly to the frame and the other warped by a crank linked to the axle, provides an escalating spring rate.

A soft-spring suspension provides a comfortable ride on a relatively even on/off road surface, but the occupants move up and down abnormally on an uneven road. The springs must be stiff enough to avoid a large deflection at any time owing to the difficulty in offering adequate clearance between the sprung portion of the automotive vehicle and the unsprung portion below the springs.

Lower roof heights make it more and more difficult to assure an adequate clearance that is indispensable for soft springs. Road-handling characteristics known as sway, or roll, also cause the vehicle's body to undergo sidewise tilting from centrifugal forces acting outward on turns.

The softer the suspension, the more the outer springs are compressed and the inner springs expanded. Front-end *'dive'* under brake action is more perceptible with soft front springs. The leaf spring, although relatively inelastic, has the distinct advantage of exactly positioning the wheel with relation to the other chassis components, both laterally and fore and aft, without the assistance of supplementary linkages.

Some vehicles delivered with rear leaf springs use a drive in which the axle housing is firmly connected to the spring seats so that the driving torque reaction, the brake torque, and the driving force are all transmitted to the spring. Those using rear coil springs with open drive shafts deal with these forces by radius rods and other linkages.

By the time that powered vehicles were developing, the springs had modified to an elliptical shape and were bolted between the axle and the chassis. The principal function of vehicular suspension at this time was to permit the four wheels to continue interaction at all time with an uneven on/off road surface and any enhancement in the comfort of the passengers was a gratuity.

Air springs submit numerous advantages over metal springs, one of the principal ones of which is the option of controlling the spring rate. Naturally, the force necessary to deflect the air unit escalates with greater deflection because the air is compressed into a smaller space and greater pressure is built up, thus progressively resisting further deflection. An air spring bellows is a column of air confined within a rubber and container similar to an automotive vehicle's wheel-tyre, or two or three wheel-tyres stacked on top of one another. The pneumatic check valves admit supplementary air to the bellows from the air-supply tank to continue vehicle height when the load is increased, and the levelling pneumatical valves vent redundant air from the bellows when the automotive vehicle rises as a result of unloading. The vehicle thus continues at a preset height irrespective of load. Even if an air spring is expandable under normal loads, it turns out to be ever stiffer when compressed under an increased load.

Air vehicular suspension was introduced on some luxury vehicles in the late 1950s, but it was dropped after several model years. Recently, innovative levelling systems have been developed for passenger vehicles, including air-adjustable rear **pneumo-mechanical** (P-M) shock absorbers; some air-spring systems operate without a **mechano-pneumatical** (M-P) compressor.

Automotive primary vehicular suspension is the term used to designate those vehicular suspension components linking axle and wheel assemblies of a vehicle's body to the frame of the vehicle. This is to distinguish them from the vehicular suspension components linking the frame and body of the vehicle, or those components located directly at the vehicle's seat, normally termed the automotive secondary vehicular suspension. There are two basic types of elements in conventional vehicular suspensions. These elements are springs and shock absorbers (dampers). The role of the spring in a vehicular suspension is to hold up the static mass of the vehicle. The function of the shock absorber (damper) is to dissipate vibration energy and control the input from the on/off road surface that is transmitted to the vehicle.

The indispensability in function and form of a suspension is the same regardless of the type of vehicle or vehicular suspension.

Primary vehicular suspensions may be divided into passive, semi-active, active and hybrid vehicular suspensions, as will be discussed next, within the context of this chapter. Vehicular suspensions are principally a series of metal bands of some sort situated between the body of the vehicle and a vehicular suspension component like an axle, *A*-arm or the like. These bands may take the form of leaf springs, coil springs or torsion bars. There are also shock absorbers (dampers) to trim down rebound, and often there are stabiliser bars to lessen the vehicle's body roll in cornering.

Considerable enhancement has been perceived in automotive vehicle riding and cornering performance in current years as a consequence of advances attained, particularly in wheel-tyre and vehicular-suspension technology. Owing to this enhancement, vehicle handling characteristics during braking have become of supplementary, rather than primary, importance. Automotive advanced technologies ought to be also applied for designing advanced adjustable shock absorbers (dampers).

ABW AWA suspension mechatronic control systems are the most essential components in driving pleasure and ride feature, and driving stability and ride comfort are the decisive factors for resolving these demands. Normally, the two do not co-exist; **smart-utility vehicles** (SUV), for driving pleasure, and luxury automotive vehicles, for ride features, each find those aims enhanced by the set up of the ABW AWA suspension mechatronic control system. An ABW AWA suspension mechatronic control system permits each wheel to act in response separately to bumps in the on/off road surface, and it intensifies the probabilities that all wheels may continue to be steadily in contact with the ground when driving on rough on/off road surfaces or when trying to avoid an accident.

Since mechatronically controlled adjustable shock absorbers (dampers) were established on the automotive market in the 1980s, active and semi-active ABW AWA suspension mechatronic control systems have been developed and used in practical applications to meet the demands of both driving pleasure and ride features [UEKI ET AL. 2004].

Active ABW AWA suspension mechatronic control systems move each wheel up and down to control vehicle-body motion in response to on/off road surface abnormalities. The mechatronic control system responds to inputs from the on/off road surface and the driver. With an active ABW AWA suspension, a vehicle may simultaneously provide the even ride of a soft suspension along with enhanced handling related with a firm suspension.

Active ABW AWA suspension mechatronic control systems use a high-pressure **mechano-fluidical** (M-F) pump or **mechano-pneumatical** (M-P) compressor or even **mechano-electrical** (M-E) generator with **fluido-mechanical** (F-M) or **pneumo-mechanical** (P-M) motors (cylinders) or tubular linear **electro-mechanical** (E-M) motors, respectively, at each wheel to position the wheels with respect to the vehicle.

Up and down motion of the wheels is actuated by mechatronically controlled fluidical or pneumatical or even electrical valves, respectively. In any mechatronic control system, sensors at each wheel detect vertical wheel position and the force of the on/off road surface acting on the wheel. Some systems use *'road preview'* sensors (microwave radar or laser) to provide data about on/off road surface abnormalities before the front wheels reach them.

Accelerometers inform the computer when the vehicle is accelerating, braking or cornering. The computer uses adaptable algorithms to continuously process data and decide the position of each wheel. Coil springs may be used at each wheel to avoid *'bottoming out'* of the suspension in case of mechatronic control system failure; they also may reduce the power necessary to hold up the sprung mass of the vehicle. The customer benefit is exceptional ride and handling, even on uneven on/off road surfaces [CARLSON 2005B].

Active tilt control winds up the stabiliser bars in the front and rear suspension to oppose vehicle-body roll while cornering. Since active control is used only when absolutely necessary, vehicle spring rates and stabiliser bar stiffness may be trimmed down, getting better normal ride characteristics. Besides, this system has the potential to escalate low-velocity, off-road traction on DBW **four-wheel driven** (4WD) vehicles. For instance, the active tilt control module collects a lateral acceleration signal from a vehicle body-mounted accelerometer. The module directs oily-fluid pressure from an E-M-F pump to F-M cylinders that substitute stabiliser bar links. During cornering, the F-M cylinders wind up the stabiliser bars that increase resistance to vehicle-body roll. The system is de-activated at slow values of the vehicle velocity to enhance driver ride comfort. Off-road traction is enhanced as a result of lower resistance from the stabiliser bars, permitting the front and rear wheels to go more smoothly along the surface of rough roads. The customer benefit is reduced vehicle-body roll when cornering and improved ride [CARLSON 2005A].

The function of an ABW AWA suspension mechatronic control system in a vehicle is to enhance ride comfort and road handling stability. An imperative consideration in ABW AWA suspension design is how to get both enhanced ride comfort and road handling stability, in view of the fact that they are as a rule in conflict. Advances in mechatronic control technology applied to vehicles, may resolve the conflict [AKATSU 1994].

One of the most examined components of future vehicles is an ABW AWA suspension mechatronic control system. Its principal actions are to [GILLESPIE 1992]:

- ❖ Afford vertical compliance so the wheels can go along the uneven on/off road surface, buffering the vehicle chassis from roughness in the on/off road surface;
- ❖ Sustain the wheels in the correct steer and camber attitudes to the on/off road surface;
- ❖ Act in response to the control forces generated by the wheel-tyres -- longitudinal (acceleration and deceleration) forces, lateral (cornering) forces, and propulsion (driving) and dispulsion (braking) torques;

- Avoid roll of the vehicle chassis;
- Retain the wheel-tyres in contact with the on/off road surface with minimal load variations.

The positive features of an ABW AWA suspension mechatronic control system imperative to the dynamics of the vehicle are first and foremost perceived in the kinematics (motion) behaviour and its response to the forces and moments that it must emit from the wheel-tyres to the vehicle chassis [SORSCHE ET AL. 1974; BASTOW 1990; MATSCHINSKY 1999; MILLIKEN AND MILLIKEN 2002].

Besides, other features taken into account in the design process are mass and size (package space), manufacturability, ease of assembly, cost and others.

Vehicular suspensions normally fall into either of two sets -- solid axles and independent vehicular suspensions. Each set can be, in practice, somewhat different, and so may be separated for that reason for examination [GILLESPIE 1992].

A solid axle is one in which wheels are placed at either end of a rigid beam so that any change the position of one wheel is transmitted to the opposite wheel [GOODSELL 1989] having an effect on them to steer and camber mutually.

Solid drives, sometimes called *'live'* axles, are used on the rear of many automotive vehicles and most trucks and on the front of many AWD trucks. Solid beam axles are normally used on the front of heavy trucks where high load-carrying capacity is required. Solid axles have the benefit that wheel camber is not influenced by the vehicle's body roll. In consequence there is little wheel camber in cornering, not including that which occurs from a slightly greater compression of the wheel-tyres on the outside of the turn. Besides, wheel alignment is without difficulty sustained, minimising wheel-tyre wear. The most important drawback of steerable solid axles is their exposure to tramp-shimmy steering vibrations [GILLESPIE 1992].

Contrary to solid axles, independent vehicular suspensions let each wheel shift vertically without having an effect on the opposite wheel. Almost all passenger vehicles and light trucks use front independent vehicular suspensions, owing to the advantages in delivering room for the ECE or ICE, and due to the enhanced resistance to steering (wobble and shimmy) vibrations. The independent vehicular suspension also has the advantage that it presents an intrinsically higher roll stiffness corresponding to the vertical spring rate.

The first independent vehicular suspension on front axles emerged in the early part of the XXth century [OLLEY 1934, 1946]. It may downgrade some of the wobble and shimmy inconveniences characteristic of solid axles (by decoupling the wheels and interposing the mass of the vehicle between the wheels).

Supplementary advantages incorporated easy control of the roll centre by selection of the geometry of the control arms, the aptitude to control tread alteration with jounce and rebound, larger suspension deflections, and greater roll stiffness for a particular suspension vertical rate [GILLESPIE 1992].

The **mechano-mechanical** (M-M) vehicular suspensions used until now characterise a compromise between necessities that are, to a degree self-contradicting; specifically, a high degree of riding comfort in opposition to precise steerability and safety in propulsion (driving). Methods for overcoming these compromises and for attaining adaptableness of M-M vehicular suspensions to loading circumstances, road roughness and handling features are accessible by innovative adjustable shock absorbers (dampers), rate-variable springs, low-cost servomechanisms, and the more or less inestimable predispositions of single-chip microprocessors [SEIFFERT AND WALZER 1991].

For better comprehension of the negative aspect let us consider the factors affecting vehicular suspension design. Decisive factors are passenger comfort, space necessity of the spring-damper system, and wheel load variation, that is, variation of the wheel-tyre to on/off road surface contact.

The dominant factor for comfort is the vehicle's vertical body acceleration; here a soft spring system with lower acceleration provides more comfort, but then, on rough road surfaces, excessive relative wheel movements occur and this tends to produce detectable variations in the normal forces between wheel-tyre and on/off road surface. Besides, soft springs necessitate an amount of wheel travel that is hardly accessible in small vehicles. For safety reasons the wheel load should be as constant as achievable during an entire trip.

Powerful shock absorbers (dampers) are necessary for small dynamic wheel-load variations. This indispensable compromise is made even more complicated as a consequence of the supplementary forces that must be controlled during lateral vehicle movements. Particular attention must be paid, in this relationship, to lateral acceleration while cornering.

Soft suspensions initiate excessive vehicle roll. And, as a final consideration, preferences concerning suspension softness in opposition to harshness fluctuate from country to country and affect the selected compromise. It is apparent that spring-damper systems have many responsibilities, the first of which is to adapt to vehicle loading.

Some innovative ABW AWA suspension mechatronic control systems may try to act in accordance with individual requirements concerning driving comfort; more complex systems may try to alter the degree of absorbing (damping) automatically for maximum comfort and for highest driving safety on rough road surfaces. The long-range objective is to produce semi- and/or active suspensions that are able to selectively absorb (dampen) *'forced vibrations'* and to offer at any time optimum comfort and maximum safety under all static and dynamic loading circumstances. To what degree the vehicular suspensions that may be described in the following represent indisputable progress is even now a matter of some scepticism among automotive vehicle specialists [SEIFFERT AND WALZER 1991].

Mechatronic control devices for automotive vehicles may be separated into four categories: passive, semi-active, active and hybrid. Passive devices, normally, are those that have preset properties and require no energy to function. On the contrary, the controllable forces generated by active devices are induced directly by energy (electrical or otherwise) put into the device. Between passive and active are semi-active devices that are passive devices with properties that are controllable by application of a small amount of energy.

Hybrid devices are combinations of the other three categories. Each of these is discussed briefly in the following paragraphs, with greater detail on semi-active devices [JOHNSON AND ELMASRY 2003].

Summing up, ABW AWA suspension mechatronic control systems that absorb vibration or movement may be categorised as follows:

- Passive;
  - No power required;
  - Directly damps vibration.
- Semi-active;
  - Requires minimal power;
  - Applies the force that changes the system;
  - Change helps dampen the vibration.
- Active;
  - Requires power;
  - Applies a force directly into the system to damp vibration;
  - Hybrid that is combinations of the other three classes.

Passive devices, such as visco-elastic shock absorbers (dampers), viscous fluid shock absorbers, friction shock absorbers, metallic shock absorbers, tuned mass shock absorbers, and tuned liquid shock absorbers may partially absorb structural vibration energy and reduce response of the vehicle body [SOONG AND DARGUSH 1997].

These passive devices are relatively simple and easily replaced. However, the effectiveness of passive devices is always limited due to the narrow frequency ranges in which they tend to be effective, the dependence of their force only on local data, and their inability to be modified if aims (or design codes) change. Active control devices, including active-mass shock absorbers (dampers) and active-tendon systems, may reduce vehicular response more effectively than passive devices because feedback and/or feedforward mechatronic control systems are used [HOUSNER ET AL. 1997].

However, large power requirements during strong vehicular response and other hazards slow down their implementation in practice. Further, active devices have the ability to inject dynamic energy into the vehicular suspension; if done improperly, this energy has the potential to cause further damage to the vehicle. In particular, this may occur when the assumptions used to design the control algorithm are incorrect or do not have a proper characterisation of the vehicular dynamics.

On the contrary, *'smart'* devices are controllable passive devices that require small amounts of power to control certain passive behaviour. Moreover, these devices cannot add energy to the vehicular mechanical system; rather, they may only (temporarily) store and dissipate energy.

Furthermore, they offer highly reliable action at a modest cost and may be considered as fail-safe in that they default to passive devices should the control hardware malfunction [DYKE ET AL., 1996].

Most vibration shock absorbers at present are passive, like the springs on an automotive vehicle that absorb the bumps in the on/off road surface.

Those shock absorbers offer a fixed amount of resistance and so do not act well when bumps are anything but average. But conceive of an automotive vehicle that senses every jiggle, large and small, and then without delay translates those movements into electrical signals that thicken or thin the **stimuli responsive fluid** (SRF) in a shock absorber (damper).

SRFs are an emerging class of XXIst century *'smart materials'* capable of reversibly transforming themselves, on a millisecond time scale, from the liquid to solid states in response to alterations in external magnetic or electrical fields. This effect initiates the mechanical properties of a device, such as an automotive shock absorber, to be continually adjusted in real-time in response to input stimuli such as uneven on/off road surfaces, thereby creating safer and more energy-efficient automotive vehicles and equipment. SRF applications in ultra-fast, computer-controlled machines and virtual reality are also emerging. SRFs are normally composed of solid, magnetically or electrically polarisable particles dispersed in a fluid medium, usually a silicon oily fluid or mineral oil.

Microcrystalline particles and derivative cellulose fibres may be formulated into **magneto-rheological fluids** (MRF) or **nano-magneto-rheological fluids** (NMRF) and/or **electro-rheological fluids** (ERF) or **giant-electro-rheological fluids** (GERF) depending on the derivation process. Thus, MRFs or NMRFs and ERFs or GERFs are smart, synthetic oily fluids changing their viscosity from liquid to semi-solid state within milliseconds if a sufficiently strong magnetic or electric field is submitted in an application. When used in appropriate **magneto-rheological** (MR) or **electro-rheological** (ER) shock absorbers, they provide an innovative potential of very fast, adaptively controllable interfaces between mechanical parts of equipment and **electronic control units** (ECU).

Some automotive OEMs have initiated a program to develop and characterise innovative polymeric materials, including biodegradable celluloses, for both MRFs or NMRFs and ERFs or GERFs. With its extreme adaptability, the MRF and ERF in the development phase could end up not only in vibration shock absorbers (dampers) but also in an innovative generation of MR or ER fluidical valves, locks, clutches, brakes or other devices in which there is a necessity for materials with variable fluidity.

The MRF or NMRF and ERF or GERF are the latest to come out of the nascent field of nanotechnology, a hybrid discipline drawing on physics and engineering in which scientists build materials from small numbers of atoms.

In the coming decade, automotive scientists and engineers have predicted, MRFs or NMRFs and ERFs GERFs that change phase in response to electrical signals may modernise engineering because they may make it possible to perform mechanical actions as quickly as an electrical switch may be turned on or off -- far faster than even the fastest M-M, F-M or P-M ABW AWA suspension mechatronic control systems may react.

An innovative step in understanding vibrations and controlling ABW AWA suspension mechatronic control systems has been made using a modelling and simulation platform for systems engineering e.g. *'AMESim'*, an efficient tool to answer typical concerns such as time response of the actuators, stability analysis of fluidical (hydraulical and/or pneumatical) or even electrical components, noise and vibrations in the fluidical (hydraulical and/or pneumatical) and electrical circuits, optimisation of drivability and comfort abilities, off-line tests of control strategies, optimisation of the network architecture, real time applications using physical models involving semi-automatic physical model simplification, development of innovative concepts with advanced technologies, increase of performance, prediction of temperatures influence, reduction of **specific energy consumption** (SEC), increase of efficiency, and so on [AMESIM 2004].

Physical and mathematical models' reliability and accuracy may be verified through continuing comparison with experimental data for various functional modes, different supply pressures and load considerations.

Mechatronic control of vertical dynamics - Conventionally, the vertical dynamics of a vehicle are mechatronically controlled using an ABW AWA suspension mechatronic control system. ABW AWA suspension mechatronic control systems are usually designed with three objectives in mind:

- To isolate the vehicle cabin from on/off road disturbances;
- To insulate the vehicle body from load disturbances (inertial loads induced by braking and cornering);
- To influence the cornering properties of the vehicle.

These requirements, referred to frequently as ride and handling, become conflicting requirements when only conventional passive suspensions are deployed (such as spring and shock absorbers), but may be satisfied independently of each other when active ABW AWA suspension mechatronic control systems are used. In fact, a well-designed ABW AWA suspension mechatronic control system should adequately address each of these design considerations.

In the present chapter, interested readers will find a concise overview of R&D works on this subject. As most readers are interested in affecting both **vehicle ride** (VR) and **vehicle handling** (VH), we consider here only R&D work on active suspensions.

As well as a large body of literature, various excellent surveys have been written on this subject.

We also refer readers to HROVAT [1997], and the handbook [GILLESPIE 1992], as entry points to this extensive subject. VR is normally associated with the perception of vehicle disturbances as experienced by vehicle passengers.

There are several components that contribute to VR; the most important of which are on/off road disturbances that are transmitted to the vehicle by means of the vehicular suspension.

From the point of view of vehicular suspension design, VR is the perception of the frequency spectrum of vehicle vibrations below 25 Hz that are produced by road roughness and on-board sources, the former being the most relevant.

On the other hand, VH refers to the manoeuvring ability of the **automotive vehicle/human driver** (AV/HD) combination [GILLESPIE 1992]. VH characteristics are greatly influenced by forces and moments generated by inertial and aerodynamic loadings such as those caused by braking, cornering and wind gusts.

Vehicular suspension hardware may range from completely passive to completely active depending on the degree of force control available to the driver, as illustrated in Figure 5.1 [Schwartzz and Rieth 2004; Illegas and Shorten 2005].

Suspension	Passive	Semi-active	Slow-active	Full-active	Full-active
Physical model					
Components	Spring with fluidic damper and stabilisator	Spring with continuously variable F-M damper < 50Hz	Fluidic energy input in parallel with F-M damper 0 -- 5 Hz	Direct fluidic energy input with F-M damper 0 -- 30 Hz	Electrical energy controlled force input with a linear E-M motor 0 -- 100 Hz

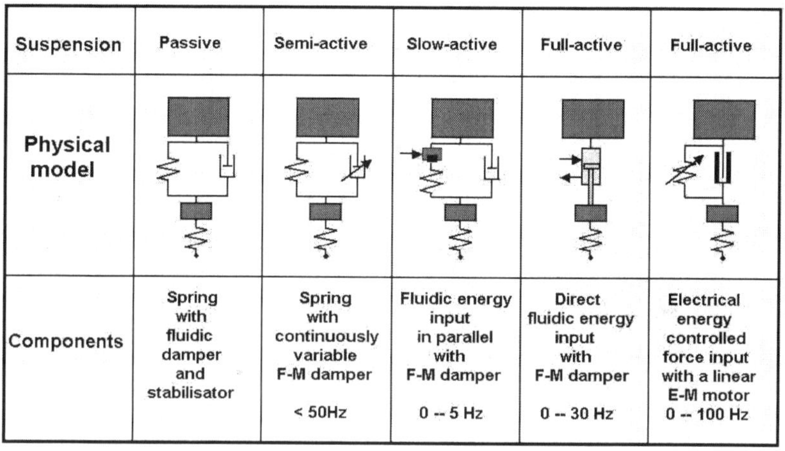

Fig. 5.1 Categories of vehicular suspensions
[Schwartzz and Rieth 2004; Illegas and Shorten 2005].

The physical model depicted corresponds to a quarter-vehicle physical model. It is composed of: a quarter of the vehicle mass called the sprunged mass m_s, a wheel mass called unsprunged mass m_u, the vehicular suspension between m_u and m_s, and the wheel-tyre represented as a spring between m_u and the uneven road. For a detailed explanation of the quarter-vehicle physical model, we refer the interested reader to Genta [1997].

The semi-active suspension illustrated in Figure 5.1 is a shock absorber (damper) with continuously variable (or adjusted) damping coefficient instead of a passive one.

As the absorbing (damping) force is proportional to the vertical velocity of the suspension displacement, semi-active suspensions may only affect transient behaviour of the suspension deflection and not its steady state.

Besides semi-active shock absorbers (dampers), there are also semi-active suspension springs where the spring stiffness alters by means of fluidical (hydraulical or pneumatical) pressure [Fischer and Isermann 2004].

Active suspensions, on the other hand, influence both steady state as well as transient behaviour of suspension deflection. These vehicular suspensions are usually categorised according to the maximum response frequency or bandwidth of the actuators that they employ.

Low bandwidth active suspensions are able to control the force at the suspension spring, and indirectly in combination with a shock absorber (damper), the suspension force.

Normally, such vehicular suspensions function using fluidical (hydraulical or pneumatical) energy and operate at frequencies below 5 Hz.

Medium bandwidth suspensions also normally employ fluidical energy and may control suspension deflection directly at up to a maximum frequency of 30 Hz.

Vehicular suspensions of the form described above have been already implemented in manufacture units. Finally, high bandwidth suspensions with a response of up to 100 Hz may be achieved with an E-M actuator [SCHWARTZ AND RIETH 2004].

The vertical movement of the automotive vehicle (heave) together with its inclinations (pitch and roll) illustrated in Figure 5.2 constitute the vehicle's vertical dynamics [VILLEGAS AND SHORTEN 2005].

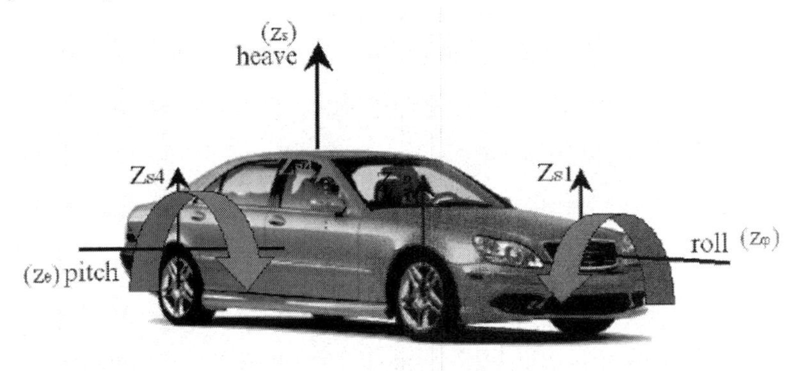

Fig. 5.2 Automotive vehicle's vertical dynamics [VILLEGAS AND SHORTEN 2005].

These are principally defined by the vehicular suspension used in the vehicle. The vehicle manufacturer may choose between a passive, semi-active or active suspension. The choice of a particular vehicular suspension depends as a rule on several issues, including energy consumption, comfort and performance requirements. As active ABW AWA suspension mechatronic control systems consume a large amount of energy [FISCHER AND ISERMANN 2004], they are used principally in high-end vehicles.

Semi-active ABW AWA suspension mechatronic control systems are favoured when low energy consumption is a priority; however, in various circumstances, a semi-active ABW AWA mechatronic control systems may yield a performance that is near to the active one.

The design of ABW AWA suspension mechatronic control systems has been an active subject of R&D works for more than half a century, and since the end of the 1980s active suspensions have been a feature of in-manufacture vehicles.

The design of AWA controllers for active and semi-active ABW AWA suspension mechatronic control systems has been advanced using various methods. These range from classical techniques to contemporary methods such as *Youla* parameterisation [SMITH AND WANG 2002].

Since the VR optimisation problem may be considered as an optimal filtering problem (where one would try to eradicate the negative effects of vibrations activated by on-off road roughness) various techniques from optimal control theory have also been used in this subject.

As previously mentioned, an extensive survey of automotive-vehicle active ABW AWA suspension mechatronic control, with particular concentration on optimal control theory, is presented in HROVAT [1997]. One of the mechatronic control concepts frequently encountered during a targeted literature survey is the *'skyhook'* and its variations. Skyhook control has the objective of absorbing (damping) the vehicle body ($\boldsymbol{m_s}$) motion from on/off road and other perturbations, as if a shock absorber (damper) is positioned between $\boldsymbol{m_s}$ and a fixed point as illustrated in Figure 5.3 [HROVAT 1997; VILLEGAS AND SHORTEN 2005].

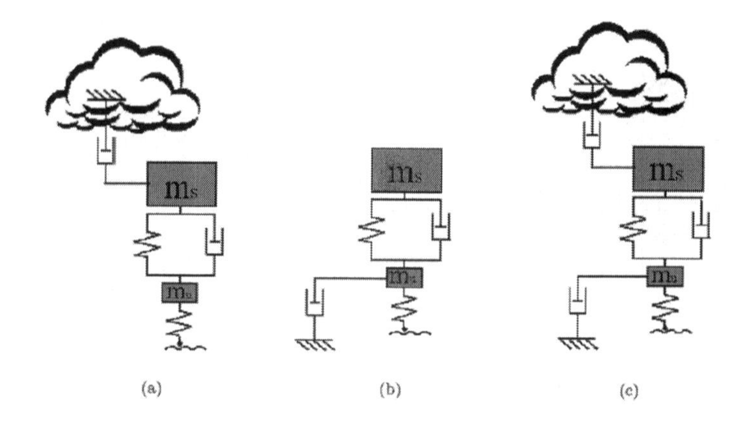

Fig. 5.3 Skyhook (a), groundhook (b), and hybrid skyhook/groundhook (c) control concepts [HROVAT 1997; VILLEGAS AND SHORTEN 2005].

Likewise, a *'groundhook'* is different from a skyhook as the shock absorber (damper) is positioned between the wheel ($\boldsymbol{m_u}$) and the fixed point. There is also the hybrid version using both skyhook and groundhook control concepts, as well as others [AHMADIAN AND GONCLAVES 2004].

Semi-active suspensions are a smart substitute for active suspensions in various circumstances with the advantage of needing an extremely small consumption of energy [FISHER AND ISERMANN 2004].

Various techniques for the design of such semi-active suspensions are reviewed in AHMADIAN AND GONCLAVES [2004]. Likewise, the limits and performance possibilities of semi-active suspensions with variable damping and stiffness coefficients are studied in FISCHER AND ISERMANN [2004].

Semi-active spring suspensions were difficult to find in the literature survey, but they are already in manufactured vehicles, for example, the electronic air suspension from Land Rover's *Range Rover*. Vehicular suspensions with fixed damping coefficients are analysed and modelled as switched linear mechatronic control systems in CORONA ET AL. [2004].

Innovative actuators termed **continuously variable dampers** (CVD) have recently become popular in semi-active suspension design. Such CVD suspensions are predominantly compatible with continuous time control design methods.

In CAPONETTO ET AL. [2003] a **fuzzy logic** (FL) mechatronic control system is presented. The motivation of this R&D work is to reduce the effect of on/off road perturbations and to concurrently complete low chassis movement and wheel -tyre movement.

A hybrid H_2/H_1 controller for a CVD-based ABW AWA suspension mechatronic control system is presented in LU AND DEPOYSTER [2002]. In this reference, using a full-vehicle physical model, an H_2 controller is used to realise an ABW AWA suspension mechatronic control system to control vertical chassis accelerations, whereas an H_1 controller is used to control vertical deflection velocity. The low cost and low power requirements of semi-active ABW AWA suspension mechatronic control systems has made their inclusion in manufactured units possible; not only high-end automotive vehicles, like Cadillac's 2002 *Seville* STS or Maserati's 2002 *Spyder* but also in small-size vehicles, like Opel's 2004 *Astra*.

Mechatronic control design techniques for active and semi-active systems may be categorised in relation to various on/off roads. A predominantly convenient method to examine the R&D work in the subject is to use the kind of automotive vehicle physical model upon which the mechatronic control design is based; namely, whether a quarter-vehicle, a half-vehicle, or a full-vehicle physical model has been used to design the mechatronic control system. Once more the interested reader should note that much of the R&D work on active suspension design is presented in HROVAT [1997].

In the past, initial attempts to design active ABW AWA suspension mechatronic control systems involved using **two-degrees-of-freedom** (2-DoF) quarter -vehicle physical models. Such designs are not only of previous concern; they are still used today. For instance, various contemporary studies [GASPAR ET AL. 2003; FIALHO AND BALAS 2002] have been based on the classical 2-DoF physical model.

Using this physical model it is possible to illustrate intuitively the skyhook and groundhook concepts used by various authors: see YAMASHITA ET AL. [1994]; ALLEYNE AND HEDRICK [1995]; AKENAGA ET AL. [2000]; AHMADIAN AND GONCLAVES [2004] and HROVAT [1997] for more details about skyhook and groundhook.

In ALLEYNE AND HEDRICK [1995], the 2-DoF physical model is used as a basis of the sliding-model mechatronic control system that is shown experimentally to have better ride performance near the chassis natural frequency.

To take into account alterations in circumstances of the on/off road surface and avoid reaching the actuator limits, a controller based on nonlinear back stepping is used in JUNG-SHAN AND KANELLAKOPOULOS [1997] and FIALHO AND BALAS [2000]. This controller is also based on a relatively simple 2-DoF physical model.

In such a reference, the vehicle suspension varies from soft (to decrease lateral acceleration) and hard (to decrease actuators displacement). Such an automotive vehicle would be comfortable on a wide variety of on/off road surfaces, and become hard on particularly bumpy on/off road surfaces.

A similar approach is used in FIALHO AND BALAS [2002]; here the authors again use nonlinear back stepping to achieve a mechatronic control system that has shown a better response to bumps in the on/off road surface than a conventional passive suspension.

In TING ET AL. [1995] the author uses an FL controller based on sliding-mode control theory to minimise spring mass acceleration, suspension deflection and wheel-tyre fluctuation, to enhance VH.

Finally, in SMITH AND WANG [2002] the author uses the 2-DoF physical model to design a linear controller that results in enhanced VH (by reducing effects of inertial perturbations such as the lateral acceleration when cornering). In this reference, on/off road surface and inertial disturbance paths are separated so responses may be adjusted independently.

Most of the recent R&D work in the subject has been based on controllers that are designed using more sophisticated vehicle physical models.

In particular, a half-vehicle physical model with **four-degrees-of-freedom** (4-DoF) is used by a number of authors; see HAYAKAWA ET AL. [1999] and SMITH AND WANG [2002].

Such physical models are used in applications where it is desired to control vehicle pitch and heave. Full-vehicle physical models with **seven-degrees-of -freedom** (7-DoF) capture more vehicle behaviours and are used when not only heave and pitch are to be controlled, but also vehicle roll and warp [SMITH AND WANG 2002].

A full-vehicle physical model is used in IKENAGA ET AL. [2000]. Here the authors demonstrate enhanced ride performance using a feedback controller. Likewise, in GASPAR ET AL. [2003], VR is enhanced (in simulation results) when compared to a passive suspension using a controller based on hybrid μ-synthesis. VH as well as VR are enhanced in automotive-vehicle tests in YAMASHITA ET AL. [1994] and HAYAKAWA ET AL. [1999] using an H_l control that is designed to be robust to parameter uncertainty.

The vehicle response to on/off road perturbations is enhanced near the chassis' natural frequency, and better handling performance is achieved by a smaller roll response to lateral accelerations.
The controller presented in HAYAKAWA ET AL. [1999] has the advantage of being simpler and easier to implement as the physical model used is decoupled into two 4-DoF physical models. In SMITH AND WANG [2002], control effort is focused on VH. A 7-DoF physical model is also used in this reference, but this physical model is later decoupled into two 4-DoF physical models and subsequently, each of these, under certain (unrealistic) assumptions, into four 2-DoF physical models. Nevertheless, the controller parameterisations designed for both, 2-DoF and 4-DoF physical models, and tested on a nonlinear mathematical model simulation, show very good rejection of inertial perturbations. Even though active suspensions require actuators for their physical implementation, various references neglect them and do not include them in the physical model to be controlled. The actuators that may be found in the literature are H-P-M and E-F-M [HROVAT 1997].

In most circumstances, E-F-M nonlinear physical models are linearised as in GASPAR ET AL. [2003]. Only in YAMASHITA ET AL. [1994] is a linearised H-P-M actuator used.

Apart from actuators, other aspects that affect vertical and lateral dynamics include elasto-kinematics and the suspension geometry. In almost every case, references that surveyed such effects were neglected.

The term *'elasto-kinematics'* refers to the influence of rubber couplings that are used to absorb (damp) vibrations coming from the wheel-tyres.

These affect the SBW AWS conversion and ABW AWA suspension mechatronic control systems. The interested reader is referred to WALLENTOWITZ ET AL. [1999] for more about this subject.

Simulation results illustrating the influence of elasto-kinematics on lateral dynamics are presented in RICHERZHAGEN [2004].

Suspension geometry on the other hand appears to have been almost completely neglected in the literature.

Finally, except for VILLEGAS-RAMOS [2004], the interested reader may find no paper that focuses on the tracking problem; rather the majority of references appear to focus on the regulation problem.

Consider an automotive vehicle's vertical dynamics physical model with 7-DoF as illustrated in Figure 5.4 and described in LINDHOLM [2003] and VILLEGAS AND SHORTEN [2005].

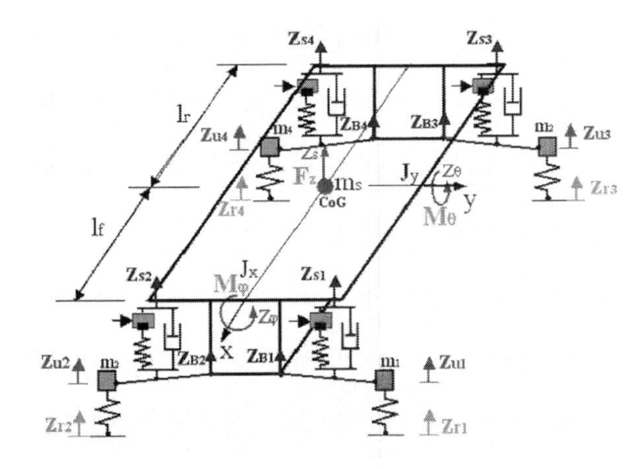

Fig. 5.4 Automotive vehicle's vertical dynamics physical model [LINDHOLM 2003 VILLEGAS AND SHORTEN 2005].

It includes the vehicle body as a stiff body and the four wheels as point masses. For the vehicle body it has 3-DoF at the **centre of gravity** (CoG), namely:

- heave z_s or vertical motion;
- pitch angle (z_θ) or rotational motion about the y-axis;
- roll angle (z_Φ) or rotational motion about the x-axis.

Each wheel position z_{ui} changes only in the vertical direction for $\boldsymbol{i}$ representing the wheel index $\boldsymbol{i} = \mathbf{1}, \mathbf{2}, \mathbf{3}$ and $\mathbf{4}$.

Let us consider that the mass of the chassis ($\boldsymbol{m_s}$), sometimes referred to as sprung mass, does not include those of the wheel-tyres ($\boldsymbol{m_i}$) also called unsprung masses.

The suspension structure is considered massless with the geometry illustrated in Figure 5.4. It has two contact points, or joints with free rotation, on the vehicle body and another at the wheel. The suspension strut at its middle is considered vertical, and a linear shock absorber (damper) is mounted in parallel to a spring and an actuator. The actuator acts directly on the spring and the latter is considered to have a linear behaviour.

The wheels are considered to move only in a vertical direction and the wheel-tyres are modelled as vertical linear springs between the wheel point mass at z_{ui} and the on/off road perturbations z_{ri}, ignoring the damping influence. Besides, their relative longitudinal and lateral motions with respect to the vehicle body are not considered. Moreover, the actuator dynamics are not considered and the input taken as the displacement of its damper piston ($\boldsymbol{u_i}$), being positive for expansion from the initial circumstance.

The outside forces acting on the automotive vehicle at the CoG are considered to be the acceleration of gravity ($\boldsymbol{g}$) for the vertical motion, and lateral and longitudinal accelerations for the roll and pitch angular motions, respectively. Nevertheless, as the vehicle is considered to be always on a flat surface, the gravity may not be considered explicitly but implicitly in the initial displacement of the linear springs. Besides, even if the vehicle is assumed to roll and pitch at the CoG, the influence of the lateral accelerations consider the roll and pitch axes to be below the CoG by the distance $\boldsymbol{h_r}$ and $\boldsymbol{h}$ respectively. It should be noted that a negative longitudinal acceleration $\boldsymbol{a_x}$, i.e. vehicle braking, may produce a positive pitch moment while a negative lateral acceleration $\boldsymbol{a_y}$, a negative roll moment because of the vehicle-body inertial behaviour to the forces generated at the wheel-tyres.

5.2 Vehicular Suspension

In order to improve the comprehensive performance of automotive vehicles in the XXIst century, vehicular suspensions featuring an active component have been developed.

The design may deal with a compilation of performance predispositions [SHARP & CROLLA 1987], but the active components adjust only the vertical force reactions of the vehicular suspensions, not the kinematics.

Active components that vary kinetic behaviour do so to steer the wheels, and would be enclosed in SBW AWS conversion mechatronic control systems [GILLESPIE 1992].

In conventional vehicle suspension design, comfort and on/off road handling are conflicting criteria.

A semi-active or an active ABW AWA suspension offers the possibility to vary the shock absorber (damper) characteristics along with the on/off road profile.

Such ABW AWA suspension mechatronic control systems have discrete settings or are limited in bandwidth up to 10 Hz [LAUWERYS ET AL. 2002].

Summing up, conventional automotive vehicle suspension designs have been a compromise between the three conflicting criteria of road holding, load carrying and passenger comfort.

The vehicle suspension must support the automotive vehicle, provide directional control during handling manoeuvres and provide effective isolation of passengers/payload from on/off road surface disturbances [WRIGHT 1984].

Good ride comfort necessitates a soft vehicular suspension, whereas insensitivity to applied loads necessitates stiff vehicular suspension.

Good handling necessitates a vehicular suspension setting somewhere between the two. Due to these conflicting demands, automotive vehicle suspension design has had to be something of a compromise, predominantly influenced by the type of use for which the automotive vehicle was designed.

Active ABW AWA suspensions are respected to be a mode of escalating the freedom one has to stipulate one by one the characteristics of load carrying, handling and ride quality.

A passive AWA suspension mechatronic control system has the aptitude to store energy by means of a spring and to dissipate it using a shock absorber (damper).

Its parameters are normally set, being selected to get a particular level of compromise between road holding, load carrying and comfort.

An active ABW AWA suspension mechatronic control system has the aptitude to store, dissipate and to set up energy for the mechatronic control system.

It may adapt its parameters conditioned on functioning circumstances and may have data other than the strut deflection to which the passive AWA suspension mechatronic control system is restricted.

5.2.1 Vehicular Suspension Categories

The miscellaneous levels of *'active'* merits in vehicular suspensions may be separated into the categories referred to below, scheduled in sequence of escalating predispositions [GILLESPIE 1992; WALKER 1997; BROGE 2000].
Passive Suspensions necessitating conventional components with spring and shock absorber (damper) special effects that are time-invariant. Passive components can only store energy for some time segment of a suspension cycle (springs) or dissipate energy (shock absorbers). No external energy is directly supplied to this category of suspension. Contemporary vehicular suspension technology refers to a range of passive activity; reactive automotive vehicle suspensions offer improved comfort and handling. All of the vehicular suspensions have some form of passive interconnection that facilitates load sharing between wheels and decouples many design and operating parameters and modes, such as cross-axle articulation and single-wheel stiffness, from roll control. Conventional passive suspensions are designed as a compromise between ride comfort and handling performance. Ride is primarily associated with the ability of a vehicular suspension to accommodate vertical inputs. Handling and attitude control relate more to horizontal forces acting through the CoG and ground level moments acting through the wheels. A low bounce frequency for maximum ride comfort normally leads to a low pitch frequency. Reactive, passive suspensions are driven by load inputs from the wheels and are regulated from within the vehicular suspension without the aid of active intervention. An early example is the Citroen *2CV*, introduced in 1948. It had the front and back suspension interconnected via two longitudinal mechanical support springs.
Self-levelling Suspensions are a variant of the passive suspension in which the most important lift component (as a rule air spring) may adjust for alterations in load. P-M vehicular suspensions that are self-levelling are used on many heavy trucks and on a few luxury passenger vehicles. A height control pneumatical valve monitors the suspension deflection, and when its mean position has oscillated from normal ride height for a predictable period of time (as usual greater than 5 s), the air pressure in the spring is adjusted to cause deflection within the indispensable range. The imperative attribute of a P-M vehicular suspension is that, as the air pressure varies with load, the air spring stiffness is altered, this maintaining a constant value of the natural frequency of the P-M vehicular suspension. Some vehicle manufacturers also invented the **fluido-pneumo-mechanical** (F-P-M) vehicular suspension in the late 1950s. In it, all four wheels are connected to one high-pressure M-F pump, so there is no requirement for individual shock absorbers (dampers). At rest, the automotive vehicle sits low to the ground. When the ECE or ICE is started, oily fluid is pumped into all suspension points. This raises the vehicle to a driveable height. The suspension has been further enhanced to set the automotive vehicle to a level that the driver desires. It may also set the height according to on/off road surface circumstances. Also, the Morris *1100*, launched in 1962, used *Moulton*'s fluido-elastic (hydro-elastic) suspension that had an oily-fluid interconnection between the front and back suspensions.

The interconnection between front and back suspensions reduced the pitching moment induced by front and back bumps.

Semi-active Suspensions consist of spring and damping components, the predispositions of which can be altered by an external mechatronic control. A signal or external power is fed to these mechatronic control systems for reasons of altering the predispositions. There are four sub-categories of semi-active ABW AWA suspensions [GILLESPIE 1992; WALKER 1997B]:

- *Slow-active* - Suspension damping and/or spring rate can be turned between some discrete levels in response to alterations in propulsion (driving) and/or dispulsion (braking) circumstances. Suspension motions, steering angle or brake pressure are normally used to activate control alterations to higher levels of damping or stiffness. Changing results within a fraction of a second provide the mechatronic control system the predisposition to control pitch, bounce, and roll motions of the sprung mass under more rough on/off road or manoeuvring circumstances. Although, the change back to softer settings results after a time delay. As a consequence the mechatronic control system does not regulate continuously during individual cycles of vehicle oscillation. Slow-active mechatronic control systems may also be termed *'adaptive'* semi-active ABW AWA suspensions.
- *Low-bandwidth* – Spring rate and/or damping are adjusted continuously in response to the low frequency sprung mass motions (1 – 3 Hz). Also known as slow-active or band-limited mechatronic control systems. In this subcategory the actuator may be placed in series with a road spring and/or a shock absorber (damper). A low band-width mechatronic control system aims to control the suspension over the lower frequency range, and specifically around the rattle space frequency. At higher values of the frequency the actuator effectively locks-up and hence the wheel-hop motion is controlled passively. With these mechatronic control systems automotive scientists and engineers may achieve a significant reduction in vehicle-body roll and pitch during manoeuvres such as cornering and braking, with lower energy consumption than a high bandwidth mechatronic control system.
- *High-bandwidth* – Spring rate and/or damping are adjusted continuously in response to both the low frequency sprung mass motions (1 – 3 Hz) and the high-frequency axle motions (10 – 15 Hz). In a high bandwidth semi-active or active ABW AWA suspension mechatronic control system automotive scientists and engineers generally consider an actuator connected between the sprung and un-sprung masses of the vehicle. An active ABW AWA suspension mechatronic control system aims to control the suspension over the full bandwidth of the mechatronic control system. In particular this means that one aims to improve the suspension response around both the *'rattle-space'* frequency (10 -- 12 Hz) and *'wheel-tyre hop'* frequency (3 – 4 Hz). The terms rattle-space and wheel-tyre-hop may be regarded as resonant frequencies of the mechatronic control system.

An active ABW AWA suspension mechatronic control system may consume a significant amount of power and may require actuators with a relatively wide bandwidth [WRIGHT 1984].

- *Preview* - These aim to increase the bandwidth of a band-limited mechatronic control system by using feedforward or knowledge of future on/off road surface inputs. Some mechatronic control systems [FOAG 1989] aim to measure on/off road surface disturbances ahead of the automotive vehicle (using perhaps a microwave radar- or laser-based sensor [PREM 1987; RADATEC 2002], and then use both standard feedback and feedforward control from the sensor to achieve a superior response. Others e.g. CROLLA AND ABDEL-HADY [1991] aim to use the data available from the front strut deflection to improve the performance of the rear suspension.

Semi-active or adaptive ABW AWA suspensions are terms usually used to describe vehicular suspensions that have some form of intelligence in the suspension's shock absorbers (dampers). Typically the damping curves may be altered such that the wheel control over the range of inputs is maximised. These vehicular suspensions also require fast-acting devices and complex mechatronic control algorithms. Semi-active actuation has received a fair amount of attention during the last decades because of its potential for reliable, low power mechatronic control. There are two distinct advantages to semi-active mechatronic control. The first of these is that, for a vehicle that is open-loop stable (i.e. stable without mechatronic control), the implementation of a mechatronic control system employing semi-active actuators physically cannot destabilise the vehicle. Thus, some questions concerning stability-robustness vanish. Secondly, semi-active actuators are unaffected by external power supply failures (as they operate on small, vehicular, **chemo-electrical/electro-chemical** (CH-E/E-CH) storage battery power making them more reliable. These two properties, combined, result in a highly appealing actuator for use in mechatronic control systems requiring extremely high reliability, such as in automotive engineering applications concerning on/off road surface response reduction. These properties have made semi-active actuators appealing in the area of vehicular suspensions, where they have exhibited favourable, reliable performance, while consuming negligible power.
Active Suspensions consist of actuators to create the considered necessary forces in the vehicular suspension. For instance, the actuators are usually F-M cylinders. External power is necessary to operate the mechatronic control system. Active suspensions may be arranged as slow-bandwidth or high-bandwidth consistent with the definitions preset above. Active ABW AWA suspensions provide independent treatment of on/off road-induced forces from vehicle body-inertia forces through active mechatronic control of some of the vehicular suspension functions. Theoretically this means that the compromise in conventional passive suspensions can be eliminated. An active ABW AWA suspension, however, usually involves a continuous power requirement, fast-acting devices, complex mechatronic control algorithms, and closed-loop control systems. The cost of these vehicular suspensions has limited their application on mass-manufactured vehicles.

The most fundamental disadvantage of active mechatronic control is that it in general necessitates an external power supply. This may present questions about both reliability and practicality.

In automotive engineering applications, an active mechatronic control system designed to protect an automotive vehicle during driving would only be as reliable as the primary energy source on which it depends. But history has shown that during driving, the **energy-and- information-network** (E&IN) is highly susceptible to destabilisation and blackouts.

On the other hand, the power demands of active mechatronic control systems for luxury and/or heavy vehicles are typically too large to meet with vehicular supplies. Thus, active mechatronic control systems are, by their very nature, inherently unreliable.

The second disadvantage of active mechatronic control is that, by designing an actuator that may accept power from an external primary energy source, the system cannot be characterised as a *'bounded energy'* E&IN.

Questions therefore quickly arise concerning the stability-robustness of such mechatronic control systems. Although significant effort has been put towards to alleviate this problem through the application of robust control theory, the nature of physical and mathematical model uncertainties in vehicles is such that significant concerns remain unresolved.

Summing up, an active suspension represents a sophisticated mechatronically-controlled active ABW AWA suspension mechatronic control system that uses powered actuators instead of conventional springs and shock absorbers. The actuators position a vehicle's wheels in the best possible manner to deal with on/off road surface disturbances and handling loads. Handling is a general term covering all the aspects of a vehicle's behaviour that are related to its directional control.

Hybrid Suspensions that are combinations of the other four categories. One compromise between active and passive actuation is found in hybrid ABW AWA suspensions. These consist of a combination of passive and active ABW AWA suspensions working in tandem. Hybrid ABW AWA suspensions are more reliable, because the passive part of the actuator may still work in the absence of power.

In this section, the word *'active'* in the context of an ABW AWA suspension mechatronic control system means it is not only a dummy passive suspension whose characteristics remain constant and the response is dependent only on the physical quantities that affect the response directly.

The active ABW AWA suspension mechatronic control systems' response depends not only on the physical quantities that affect the response directly, but also on physical quantities that do not affect the response directly.

A physical quantity that affects the response of the active ABW AWA suspension mechatronic control system directly is, for instance, the shock absorber (damper) velocity, while the vehicle-body's roll velocity may be used as an example of a physical quantity that has no direct effect on the function of the suspension.

A controller, the *'intelligence'* of the active ABW AWA suspension mechatronic control system, characterizes the latter.

The idea of a passive and/or active ABW AWA suspension mechatronic control system may be more easily appreciable with the structural and functional block diagrams shown in Figure 5.5 [HYVÄRINEN 2004].

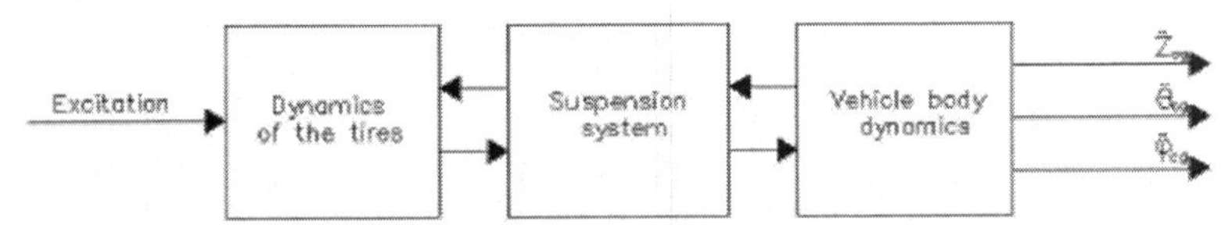

Fig. 5.5 A passive AWA suspension mechatronic control system [HYVÄRINEN 2004].

From the structural and functional block diagram shown in Figure 5.5, it may be observed that a passive AWA suspension mechatronic control system's response to excitation is affected only by the excitation and the state variables of the system that have a direct affect on the function of the passive ABW AWA suspension mechatronic control system.

A structural and functional block diagram of an active ABW AWA suspension mechatronic control system is shown in Figure 5.6 [HYVÄRINEN 2004].

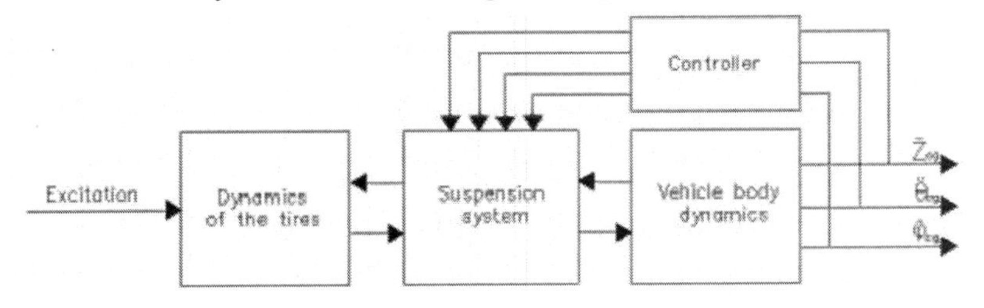

Fig. 5.6 An active ABW AWA suspension mechatronic control system [HYVÄRINEN 2004].

From Figure 5.6 it may be observed that the function of the suspension-system block is affected also by indirect physical quantities, for instance, second derivatives of the heave z, pitch θ and roll φ. A common way of implementing the actuator in an active ABW AWA suspension mechatronic control system is to use the variable absorbing (damping) in which context the semi-active ABW AWA suspension mechatronic control system is activated. Another way of affecting the function of the vehicular suspension is to create counter-force or counter-moment with the damping system in which context the active ABW AWA suspension mechatronic control system is activated [HYVÄRINEN 2004].

The idea of an active ABW AWA suspension mechatronic control system in automotive vehicles is from the past, probably from the 1930s.

The most significant progress in this area began at the end of the 1970s, with the active ABW AWA suspension used in the *Formula 1* racing cars of that era [DIXON 2000]. The first vehicles with adjustable damping characteristics appeared in the beginning of the 1980s [CITROEN 2003A].

In the beginning, the adjustable damping mechatronic control systems could be disregarded as adaptive suspension mechatronic control systems, since their reaction times were below the natural frequencies of the vehicle. Nowadays, almost every major vehicle manufacturer and supplier has some type of an active ABW AWA suspension mechatronic control system commercially available. They vary from the simple manual selection between a soft and firm damping setting to a fully automatic tandem active/passive system [SACHS 2003A; DELPHI 2003A].

Because of the growing popularity of modern off-road vehicles, namely SUVs, there is also an active damping mechatronic control system suitable for them on the market [DELPHI 2003B].

Adjustable shock absorbers (dampers) have also been developed for commercial automotive vehicles, but they have not come into mass manufacture yet [SACHS 2003B].

The wheel-tyre acts as a connecting link between the on/off road surface and the automotive vehicle and has an outstanding influence on vehicle handling and comfort. Still the wheel-tyre is the most difficult suspension factor to control. The complexity of the wheel-tyre's behaviour has been pointed out in many forms of wheel-tyre physical and mathematical models used in computer simulations [BLUNDELL 1999B; LEE 1997].

Despite this fact, the ride and handling properties of an automotive vehicle may be changed within a wide range with an active ABW AWA suspension mechatronic control system for both on-road and/or off-road automotive vehicles.

The basic dynamic behaviour still remains the same, as long as the major design consists of a vehicle's body equipped with two or more axles. It is useful to distinguish among several categories of ABW AWA suspension mechatronic control systems currently being used in practice.

A purely active ABW AWA suspension mechatronic control system has the basic configuration as shown schematically in Figure 5.7a.

It consists of (a) sensors located about the automotive vehicle's ABW AWA suspension to measure either external on/off road surface excitations, or vehicular response variables, or both; (b) **electronic control units** (ECU) to process the measured data and to compute indispensable control forces necessary based on a given control algorithm; and (c) actuators, usually powered by external sources, to generate the necessary absorbing (damping) forces.

When only the vehicular response variables are measured, the mechatronic control configuration is referred to as feedback control since the vehicular response is continually supervised and this data is used to create recurrent corrections to the related control forces. A feedforward control results when the control forces are adjusted only by the measured excitation that may be attained, for on/off road surface inputs, by measuring accelerations at the vehicular base.

In the case where the data on both the response quantities and excitation are used for mechatronic control design, the term feedback-feedforward control may be used.

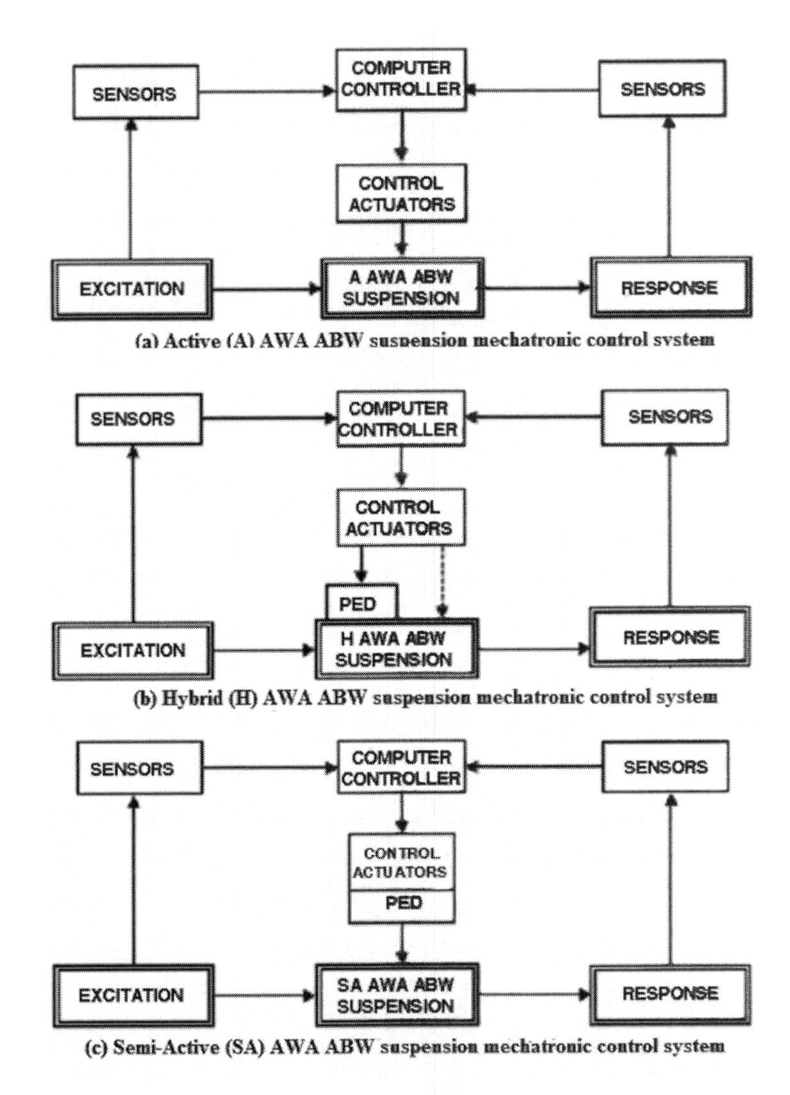

Fig. 5.7 Various ABW AWA suspension mechatronic control systems: PED – passive energy dissipation.

The term hybrid mechatronic control normally applies to a mutual passive and/or active ABW AWA suspension mechatronic control system as shown in Figure 5.7b. A fragment of the mechatronic control aim is realised by the passive mechatronic control system, involving an appropriately designed hybrid mechatronic control system in which less power and mechatronic control resources are necessary. Comparable control resource savings may be realised using the semi-active ABW AWA suspension mechatronic control system depicted in Figure 5.7c, where the mechatronic control devices do not deliver mechanical energy directly to the vehicle; hence bounded-input/bounded-output stability is assured.

Semi-active mechatronic control devices are often considered as controllable passive ones. A countenance advantage of hybrid and semi-active ABW AWA suspension mechatronic control systems is that, in the circumstances of a power failure, the passive components of the mechatronic control still present some degree of safety, unlike an active ABW AWA suspension mechatronic control system.

Various categories of advanced ABW AWA suspension mechatronic control systems have appeared recently in prototype and manufactured automotive vehicles [ELBEHEIRY AND KARNOPP 1996].

Unlike classical passive suspension mechatronic control systems, these advanced ABW AWA suspension mechatronic control systems often depend on applying a control force as a function of measured dynamic vehicle parameters. Some of these mechatronic control systems require a power supply. For instance, a large amount of power is required for the operation of so-termed broadband

For active ABW AWA suspension mechatronic control systems, in comparison, little power is required for the operation of variable shock absorber (damper) suspensions such as semi-active shock absorbers (dampers) or mechatronically adjusted shock absorbers.

In most cases, active ABW AWA suspensions provide theoretically superior performance with either broadband or limited-band actuators. This is at the expense of high cost, construction complexity and possible stability problems when components fail. Therefore, active ABW AWA suspensions remain limited to expensive automotive vehicles.

Semi-active ABW AWA suspension mechatronic control systems are capable of providing most performance features of active ABW AWA suspensions with cost and complexity reduced to a great extent [KARNOPP 1992].

Moreover, the semi-active ABW AWA suspension mechatronic control system has no inherent stability problems. The concept of active absorbing (damping) is presented in KARNOPP [1983, 1987].

The idea is use of actuators capable of generating absorbing (damping) forces as functions of absolute vehicle velocity (as if there was skyhook damping) as well as passive forces related to the velocity across the shock absorber (damper).

The application of mechatronic control theory to the design of ABW AWA suspension mechatronic control systems has been the subject of many publications during the past four decades. A collection of methods and techniques of this theory applicable to the design of ABW AWA suspension mechatronic control systems has been reported in HEDRICK [1973] and KARNOPP [1973].

Numerical application of the **linear-quadratic-*Gaussian*** (LQG) criterion to the design of active ABW AWA suspension mechatronic control systems has been presented in THOMPSON [1976].

Further developments and investigations of the same problem are found in HAC [1985]; WILSON ET AL. [1986]; CHALASANI [1987] and ELMADANY [1990]. Since some state variables are difficult to measure or estimate, many algorithms on the basis of mechatronic control techniques have been developed and applied to the design of suboptimal or limited-state controllers of active ABW AWA suspension mechatronic control systems [THOMPSON 1984; THOMPSON ET AL. 1984; WILSON ET AL. 1986].

Also, optimal mechatronic control theory methods have been extended to the design of ABW AWA suspension mechatronic control systems for physical and mathematical models of the automotive vehicle excited by non-stationary on/off road surface inputs during variable velocity runs [RAJU AND NARAYANAN 1991], for vehicle stability investigations [PALKOVICS AND VENHOVENS 1992] and for non-linear controllers [GORDON ET AL. 1991].

Some invariant properties of all ABW AWA suspension mechatronic control systems that do not include reaction mass are discussed in HEDRICK AND BUTSUEN [1988].

When a physical and mathematical model of the automotive vehicle is considered linear and under the assumption that the on/off road surface input statistics follows a *Gaussian* probability distribution, the automotive-vehicle response is also *Gaussian* [ELBEHEIRY AND KARNOPP 1996].

In this manner, the design of passive suspensions with the variance of suspension deflection kept within specified values is investigated in HAC [1985]. An innovative generalised algorithm for solving the limited-state regulator problem may be developed. The variance of suspension deflection is constrained to specific values for various disturbance intensities. A generalised performance index formulated as a weighted sum of variances of sprung mass acceleration, suspension travel, wheel-tyre deflection, and the suspension control force may be considered. Comparison may be made among different ABW AWA suspension mechatronic control systems in terms of **root-mean-square** (RMS) values of important physical variables, frequency response plots, and eigenvalue loci [ELBEHEIRY AND KARNOPP 1996].

The active ABW AWA suspension mechatronic control system provides much better vehicle-body isolation than the other categories with or without equality constraint. Finally, the active ABW AWA suspension mechatronic control system requires minimum suspension control force to maintain constant suspension deflection compared with the other categories of suspension above mentioned [ELBEHEIRY & KARNOPP 1996].

In KOSLIK ET AL. [1998] an active ABW AWA suspension mechatronic control system for improving the ride comfort and safety of a tractor is investigated. The underlying planar dynamic tractor model and a suitable objective for optimal suspension are introduced. The problem of optimal active suspension leads to a linear-quadratic optimal control problem.

Classical LQG theory provides a closed-loop control for the steady state problem that is optimal only for an initial disturbance input from the on/off road surface. A direct transcription method can handle more general disturbances and physical and mathematical models but provide only an open-loop solution, where the time history of the optimal control is given along the optimal trajectory for one type of deterministic disturbance and initial value only.

Simulation results for two different road disturbances are given comparing both approaches. It is well known that compromise between ride comfort and handling performance has to be made to design a passive suspension of an automotive vehicle [ROH AND PARK 1999].

To overcome this problem, some automotive scientists and engineers have proposed the use of active ABW AWA suspensions. Unlike passive suspensions that may only store or dissipate energy, active ABW AWA suspensions may continually change the energy flow to or from the mechatronic control system when required.

Furthermore, characteristics of active ABW AWA suspensions may adapt to instantaneous changes in driving conditions detected by sensors. As a result, an active ABW AWA suspension may improve both ride comfort and handling performance to satisfactory levels.

It has been proposed in BENDER [1968] that performance of active ABW AWA suspension may be further improved if knowledge of the on/off road surface in front of the actively controlled axles, i.e., preview data, are used in the mechatronic control strategy. With preview data, one may, for example, prepare the automotive vehicle for an approaching on/off road surface input and pass through abrupt on/off road obstacles without severe impacts.

There are two possible modes to obtain preview data. One is to use a *'look-forward'* sensor and the other is to estimate on/off road profile from the response of the front wheels by assuming that on/off road surface inputs at the rear wheels are the same as those at the front wheels except for time delays. In ROH AND PARK [1999], the look-ahead preview control is considered.

In an automotive vehicle active ABW AWA suspension mechatronic control system, for practical reasons one cannot measure all the state variables and therefore, data on them are incomplete. All the measurement signals including the preview sensor signal are assumed to be contaminated by sensor noises.

To realise preview control in the automotive vehicle active ABW AWA suspension, one needs an optimal filter that may filter out the sensor noises from the preview sensor signal as well as a state estimator that may minimise estimation errors due to sensor noises. Thus, to obtain an optimal preview controller for a mechatronic control system with incomplete and noisy measurements, one needs to solve a stochastic optimal preview regulator problem with incomplete and noisy measurements, or simply the stochastic optimal, output feedback, preview regulator problem.

In YOSHIMURA AND EDOKORO [1993], the authors changed the problem into LQP form by augmentation of the dynamics of the original mechatronic control system and the on/off road surface inputs.

In these authors' formulation the correlation between the on/off road surface inputs is considered only in the estimation scheme but not in the control scheme. The controller does not use the preview data estimated in the previous steps that is said to be characteristic of preview control.

In HAC [1992], the author also solved the problem by the variational approach. The author derived the optimal control and estimation schemes independently and then showed that a separation principle may satisfy his solution to prove its optimality. However, the author did not consider the optimal estimation of the on/off road surface input from preview sensor signals; rather, the author use the delayed raw preview sensor signals as the estimated on/off road surface input. Therefore, the author's solution is not optimal if the preview sensor signal is corrupted by measurement noise.

In LOUAM ET AL. [1988], the authors derived the optimal preview regulator assuming availability of exact data on the state and the on/off road surface inputs. The authors transformed the preview regulator problem into the LQG problem by introducing a state vector, whose dynamics is a combination of those of the original mechatronic control system and previewed on/off road surface input.

The resulting control input may be given as the feedback input driven by the augmented state vector that is different from the structure of the conventional preview controller consisting of feedback and feedforward parts.

In KOSLIK [1997], it is written that recent research in automotive engineering has proven the superiority of active ABW AWA suspension mechatronic control systems for vehicles compared to conventional type passive suspension mechatronic control systems. They are superior both with respect to ride comfort and overall stiffness to resist body forces. As a rule, linear multi-body systems, so-called quarter-vehicle physical models, and **quadratic-cost-functionals** (QCF) are considered to obtain an optimal feedback control law for active ABW AWA suspension by the well-known solution of an algebraic matrix *Riccati* equation. Furthermore, the on/off road surface disturbance is assumed to be a step function. To enable (almost) optimal suspension of arbitrary disturbances in real-time an *'adaptive critic'* method is introduced. It is based on approximate dynamic programming that avoids the so-termed *'course of dimensionality'* by using **neural networks** (NN) as generalising function approximators for the value function and the feedback control. The network approximations are called critic network and action network, respectively. Radial basis function networks are chosen for approximation because locally tuned processing units are especially suited for incremental adaptation. Q learning offers network training that is a model-free adaptive critical method. Therefore, both non-linear systems and systems without full information may be considered. The latter point is important since it is very expensive to get full state information for realistic vehicle models. For the linear-quadratic control problem the proposed method may be investigated and the quality of approximation is compared to the *Riccati* solution. Extensions to problems with more general disturbances, more realistic cost functionals and non-linear dynamical systems may be discussed.

In ROH AND PARK [1999], the solution for the stochastic optimal output feedback; the preview regulator problem has been derived assuming incomplete and noisy measurements of the state variables and the on/off road surface input.

Augmenting dynamics of previewed on/off road surface input with the original mechatronic control system dynamics may solve the problem based on the LQG framework. The resulting compensator may be divided into the controller and the estimator part. While the controller may be identical to that in LOUAM ET AL. [1988], the estimator may be unique in its optimality under the noisy measurement condition. The proposed compensator may have high order dynamics compared with the original mechatronic control system that makes it impractical to implement in real applications. The compensator structure may be transformed into a combination of the feedback part with the same order of the original mechatronic control system and the feedforward part based on the measured on/off road surface inputs only. This transformation explicitly shows a relation between the conventional and the present preview controls [ROH AND PARK 1999].

In CHANTRANUWATHANA AND PENG [2004], the **modular-adaptive-robust-control** (MARC) technique is applied to design the force loop controller of an active F-M ABW AWA suspension mechatronic control system. A key advantage of the modular design approach lies in the fact that the adaptation algorithm may be designed for explicit estimation convergence. The effect of parameter adaptation on force tracking performance may be compensated for and thus it is possible to guarantee certain control performance.

Experimental results from a quarter-vehicle active suspension rig show that when realistic external disturbances and measurement noises exist, the modular design achieves a better estimate than the non-modular **adaptive-robust-control** (ARC) design. The improved estimation may be found to result in control signals with slightly lower magnitude while maintaining similar tracking performance.

Some automotive scientists and engineers in recent years have still concentrated on an active ABW AWA suspension mechatronic control system. In these researches, a variety of physical and mathematical models including a quarter-, half- and/or full-vehicle model has been considered. Although experiments show the importance of a non-linear behaviour of the actuator in determination of a suitable trade-off in an active ABW AWA suspension mechatronic control system, in some publications, no attention is paid to the non-linear behaviour of the actuator [YAMASHITA ET AL. 1994; WANG ET AL. 2001].

JAMEI [2003] describes the development of a novel generic methodology for eliciting optimal fuzzy rules of the *Mamdani* type using **symbiotic evolution** (SE). In this evolutionary computation-based approach a randomly selected group of rules sets up a fuzzy inference system, then based on the system's performance a proportional score is allocated to each contributing rule.

Subsequently, the overall fitness of all rules in the population is calculated and on the basis of this fitness the rules are selected to reproduce and survive to the next generation. In contrast to the conventional **genetic algorithm** (GA) based fuzzy rule generation algorithms, in this proposed approach, the rules are evolved from one generation to another and not the rule-bases.

This algorithm is implemented in two versions, namely the **self-organising symbiotic evolution** (SOSE) and the **self-adaptive symbiotic evolution** (SASE) methods.

In the SOSE method the membership functions' parameters are fixed and the method generates only the fuzzy rules, however, in the SASE method, the algorithm optimises both the inference system's structure and the membership functions' parameters.

In order to evaluate the capabilities of the proposed method, it was applied successfully to the design of active suspension systems. In this investigation, the **bond graphs** (BG) method is used to model the non-linear quarter- and half-vehicle models using parameters that relate, for example, to a Ford *Fiesta* MK2.

Simulation results proved that SE, coupled with FL to form a hybrid structure could be very successful in the design of an efficient active suspension system in terms of performance and size of the fuzzy rule-base. The carried-out investigations demonstrated that the obtained optimal fuzzy controllers not only perform well for the training road surfaces but also in the face of unseen road surfaces. Moreover, these obtained controllers demonstrated a robust behaviour in the presence of uncertainties in the system's parameters. Almost all fuzzy rule-base generation algorithms produce rule-bases with redundant and overlapped membership functions that limit their interpretability and elegance. This problem is also addressed by applying an algorithm to merge any similar membership functions.

It is shown that the applied algorithm leads generally to a more transparent and more interpretable rule-base with a minimum number of membership functions and a reduced number of elicited fuzzy rules. In addition, a new rule post-processing approach is proposed for recovering any lost performance following the above membership functions merging.

Furthermore, in order to improve the rule-generation process, the fuzzy rules' average activating strength is used as a parameter that gives an indication of the rules activity in the rule-base and hence helps in the rules' fitness assignment.

Preliminary results following the application of this method are encouraging and further experiments are underway to validate this method. It is worth noting that all procedures, simulations, and approximate reasoning computations are implemented in a versatile C^{++} environment that features a variety of GAs. The use of such environment reduces significantly all incurred computation times.

In GRUNDLER [1997], the innovative method for complex processes control with the coordinating control unit based upon a GA has been described.

Minimal energy spent criteria, restricted by given process response limitations, has been applied, and improvement in relation to other known optimising methods has been found.

Independent and non-coordinating PID and fuzzy regulator parameter tuning has been performed using a GA and the results achieved are the same or better than with traditional optimising methods, while at the same time the proposed method can be easily automated.

In some researches, use of a back-stepping mechatronic control may be used to investigate a non-linear behaviour of the actuator in different on/off road surface circumstances. A non-linear H_∞ control may be another design approach for a quarter-, half- and/or full-vehicle model. Also, usage of the cascade feedback structure may be another way for investigation of F-M, P-M or E-M actuators.

In SHARIATI ET AL. [2004] a decentralised robust H_∞ control may be designed for a half-vehicle model considering non-linear behaviour of the F-M actuator. To remedy this drawback, a cascade feedback structure may be used. Using this structure, the fact that the behaviour of the active ABW AWA mechatronic control system may be significantly linearised makes it plausible to determine the linear physical and mathematical model for it, in addition to formulate a minimum bounded norm multiplicative uncertainty description.

The two H_∞ controllers have been designed for the above structure, considering nominal performance for the first case and robust performance in the second one.

The solution of a mixed sensitivity problem may significantly reduce automotive vehicle vibration in the human being sensitivity frequency range.

Statistical analysis of the simulation results using random input as on/off road surface roughness illustrates that the proposed strategy may provide a suitable trade-off between road comfort and on/off road holding, despite the actuator's non-linear behaviour.

TRUSCOTT AND WELLSTEAD [1994] describes the development of an adaptive control algorithm for active ABW AWA suspension mechatronic control systems based on optimal regulated methods. The aim is to design an algorithm that may automatically tune at start-up to changed vehicle conditions and adaptively retune to changes in driving conditions (in particular on/off road surface generated disturbances). The proposed algorithm is a self-tuning regulator based on generalised **minimum-variance** (MV) control. Simulation results obtained for 3-DoF quarter vehicle-suspension demonstrate potential benefits of fully adaptive control in vehicular suspensions.

5.2.2 Vehicular Suspension Functions

The motivation for automotive vehicle designers to give their attention to semi-active and/or active ABW AWA suspensions develops from a predisposition for innovations to vehicles ride performance with no compromise (and conceivably progress) in handling stability. The modes of performance that may be enhanced by active mechatronic control are [GILLESPIE 1992]:

- *Ride Control* – Ride enhancements can be achieved by various methods. The mechatronic control system may detect and control pitch and bounce motions of the vehicle body directly. Ride enhancements are also attained indirectly when active mechatronic control is relevant to the modes explained below.

Suspension predispositions that optimise ride always decay performance in other modes, thus dictating a compromise in design. With active ABW AWA suspension, on the other hand, the mechatronic control may be relevant only during the manoeuvre, and ride performance demand is not compromised during other modes of travel. Particularly, the suspension may be enhanced for optimal ride performance during steady, straight-ahead travel, and ride separation predispositions, advanced to that achieved with only passive components, may be attained without compromise of handling behaviour.

- *Height Control* – Mechatronic control of vehicle height presents some advantages in performance. By adjusting to maintain a constant height despite alterations in load or aerodynamic forces, the suspension can always function at the design ride height, giving maximum stroke for negotiating bumps, and eradicating alterations in handling that would occur from function at other than the design ride height. A height control may let down the vehicle for trimmed down drag at high values of the vehicle velocity or adjust the pitch attitude to adapt aerodynamic lift. Height can be lifted up for greater ground clearance and suspension stroke on bad roads. Height altitude can also be appropriate for exchanging wheel-tyres and to cause clearance for wheel-tyre chains.
- *Roll Control* – Roll control in cornering is enhanced by escalating damping or exerting anti-roll forces on the suspension during cornering. Vehicle velocity, steer angle, steer rate and/or lateral acceleration may be detected to choose when roll control is right and proper. With the operation of active force-generating components it is possible to eradicate roll in cornering completely, and thereby eradicate any roll-induced understeer or oversteer effects from the suspensions. Besides, the roll moments may be selectively relevant at either the front or rear axles to adjust the understeer gradient by action of the alteration in cornering stiffness as a result of lateral load shift.
- *Dive Control* – Mechatronic control of dive (forward pitch) during dispulsion (braking) can be enhanced by greater than ever damping or exerting anti-pitch forces in the suspension during dispulsion (braking). Mechatronic control may be triggered by the brake light signal, brake pressure and/or longitudinal acceleration. Dive control in an active suspension replaces the necessity to design anti-dive geometry in the suspension linkages.
- *Squat Control* – Mechatronic control of squat (rearward pitch) during acceleration can be enhanced by greater than ever damping or exerting anti-pitch forces in the suspension during acceleration. Mechatronic control may be stimulated by the throttle position, gear selection and/or longitudinal acceleration. Squat control in an active ABW AWA suspension relieves the need to design anti-squat geometry in the suspension linkages of drive wheels, and may overcome the squat or lift action on the non-driven wheels.

- *On/Off Road Holding* – Additionally to mechatronic control of vehicle-body motion during manoeuvres in the modes explained above, active suspensions have the aptitude to get better on/off road holding by decreasing the dynamic variations in wheel loads that are activated from on/off road roughness. Normally, cornering performance is enhanced when dynamic load variations are minimised. The on/off road damage made by automotive vehicles, mostly heavy trucks, is also trimmed down by minimising dynamic wheel loads.

5.2.3 Vehicular Suspension Performance

Normally, the semi-active and/or active ABW AWA suspension have the greatest potential to attain optimum performance in the modes explained above, but at a discontent in mass and size, complexity and reliability as well as cost. For this reason the challenge to vehicle designers is to attain the advantages of active mechatronic control with a minimum of hardware.

Table 5.1 characterises the comparative performance that can be achieved with each level of convolution in the design [GILLESPIE 1992].

Table 5.1 Performance Potential of Various Types of Suspensions [GILLESPIE 1992]

	Suspension type	**Passive**	**Self-levelling**	**Semi-active**	**Active**
Performance Mode	Ride	Performance is a compromise between all modes	High	Medium	High
	Height		High		High
	Roll			Low	High
	Dive			Low	High
	Squat			Low	High
	On/off road-loading			Medium	High

With semi-active ABW AWA suspensions, even slow-active variable damping permits enhancement in roll, dive and squat control in conjunction with ride and on/off road holding. Variable stiffness may offer similar advantages, albeit greater cost because of the necessity to use air springs or adjustable mechanical springs. With low-bandwidth stiffness or damping control a more responsive suspension can be realised, High-bandwidth control is effective for retaining the constant wheel loads advantageous to handling, but there is little supplementary ride advantage from a high-bandwidth mechatronic control system.

Only with an active ABW AWA suspension may the widest range of enhancements be acquired in all performance modes. The performance of an active ABW AWA suspension optimised for ride compares with that of a passive suspension by much enhanced control of the vertical, pitch and roll motions at the sprung mass resonant frequencies [CHALASANI 1986a, 1986b].

Figure 5.8 contrasts the *Bode* diagram's response behaviour in these three modes for the two categories of suspensions [CHALASANI 1986A, 1986B; GILLESPIE 1992].

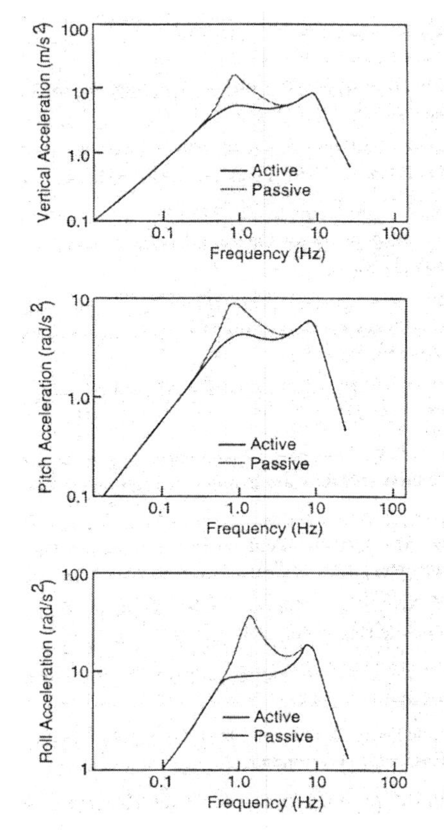

Fig. 5.8 Comparison of the *Bode* diagram's responses for passive and active suspensions [CHALASANI 1986B].

Whereas the passive suspension demonstrates sprung mass resonance near 1 Hz in the vertical, pitch and roll directions, a much-reduced response emerges with the active ABW AWA suspension. For practical purposes, control forces generated in the active ABW AWA suspension may effectively damp the sprung mass motions in these directions that are detected by accelerometers.

With mechatronic control features optimised for ride, there is no significant alteration in response at the un-sprung mass resonant frequency near 10 Hz. This is trimmed down by the fact that for the suspension to exert control forces that may lower un-sprung mass motions; those forces must be acted upon in response to the sprung mass, thus escalating the ride vibrations.

Handling is influenced by mechatronic control system response at the wheel hop frequency as a result of the related load vibrations on the wheel-tyres.

Since the *Bode* magnitude plots of the passive and active ABW AWA suspensions are equal in this bandwidth, little handling advantage is seen. To improve handling, the mechatronic control system design should be altered to trim down wheel hop response, even if some discontent in ride must be anticipated [CHALASANI 1986B; GILLESPIE 1992].

5.3 Passive Suspension

5.3.1 Foreword

The modern automotive vehicle has come a long way since the days when *'just being self propelled'* was enough to satisfy the vehicle's driver. Improvement in vehicular suspension, increased strength and durability of components, and advances in wheel and tyre design and construction have made large contributions to riding comfort and driving safety.

On a summer day in 1904 a young man by the name of *William Brush* helped bring about the modern vehicular suspension system. Driving his brother *Alanson*'s *Crestmobile*, *Brush* was rolling along too fast for the unpaved roads of the day and went into a curve at 48 km/h (30 mph). The vehicle's right front wheel skittered onto the dirt shoulder and whammed into a deep rut. Almost at once, the wheel started to shimmy violently. The undulations of the jarred right front elliptic leaf spring had sent shock waves across the solid *I*-beam axle to the left side of the vehicle. This set the entire front of the vehicle to vibrating furiously. *Brush* was caught unawares and lost control. The vehicle crashed through a barbed wire fence, hit a ditch and overturned in a cow pasture. Several hours later young *William* fussed up to *Alanson*, whose demeanour switched from stern to thoughtful, since he was trying to design a better automotive vehicle. That vehicle, dubbed the *Brush* Two-Seat *Runabout*, finally appeared in 1906. It featured a revolutionary vehicular suspension that incorporated two innovations never before assembled together: front coil springs and devices at each wheel that dampened spring bounce - shock absorbers -- mounted on a flexible hickory axle. Principally, conventional passive ABW AWA suspension (Fig. 5.9) refers to the use of front and rear springs to suspend a vehicle's frame, body, engine and powertrain above the wheels [DAS 2006].

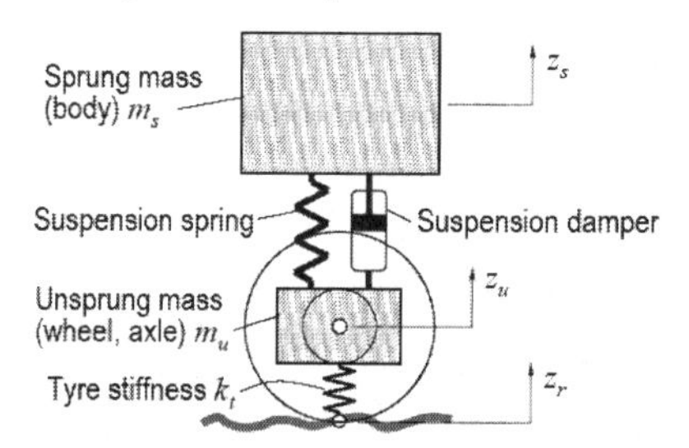

Fig. 5.9 Physical model of the conventional passive ABW AWA suspension [DAS 2006].

These relatively heavy assemblies constitute what is known as *'sprung masses. 'Unsprung mass'* on the other hand, includes wheels and tyre, brake assemblies and other structural members not supported by the springs.

The springs used in today's automotive vehicles are engineered in a wide variety of types, shapes, sizes, rates and capacities. Types include leaf springs, coil springs, air springs and torsion bars. These are used in sets of four per vehicle, or they are paired off in various combinations and are attached to the automotive vehicle by a number of different mounting techniques [AUTO PRO 2000].

Front suspension kinds -- There are two kinds of FWA suspension in general use: the independent system and the solid axle system. Independent RWA suspension became popular on the rough, twisty roads of Europe because it can offer improved ride and handling. The cheapest method is the swing axle, for which early **Volkswagen** (VW) vehicles were infamous. The differential is bolted to the frame, with constant-velocity joints on each side. However, as the wheels bounce over bumps, the wheel-tyre camber and rear track change radically, causing some handling quirks. In extreme manoeuvres, an outside wheel may actually tuck under the vehicle, causing it to flip. Axles with joints at both ends do a better job of keeping the wheels upright in a turn, and an amazing variety of control arms have been used to meet this end. Trailing arms, once popular, sometimes allowed trailing throttle oversteer - lift driver's foot off the acceleration (gas) pedal in a turn and the rear wheels shift slightly, throwing the vehicle into a skid. Modern designs use up to six control links at each wheel to prevent such erratic behaviour as bump steer and trailing throttle oversteer. Independent vehicular suspensions usually operate through heavy-duty coil springs or torsion bars and direct, double acting shock absorbers. In solid axle construction, the axle beam and wheel assemblies are connected to the vehicle by leaf springs and direct or indirect shock absorbers. With the solid axle set-up, the steering knuckle and wheel spindle assemblies are connected to the axle beam by bronze-bushed kingpins, or spindle bolts that provide pivot points for each front wheel. Modern independent FWA suspension systems use ball joints, or spherical joints, to accomplish the purpose. In operation, the swivelling action of the ball joints allows the wheel and spindle assemblies to be turned left and right and to move up and down with changes in road surface [AUTO PRO 2000].

Leaf spring -- Leaf springs in one form or another have been used since the Romans suspended a two-wheeled vehicle called a *Pilentum* on elastic wooden poles. The first steel spring put on an automotive vehicle was a single flat plate installed on carriages by the French in the 18th century. In 1804 *Obadiah Elliot* of London invented the venerable leaf spring that some automotive vehicle manufacturers still use in RWA suspensions today. He simply piled one steel plate on top of another, pinned them together and shackled each end to a carriage. Front leaf, or plate, springs are used in conjunction with solid axle beams in most truck applications. Rear leaf springs are used on trucks and some passenger vehicles. Single leaf or multi-leaf springs are usually mounted longitudinally over the front axle beam or under the rear axle housing. The spring centre bolt fastens the leaves together, and its head locates the spring in the front axle beam or saddle on the rear axle housing. *U*-bolts clamp the spring firmly in place and keep it from shifting. Eyebolts, brackets and shackles attach it to the frame at each end.

In many cases, leaf springs are used at the rear of the vehicle in combination with another type of spring in front. Chrysler, for example, uses leaf springs at the rear, torsion bars in front. For many years, *Ford* used leaf springs at rear, coil springs in front. Now, full-size vehicles have coil spring suspension, front and rear. Ford's small vehicles have coil springs in front; leaf springs at rear. Buick still uses coil spring all around. In some European and Pacific Asian vehicles, torsion bars are used front and rear; in others, leaf springs are mounted crosswise for use with independently suspended wheels. Rear leaf springs in American vehicles generally are placed parallel to the frame to absorb the torque of the driving wheels, while the front half of each leaf spring acts like a radius rod or control arm to transmit the driving force from the rear wheels to the frame. With this suspension set-up, the leaf springs also serve as a stabiliser in the side sway of the chassis [AUTO PRO 2000].

Coil springs -- Some European automotive vehicle manufacturers had tried coil springs, with *Gottlieb Daimler* in Germany being the leading exponent. However, most vehicle manufacturers stood fast with leaf springs: They were less costly, and by simply adding leaves or changing the shape from full elliptic to three-quarter or half elliptic, the spring could be made to support varying masses. The coil spring is not a spring chicken, either. The first patent for such a spring (British patent No. 792) was issued to *R. Tredwell* in 1763. The main advantage of coil springs was that they did not have to be spread apart and lubricated periodically to keep them from squeaking, as leaf springs did. *Henry Ford*'s 1908 *Model T* Ford featured old-fashioned leaf springs with a novel twist - he used only one spring at each axle, mounted transversely, instead of one at each wheel. Ford's adaptation of high-strength vanadium steel from a French racing car allowed him to save mass and cut costs in many areas of the *Model T* without compromising its durability. With the exception of a vehicle here and there, independent coil spring front suspension remained in limbo for 25 years after the introduction of the *Brush Runabout*. Then suddenly in 1934, General Motors, Chrysler, Hudson and others reintroduced coil spring front suspension, this time with each wheel sprung independently. In that year, most vehicles started using F-M shock absorbers and balloon (low-pressure) wheel-tyres. Coupling a solid front axle with shock absorbers and these wheel-tyres really aggravated front-end shimmy. Suspending each wheel individually lessened the effects of spring bounce. Not all vehicles used coil springs at first. Some had independently suspended leaf springs. But soon after World War II, all vehicle manufacturers switched to coil springs for the front wheels. Buick became the first U.S. vehicle manufacturer to use back-end coil springs in 1938. Vehicle manufacturers have switched back and forth from model to model between leaf and coil springs since then. Generally, large, heavy vehicles are equipped with leaf springs, while small light vehicles have coil springs. Many independent FWA suspension systems incorporate compression-type coil springs mounted between the lower control arms and spring housing in the frame. Others have the coil springs mounted above the upper control arms, compressed between a pivoting spring seat bolted to the control arm and a spring tower formed in the front-end sheet metal.

Generally, the upper control arm pivots on a bushing and shaft assembly that is bolted to the frame. The lower arm pivots on a bushing and shaft assembly or on a bolt cross frame member. When the lower control arm is not the A-frame type, it is supported by a strut that runs diagonally from the lower control arm to a bracket attached to the frame. On some models, this strut serves as a support; on others, it provides a means of adjusting caster. Stabilisers or sway bars are used in conjunction with front suspension on many vehicles to dampen rod shocks and minimise road sway. These bars are bracketed to the frame front cross member and extend from one lower control arm to the other. Actually, the **short-long arms** (SLA) system of front suspension has been adopted almost universally for passenger vehicles. The proportional lengths of the upper and lower control arms (and their engineered placement) are designed to keep the rise and fall of each front wheel in a vertical plane. With this arrangement, changes in wheel angularity, mass balance and tyre-scuffing tendencies are negligible when compared with solid axle suspension. Another coil spring set-up that is gaining application in small vehicles is *MacPherson* strut suspension. It combines coil spring, shock absorber and strut in a single assembly. When coil springs are used in both front and rear suspension, three or four control arms are placed between the rear housing axle and the frame to carry driving and braking torque. The lower control arms pivot in the frame members and sometimes support the rear coil spring to provide for up and down movement of the axle and wheel assembly. In addition, a sway bar (track bar) is usually attached from the upper control arm to the frame side rail to hold the rear axle housing in proper alignment with the frame and to prevent side sway of the vehicle body. However, if the rear coil springs are mounted between the frame and the swinging half axle, the independently suspended rear wheels have a sturdy axle housing attached to the differential housing that, in turns, is bolted to the frame [AUTO PRO 2000].

Torsion bars -- The first automotive vehicle to use torsion bar suspension was the 1921 Leyland. Most of the credit for the wide acceptance of torsion bars in Europe goes to Dr *Ferdinand Porsche* who made it standard on most of his vehicles, beginning with the 1933 VW prototypes.

By 1954, 21 European vehicle manufacturers had equipped their automotive vehicles with torsion bars. By contrast, in America, only Chrysler went the torsion bar route on its large-sized automotive vehicles. Despite its excellent ride qualities, high cost has limited its acceptance in this country.

A renowned British surgeon, who had been knighted by *Queen Victoria*, was convinced of a direct relationship between sound health and driving an automotive vehicle. Dr *William Thomson*'s observations were made in a 1901 edition of the *Journal of Medicine*: *"I have found my drives to improve my general health"*, Sir *Thomas* stated. *"The jolting which occurs when a motor car is driven at fair speed conduces to healthy agitation that acts on the liver. This aids the peristaltic movements of the bowels and promotes the performance of their functions"*. Vehicle manufacturers either did not read Sir *Thomas*'s report or did not care for his views, because soon afterward they began using shock absorbers to suppress vehicular jolting.

Since early automotive vehicles were limited to much the same vehicle velocity (speed) as carriages, leaf springs for them could be made of the right proportion to provide relatively jolt-free rides.

As roads were improved and values of the vehicle velocity shot up, a 1909 edition of *Automobile Engineering* noted: *"When springs are made sufficiently stiff to carry the load properly over the small inequalities of ordinary roads, they are too stiff to respond readily to the larger bumps. The result is a shock, or jounce, to the passengers. When the springs are made lighter and more flexible in order to minimize the larger shocks, the smaller ones have too large an influence, thus keeping the [vehicle] body and its passengers in motion all the time. These two contradictory conditions have created the field for the shock absorber"*. Although torsion bars were and are used extensively on European vehicles, this kind of vehicular suspension system received only token attention from the U.S. vehicle manufacturers until Chrysler developed their system in the early 1950s. Before that only a few buses, trailers and racing cars were equipped with torsion bar suspension. Basically, torsion bar suspension is a method of using the flexibility of a steel bar or tube twisting lengthwise to provide spring action. Instead of the flexing action of a leaf spring, the torsion bar twists to exert resistance against up and down movement. For example, an independently suspended front system with torsion bars mounted lengthwise would have one end of the bars anchored to the vehicle's frame and the other end attached to the lower control arms. With each rise and fall of a front wheel, the control arm pivots up and down, twisting the torsion bar along its length to absorb road chock and cushion the ride. Chrysler automotive vehicles are equipped with left and rights, non-interchangeable, front torsion bars with hex-shaped ends. In position, the bars extend from hex-shaped rear anchors in the frame cross members to hex-shaped holes in the front lower control arm, adjusting bolts are provided at the front mounting to increase or decrease torsion bar twist and thereby control front suspension height. Over the years, Chrysler has made many improvements in the system, including lengthening the torsion to lower the spring rate; adding a removable rear anchor cross member that rubber-isolated from the frame; devising a plastic plug and a balloon seal for the rear anchor. Oldsmobile *Tornado* and Cadillac *Eldorado* front wheel drive automotive vehicles also use lengthwise mounted torsion to support the front and to provide for high adjustment. Torsion bars can also be used laterally to provide spring action for front and/or rear wheel independent suspension system. Older VW vehicles offer a unique torsion bar arrangement with all four wheels independently suspended, but with two different torsion bar setups in use. At the front, two laminated square torsion bars behind separate axle tubes are anchored at the centre to counteract twisting and lateral movement. Each has a lever or torsion arm attached to its outer end. Ball joints connect the torsion arms to the steering knuckle. The wheel spindle trails behind the axle and tends to swing in an arc when moved up and down by road irregularities. For instance, at the rear, VW used one short, round torsion bar on each side. These bars are splined at each end and anchored in the centre of the frame cross member. The outer ends of the torsion bars carry the spring plates to which the wheels are attached.

Here, too, the wheels follow behind the torsion bar on *'trailing arms'* [AUTO PRO 2000].

Air suspension – The Cowey Motor Works of Great Britain introduced air suspension that *Lincoln* ballyhooed for some models introduced +in 1984 in 1909. It did not work well because it leaked. The first practical air suspension was developed by *Firestone* in 1933 for an experimental vehicle termed the *Stout -Scarab*. This was a rear-engine vehicle that used four rubberised bellows in place of conventional springs. Small M-P compressors attached to each bellow supplied air. As people might imagine, the air bag suspension was an expensive setup -- still is, in fact. Air suspensions are designed to cushion the ride and keep the car, bus or truck level fore and aft and at a constant height regardless of load. Air suspension was introduced on many luxury vehicles in the late 1950s, but it was dropped after one or two model years. Recently, however, new levelling systems have been researched and developed for passenger automotive vehicle use, including air-adjustable rear shock absorber. A typical air suspension system consists of an ECE- or ICE-driven M-P compressor, air supply pneumatical tank, filter or condenser, pneumatical valves, piping, controls and air springs or bellows. In operation, the air compressor maintains a constant pressure in the air supply pneumatical tank. Air is piped to the control pneumatical valves that feed air to each spring as needed. Air pressure is automatically increased on either side or at front or rear as required to keep the vehicle level and to keep any desired height from the road (within the limits of the system) [AUTO PRO 2000].

Automatic level control - Air springs are not used in Cadillac's automatic level control system. Rather, the rear shock absorber extends or compresses to bring the rear of the vehicle to the same level as the front. This automatic system is used on M-P compressors, pneumatical reservoir tank assemblies, air pressure regulators, hoses, flexible airlines, height control pneumatical valves and special shock absorbers. Older Ford vehicles used one of two levelling systems that use air bags in conjunction with rear coil springs. The automatic system consists of air reservoir and M-F pump, levelling fluidical valve, air bags, nylon tubing and metal fittings and connectors. The manual system has similar air bags connected to lines leading to the trunk of the automotive vehicle where a pneumatical valve connection permits levelling by application of air under pressure from an outside compressed air source. Late model Ford automotive vehicles carry an option for a rear suspension automatic load-levelling system that functions only after a load, approximately 1.78 kN (400 lbs.), is added to the automotive vehicle. When the load lowers the automotive vehicle to a specific level, an air sleeve rear shock absorber inflates and extends, raising the automotive vehicle to design height. When the load is removed, the air sleeve shock absorbers deflate and lower the automotive vehicle to design height. Ford's rear suspension automatic load-levelling system includes an M-P compressor and air bleed pneumatical valve, a pressure reservoir pneumatical tank and height control pneumatical valve and link, three flexible nylon airlines and a rubber vacuum hose. The M-P compressor maintains 464 or 888 kPa (60 or 125 psi) in the system. The air bleed pneumatical valve permits air to bleed for testing and servicing or for hooking up a trailer.

The pneumatical pressure tank stores high-pressure air. The height control pneumatical valve and link maintain vehicle design height at the rear. The lever on the height control pneumatical valve is attached to the rear suspension upper arm by means of the link. The control pneumatical valve senses riding height and either admits or exhausts height pressure air to or from the air sleeve shock absorber. A time delay mechanism in the height control valve prevents the transfer of air when the lever is moving during normal ride motions [AUTO PRO 2000].

Shock absorber -- The first recorded use of a crude shock absorber (damper) is the invention by one *A. Gimmig* in 1897. He attached rubber blocks to the top of each leaf spring. When the suspension was compressed sufficiently, the rubber bumpers hit bolts that were attached to the frame. Rubber bump stops are still used in many modern suspensions, but their effect on ride control is minimal. The first true shock absorbers were fitted to a racing bicycle in 1898 by a Frenchman named *J. M. M. Truffault.* The front fork was suspended on springs, and incorporated a friction device that kept the bike from oscillating constantly. In 1899, an American automobile enthusiast named *Edward V. Hartford* saw one of *Truffault*'s bikes win a marathon race at Versailles France. *Hartford* immediately recognized the automotive potential of the friction device. *Hartford* and *Truffault* got together and by the next year *Hartford* had outfitted an Oldsmobile with a variation of *Truffault's* device. This first automotive vehicle shock absorber consisted of a pair of levers that were hinged together with a pad of rubber placed at the pivot point. One of the lever arms was attached to the frame, while the other was bolted to the leaf spring. A bolt placed at the hinge point could be tightened or loosened to increase or decrease the friction, providing a stiffer or softer ride.

The *Truffault-Hartford* unit was, therefore, not only the first automotive shock absorber, but also the first adjustable “shock”. *Hartford* brought the car to America, where he opened his own plant, the Hartford Suspension Co., in Jersey City, New Jersey. His first big contract came from *Alanson P. Brush,* who installed shock absorbers along with front coil springs on the 1906 *Brush Runabout.*

The ride given by the automotive vehicle was called *'magnificent'* in a critique written by *Hugh Dolnar* for *Cycle and Automobile Trade Journal.* From then on shock absorber designs came fast and furious.

Among them:

- *Gabriel Snubbers* -- This consisted of a housing that contained a belt wound into a coil. It was kept under tension by a spring. The housing was fastened to the frame and the outer end of the belt was attached to the axle to limit the degree of rebound from a jolt. The Gabriel Co. started operation in 1906 making accessory auto horns. The founder, *Claude H. Foster*, named his firm after the horn-tooting angel *Gabriel.* When the pushbutton horn came along in 1914, it killed the *Gabriel* and all other body-mounted horns. *Foster* looked for a product to keep his company in business and came across the *Snubber*.
- *Equalizing springs* -- These were auxiliary coil springs used in addition to the leaf spring. Since each spring had a different harmonic frequency, they tended to cancel out one another's oscillations. But they also added to ride harshness and soon fell out of favour.

- *Air springs* -- Air springs combine spring and shock absorbing action in one unit and were often used without metal springs. The Cowey Motor Works of Great Britain developed the first one in 1909. It was a cylinder that could be filled with air from a bicycle pump through a pneumatical valve in the upper part of the housing. The lower halves of the cylinder contained a diaphragm made of rubber and cord that, because it was surrounded by air, acted like a pneumatical tyre. Its main problem was that it often lost air. The newest air spring, developed by Goodyear, is found on some late-model Lincolns. Like the ones that have preceded them, these ride-on-air units are more costly than conventional springs and hydraulical shock absorbers.
- *F-M shock absorbers* -- *M. Houdaille* of France gets credit for designing the first workable F-M shock absorber in 1908. F-M shocks dam spring oscillations by forcing oily fluid through small passages. In the popular tubular shock, a piston with small orifices is attached to the chassis and a cylindrical oily fluid reservoir is attached to the suspension or axle. As the suspension moves up and down, the piston is forced through the oily fluid, resisting the action of the spring. One-way fluidical valves allow different orifices to be used to control suspension jounce and rebound. This is termed a double-acting shock. The latest wrinkle is to add a pneumatical chamber of compressible gas at one end of the oily fluid reservoir to cushion the damping action. *Monroe* built the first original equipment hydraulic shocks for Hudson in 1933. By the late 1930s the double-acting tubular shock absorber became common on automotive vehicles made in the USA. In Europe, lever-type F-M shocks prevailed into the 1960s. They resembled the *Hartford* friction shock, but used oily fluid instead of a friction pad.
- *MacPherson struts* -- With the advent of FWD automotive vehicles, manufacturers in the 1970s and 1980s started using *MacPherson* struts. *MacPherson*, a General Motors engineer, developed this unit in the 1960s. It combines the coil spring, F-M shock absorber, and upper suspension arm into a single compact device. The main advantage is that it permits the necessary space for positioning the front-drive transaxle.

For instance, several Japanese automotive vehicles now feature struts with shock valving that may be adjusted from soft to firm by E-M motors while the vehicle is moving. The driver has a choice of three settings, but a signal from the speedometer usually overrides the manual control at highway values of the vehicle velocity to set the shocks on firm. The Nissan *Maxima* for 1985 sold in Japan had mechatronically controlled shocks that automatically provided a soft, medium, or firm ride depending upon on/off road surface circumstances, vehicle velocity (speed), and driving style. A sonar unit under the bumper monitored the on/off road surface, while other sensors checked vehicle velocity, acceleration, steering angle, and brake use. Data were fed to a **central processing unit** (CPU) that decided if you were driving gently or aggressively, and then activated shafts in the shock absorbers that altered the size of fluid passages.

The Lotus active suspension does away with springs and shock absorbers altogether. Eighteen motion-sensing transducers send data to four computer-controlled F-M RAMs. The system distinguishes roll, dive, jounce, and bump.

Fluidical valves in the RAMs adjust the ride accordingly. These fluidical valves may alter position as much as 250 times per second. For instance, the Lotus active suspension has the uncanny ability to keep a vehicle level in a tight turn or even bank it toward the inside of the turn, rather than leaning to the outside as other automotive vehicles do.

A wide variety of shock absorbing devices have been used to control spring action. Today, however, direct double – acting, *'telescoping'* F-M shock absorbers have almost universal application. At the front, each shock absorber often extends through the coil spring from the lower control arm to a bracket attached to the frame.

On Chrysler automotive vehicles with torsion bar suspension, the front shock absorbers attach to the lower control arm and mount to a bracket on the frame. In the case of high-mounted coil springs, each front shock absorber extends from the upper control arm to a platform mounted in the spring tower or to a bracket on the wheel housing in the engine compartment. At the rear, the lower end of the shock absorber usually is attached to a bracket welded to the axle housing. The upper end is fastened to the frame or to the coil spring upper seat that is integral with the vehicle frame or body.

On vehicles with rear leaf springs the rear shock absorbers generally extend from a stud attached to the spring U-bolt mounting bracket to the frame cross member. Quite often the rear shock absorbers are mounted at an angle to assist in restricting lateral movement as well as vertical movement.

Some Oldsmobile *Tornado* and Cadillac *Eldorado* automotive vehicles use four rear shock absorbers to give better control. And, for this same reason, some Chevrolet automotive vehicles have *'bias-mounted'* rear shock absorber. The cub -side unit is mounted in front of the axle housing.

Some Ford vehicles also feature this arrangement. The operating principle of direct-acting F-M shock absorbers consists of forcing oily fluid through restricting orifices in the fluidical valves. The restricted flow serves to slow down and control the rapid movement of the vehicle springs as they react to on/off road surface irregularities.

Generally spring loaded fluidical valves control oily fluid flow through the piston. The F-M shock absorber automatically adapts itself to severity of the shock. If the axle moves slowly, resistance to the flow of the oily fluid may be light. If axle movement is rapid or forceful, the resistance is much stronger since more time is required to force the oil through the orifices. By these fluidical actions and reactions, the shock absorbers permit a soft ride over small fluidical valves and provide firm control over spring action for cushioning large bumps. The double-acting units operate efficiently in both directions. Spring rebound may be almost as violent as the original action that compressed the shock absorber [AUTO PRO 2000].

Conventional passive suspensions exhibit a significant compromise between ride comfort (bounce and single wheel stiffness), vehicle handling (roll stiffness) and wheel loading during articulation (articulation stiffness). These suspensions do not allow for decoupled roll and bounce damping modes, further hindering the ability to tune a well balanced, safe and capable suspension for all driving conditions both on/off road surface.

Innovative passive suspensions are unique in that they may provide reduced articulation stiffness, increased roll stiffness; high levels of comfort and mode decoupled roll and bounce damping all in a passive suspension. All these parameters are independently tuneable.

Innovative passive suspensions are not active; they typically require no F-M or P-M or even E-M motors, M-F pumps or M-P compressors or even M-E generators, respectively, as well as computers.

Nonetheless such suspensions may be easily equipped to provide load levelling and ride height adjustment and may also be adapted to include active or semi-active mechatronic control features to enhance their inherent passive functionality.

Innovative passive suspensions have been applied and proven in a diverse range of automotive vehicles; from passenger vehicles, light SUVs, full size SUVs and motor sport applications to military and heavy trucks, vehicle manufacturers continue to develop advanced technologies for a wider range of market applications.

5.3.2 Passive F-M or P-M Shock Absorber Suspension Mechatronic Control System

A conventional shock absorption component of a suspension mechatronic control system consists of a totally enclosed F-M or P-M shock surrounded by a large spring. This kind of suspension is termed passive F-M or P-M absorber suspension because there is nothing occurring to alter the properties of the self-regulating mechatronic control system as it is moved by shocks to the wheels.

A passive F-M or P-M shock absorber suspension mechatronic control system dissipates the energy of shock by compressing the F-M or P-M shock absorber and the spring. The restorative force works to put the wheel back on the on/off road surface as soon and efficiently as feasible. The downside of a P-M shock absorber suspension mechatronic control system is the energy that is not dissipated by the spring and F-M or P-M shock absorber is transferred to the vehicle body of the automotive vehicle and subsequently, the drivers. Hitting a velocity bump or any object in the on/off road surface fast enough may send an automotive vehicle into the air. Passive F-M or P-M shock absorber suspension mechatronic control systems are self-regulating mechatronic control systems that are constructed so that changes in attitude or height of the vehicle body relative to the on/off road surface are compensated for without the need for means specifically provided for detecting such alterations. Important examples are interconnected F-M shock absorber suspension mechatronic control systems where the interconnections are such that movement of one F-M shock absorber forces oily fluid through pipe work to an opposing shock absorber e.g. on the other side of the vehicle, to cause movement of that F-M shock absorber in the same, or opposite sense of direction.

Simple mechanical linkages may also achieve compensatory movement of this type. Also included here are mechatronic control systems of torsion bars, usually but not necessarily linked, specially designed for the purpose of controlling roll, pitch and so on. This may sometimes form the sole means of suspension.

A passive F-M or P-M shock absorber suspension mechatronic control system is one in which the characteristics of the components (coil springs and shock absorbers) are fixed.

The designer of the suspension according to the design goals and the intended application determines these characteristics. Passive F-M or P-M shock absorber design is a compromise between vehicle handling and ride comfort, as shown in Figure 5.10 [SIMON 2001].

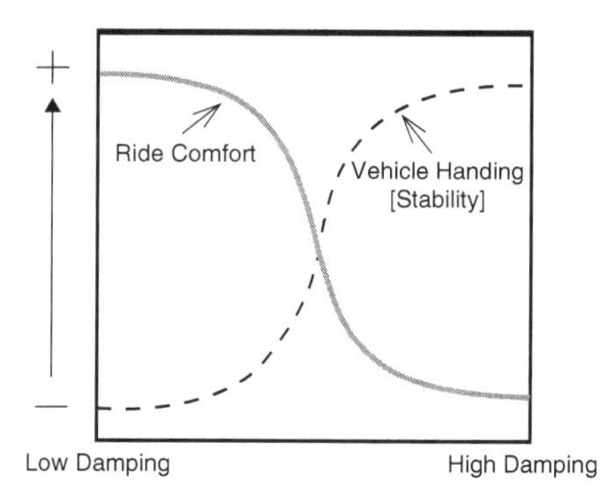

Fig. 5.10 Damping compromise for the passive shock absorbers [SIMON 2001].

A heavily damped passive F-M or P-M shock absorber suspension may yield good vehicle handling, but also transfers much of the on/off road surface input to the vehicle body. When the vehicle is moving at low values of the vehicle velocity on a rough on/off road surface or at high values of the vehicle velocity in a straight line, this may be perceived as a harsh ride.

The drivers may find the harsh ride objectionable or that it damages cargo. A lightly damped passive F-M or P-M shock absorber suspension may yield a more comfortable ride, but may significantly reduce the stability of the vehicle in turns, lane change manoeuvres, or in negotiating an exit ramp.

Good design of a passive F-M or P-M shock absorber suspension may to some extent optimise ride and stability, but cannot eliminate this compromise.

Passive F-M or P-M Shock Absorber Suspension Mechatronic Control System Arrangements - Fundamentally, all contemporary passive suspensions provide support springs roll control and damping. All contemporary suspensions use some form of F-M, P-M or E-M shock absorber (damper) e.g. telescopic or inline.

For support springs, contemporary pneumatical suspensions may contain conventional mechanical springs such as leaf, coil, or torsion; F-P-M springs that contain nitrogen over oily-fluid; air springs; or a combination of any two types above.

For roll control, suspensions may contain conventional torsion-type roll control springs with the front and rear torsion springs interconnected by F-M cylinders, F-P-M springs, or a combination of the two (Fig. 5.11) [BROGE 2000].

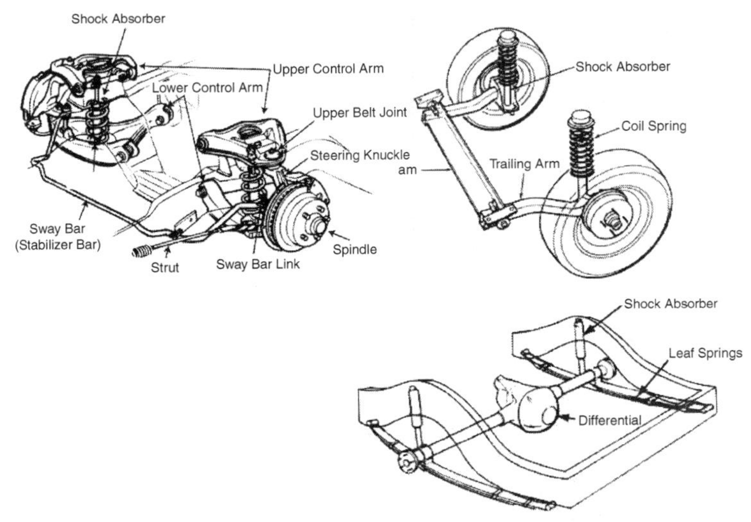

Fig. 5.11 One version of contemporary passive suspension
[Tenneco Automotive; BROGE 2000 – Bottom left corner image].

Suspension stiffness is used to describe the handling performance of a vehicle (Fig. 5.12) [BROGE 2000].

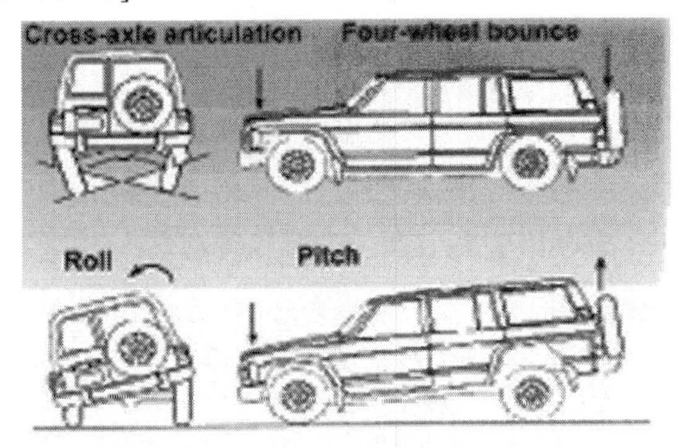

Fig. 5.12 Ride and handling parameters of a vehicle [BROGE 2000].

Roll stiffness is the stiffness of the suspension supporting the sprung mass when the wheels on one side of a vehicle have an increase in load and the wheels on the other side have a decrease in load.

Pitch stiffness is the stiffness of the suspension supporting the sprung mass when the wheels on the front of a vehicle have an increase in load and the wheels on the back have a decrease in load.

During cornering (vehicle roll) the front wheels and the back wheels experience a separate moment caused by the respective vertical force couple at the wheels.

The reaction to these two moments defines the roll stiffness of the front wheels, the back wheels, and the whole vehicle. In the past, some vehicle manufacturers have increased the size of the stabilizer bars to reduce the risk of rollover in vehicles, though the larger bars brought about a stiff uncomfortable ride.

For instance, the Kinetic™ Suspension Technology may cut and reconnect the bars through F-M cylinders. The left and right sides of the stabilizer bar rotate relative to one another.

Then the front and rear F-M cylinders are connected, allowing fluid to move back and forth in response to wheel motion. This shift redistributes road forces through the entire suspension system to absorb shock and equalise load on the wheels (Fig. 5.13) [BROGE 2000].

Fig. 5.13 Kinetic™ suspension-equipped two-wheel-drive (2WD) vehicle can perform a conventionally equipped four-wheel-drive (4WD) vehicle [Tenneco Automotive].

The results shown in Figure 5.14 were created by measuring the four vertical wheel loads while raising the front right wheel at low velocity, effectively giving static results [BROGE 2000].

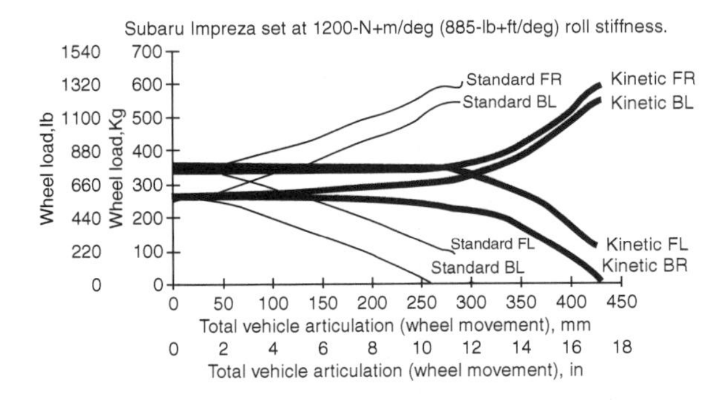

Fig. 5.14 Articulation wheel loads: a *Standard* suspension vs. a *Kinetic* suspension [BROGE 2000].

The difference between the front and rear load changes is due to the test creating a small roll moment as well as pure cross-axle articulation. As the wheel movement in the conventional vehicle is increased, the support springs deflect and the roll stabilizer bars wind up, and the wheel loads change almost linearly.

When the load on the back right wheel reaches zero the wheel lifts from the ground, leaving the vehicle balanced on three points. When the vehicle fitted with the suspension is articulated, the support and roll stabilisation systems are initially not subject to any wind up and the wheel loads remain almost constant, providing free articulation (Fig. 5.15) [BROGE 2000].

Fig. 5.15 Jeep *Grand Cherokee* with the Kinetic™ suspension has a free articulation limit of 55.0 cm (22 ") *an*d an ultimate articulation limit of 77.5 cm (31 "), compared to the standard automotive vehicle's ultimate limit of 47.5 cm (19 ") [Tenneco Automotive; BROGE 2000].

This is continued until the relevant bump stop is compressed, at which point the wheel loads alter relatively quickly. Nearly all the rear suspension move occurs before the rear wheel lifts off the ground.

Kinetic™ Suspension Technology also investigates alarms about rollover accidents that OEMs quote as causing more than 165,000 deaths or injuries per year in the USA, with the death rate in SUV rollovers having doubled since 1988 [BROGE 2000].

When a vehicle is cornering or involved in an avoidance manoeuvre, a kinetic suspension responds instantly to reduce the risk of rollover.

For instance, oily fluid in the fluidical line stops flowing causing the stabilizer bars to reconnect for optimum stability.

Conventional roll stabiliser bars fitted to most modern automotive vehicles act well to trim down vehicle-body roll during cornering but also greatly increase single wheel and articulation stiffness that downgrade ride comfort and the automotive vehicle's aptitude to continue equal wheel loading on uneven on/off road surfaces.

Vehicular suspension designers struggle to find a balance between these ride parameters as each compromises the other. For instance, the Kinetic™ **reverse function stabiliser** (RFS) (Fig. 5.16) may retain conventional bounce and damping methods and operates by splitting the roll stabiliser bars using a simple cradle device incorporating a double-acting F-M cylinders [KINETIC 2005].

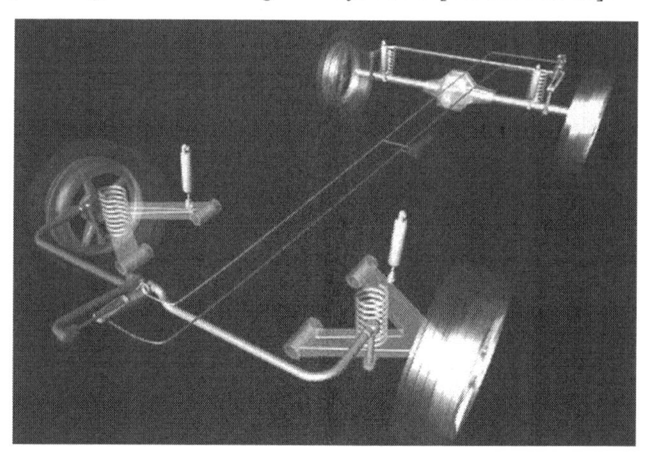

Fig. 5.16 Reverse function stabiliser (RFS)
[Kinetic™ Suspension Technology; KINETIC 2005].

These F-M cylinders are connected using kinetics unique to the proprietary RFS arrangement. This arrangement passively frees the bars to allow articulation and single wheel type movements but may not permit vehicle-body roll. This roll stiffness and *'articulation looseness'* are achieved simultaneously with neither mode affecting the other. This system requires no computers, M-F pumps or M-P compressors or even M-E generators as well as F-M or P-M or even E-M motors, respectively.

The H2 is a passive suspension (Fig. 5.17) that replaces existing shock absorbers (dampers) with simple double-acting F-M cylinders [KINETIC 2005; WILDE 2005; WILDE ET AL. 2005].

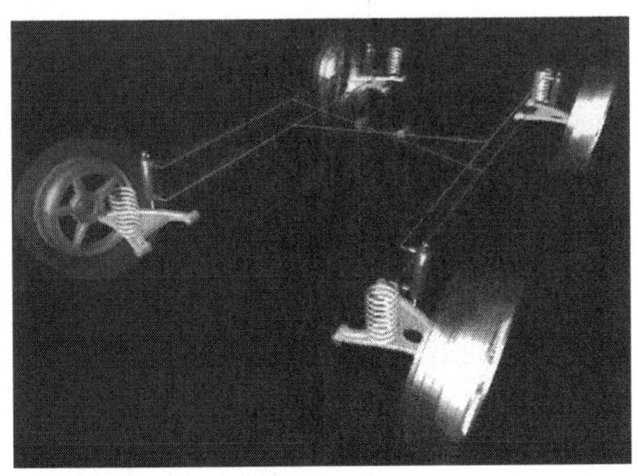

Fig. 5.17 H2 passive suspension [Kinetic™ Suspension Technology; KINETIC 2005].

Each cylinder performs the normal wheel/body damping functions using conventional or advanced damping technology incorporated within the cylinders. All cylinder chambers (eight chambers in total – two per wheel) are interconnected using relatively small fluidical (hydraulical) lines. The interconnection provides extremely high levels of roll stiffness while passively maintaining low levels of single wheel and articulation stiffness. The interconnection also provides the opportunity to incorporate roll damping independently of wheel/ body damping. The system also incorporates small accumulators at various positions depending on the design specification and vehicle parameters.

The X is Kinetic's most sophisticated and complex pneumatical suspension (Fig. 5.18) providing a very high degree of functionality and performance [KINETIC 2005; WILDE ET AL. 2005].

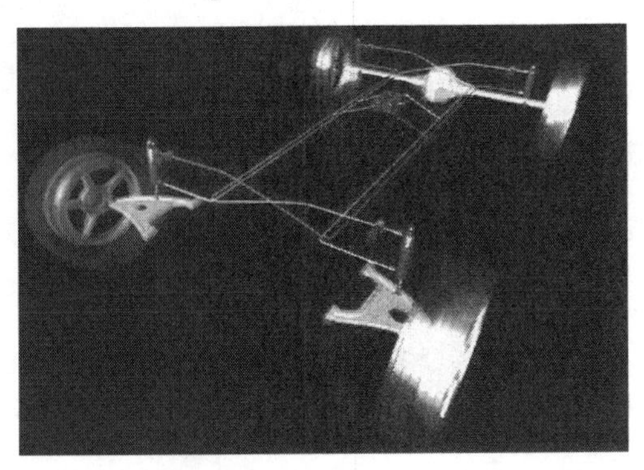

Fig. 5.18 X passive suspension [Kinetic™ Suspension Technology; KINETIC 2005].

For a comparison, the passenger automotive vehicle Kinetic™ Subaru *Imprezza* (blue) showing increased roll stiffness and reduced articulation stiffness providing enhanced handling on all road surfaces is exemplified in Figure 5.19 [KINETIC 2005].

Fig. 5.19 Comparison of two the same passenger automotive vehicles: Shubaru *Imprezza* with the conventional passive suspension (red) and Kinetic™ passive (P) suspension (right) [Kinetic™ Suspension Technology; KINETIC 2005].

The comparison of the Kinetic™ suspension's features and benefits to those of a conventional suspension is shown in Table 5.2 [KINETIC 2005].

Table 5.2 The comparison of the Kinetic™ suspension's features and benefits to those of a conventional suspension [KINETIC 2005].

Features & Benefits	Kinetic	Conventional Suspension
Decoupled roll, articulation & bounce Stiffness	yes	no
*Decoupled roll and bounce Damping	yes	no
**Load Dependent Roll Stiffness	yes	no
*~Height Control	yes	no
*~Load Levelling	yes	no

*** inherent in X systems, optional on H2 & RFS systems;**
**** inherent in X systems, optional on H2 system;**
***~ inherent in X & H2 systems.**

Design of a passive suspension mechatronic control system is equivalent to design of a controller with structural static output feedback. The feedback gains are composed of the stiffness and damping parameters associated with each wheel, the closed loop inputs are the on/off road surface disturbances, and the closed-loop outputs are the performance indices [ZUO AND NAYFEH 2003A].

The shock absorbers (dampers) are approximated as linear dashpots, ignoring the asymmetry in the jounce and rebound. The wheel-tyre is physically modelled as a linear spring with or without some small damping. The wheel is taken as a one **degree-of-freedom** (DoF) mass, and the vehicle body is physically modelled as a rigid body.

Figure 5.20 shows a classic full-vehicle (four-wheel automotive vehicle) physical model with **eight-degrees-of-freedom** (8-DoF) [ZUO AND NAYFEH 2003A].

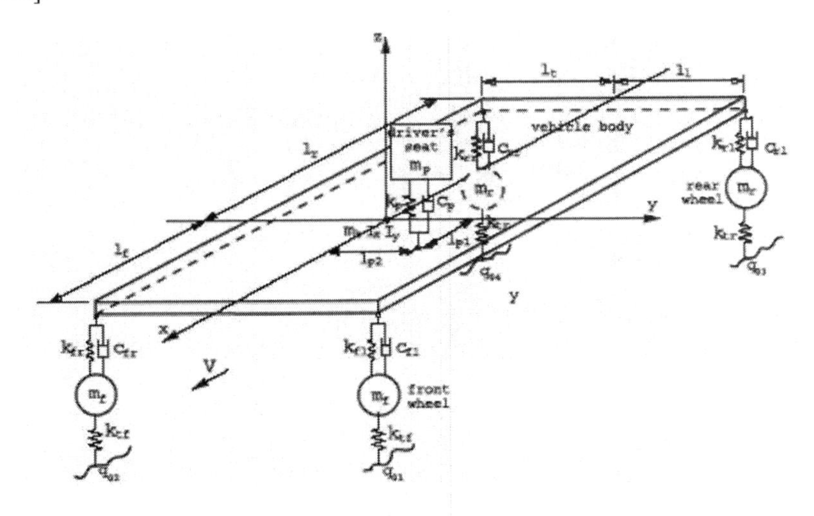

Fig. 5.20 A classic full-vehicle (four-wheel vehicle) physical model [ZUO AND NAYFEH 2003A].

Design of a passive mechanical system composed of masses, springs, and shock absorbers (dampers) may be cast as a decentralised control problem [ZUO AND NAYFEH 2002] by recognising that springs and shock absorbers (dampers), respectively, feed back the relative displacements and velocities locally. Replacing the forces generated by the passive elements with a control force vector $\{u\}$, one can formulate a mathematical model and write the *Euler-Lagrange* second-order differential equation system of dynamics governing vibration of the automotive vehicle as follows:

$$\{M_q\}\frac{d}{dt^2}\{q\}+\{C_q\}\frac{d}{dt}\{q\}+\{K_q\}\{q\}=\{B_u\}\{u\}+\{B_p\}\{q_0\}+\{B_v\}\frac{d}{dt}\{q_0\} \qquad (5.1)$$

where: $\{M_q\}$, $\{C_q\}$ and $\{K_q\}$ are the mass, damping, and stiffness matrices, respectively;

$\{q_0\}$ and $d/dt\{q_0\}$ are the disturbance vectors arising from the on/off road surface displacement and velocity;

$\{B_u\}$, $\{B_p\}$ and $\{B_v\}$ are constant matrices with appropriate dimensions.

Defining the state variables as

$$\langle \bar{x} \rangle = \left\langle \begin{matrix} \{q\} \\ \frac{d}{dt}\{q\} - \{M_q\}^{-1}\{B_v\}\{q_0\} \end{matrix} \right\rangle, \tag{5.2}$$

one can write the *Euler-Lagrange* second-order differential equation system (5.1) of dynamics as

$$\frac{d}{dt}\langle \bar{x} \rangle = \left\langle \begin{matrix} 0 & I \\ -\{M_q\}^{-1}\{K_q\} & -\{M_q\}^{-1}\{C_q\} \end{matrix} \right\rangle \langle \bar{x} \rangle + \left\langle \begin{matrix} \{M_q\}^{-1}\{B_v\} \\ \{M_q\}^{-1}(\{B_p\} - \{C_q\}\{M_q\}^{-1}\{B_v\}) \end{matrix} \right\rangle \langle q_0 \rangle + \left\langle \begin{matrix} 0 \\ \{M_q\}^{-1}\{B_u\} \end{matrix} \right\rangle \langle u \rangle$$

$$\equiv \langle \bar{A} \rangle \langle \bar{x} \rangle + \langle \bar{B}_1 \rangle \langle q_0 \rangle + \langle \bar{B}_2 \rangle \langle u \rangle .$$

(5.3)

Based on the geometry of the automotive vehicle, ones write the vector of *'measured'* outputs – the relative displacements and velocities at the suspension connections – as a linear combination of the states and inputs; that is,

$$\langle y \rangle = \langle \bar{C}_2 \rangle \langle \bar{x} \rangle + \langle \bar{D}_{21} \rangle \langle q_0 + \langle \bar{D}_{22} \rangle \langle u \rangle \rangle \tag{5.4}$$

in which the matrix $\langle \bar{D}_{21} \rangle$ turns out to be zero naturally in passive suspension design.

Similarly, one can write the vertical velocities of the driver and vehicle body, suspension deformation, and dynamic contact force as an output vector in the following form:

$$\langle \bar{z} \rangle = \langle \bar{C}_1 \rangle \langle \bar{x} \rangle + \langle \bar{D}_{11} \rangle \langle q_0 \rangle + \langle \bar{D}_{12} \rangle \langle u \rangle . \tag{5.5}$$

The forces generated by the suspension springs and shock absorbers (dampers) are determined from $\langle y \rangle$ according to

$$\langle u \rangle = \langle F_d \rangle \langle y \rangle \tag{5.6}$$

where the *'feedback gain'* $\langle F_d \rangle$ is a decentralized (block –diagonal) matrix composed of the suspension parameters to be optimised.

For instance, for a full-vehicle (four-wheel vehicle) physical model, the feedback gain takes the following form:

$$\langle F_d \rangle = \left\langle \begin{matrix} k_{fl} & c_{fl} & & & & & & \\ & & k_{fr} & c_{fr} & & & & \\ & & & & k_{rl} & c_{rl} & & \\ & & & & & & k_{rr} & c_{rr} \end{matrix} \right\rangle \tag{5.7}$$

where k_{fl} and c_{fl} denote, respectively, the stiffness and damping of the suspension at the front-left wheel, k_{fr} and c_{fr} denote those at the front-right wheel, and so on. To preserve automotive vehicle symmetry, one constrains the suspension parameters corresponding to the left and ride sides of the automotive vehicle to be equal:

$$k_{fl} = k_{fr}\,,\quad k_{rl} = k_{rr}\,,\quad c_{fl} = c_{fr}\,,\quad \text{and}\quad c_{rl} = c_{rr}\,. \tag{5.8}$$

The disturbances acting on the automotive vehicle's passive suspension mechatronic control system include on/off road surface irregularities, braking forces, acceleration forces, inertial forces on a curved track, and payload changes. Among them, on/off road surface roughness is the most important disturbance to either the rider or the automotive vehicle structure itself [ELBEHEIRY ET AL. 1995].

Many on/off road surface profiles have been measured, and several physical road models have been discussed in SHARP AND CROLLA [1987] and HROVAT [1997].

In the context of vibration, the on/off road surface roughness is typically represented as a stationary *Gaussian* stochastic process of a given displacement **power spectral density** (PSD) in [m^2/(cycle/m)]:

$$S_{psd}(\nu) = G_r\, \nu^{\beta} \tag{5.9}$$

where ν is the spatial frequency, G_r is the road-roughness coefficient, and the exponent β is commonly approximated as – 2.
The *International Organization for Standardization* (ISO) suggests a road classification scheme based on the value of G_r as shown in Table 5.3 [SHARP AND WILSON 1990; TAGHIRAD AND ESMAILZADEH 1998].

Table 5.3 Road-roughness coefficients G_r [$m^2/(cycle/m)$] classified by ISO [SHARP AND WILSON 1990].

Road Class	A very good	B good	D average	C poor	E very poor	G	F
G_r ($\times 10^{-7}$)	4	16	64	256	1024	4096	16384

When an automotive vehicle is driven at a constant value of the vehicle velocity (speed) V, the temporal excitation frequency ω and the spatial excitation frequency ν are related by $\omega = 2\pi\nu V$. And the displacement PSD in terms of temporal frequency may be obtained by using $S(\omega)\, d\omega = S(\nu)\, d\nu$ [SHARP AND CROLLA 1987]:

$$S_{psd}(\omega) = \frac{2\pi G_r V}{\omega^2}\,. \tag{5.10}$$

But the road-roughness characteristic given by expression (5.9) is not valid at very low values of the spatial frequency.

Thus a cut-off v_0 between 0.001 and 0.04 cycle/m is used to limit the displacement to be finite at vanishingly small values of the spectral frequency, and one modifies expression (5.10) to become as follows:

$$S_{psd} = \frac{2\pi G_r V}{\omega_0^2 + \omega^2} \tag{5.11}$$

where $\omega_0 = 2\pi v_0 V$.

This displacement disturbance to the automotive vehicle's wheel-tyre may be represented by a white noise signal $w(t)$ passing through a first-order filter given by

$$G(s) = \frac{\sqrt{2\pi G_r V}}{\omega_0^2 + \omega^2} \ . \tag{5.12}$$

For a full- or half-vehicle physical model, the automotive vehicle may be excited by the on/off road surface irregularities at more than one location. The excitations at the left and right wheels are correlated at low values of the frequency and uncorrelated at high values of the frequency [ZUO AND NAYFEH 2003A].

A two-dimensional road-roughness physical model is proposed in RILL [1983] and has been used in CROLLA AND ABDEL-HADY [1991 and ELBEHEIRY ET AL. [1996] in active ABW AWA suspension design.

In CROLLA AND ABDEL-HADY [1991] the authors find that the correlation of excitations acting on the left and right wheels is not important for design; therefore it may be ignored. For most multi-axle automotive vehicles, the distances between the wheels at different axles differ little. Hence it is reasonable to take the excitations at the rear axle as a pure delay of those at the front axle, with a delay time $t_0 = L/V$, where L is the distance between axles. The passive suspension mechatronic control system plays an important role in determining many aspects of the performance of an automotive vehicle; especially ride comfort, road handling, and the vehicle attitude. Ride comfort is measured by a specific index that depends on the acceleration level, frequency, direction and location.

The ISO 2631 standard [ISO 2631 1997; ZUO AND NAYFEH 2003B] specifies a method of evaluation of the effect on humans of exposure to vibration, by weighting the **root-mean-square** (RMS) acceleration with human vibration-sensitivity curves. The frequency-weighting curve for vertical acceleration (measured at the seat surface) is shown in Figure 5.21.

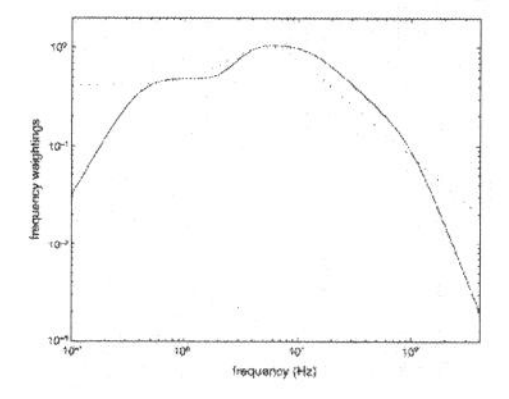

Fig. 5.21 Human vibration sensitivity weighting curve for vertical acceleration: ISO 2631-1(solid); approximate second-order filter (dotted) [ZUO AND NAYDEH 2003A].

A second-order shape filter of the form

$$H_{2631}(s) = \frac{50s2 + 500}{s^2 + 50s + 1200} \tag{5.13}$$

has been used in ELMADANY [1987] and ELBEHEIRY ETAL. [1996] to approximate this ISO weighting curve and is plotted in Figure 5.16. It is also possible to design other filters to better approximate the ISO weighting curve [ZUO AND NAYDEH 2003A, 2003B].

One measure of road handling is obtained from the dynamic contact forces between the wheel-tyres and the on/off road surface, because maintenance of large contact forces is necessary to maintain the traction required for acceleration, braking and steering. The dynamic contact forces may be calculated from the deformation and stiffness of each wheel-tyre. For a full-vehicle physical model, the attitudes include velocities of heave, pitch, and roll of the vehicle body. The working space of a passive suspension mechatronic control system is bounded by stops, so a small suspension deformation is preferred. The vector of cost outputs should include all of the performance requirements: the weighted acceleration of the driver or passenger, vehicle body attitude, dynamic contact force of the wheel-tyres, and suspension deformation.

The full-vehicle physical model and road physical model, as well as performance requirements may be combined to formulate a decentralised control problem for design of passive suspension mechatronic control systems as shown in Figure 5.22 [ZUO AND NAYFEH 2003A].

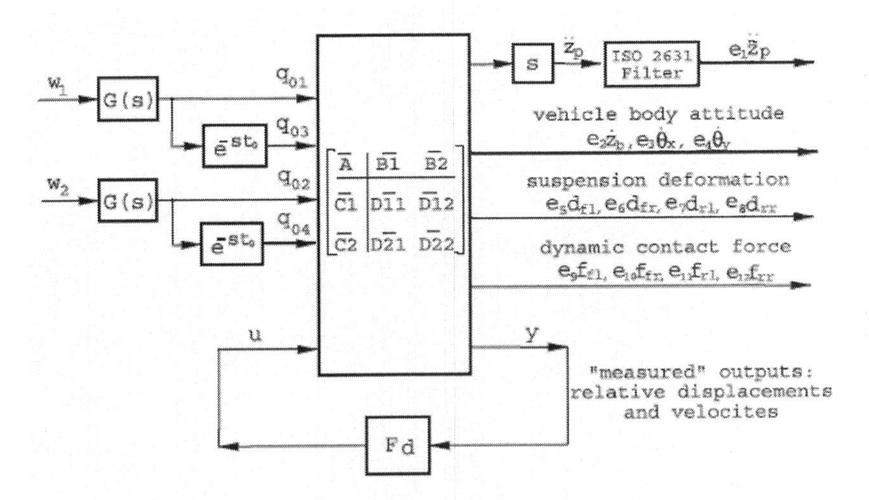

Fig. 5.22 Structural and functional block diagram of a passive suspension mechatronic control system [ZUO AND NAYFEH 2003].

Using *Pade* expressions to approximate the delay, one obtains a generalised **linear time-invariant** (LTI) system, represented by a physical and mathematical model including all of the shape filters and weighting factors, namely:

$$\frac{d}{dt}\langle x\rangle = \langle A\rangle + \langle B_1\rangle\langle w\rangle + \langle B_2\rangle\langle u\rangle, \quad \langle z\rangle = \langle C_1\rangle\langle x\rangle + \langle D_{12}\rangle\langle u\rangle, \qquad \langle y\rangle = \langle C_2\rangle\langle x\rangle \tag{5.14}$$

where the on/off road surface inputs are replaced by a white-noise input vector $\langle w\rangle$, the state vector $\langle x\rangle$ is obtained by augmentation of vector $\langle \overline{x}\rangle$ from matrix equation (5.2) to include the state necessary for incorporation of the shape filters, and likewise, the coefficient matrices $\langle A\rangle$, $\langle B_1\rangle$, $\langle B_2\rangle$, $\langle C_1\rangle$, $\langle D_{12}\rangle$ and $\langle C_2\rangle$ are obtained by augmentation of the corresponding (barred) matrices in matrix equations (5.3) – (5.5).

For instance, the task is to design the decentralised controller law $\langle u\rangle = \langle F_d\rangle\langle y\rangle$ in order to minimise the H2 norm of the closed-loop system from $\langle w\rangle$ to $\langle z\rangle$ where $\langle F_d\rangle$ is the block diagonal form given by expression (5.7) subject to the symmetry constraints given by expressions (5.8).

For instance, in ZUO AND NAYFEH [2003A], the authors show that H2 optimisation is equivalent to minimisation of a RMS value of the response to a random road excitation.

The *'closed-loop'* passive suspension mechatronic control system from $\langle w\rangle$ to $\langle z\rangle$ may be written in the standard form as follows:

$$\frac{d}{dt}\langle x\rangle = \langle A_c\rangle\langle x\rangle + \langle B_c\rangle\langle w\rangle, \qquad \langle z\rangle = \langle C_c\rangle\langle x\rangle + \langle D_c\rangle\langle w\rangle \tag{5.15}$$

where

$$\begin{Vmatrix} \langle A_c\rangle & \langle B_c\rangle \\ \langle C_c\rangle & \langle D_c\rangle \end{Vmatrix} = \begin{Vmatrix} \langle A\rangle + \langle B_2\rangle\langle F_d\rangle\langle C_2\rangle & \langle B_1\rangle \\ \langle C_1\rangle + \langle D_{12}\rangle\langle F_d\rangle\langle C_2\rangle & \langle D_{12}\rangle\langle F_d\rangle\langle C_2\rangle \end{Vmatrix}. \tag{5.16}$$

5.3.3 Passive F-P-M Suspension Mechatronic Control Systems

Passive **fluido-pneumo-mechanical** (F-P-M) suspension mechatronic control systems are self-regulating mechatronic control systems that are constructed so that changes in attitude or height of the vehicle body relative to the on/off road surface are compensated for without the need for means specifically provided for detecting such changes. Important examples can interconnect F-P-M suspension mechatronic control systems where the interconnections are such that movement of one suspension unit forces fluid through pipe work to an opposing unit, e.g. on the other side of the automotive vehicle, to cause movement of that unit in the same, or opposite sense of direction.

A tightly sealed amount of gas (air) is used in the F-P-M suspension mechatronic control system. The oily-fluid and gas (air) are isolated by a rubber diaphragm as shown in Figure 5.23 (a) [AKATSU, 1994].

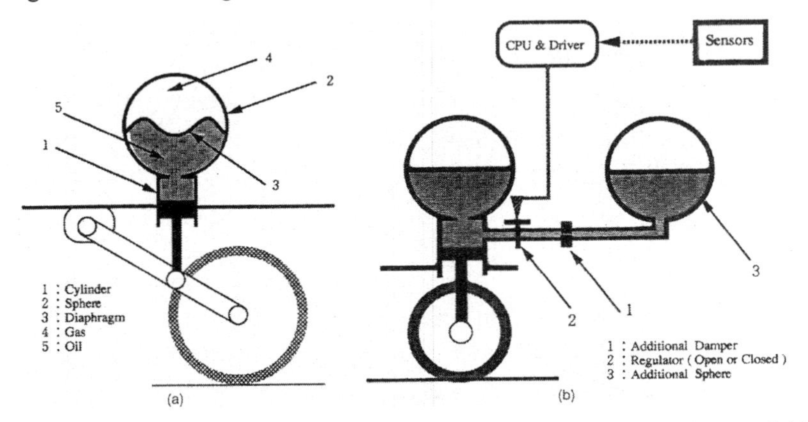

Fig. 5.23 Passive F-P-M suspension mechatronic control system: (a) uncontrollable; (b) controllable [AKATSU, 1994].

The mechanical springs are replaced by gas (air) ones. The orifice integral with fluidical valves attains the F-P-M shock absorber's damping mechanism [AKATSU 1994]. The F-P-M shock absorber (actuator) is a popular choice [WILLIAMS 1994], giving both a low frequency active element and a high frequency passive element in one unit.

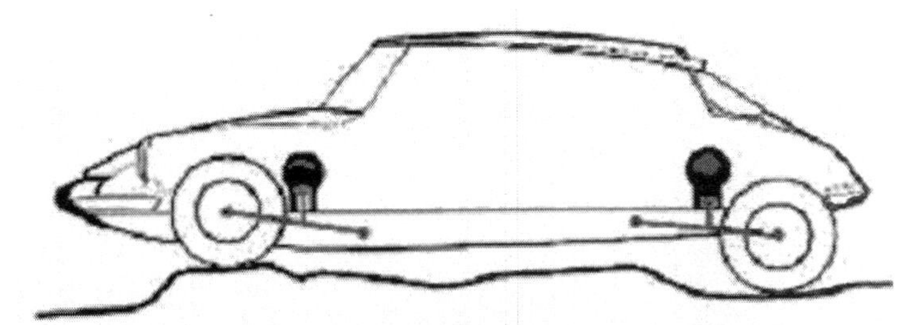

Fig. 5.24 Passive F-P-M suspension mechatronic control [Citroën; CITROËN 2001].

The passive F-P-M suspension mechatronic control system (Figure 5.24) showing the spheres with gas (nitrogen) are represented in gray and the oily-fluid in black [CITROËN 2001].

Passive F-P-M Suspension Mechatronic Control System - As shown in Figure 5.23 (b) by appendage of an additional sphere to the F-P-M suspension mechatronic control system, a controllable F-P-M suspension may be created.

If the regulator is closed, the F-P-M suspension mechatronic control system is in a firm mode. If the regulator is open, the spring constant of the F-P-M suspension mechatronic control system turns out to be lower by escalating the total volume of the sphere, and the total absorbing (damping) force is trimmed down. Relying on the use of sensors that sense vehicle propulsion (driving), dispulsion (braking) and conversion (steering) as well as on/off road circumstances, this mechatronic control system may alter the regulator characteristics so as to attain both high-quality ride comfort and handling stability [AKATSU, 1994].

5.4 Self-Levelling Suspension

5.4.1 Foreword

Self-levelling suspension mechanisms take and keep possession of first place among the various potentials to further enhance a vehicle chassis that has already been fine tuned by design and is in manufacture.

Such mechanisms are most indispensable for automotive vehicles of the low- and/or medium-mass category seeing that here the ratio of probable supplementary loading in opposition to normal vehicle mass is unpredictably relatively high.

A self-levelling suspension mechatronic control system combines the passive spring element found in a passive suspension with a shock absorber (damper) element whose characteristics may be adjusted by the driver.

As shown in Figure 5.25 the drivers may use a selector device to set the desired level of absorbing (damping) based on their objective feel [BARAK 1989].

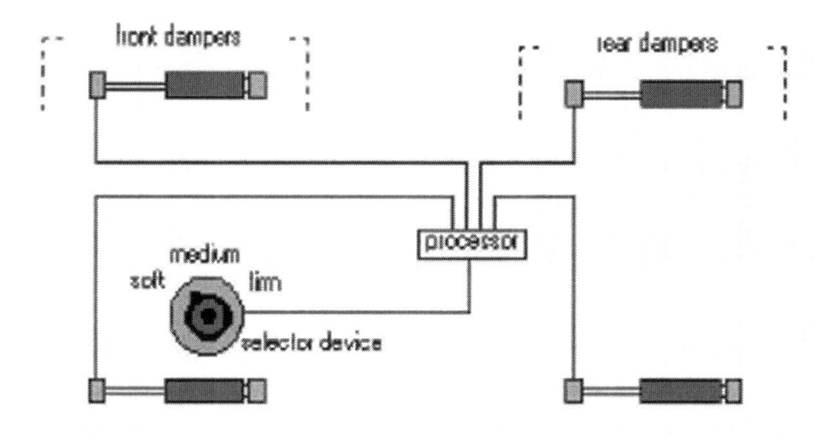

Fig. 5.25 Self-levelling suspension mechatronic control system [BARAK 1989].

This mechatronic control system has the advantage of allowing the driver to occasionally adjust the shock absorbers according to the on/off road surface characteristics. It is however, unrealistic to expect the driver to adjust the self-levelling suspension mechatronic control system to respond to time inputs such as potholes, turns, or other common on/off road surface inputs.

In some circumstances, self-levelling suspension mechanisms are already technologically advanced.

They contain for the most part conventional F-P-M or only P-M components set up parallel to the standard suspension spring that holds up the vehicle body.

Thus the normal spring travel is preserved even at increased static loads, in view of which, softer springs that enhance comfort contrasted to the standard suspension can be used.

An attractive and comparatively low-cost design variant of the vibration shock absorber (damper) with self-pumping levelling feature is shown in Figure 5.26 [SEIFFERT AND WALZER 1991].

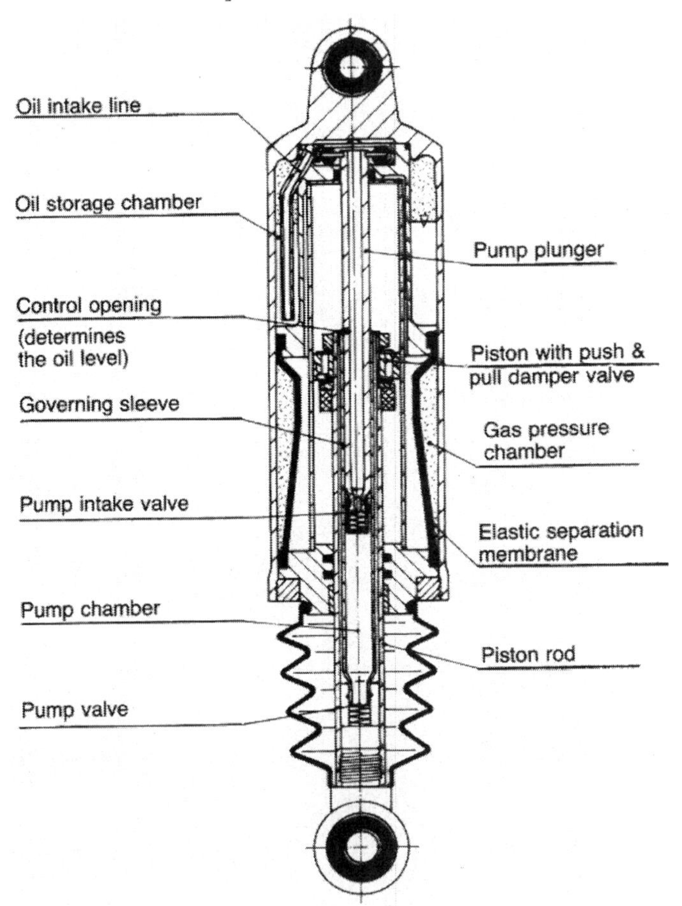

Fig. 5.26 Vibration shock absorber *'NIVOMAT'* – with self-pumping levelling feature [Boge; SEIFFERT AND WALZER 1991].

This damper does not necessitate an independently driven M-F pump; the necessary oily-fluid pressure develops by itself as the vehicle's body moves up and down. But not many preliminary actions are effective enough to attain this on their own. [SEIFFERT AND WALZER 1991].

Summing up, in a P-M and F-P-M suspension, the vehicle's body may be suspended at a constant height from the on/off road surface, maintaining a low spring constant. The benefits of a self-levelling mechatronic control system are [AKATSU, 1994]:

- Maintaining a low spring rate to attain high-quality ride comfort regardless of load circumstances;
- Escalation in vehicle body height on rough road surfaces;
- Altering spring rate and absorbing (damping) force compliant with propulsion (driving) and/or dispulsion (braking) circumstances and on/off road surfaces.

5.4.2 Self-Levelling Suspension Mechatronic Control System Arrangement

The self-levelling suspension mechatronic control system is shown in Figure 5.27. It contains eight sensors, a mode-select switch, air spring/shock absorber (damper) units on four wheels, P-M actuators to operate the alteration pneumatical valves in the unit, an **electro-mechano-pneumatical** (E-M-P) compressor unit and five height-control pneumatical valves for air springs and ECU.

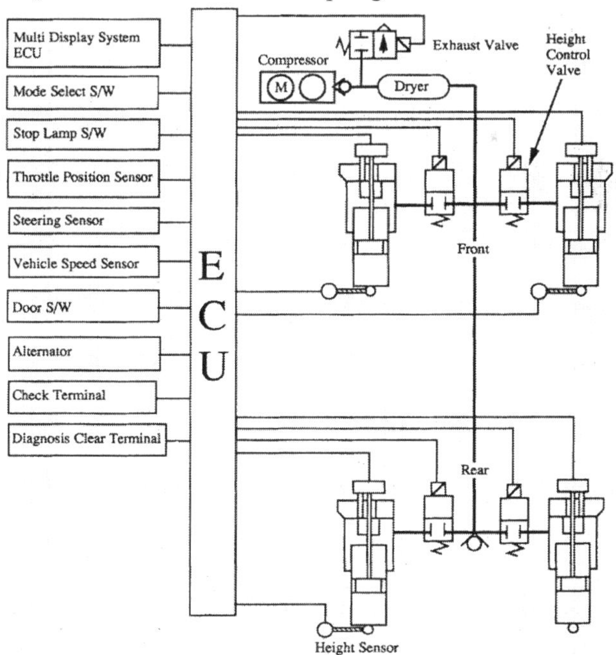

Fig. 5.27 Principal components of a self-levelling P-M ABW AWA suspension mechatronic control system [AKATSU, 1994]

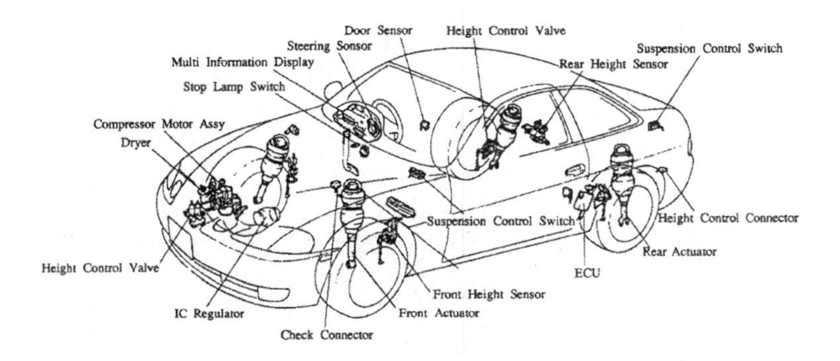

Fig. 5.28 Configuration of a self-levelling P-M suspension mechatronic control system [AKATSU, 1994].

The self-levelling suspension mechatronic control system configuration is shown in Figure 5.28 [AKATSU, 1994].

5.4.3 Self-Levelling Suspension Mechatronic Control System Components

The arrangement of the air suspension unit contains a shock absorber (damper), a pneumatical piston nearby the shock absorber, major- and minor-air chambers, a rolling diaphragm, and pneumatical valves that adjust the suspension stroke.

The P-M actuator uses a DC-AC commutator motor that has two shafts to function the pneumatical valves for the air spring and the shock absorber (damper).

The sector gear decreases the rotation of the DC-AC commutator motor and functions the rotary pneumatical valve to alter the absorbing (damping) force. Simultaneously, another gear connected with the sector gear functions the pneumatical valve to adjust the spring rate [AKATSU, 1994].

5.4.4 Self-Levelling Suspension Mechatronic Control System Function

A self-levelling suspension mechatronic control system may adjust the spring rate and the absorbing (damping) force into three levels and vehicle height levels of low-, normal-, or high-level may be selected.

One of the self-levelling suspension mechatronic control logics is exposed in Table 5.4 [AKATSU, 1994].

Table 5.4 Self-Levelling Suspension Mechatronic Control Logic of Air Suspension System [AKATSU 1994].

		Spring rate, damping force					
		Soft mode			Medium mode		
Function	Operating condition	Soft	Medium	Firm	Soft	Medium	Firm
Antiroll	Rapid steering	○ →	○ →	○		○ →	○
Antidive	Braking at V* > 60 km/h	○ →	○ →	○		○ →	○
Antisquat	Rapid starting at V < 20 km/h	○ →		○		○ →	○

V* – vehicle velocity (speed).

This is a mechatronic control that alters the suspension features in response to vehicle velocity and on/off road circumstances. The spring's rate/absorbing force and the vehicle height is controlled autonomously compatible with anyone's control logic.

Self-levelling suspension mechatronic control systems do not necessitate much energy to control vehicle height. They control both spring rate and absorbing (damping) force. As a consequence of maintaining the low spring rate, self-levelling suspension mechatronic control systems can offer both high-quality ride comfort and handling stability. As an example, the AiROCK™ self-levelling suspension mechatronic control system may operate to both improve highway operation and drastically enhance off-road control. This self-levelling suspension mechatronic control system is automated.

With its harness tied into the vehicle-velocity sensor, the ECU may monitor and adjust the ride height to always ensure safe driving conditions. There are three driving modes that may be displayed by the interface.

By understanding the three modes, drivers are well on their way to understanding the logic behind the mechatronic control system and have a good understanding of how drivers want to have their interface customised.

The following is an explanation of three driving modes [AIROCK 2006]:

- *Off Road mode* that a vehicle both starts and stops in; the vehicle is automatically entered in this mode between 0 – 32 km/h (0 – 20 mph); in this mode drivers have complete control of the height, pitch and roll of the vehicle;
- *Highway mode* is automatically entered when drivers are travelling between 32 – 80 km/h (20 – 50 mph); in this mode the vehicle returns to its predetermined *'Ride Height'*; this mode assures that the suspension and steering geometry may be correct during high values of the vehicle velocity; during highway mode drivers have the option to adjust their ride height by a small amount only; the mechatronic control system may compensate for vehicle-body roll by inflating the air springs as it detects such a need;

❖ Freeway mode is automatically entered once drivers are travelling faster than 80 km/h (50 mph); highway mode may be reengaged after drivers slow to approximately 56 km/h (35 mph); while the vehicle is in this mode, it may remain in its predetermined ride height (that is the same height as highway mode) and may not let drivers adjust it at all; this mode assures that the suspension and steering geometry may be correct during high values of the vehicle velocity; although the mechatronic control system will compensate for vehicle-body roll in this mode, it will do so less often since turns on freeways are far more gradual than on smaller streets.

Fig. 5.29 Button Menu Up/Down: Y – Select; X – Cancel [AIROCK 2005].

During configuration of a driver interface, highway and freeway modes are referred to as *'Right Height'*, and off road mode may be referred to as *'Off-road Height'*. It is important that the suspension and steering geometry is dialled in while in the drivers predetermined *'Ride Height' mode* to ensure proper handling; as drivers adjust their height value, the toe function may change.

The toe should be adjusted at *'Ride Height'*; if the driver changes *'Ride Height'* later on, the toe value must be readjusted to ensure proper handling; it is also important to have the proper speedometer gear installed so that the computer may accurately judge the vehicle's velocity.

Now that we have clarified the three different driving modes, let us move on to the setup button menu. This button menu is where all of the user configurations (Fig. 5.29) are displayed and adjusted. Understanding how this menu acts is important during initial set-up, troubleshooting, and making changes later on.

Pressing "*CHECK*" may allow the user to swap the operations of the *'Up'* and *'Down'* buttons from one touch all up and all down to pitch front up and pitch front down respectively. The two functions that are not on a single button push are still available as a two-button push. These two button push drivers add the up button and left button together -- or the down button and the right button together.

5.5 Semi-Active Suspensions

5.5.1 Foreword

A semi-active ABW AWA suspension (Fig. 5.30) mechatronic control system is much less complex than an active ABW AWA suspension mechatronic control system, and may be manufactured at a lower cost [DAS 2006].

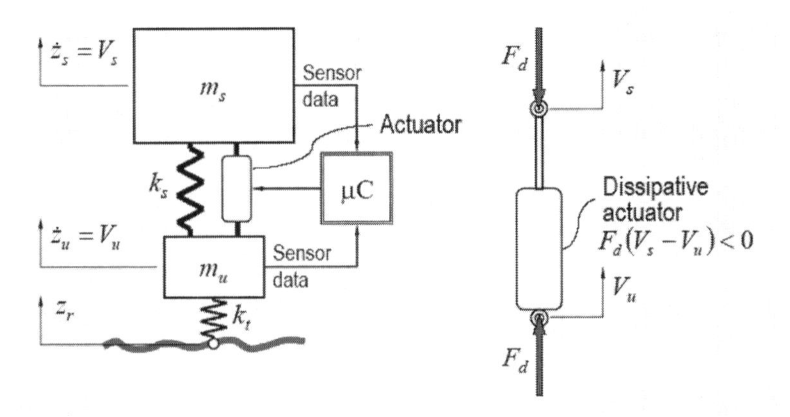

Fig. 5.30 Physical model of a semi-active ABW AWA suspension – dissipative forces only [DAS 2006].

The semi-active ABW AWA suspension mechatronic control system approaches the performance capability of the active system, even though it requires less power to function. A conventional suspension spring is used, but in conjunction with an F-M or even E-M actuator rather than a shock absorber (damper).

The absorbing (damping) force absorbed by the actuator is modulated in accordance with the current operating circumstances. This type of mechatronic control system is a closed-loop process.

A number of sensors are required to measure the vehicle's attitude incorporating, position sensors between the sprung and unsprung masses, vehicle velocity (speed) sensors, and steering sensors. In concept design, the *'skyhook'* theory is relevant.

Semi-active suspension systems may be categorised into either high-frequency bandwidth or low-frequency bandwidth mechatronic control systems.

High-frequency bandwidth mechatronic control systems concentrate on improving the suspension response of 10 -- 12 Hz; this range is known as the *'rattle space'* frequency. Mechatronic control systems in this category also improve the *'wheel-tyre hop'* frequency that is in the 3 -- 4 Hz region.

However, the majority of semi-active ABW AWA suspension mechatronic control systems are low bandwidth ones that only cover the *'wheel-tyre hop'* frequency. Ride comfort and road handling performance of a vehicle may be primarily set up by the damping characteristic of the shock absorbers (dampers). Passive shock absorbers have a preset damping characteristic set up by their design.

Semi-active and active ABW AWA suspension mechatronic control systems present the preference to vary online the damping characteristic derived from measurements of the vehicle motions, *e.g.* by controlling the limitation of the shock absorber fluidical or pneumatical or even electrical valves, respectively. For instance, an adaptable model-free control system configuration may be developed for a semi-active ABW AWA suspension mechatronic control system. The mechatronic control system configuration may be derived from physical insights in the vehicle and semi-active ABW AWA suspension dynamics that may be used to linearize and decouple the mechatronic control system. It does not necessitate a physical model of its dynamics, and accordingly, may be appropriate to any semi-active or active ABW AWA suspension mechatronic control system and any class of the automotive vehicle [SWEVERS ET AL. 2005].

A static decoupling may be used to decouple the semi-active ABW AWA suspension mechatronic control system into its modal motions heave, roll and pitch that are then controlled by modal (diagonal) controllers that may be composed of several feedback and feedforward modules, each engaging in a particular ride comfort or road handling area. The feedback controller may be derived from the skyhook principle, with enhancements that are adapted automatically in relation to the on/off road surface excitation. Steering angle, throttle, and brake pedal feedforward control lean during cornering, and pitch during acceleration and deceleration, respectively [SWEVERS ET AL. 2005].

Semi-active ABW AWA suspensions were first proposed in the early 1970s [KARNOPP ET AL. 1974; CROLLA AND ABDEL-HADY 1991]. In this category of the mechatronic control system, the conventional spring element is retained, but the shock absorber (damper) is replaced with a controllable shock absorber (damper) as shown in Figure 5.31 [SIMON 2001].

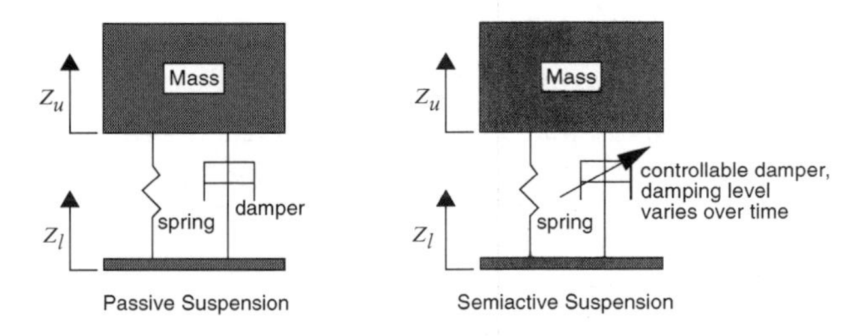

Fig. 5.31 Passive and semi-active suspensions [SIMON 2001].

Whereas an active ABW AWA suspension mechatronic control system requires an external energy source to power an actuator that controls the vehicle, a semi-active ABW AWA suspension mechatronic control system uses external power only to adjust the damping levels, and operate an embedded controller and a set of sensors. The controller determines the level of absorbing (damping) based on a control strategy, and without driver intervention adjusts the shock absorber (damper) to achieve that damping.

One of the most common semi-active control policies is skyhook control that adjusts the damping level to emulate the effect of a shock absorber (damper) connected from the vehicle to a stationary ground as shown in Figure 5.32 [SIMON 2001].

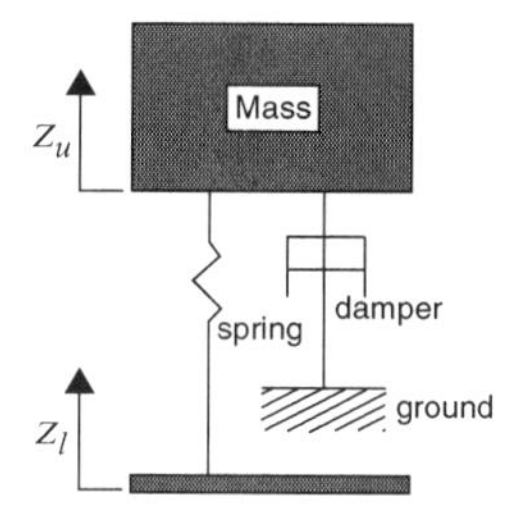

Fig. 5.32 Quarter-vehicle physical model with a skyhook shock absorber [SIMON 2001].

A mathematical model of skyhook control may be formulated as [SIMON 2001]:

$$\dot{z}_u * (\dot{z}_u - \dot{z}_l) \geq 0 \qquad C = high\ damping$$

$$\dot{z}_u * (\dot{z}_u - \dot{z}_l) < 0 \qquad C = low\ damping$$

where: $\dot{z}_u$ - the velocity of the upper mass;

$\dot{z}_l$ - the velocity of the lower mass;

C - the damping constant.

This type of skyhook control is termed *'on-off'* or *'bang-bang'* control since the shock absorber (damper) switches back and forth between two possible absorbing (damping) states.

When the upper mass is moving up, and the two masses are getting closer, the damping constant of zero is not practical and a low damping constant should ideally be zero.

Due to the physical limitations of a practical shock absorber, a damping constant of zero is not practical and low damping constant may be used.

When the upper mass is moving down and the two masses are getting closer, the skyhook control policy ideally calls for an infinite damping constant.

An infinite damping constant is not physically attainable, so in practice, the adjustable damping constant may be set to a maximum.

The effect of the skyhook control scheme is to minimise the absolute velocity of the upper mass, as it is shown in Figure 5.33 [SIMON 2001].

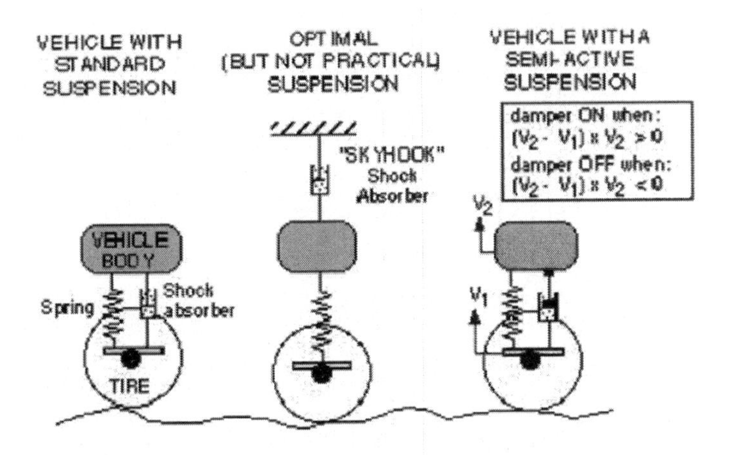

Fig. 5.33 Skyhook control policy [SIMON 2001].

It has been shown that a continuously variable semi-active ABW AWA suspension mechatronic control system is able to achieve performance comparable to that of an active ABW AWA suspension mechatronic control system [KARNOPP ET AL. 1974]. It is also possible to develop a control policy in which the shock absorber (damper) is not just switched between high and low state, but has an infinite number of positions in-between. This type of mechatronic control system is termed a continuously variable semi-active ABW AWA suspension mechatronic control system. The ranges of damping values used in these two mechatronic control systems are illustrated in Figure 5.34 [SIMON 2001].

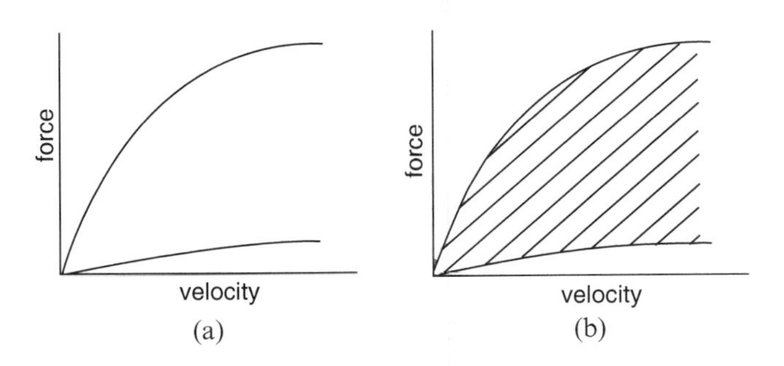

Fig. 5.34 Range of damping force: on-off semi-active damping; (b) continuously variable SA damping [SIMON 2001].

Further research indicated that performance of an *'on-off'* SA ABW AWA suspension mechatronic control system would be very close to the performance of a continuously variable semi-active ABW AWA suspension mechatronic control system [FODOR AND REDFIELD 1995].

In the case that the controllable shock absorber (damper) necessary in a semi-active ABW AWA suspension fails, it may simply revert to a conventional shock absorber. Semi-active ABW AWA suspension mechatronic control systems not only have a less dangerous failure mode, but are also less complex, less prone to mechanical failure, and have much lower power requirements compared to active ABW AWA suspension mechatronic control systems [SIMSON 2001].

Semi-active ABW AWA suspension mechatronic control has been developed as a compromise between passive and active control. Instead of opposing a primary disturbance, as is the case with active control, a semi-active control scheme applies a secondary force to the semi-active ABW AWA suspension mechatronic control system. Semi-active control has recently been an area of much interest because of its potential to provide the adaptability of active means without requiring a significant external power supply for actuators. A semi-active control system cannot provide energy to a system comprising the vehicle body and actuator, but it may achieve favourable results by altering the properties of the system, such as stiffness and damping. Unlike an active ABW AWA suspension mechatronic control system, the control forces developed are related to the motion of the vehicle body. Furthermore, the stability of the semi-active ABW AWA suspension mechatronic control system is guaranteed as the control forces typically oppose the notion of the vehicle body. A semi-active ABW AWA suspension consists of a spring and a shock absorber (damper) but, unlike a passive suspension system, the value of the damping coefficient may be controlled and updated. In some types of semi-active ABW AWA suspensions it may also be possible to control the elastic constant of the spring. The use of semi-active ABW AWA suspension mechatronic control systems is gaining more and more automotive market share in the last years because it provides a valid trade-off between purely active ABW AWA suspension mechatronic control systems and passive suspension systems.

Many efficient and innovative contributions in this area have been recently proposed in GÖRING ET AL. [1993]; KITCHING ET AL. [2000]; CAPONETTO ET AL. [2001]; SAMMLER [2001] and GIUA ET AL. [2004].

A semi-active ABW AWA suspension mechatronic control system is nowadays normally set up on high-end manufacture sports automotive vehicles and incorporates damping control on all four corners. The mechatronic control function of semi-active ABW AWA suspension is to counteract for heave, roll and pitch. This control is realised by coupling a microprocessor with four shock absorbers (dampers) that have a constantly adjustable (and controllable) damping coefficient. Sometimes, these shock absorbers (dampers) are coupled with pneumatical springs to afford ride height/ levelling control.

Eventual driver advantages of the semi-active ABW AWA suspension mechatronic control system incorporate [FREESCALE 2005]:

- An adaptable ride, most advantageous for comfort or handling performance;
  - Suspension may automatically adjust compatible with on/off road surface circumstances;
  - The driver has the preference to opt for the firmness of the semi-active ABW AWA suspension mechatronic control system.
- There is no alteration in size from conventional passive suspensions.

During the past 30 years, lots of different semi-active ABW AWA suspension mechatronic control systems have been examined. The major function of all these systems is to choose the optimum absorbing (damping) force for various propulsion (driving) and/or dispulsion (braking) circumstances. The first function of a shock absorber (damper) is to control vehicle movement at variance with inertial forces, such as roll when the vehicle turns and pitch when it is braked. The second function -- is to avoid vehicle vibration generated by on/off road surface inputs. To fulfil both functions it is indispensable to control absorbing (damping) forces. There are three indispensable components of an ABW AWA suspension mechatronic control system: an absorbing/damping control set-up (actuator), sensors, and software (control strategy). Optimum value of the absorbing (damping) force should be adjusted for various propulsion (driving), dispulsion (braking) or conversion (steering) circumstances in order to better ride comfort and handling stability [AKATSU 1994].

A notable innovation on the automotive market was a vibration damping mechatronic control system with variable absorbing (damping) force. The purpose here is to acquire maximum comfort under all propulsion (driving), dispulsion (braking) or conversion (steering) circumstances and the most risen and fallen on/off road surfaces without loss of handling features or safety measures. For instance, Figure 5.35 presents an illustration of such a mechatronic control system in which electromagnetic servo fluidical valves are used to control the features of the vibration shock absorbers (dampers) on the front and rear axles [SEIFFERT AND WALZER 1991].

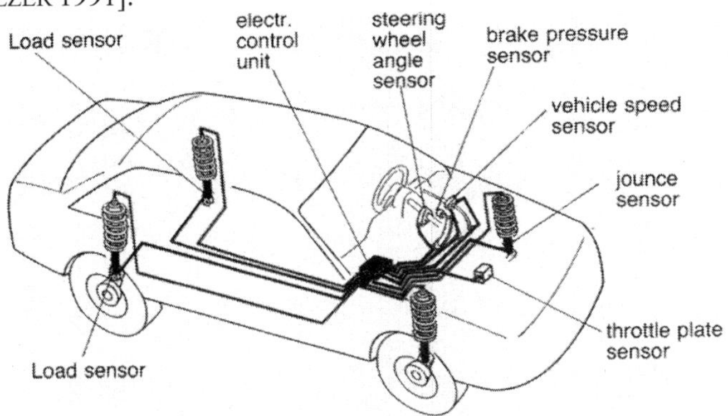

Fig. 5.35 Schematic of an automotive vehicle shock absorber system for variable damping characteristics: blue lines – sensor signals that go to the ECU; red lines – control signals for the fluidical valves in the shock absorbers [SEIFFERT AND WALZER 1991].

An acceleration transducer on one of the front suspension struts estimates the vertical vehicle body acceleration that is caused by on/off road irregularities.

Lateral forces resulting from obstacle avoidance manoeuvres, lane changes and cornering are sensed by means of a steering wheel angle sensor.

Conditional on the transmitted signal, the mechatronic control system chooses the optimum shock absorber (damper) adjustment. High values of the vertical acceleration create a soft shock absorber (damper) characteristic while rapid steering wheel movements cause a short-term recourse to hard turn characteristics in order to avoid roll development [SEIFFERT AND WALZER 1991].

In the case of steady cornering at high velocity, a mutual vehicle velocity and steering angle signal escalates the absorbing (damping) force. Pitch motions generated by sudden intense brake manoeuvres are admitted by the mechatronic control system, for example, through a signal from the brake pressure pickup that also turns the shock absorbers (dampers) for the short term to *'hard'*. The hard shock absorber (damper) characteristic must be activated to absorb (dampen) vibrations in an aperiodic mode. So as to attain harmonic vibration behaviour, the time interval from signal pickup to adjustment of the absorbing force must not surpass 20 ms [SEIFFERT AND WALZER 1991].

5.5.2 Shock Absorber Suspension Mechatronic Control System Arrangement

One of the ABW AWA suspension mechatronic control system arrangements is shown in Figure 5.36. This mechatronic control system uses five sensors, counting a supersonic on/ off road sensor, to sense all propulsion (driving), dispulsion (braking) or conversion (steering) circumstances.

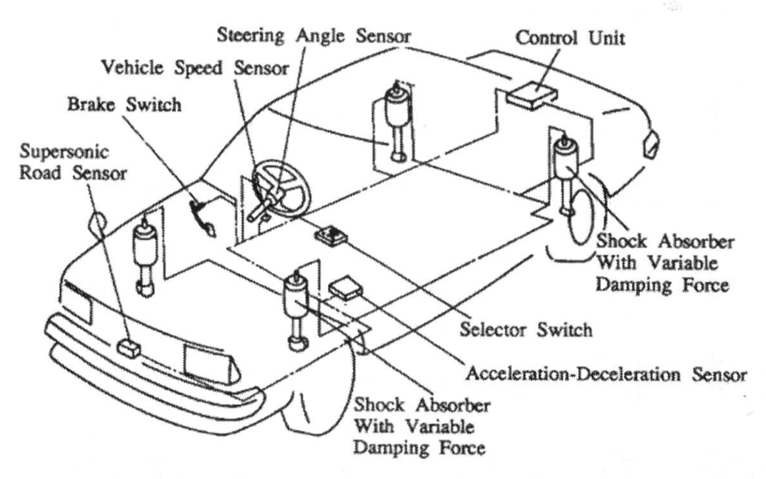

Fig. 5.36 Principal components of an ABW AWA suspension mechatronic control system [AKATSU, 1994].

Control signals are transmitted to regulate the absorbing (damping) force of the variable shock absorbers (dampers) to optimum values.

A major benefit to this kind of mechatronic control system is that, because of the use of an on/off road sensor, it can offer optimum control in relation to the actual on/off road circumstances. This mechatronic control system includes three discrete shock absorber (damper) characteristics [AKATSU 1994].

Sensors used are: a vehicle-velocity sensor, a steering-angle sensor, acceleration and/or deceleration sensor, a brake sensor, and a supersonic sensor to sense on/off road circumstances. For instance, one shock absorber suspension mechatronic control system uses four piezoelectric sensors and four piezoelectric actuators on each wheel in order to adjust the absorbing (damping) forces as quickly as possible. This mechatronic control system includes two discrete shock absorber (damper) characteristics. Sensors used are: four piezoelectric sensors on each wheel, a stop-lamp switch, a steering-angle sensor, and a vehicle-velocity sensor.
Actuator and Sensor Discrete damping control P-M actuators normally use an incorporated DC-AC commutator motor to adjust the absorbing (damping) force. This **electro-mechanical** (E-M) motor switches a rotary pneumatical valve to choose the orifice diameter for three different absorbing (damping) levels: soft-, medium- and/or hard-level. The halt position of the rotary pneumatical valve is monitored by encoder signals. Another E-M actuator is a piezoelectric one, for example, containing 88 piezoelectric elements, and it is embedded in the piston rod of a shock absorber (damper). When a high voltage (for example 0.5 kV) is applied to the piezo-electric E-M actuator, it elongates about 50 μm with reverse piezoelectric effect. Elongation in the piezoelectric E-M actuator causes the plunger pin to be pushed out through the displacement F coupling unit.

As a result, the plunger pin moves down to open the bypass of the absorbing force switching pneumatical valve. The effect is a soft absorbing/damping force [AKATSU 1994].

A sensor using a supersonic wave may sense the on/off road surface. The vehicle height from on/off road to vehicle body is estimated on account of reflection time T.

Analysing the pattern of adjustments in vehicle height involves deduction of the on/off road circumstance. A sensor using the piezoelectric effect may be embedded in the piston rod of a shock absorber (damper). The sensor creates an electric charge in proportion to axial force from the on/off road surface.

5.5.3 Shock Absorber Suspension Mechatronic Control System Function

The shock absorber suspension mechatronic control system of Figure 5.36 is described in Table 5.5. The incidence of roll and pitch can be estimated from an assortment of sensors. Single bumps or dips are sensed from alterations in vehicle height [AKATSU 1994].

Table 5.5 Absorbing (Damping) Force Control [AKATSU 1994].

Control objectives		Sensors used: Vehicle speed	Steering angle	Accel/ decel	Brake	Road condition	Damping force Front / Rear
Roll	Roll reduction for quick steering operation	○	○				Hard / Hard
Pitch	Reduction of nose diving by braking				○	○	Hard / Hard
	Reduction of pitching when accelerating and decelerating	○		○			Medium / Medium
Bouncing	Reduction of light, bouncy vibrations in bottoming	○				○	Medium / Medium
	Reduction of light, bouncy vibrations in bouncing on a heaving road	○				○	Medium / Medium
Road holding performance	Road holding performance improvement when running on rough roads	○				○	Medium / Medium
Others	Stability improvement at high speed	○					Medium / Soft
	Prevention of shaking when stopping and rocking when passengers exit or enter	○					Hard / Hard

The outline of the on/off road surface deduction logic is described in Table 5.6.

Table 5.6 On/Off Road Surface Deduction Logic [AKATSU 1994].

High frequency components	**Low frequency components**	
	Small	**Large**
Small	Smooth road damping force control unnecessary	Heaving road
Large		Rough road

The structural and functional block diagram of the shock absorber suspension mechatronic control system using four piezoelectric sensors and four piezoelectric E-M actuators is shown in Figure 5.37 [AKATSU 1994].

The absorbing (damping) force resulting in a shock absorber (damper) intensifies as soon as the tyre-wheel goes up or down in relation to roughness of the on/off road surface.

All piezo-electric sensor outputs to the ECU continuous signals imitating absorbing (damping) force differential.

If the value of absorbing (damping) force differential surpasses a prearranged level, the mechatronic control system changes from firm mode to soft mode.

The shock absorber mechatronic control system is designed so that the firm mode reinstates instantaneously after the vibrations by reason of poor on/off road surface cease.

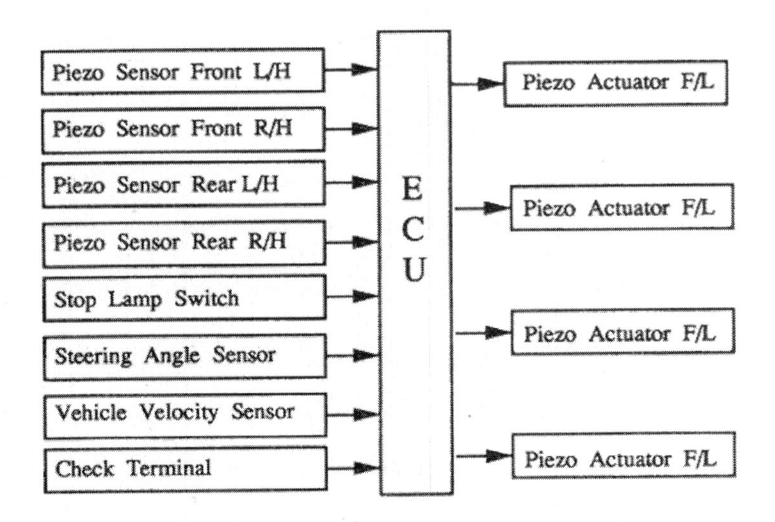

Fig. 5.37 Structural and functional block diagram of the ABW AWA suspension mechatronic control system [AKATSU, 1994]

Many of the restrictions and disadvantages of the conventional shock absorber (damper) can be eradicated by the ABW AWA suspension mechatronic control systems. They contain semi-active suspension systems that are appropriate to enduing both ride comfort and super handling stability [AKATSU, 1994].

5.5.4 Types of Semi-Active Devices

Different types of semi-active devices have been developed recently. One type is the semi-active shock absorber (damper), such as variable-orifice shock absorbers, controllable fluid shock absorbers, and controllable friction devices.

Variable-orifice shock absorbers (dampers) use an **electro-mechanical** (E-M) variable orifice to alter the resistance to flow in a conventional liquid fluid.

Controllable fluid shock absorbers are passive F-M shock absorbers containing an oily fluid, such as MRF or ERF, with controllable yield stress [SPENCER ET AL. 1997].

Another type of semi-active device is a semi-active stiffness device such as those published in KOBORI AND TAKAHASHI [1993]; PATTEN ET AL. [1999] and YANG ET AL. [1996]. They are on-off F-M devices capable of providing mainly variable damping and limited variable stiffness capability.

In the 1990s, some scientists introduced a variable stiffness device that consists of four sets of spring elements and telescoping tube elements. Varying the position of the springs with a servomotor produces continuously variable stiffness [NAGRARAJAIAH AND MA 1996].

Controllable fluid shock absorbers use oily fluids with properties that may be modified by some influence.

MRFs or ERFs change their properties in the presence of a magnetic or electric field, respectively. These fluids were originally developed in the 1940s [RABINOW 1948; WINSLOW 1949], but few applications were foreseen at that time.

While ERFs showed early promise for automotive applications, see e.g. EHRGOTT AND MASRI [1992], most of the attention of the vehicular control community has shifted to using MRFs due to their insensitivity to impurities, relatively constant behaviour over a wide range of operating temperatures, and the low voltage required to activate them [SPENCER ET AL. 1997].

MR shock absorbers (dampers) typically consist of an F-M cylinder containing micron sized magnetically polarized particles suspended within a MRF. In the presence of magnetic field, the particles polarise and form particle chains that resist MRF flow.

By varying the magnetic field, the mechanical behaviour of an MR shock absorber (damper) may be modulated. Since MRFs may be changed from a viscous oily fluid to a yielding semisolid one within milliseconds and the resulting absorbing (damping) force may be considerably larger with a low-power requirement, MR shock absorbers are applicable to automotive vehicles.

5.5.5 Semi-Active ABW AWA Suspension Design Challenges

Semi-active ABW AWA suspension requires innovative mechanical hardware and sensor input requests to function within its mechatronic control system. The performance of the semi-active ABW AWA suspension mechatronic control system depends in some measure on the response time of the shock absorbers (dampers) and their damping coefficient range.

Mechanical Hardware Necessitated Adjustable damping coefficient shock absorbers (dampers), with a response time less than *10 ms* are necessary for the most rudimentary of semi-active ABW AWA suspension mechatronic control systems. These shock absorbers (dampers) can be F-M or P-M or even E-M and are as a rule controlled with electromagnetic switches or M-F pumps or M-P compressors or even M-E generators.

Sensor Input Necessitated semi-active ABW AWA suspension mechatronic control systems are practically inestimable in the amount of input variables for which they are proficient to counteract, thus, a mechatronic control system designer can establish the amount of control a system submits. Semi-active ABW AWA suspension mechatronic control systems can monitor for [FREESCALE 2005]:

- Vehicle velocity (speed);
- Brake condition;
- Vertical acceleration;
- Lateral acceleration;
- Steering angle position;
- Steering angle velocity;
- Vehicle level position.

These innovative mechatronic control system requests must be encountered using high-end components at very economical costs to substitute conventional, profitable technology, while sustaining exact dependability in the automotive circumstances.

5.5.6 Semi-Active F-M ABW AWA Suspension Solution

The semi-active F-M ABW AWA suspension mechatronic control system is referred to as a **real-time damping** (RTD) system in the onboard diagnostics. This mechatronic control system controls absorbing (damping) forces in the front struts (Fig. 5.38) and rear shock absorbers (dampers) in response to various on/off road surface and propulsion (driving) or dispulsion (braking) or even conversion (steering) circumstances.

Fig. 5.38 Semi-active F-M ABW AWA suspension [ZF].

The semi-active F-M ABW AWA suspension mechatronic control system also changes shock and strut absorbing (damping) forces in 10 – 12 ms, whereas other suspension absorbing (damping) systems require a much longer time interval to change absorbing (damping) forces. It requires about 200 ms to blink one's eye. This gives us some idea how quickly the ABW AWA suspension mechatronic control system reacts.

The semi-active F-M ABW AWA suspension mechatronic control system's ECU receives inputs regarding vertical acceleration, wheel-to-body position, angular velocity of wheel movement, vehicle velocity (speed) and, lift/dive (Fig. 5.39) [KNOWLES 2002].

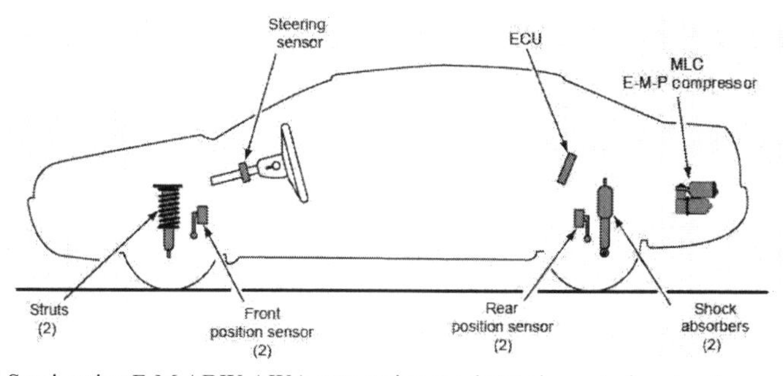

Fig. 5.39 Semi-active F-M ABW AWA suspension mechatronic control system's components [KNOWLES 2002].

For instance, the DBW AWD suspension mechatronic control system's ECU evaluates these inputs and controls with an electromagnetic solenoid in each shock or strut to provide suspension-damping control. The solenoids in the shocks and struts may react much faster compared with the conventional strut F-M actuators. This ECU may also control the vehicle-velocity dependent **all-wheel-steered** (AWS) **steer-by-wire** (SBW) conversion mechatronic control system and the **level control** (LC). The LC maintains the rear suspension trim height regardless of the rear suspension load. The semi-active F-M SBW AWS conversion mechatronic control system is similar to the **variable orifice** (VO) R&P steering gears [KNOWLES 2002]. Some automotive OEMs have developed ABW AWA suspension mechatronic control systems in response to concerns about SUVs that roll over more easily compared with automotive vehicles because of the SUV's higher CoG. These active roll mechatronic control systems were developed in response to this concern. For instance, this mechatronic control system contains an ECU, accelerometer, vehicle-velocity (speed) sensor, oily-fluid reservoir, E-M-F pump, oily-fluid pressure control fluidical valve, directional control fluidical valve, and an F-M actuator in both the front and rear stabiliser bars (Fig. 5.40) [KNOWLES 2002].

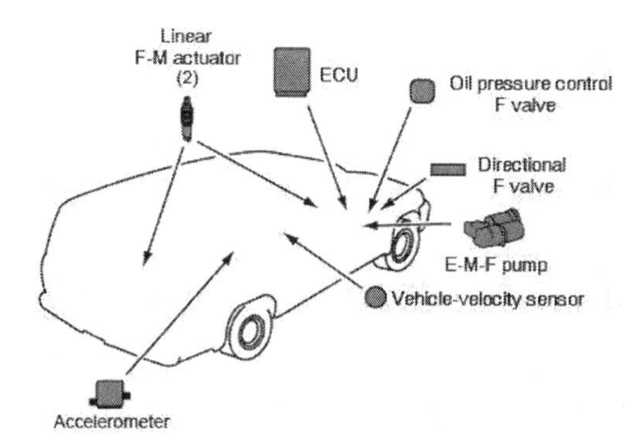

Fig. 5.40 Active roll mechatronic control system components [KNOWLES 2002].

The accelerometer and vehicle-velocity sensor may be used as the power steering E-M-F pump. The active roll mechatronic control system has not been used as standard or optional equipment to date.

When an OEM installs this mechatronic control system on automotive vehicles, it may be integrated with other mechatronic control systems such as the **anti-lock braking system** (ABS), **traction control system** (TCS) and **stability control system** (SCS).

When the automotive vehicle is driven straight ahead, the active roll mechatronic control system does not supply any oily-fluid pressure to the linear F-M actuators in the stabiliser bars. Under this circumstance, both stabiliser bars move freely until the linear F-M actuators are fully compressed. This action provides improved individual wheel bump performance and better ride feature.

If the automotive vehicle's chassis begins to roll while cornering, the ECU operates the directional fluidical valve so it supplies oily-fluid pressure to the linear F-M actuators in the stabiliser bars. This action stiffens the stabiliser bar and reduces vehicle-body roll (see Fig. 5.41) [KNOWLES 2002].

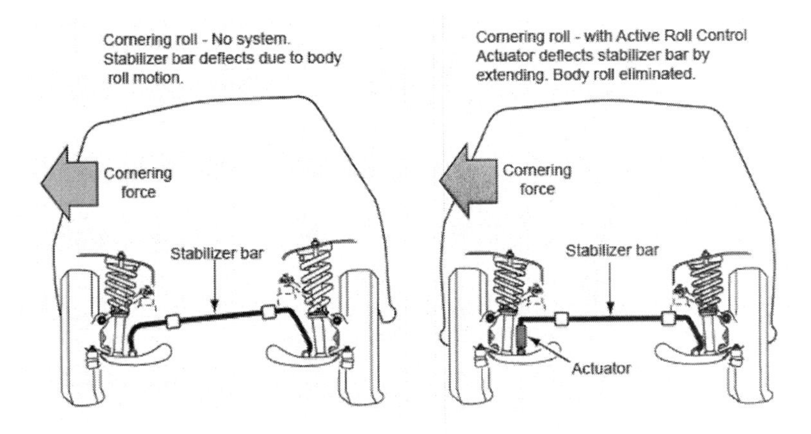

Fig. 5.41 Active roll mechatronic control system functions while cornering [KNOWLES 2002].

The active roll mechatronic control system increases safety by reducing the vehicle's body lean that decreases the possibility of an automotive vehicle rollover. Semi-active F-M shock absorbers (dampers) that are commonly used in semi-active F-M ABW AWA suspension mechatronic control systems generally consist of a F-M shock absorber combined with an external bypass loop containing a servo fluidical valve. The amount of oil passing through the bypass loop is varied to control the behaviour of the shock absorber that essentially behaves as a variable force device with hysteretic type of absorbing (damping). The control fluidical valve is a normally closed direct-drive servo fluidical valve that offers fail-safe function. In case of a power loss, the fluidical valve is fully closed and therefore, the semi-active F-M shock absorber behaves as a passive device with high damping characteristics.

As an example, the construction of the semi-active F-M shock absorber is shown in Figure 5.42 [UNSAL 2002].

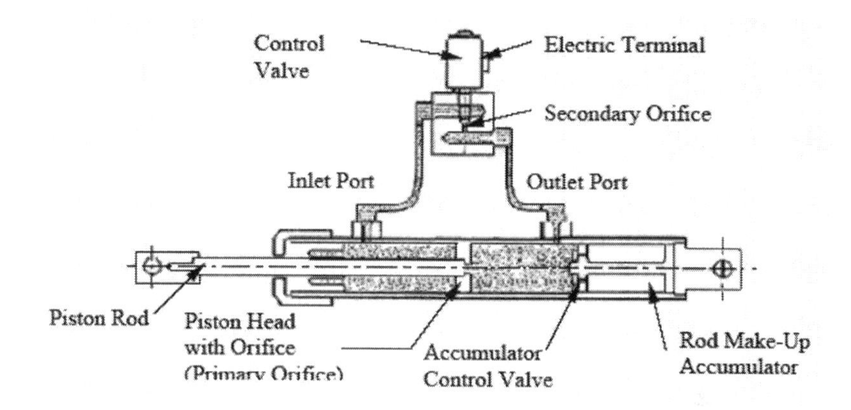

Fig. 5.42 Construction of the semi-active F-M shock absorber [UNSAL 2002].

As another example, the construction of the semi-active F-M shock absorber with the electromagnet solenoid fluidical valve is frequently used in heavy vehicles.

The main feature of this semi-active F-M ABW AWA suspension mechatronic control system is the structure of the fluidical valves regulating the damping coefficient that should be extremely rapid and precise, and at the same time should be capable of taking the stress due to the high values of the oily-fluid pressure within the F-M shock absorber (damper) [GIUA ET AL. 2004].

The fluidical circuit of the F-M shock absorber with the solenoid fluidical valve is shown in Figure 5.43 (a) [KITCHING ET AL. 2000, GIUA ET AL. 2004].

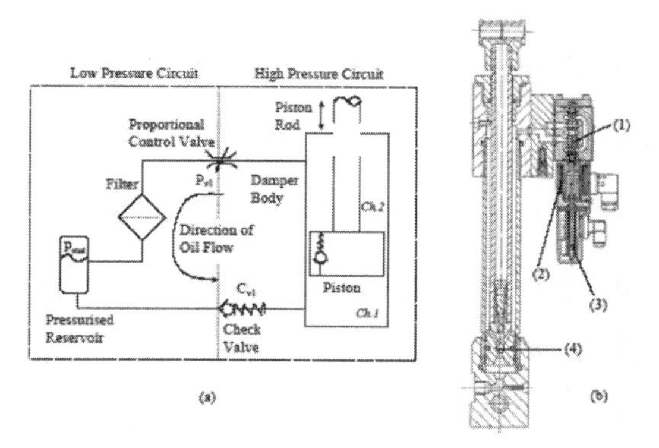

Fig. 5.43 The fluidical circuit of the semi-active F-M shock absorber (a); a cross-section view (b) of the SA F-M shock absorber (fluidical reservoir, fluidical filter and check fluidical valve are not shown) where: (1) is the spool, (2) is the solenoid, (3) is the position transducer and (4) is the piston check fluidical valve [KITCHING ET AL. 2000].

A high-speed solenoid fluidical valve P_{v1} controls the oily-fluid pressure drop through the fluidical circuit, thereby appropriately updating the damping coefficient. A fluidical reservoir accommodates the oily-fluid displaced by the volume of the piston rod in the high-pressure fluidical circuit and is pressurised with nitrogen gas (at a pressure p_{stat}) to prevent cavitations occurring. Note that the flow of oily-fluid always occurs in the same direction, thus allowing keeping the structure of the system simple and compact. Moreover, the presence of the check fluidical valve and the solenoid fluidical valve at the opposite sides of the fluidical circuit create a physical separation among the low-pressure circuit and the high-pressure circuit.

A cross section view of the semi-active F-M shock absorber is shown in Figure 5.43 (b) where for simplicity the fluidical reservoir, the fluidical filter and the check fluidical valve are not shown. It is possible to distinguish three main parts that constitute the electromagnet solenoid fluidical valve: the fluid part, the electrical part and the position transducer. In the fluidical part of the fluidical valve it occur the variation of the cross section that allows updating the actual value of the damping coefficient. This updating can be realised thanks to a small F-M cylinder that translates in the direction of its own axis. The electrical part is basically constituted by an electromagnet solenoid that can modify the position of the F-M cylinder by simply applying an axial force.

Finally, the position transducer is particularly useful when the damping coefficient requires to be updated at a very high frequency. Note that the feedback control of the cylinder position is also necessary due to the disturbance on the cylinder produced by the flow of oil in the fluidical part of the fluidical valve.

Summarising, such a semi-active F-M ABW AWA suspension mechatronic control system requires two different mechatronic control devices [KITCHING ET AL. 2000]. The first one is used to determine the position of the cylinder on the base of the difference among the target force and the actual force produced by the semi-active F-M shock absorber. Finally, the goal of the second mechatronic control device is that of modifying the intensity of the electrical current through the electromagnet solenoid so as to reduce the difference among the required and the actual position of the cylinder [GIUA ET AL. 2004].

A real existing F-M shock absorber's static characteristics *'force-velocity'* is shown in Figure 5.44 [SAMMLER ET AL. 2002; GIUA ET AL. 2004].

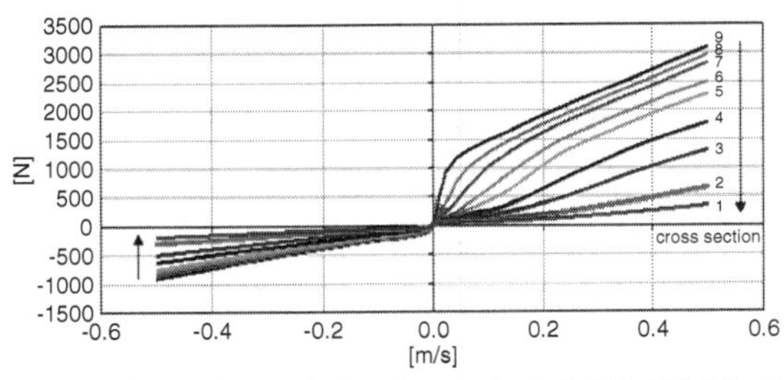

Fig. 5.44 The non-linear characteristics *'force-velocity'* of the F-M shock absorber with the electromagnet solenoid fluidical valve [SAMMLER ET AL. 2002].

Using standard notation, positive values of the deformation velocity correspond to positive values of the force.

Each curve is parameterised by the position of the cylinder and consequently by the value of the opening section where the oily fluid flow from one chamber to the other one within the F-M shock absorber body.

Figure 5.45 shows the major components for semi-active F-M ABW AWA suspension mechatronic control systems [UEKI ET AL. 2004].

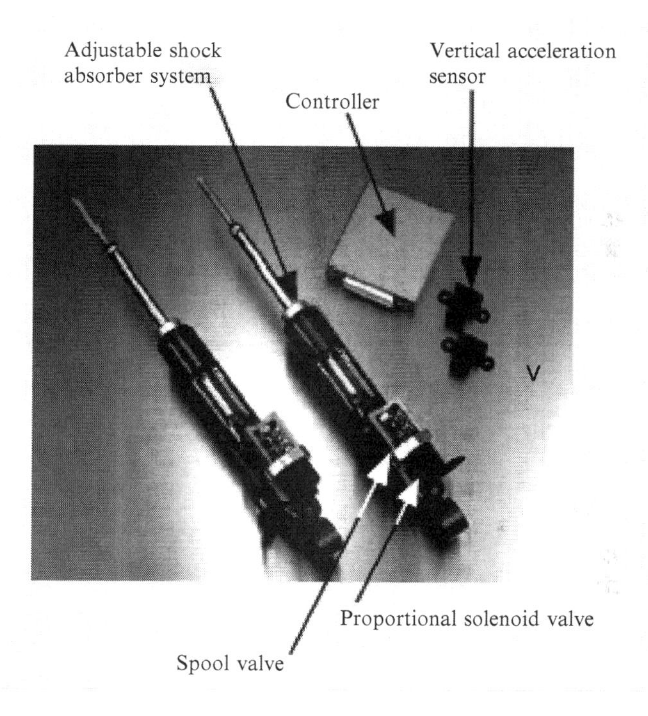

Fig. 5.45 Major components of the semi-active F-M ABW AWA suspension mechatronic control system: it illustrates the adjustable shock absorber (damper), proportional electromagnetic solenoid and spool fluidical valves that shift the absorbing (damping) force [Hitachi, Ltd.].

The controller gains knowledge of vehicle vibration founded on the data from the vertical acceleration sensor bedded in the vehicle, estimates the most favourable absorbing/damping force to circumvent vibration, and controls the adjustable shock absorbers (dampers) to realise better ride feature.

Besides, the controller estimates vehicle cornering founded on data about wheel angle and driving vehicle velocity, and controls the absorbing (damping) force to accomplish better driving stability.

A semi-active F-M ABW AWA suspension mechatronic control system (Fig. 5.46) enable smooth and reliable control with the adoption of **continuously variable-damping- force shock absorber** (CVSA), a damping-force inversion characteristic and unique control logic [HITACHI 2004].

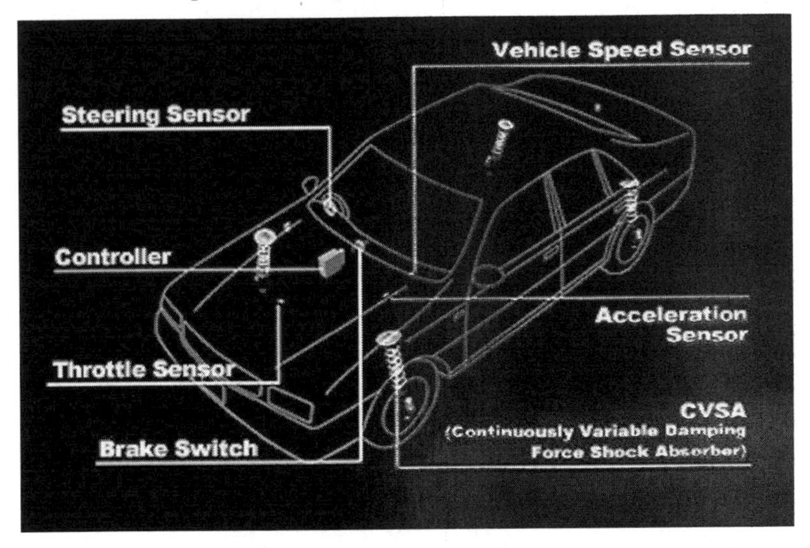

Fig. 5.46 Layout of the semi-active F-M ABW AWA suspension [Hitachi, Ltd; Hitachi 2004].

The CVSAs are shown in Figure 5.47 [Hitachi 2004].

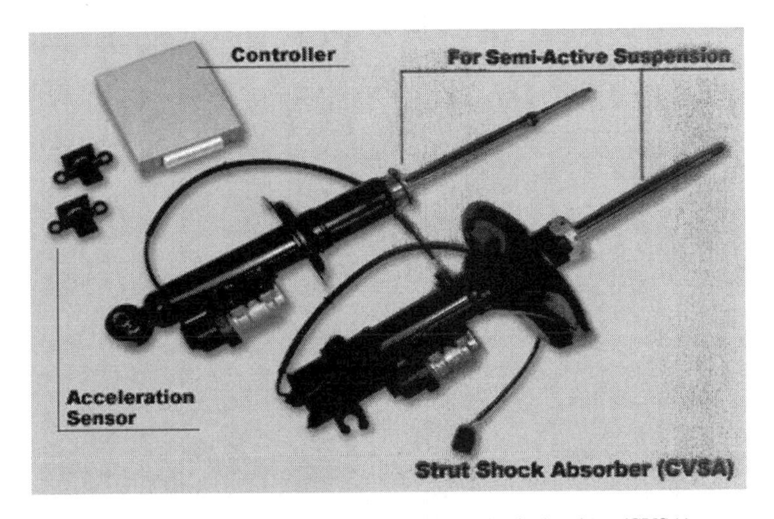

Fig. 5.47 Continuously variable damping force shock absorbers (CVSA) [Hitachi, Ltd; HITACHI 2004]

A semi-active F-M ABW AWA suspension mechatronic control system has the following features [HITACHI 2004]:

- ❖ Absorbing (damping) force characteristic of rebound/compression inversion type;
- ❖ Continuously variable absorbing (damping) force (adoption of proportional solenoid);
- ❖ Wide absorbing (damping) force variable width.

A semi-active F-M ABW AWA suspension mechatronic control system suppresses the exchange between comfort and handling. It continually adjusts damping levels compatible with on/off road surface circumstances and vehicle dynamics, such as vehicle velocity (speed), turning and cornering as well as driver inputs. It demands shock absorbers that include dampers and continuously variable electromagnet solenoid fluidical valves. These fluidical valve adjustments may occur very rapidly -- as a rule within 10 ms - to generate different absorbing (damping) forces. This adjustment velocity (speed) affords active control of wheel resonant vibrations up to 20 Hz. Therefore, wheel hop frequency control is attained as well as vehicle-body frequency control [TEN 2004].

A **servo-valve optimised active damper suspension** (SOADS) mechatronic control system (Figure 5.48) was the first foray into a computer-controlled suspension for the *Hummer* termed in U.S. military parlance **high mobility multipurpose wheeled vehicle** (HMMWV) [MILLENWORKS 2005A].

Fig. 5.48 HMMWV SOADS mechatronic control system [MILLENWORKS 2005A].

The aim was to develop a semi-active F-M ABW AWA suspension mechatronic control system capable of bolt-on retrofit to the HMMWV platform.

The mechatronic control system was designed to replicate the benefits of an active ABW AWA suspension without the associated cost, complexity, power use, and poor failure modes.

In the SOADS mechatronic control system, a **digital signal processor** (DSP) modulates special F-M shock absorbers (Fig. 5.49) in real time in response to instantaneous vehicle behaviour [MILLENWORKS 2000A].

Fig. 5.49 Special F-M shock absorber for the SOADS mechatronic control system [MILLENWORKS 2005A].

The modulation is primarily accomplished using a precision, high-speed, electronically controllable servo-valve.

The SOADS system proved to be a very effective enhancement for the HMMWV platform, demonstrating higher speed over given terrain, increased stability, and reduced shock loading to the vehicle and occupants.

At the heart of the semi-active F-M ABW AWA suspension mechatronic control system is an ECU that processes driver inputs and data from sensors positioned at considerable locations on the automotive vehicle.

The sensors incorporate three accelerometers located on the vehicle body and four suspension-position sensors that deliver data to the ECU on steering-wheel angle, vehicle velocity (speed), brake pressure and other vehicle-chassis control data.

The ECU employs control software that processes the sensor data in real time and transmits signals that adjust individually the damping level of each shock absorber valve.

Adjustable F-M shock absorbers (dampers) permit a large separation between maximum and minimum damping levels and adjust without delay to guarantee riding comfort and firm, safe vehicle control.

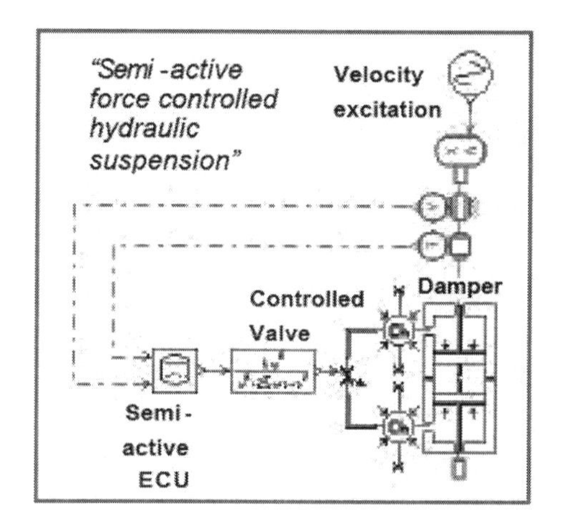

Fig. 5.50 Quarter-vehicle physical model of the semi-active F-M 4WA ABW suspension mechatronic control system [AMESIM 2004].

In Figure 5.50 is shown a quarter-vehicle physical model of the semi-active F-M ABW AWA suspension mechatronic control system [AMESIM 2004].

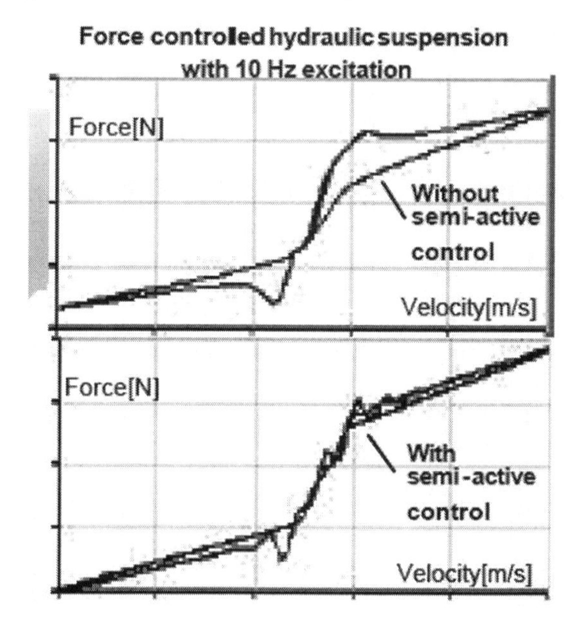

Fig. 5.51 Absorbing (damping) force-velocity profile of the semi-active F-M ABW AWA suspension: (above) without semi-active control; (below) with semi-active control [AMESIM 2004].

Absorbing (damping) force-velocity profile of the semi-active F-M ABW AWA suspension is drawn in Figure 5.51 [AMESIM 2004].

5.5.7 Semi-Active P-M ABW AWA Suspension Solution

The semi-active P-M ABW AWA suspension mechatronic control system is a completely operating ABW AWA suspension mechatronic control system that assumes control of the complete static and dynamic supportive elements by means of four air bellows positioned correspondingly at the wheels. The level control for the ride-height control system is attained with the assistance of air bellows: if the M-P compressor supplies air to one of the air bellows, the automotive vehicle height escalates at the individual wheel. On the contrary, the height level declines when air is released from one of the air bellows by means of the levelling discharge pneumatical valve.

The mechatronic control system is prepared with a **central storage unit** (CSU) as a compressed air reservoir. This raises the adjustment speed when the automotive vehicle gets higher and permits for mechatronic control even when the ECE or ICE is at a standstill. Besides the semi-active P-M ABW AWA suspension mechatronic control system is combined with an automatic damper adjustment that acts according to the *'skyhook'* algorithm.

Four control pneumatical valves may be turned on or off per wheel for supplementary air volume, permitting for the semi-activity of the P-M shock absorber (damper) adjustment with concurrent roll stabilisation. Moreover the semi-active suspension mechatronic control system is combined with an automatic damper adjustment that acts in relation to the *'skyhook'* algorithm.

A semi-active P-M ABW AWA suspension mechatronic control system continuously adjusts absorbing (damping) levels according to on/off road surface conditions and vehicle dynamics, such as vehicle velocity, turning and cornering as well as driver inputs. It may use P-M shock absorbers and continuously variable electromagnetic pneumatical valves.

As a result, wheel-tyre hop frequency control is attained in addition to body frequency control [TEN 2005].

The semi-active P-M ABW AWA suspension uses compressed air in bellows-like springs in each corner, to maintain constant ride height regardless of load.

On the highway, the automotive vehicle is lowered by about 2.5 cm (1 inch) that improves aerodynamics and thus fuel kilometreage (mileage).

The semi-active P-M ABW AWA suspension also incorporates the adaptive absorbing (damping) mechatronic control system that allows a driver to adjust the shock absorbers (dampers) according to on/off road surface circumstances, load and vehicle velocity (speed). This is not the same as the more complex semi-active ABW AWA suspension that is optional on the automotive vehicles, and which virtually eliminates vehicle-body roll in cornering, squat under acceleration, and dive during braking. A semi-active P-M ABW AWA suspension mechatronic control system necessitates high-performance control architecture.

Digital signal processor/micro-controller unit (DSP/MCU) tools exceed this requirement including a hybrid core, **A-D converters** (ADC), **pulse width modulators** (PWM), timers, and on-board flash.

The semi-active P-M ABW AWA suspension mechatronic control system that is shown in Figure 5.52 incorporates DSP and MCU functionalities.

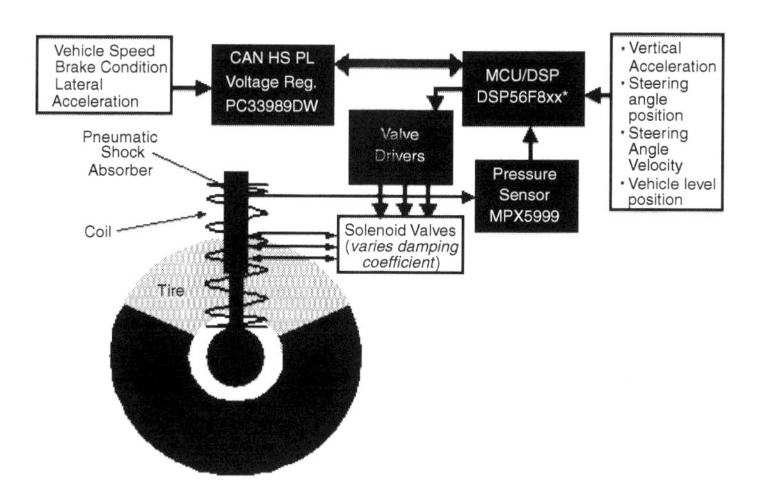

Fig. 5.52 Structural and functional block diagram of the semi-active P-M ABW AWA suspension mechatronic control system [FREESCALE 2005]

This solution deals with design challenges with ADCs for multiple sensor inputs and quad decoder for use with quad encoder equipped steering systems [FREESCALE 2005].

The semi-active P-M ABW AWA suspension mechatronic control system also reciprocates with PWMs for controlling fluidical-valve F-M actuators and has a 60 MHz process to guarantee high-speed DSP/MCU response to mechatronic control system demands. Not only does this semi-active P-M ABW AWA suspension mechatronic control system afford high-speed processing for automotive frequency range and a low cost MCU/DSP, it also offers the most cost-effective automotive frequency range of 60 MHz/MIPS MCU/DSP that is insert its aptitude to control semi-active ABW AWA suspensions. The semi-active P-M ABW AWA suspension mechatronic control system reaches an innovative level in ride and handling control.

At the touch of a button a driver may choose from three driving modes – two for sport and one for comfort. A self-levelling mechatronic control system can even raise or lower the ride height -- automatically, or at the driver's pushbutton command.

The sport mode using a switch on the console may be chosen for even sharper handling response and virtually flat cornering. The ride height may even be raised or lowered for greater clearance or enhanced aerodynamic efficiency.

The modern semi-active P-M ABW AWA suspension mechatronic control system is all-aluminium in order to reduce unsprung mass to a minimum.

Unsprung mass includes the wheel and tyre, brake rotor and calliper, and the suspension arms: in other words, all the components that move up and down with the road surface and are not supported by the springs. The lighter mass of these components keeps the wheel-tyres in better contact with the on/off road surface by lessening the inertia that creates a bouncing action on rough on/off road surfaces. By avoiding this tendency of heavy suspension components to bounce, the automotive vehicle maintains more consistent wheel-tyre grip.

Semi-active P-M ABW AWA suspension is also available on the rear axle as an option and serves to keep the automotive vehicle level regardless of load.

Another suspension feature, **active roll stabilisation** (ARS), compensates for vehicle body sway by adding semi-active P-M ABW AWA suspension control to the mix. A third system, **electronic damping control** (EDC) provides variable shock damping under computer control for no-compromise ride and handling over a variety of on/off road surface conditions. EDC is available as part of an optional **adaptive ride package** (ARP) and offers *'stepless'* control of the P-M shock absorbers from full soft to full firm.

The most interesting of these suspension goodies has to be the ARS mechatronic control system that consists of front and rear active anti-roll bars. These are sway bars that are cut in half and attached at the centre to a control unit that uses oil pressure to apply a reverse-twist to the bar under computer control, in order to compensate for vehicle-body lean during turns. This produces near flat cornering up to about half a *G* with no compromise to the superb ride. Beyond that, the system allows some lean to creep in so the driver may feel the feedback and realise that he or she is approaching cornering limits. There are still limits.

Automotive scientists and engineers have not yet found a way to disobey the laws of physics.

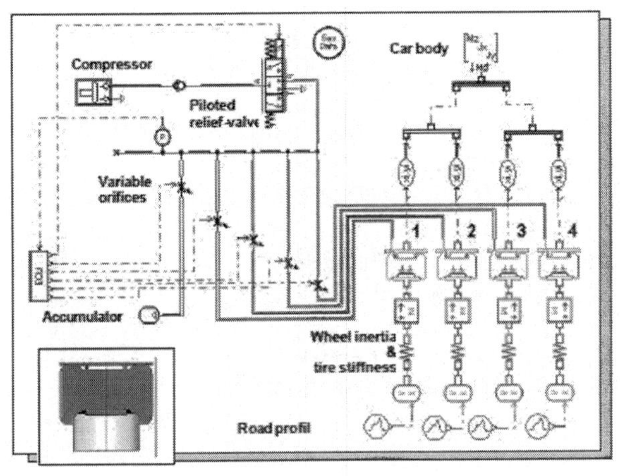

Fig. 5.53 Full-vehicle physical model of the semi-active P-M 4WA ABW suspension [AMESIM 2004].

In Figure 5.53 is shown a full-vehicle physical model of the semi-active P-M ABW AWA suspension mechatronic control system [AMESIM 2004].

5.5.8 Semi-Active E-M ABW AWA Suspension Solution

In the 2010s, however, an alternative opportunity was exposed by OEMs when it pulled the veils of secrecy off a semi-active ABW AWA suspension that exchanges an F-M or a P-M actuator for a linear E-M motor as shown in Figure 5.54 [McCormick 2004; Eisenstein 2005; Lotus 2005].

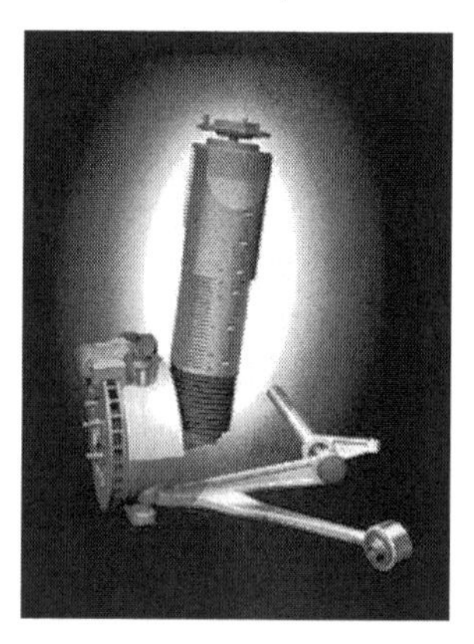

Fig. 5.54 Semi-active E-M ABW AWA suspension (with the linear E-M motors) that is termed *'Bose® suspension'*: front corners contain E-M uprights and custom lower control arms [Eisenstein 2005; Lotus 2005].

Each unit resembles a conventional strut, minus coil spring. Instead, automotive vehicle mass is supported by a torsion bar integrated with a two-piece lower control arm.

Power amplifiers using switching amplification technology feed energy to the linear E-M motors under the mechatronic control of algorithms.

The algorithms in turn, respond to data from sensors, continually reporting on the dynamic state of the automotive vehicle.

Linear M-E motors are proving simple and reliable in other areas of industry and in this instance, provide one benefit that no fluidical (hydraulical and/or pneumatical) active unit could do; they can regenerate energy.

When the mover of the linear E-M motor is active, then naturally, it consumes electrical energy.

But when it is passive and being moved in bump or rebound by the on/off road surface, then it can generate electricity that could be stored in a CH-E/E-CH storage battery or better still, an ultracapacitor.

This is a crucial aspect of the design and the designers claim the energy consumption is less than 30% of a conventional automotive vehicle air conditioning system.

However, that's still more than a spring and shock absorber (damper) unit and ultimately the benefits of using such a system may be weighed not just against the cost of the components, but against the cost of improving powertrain efficiency by an equivalent amount, if the automotive vehicle's efficiency is to remain in balance.

Assuming everyone's happy on that front, the linear E-M motor could conceivably crop up in other chassis applications too. SBW AWS conversion may be in its infancy but it has many supporters.

The problems of doing away with a mechanical link between driver and wheels stood in the way for some time until vehicle manufacturers devised the **active front steering** (AFS) system.

A rotary E-M motor and epicyclical gearbox spitting the steering column meant that steering angle could be changed independently of driver input while retaining a mechanical link [McCormick 2004; Lotus 2005].

It's clever stuff, although automotive scientists and engineers are still working on the principal of electric steering racks to provide full SBW AWS conversion and do away with the packaging and safety constraints of a conventional steering column. In that case, linear E-M motors could have a role to play.

The same electromagnetic forces used to apply torque in a conventional rotary E-M motor produces linear force.

A linear E-M motor can deliver accuracy measured in microns, produce both miniscule and massive forces, are fast and don't wear out [Lotus 2005].

Another M-E shock absorber (damper) uses a ball-screw mechanism to transform vibration motion between a vibrating vehicle body and an of/off road surface into rotational motion. The structure of the E-M shock absorber is shown in Figure 5.55 [Unsal 2002].

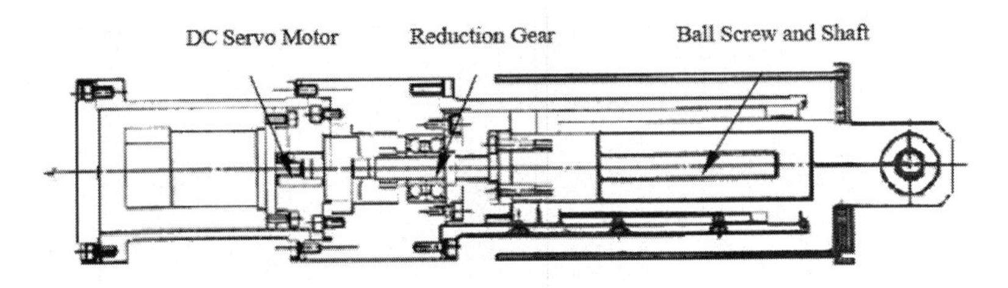

Fig. 5.55 Structure of the E-M shock absorber [Unsa l 2002].

A DC servomotor, that is, the rotary DC-AC/AC-DC commutator servo motor/ generator is driven by this rotational motion through a speed-increasing gear that in turn generates an inverse induced voltage of rotation.

This induced voltage of rotation is used to create the absorbing (damping) force of the E-M shock absorber.

The resistance between the terminals of the DC-AC/AC-DC commutator servo motor/ generator may be varied to adjust the level of the induced voltage of rotation and therefore, the absorbing (damping) force [Iiyama et al. 1998].

For instance, Showa Corporation has publicly presented a prototype of a semi-active E-M ABW AWA suspension at the *39th Tokyo Motor Show*, Japan [Takano 2004].

When used with a certain mechatronic control technology, the suspension may eliminate the necessity for springs.

The semi-active E-M ABW AWA suspension on display (Fig. 5.56) employs an actuator unit including a rotary DC-AC commutator motor and ball screw [TAKANO 2004].

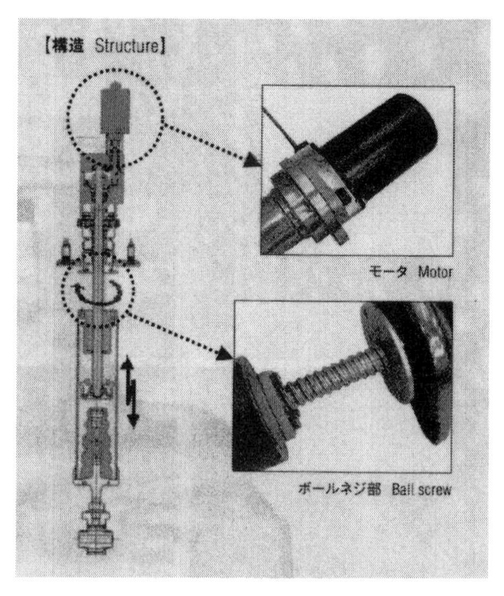

Fig. 5.56 Structure of the semi-active E-M ABW AWA suspension's actuator [TAKANO 2004].

The unit enables semi-active control to adjust damping force or active control to control rolling or braking dive by using a shock absorber (damper) that moves vertically in advance.

According to a *Show* spokesperson, the company expects that *"as the driving population and environment become diverse, just adopting semi-active control able to adjust damping force may be a great added value"*.

In addition to testing the suspension alone, the company has test-driven a *Step Wagon* automotive vehicle made by Honda Motor Co., Ltd. equipped with this suspension to collect driving data [TAKANO 2004].

5.5.9 Semi-Active MR ABW AWA Suspension Solution

Some automotive OEMs introduce a revolutionary innovative suspension technology. It is semi-active **magneto-rheological** (MR) ABW AWA suspension mechatronic control system, and it may replace the semi-active F-M or P-M ABW AWA suspension mechatronic control system beginning in the 2010's. Although semi-active F-M or P-M ABW AWA suspension mechatronic control systems with variable dampers (shock absorbers/struts) have existed in the past, this is the first one to dispense with electromagnet solenoid fluidical or pneumatical valves and small moving parts. Conventionally, dampers (shock absorber/struts) have relied on the movement of a piston through an oily-fluid- or air-filled chamber.

Piston movement is resisted and controlled by fluidical or pneumatical valves that limit the amount of oily-fluid or air (gas) that can flow past the piston. In more sophisticated semi-active F-M or P-M ABW AWA suspension mechatronic control systems, computer controlled fluidical or pneumatical valves, respectively, are used to vary the flow rate. Currently, the semi-active MR ABW AWA suspension mechatronic control system eliminates even the fluidical valves. The principle being used sounds like it's taken from the pages of science fiction.

Semi-active MR ABW AWA suspension mechatronic control systems are becoming more popular because they offer both the reliability of passive systems and the versatility of active control without imposing heavy power demands. It has been found that, **magneto-rheological fluids** (MRF), **nano-magneto-rheological fluids** (NMRF) or **magneto-rheological elastomers** (MRE) may be designed to be very effective vibration control actuators. The MR shock absorber (damper) is a semi-active control device that uses MRFs. NMRFs or MREs to produce a controllable absorbing (damping) force [LAI AND LIAO 2002].

A MR shock absorber is a recently invented E-M semi-active mechatronic control device that holds promise for applications to an automotive vehicular vibration. However, non-linear characteristics of MR shock absorbers are delaying practical installation. Using a dynamic **neural network** (NN), that is applicable for an on-line mechatronic control environment, it may provide a linearization scheme for MR shock absorber behaviour.

As recently as 2010, automotive scientists and engineers have experienced very rapid advancements of mechatronics technology in the automotive industry. The rate of mechatronic developments promise to intensify each year. Mechatronics concerns all subjects pertaining to automotive vehicles and their incorporation of semi-active MR ABW AWA suspension mechatronic control systems. In this innovative mechatronic control system, the shock absorbers (dampers) or struts do not have any electromagnetic solenoids or fluidical or pneumatical valves. Instead of these components, the shock absorbers (dampers) or struts are filled with an MRF, NMRF or MRE. MRFs and NMRFs are a class of controllable fluids whose rheological properties may be rapidly varied by the application of a magnetic field. MRFs are suspensions of micron-sized, magnetically polarisable micro-particles in oily-fluid or other liquid [CARLSON 2001]. The flow ratio of MRFs may be varied by the application of a magnetic field as shown in Figure 5.57 [GILBERT AND JACKSON 2002].

Fig. 5.57 Magneto-rheological fluid (MRF) at reference (left) and activated states (right) [Lord Corp. & Rheonetic™; GILBERT AND JACKSON 2002].

Under normal conditions, MRF is a free-flowing liquid having a consistency similar to that of motor oil. Exposure to a magnetic field transforms the oily fluid into a plastic-like solid in milliseconds. Removal of the field allows the MRF to return to its original free-flowing liquid state.

The degree of change in a MRF depends on the magnitude of the applied field. MRFs develop a yield strength that scales with the applied magnetic field strength.

Stable, robust, high-strength MRFs have recently been developed that provide the enabling technology to realise the benefits of controllable fluids in many practical, real world applications. MRF is named for rheology, the science dealing with the deformation and flow of matter.

MRF consists of micro-spherical iron particles suspended in a synthetic hydrocarbon base fluid. It can change mineral-oil consistency to jelly-like consistency within 1 ms, when a magnetic field is applied [GILBERT AND JACKSON 2002].

Without a magnetic field the fluid is not magnetised and the iron particles are scattered randomly. But when the magnetic field is turned on, the metal particles align into fibrous structures, changing the fluid rheology. This essentially instantaneous thinning or thickening of the fluid regulates the damping properties of the struts. MRFs respond to an applied magnetic field with a change in their viscosity; in essence they stiffen from liquid to solid (Figure 5.58).

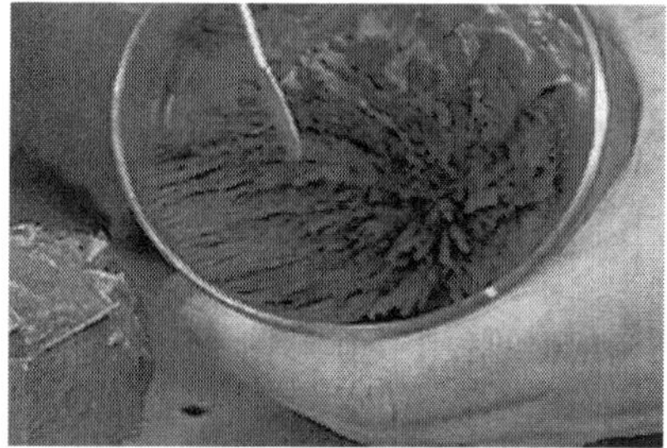

Fig. 5.58 Oily fluid full of tiny iron particles before being near a magnet (above); and after [MIT Hatsoupoulos Microfluids Lab.; GILBERT AND JACKSON 2002].

The essential characteristic of these oily fluids is their ability to reversibly change from a free-flowing, linear, viscous liquid to a semi-solid with controllable yield-strength in fractions of a second when a magnetic field is applied.

An increase in the power of the magnetic field increases the stiffness of the oily fluid, allowing a corresponding linear increase or decrease in viscosity with magnetic field. In their simplest state, MR fluids are simply iron filings or spheres in an oily paste, with particle sizes of the order of 1 – 10 μm.

The oily fluid is normally a silicone oil or corn oil, to prevent rusting, and is *'mayonnaise thick'* to avoid dripping. Varying the morphology of the particles also changes the ultimate yield strength of the fluid, with faceted platelets or donut shaped particles allowing for easier stacking of particles and therefore greater strength.

Although not a new discovery (MRFs were first reported by *Jacob Rainbow* at the U.S. National Bureau of Standards in the late 1940s), the applications for this material fall into three distinct categories, vibration damping, highly accurate polishing tools of high precision optical lenses and controlled movement levers for fitness equipment.

The use of MRF in the suspension of automotive vehicles, show how more recent work has harnessed the properties of these fluids for motion control and damping, proving effective alternatives to an E-M valve that has no small moving parts.

The diagram shown in Figure 5.59 shows how the alignment of the magnetic particles in the MRF increases the yield stress over the shear stress, thus *'stiffening'* the ride. This change from mobile to stiff is reversible and can occur hundreds of times a second.

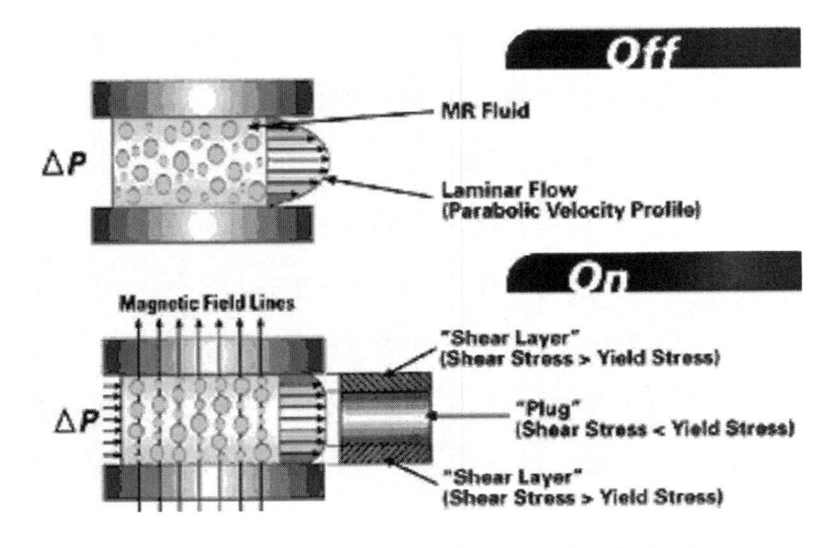

Fig. 5.59 Schematic of the stiffening process in an automotive vibration dampener [Delphi Inc.; GILBERT & JACKSON 2002]

In the *'off'* state, the **magneto-rheological** (MR) fluid is not magnetised and the iron particles are dispersed randomly; however, in the *'on'* state, the applied magnetic field aligns the particles into fibrous structures, changing the fluid rheology and thus regulating the damping properties of the mono-tube struts.

Due to the fact that the MRF is a synthetic oily fluid containing suspended iron particles, each MR shock absorber (damper) or strut contains an exciter-coil winding that is energised by the ECU.

The piston is fitted with an electromagnet; the exciter-coil winding is imaged in Figure 5.60 [GILBERT AND JACKSON 2002].

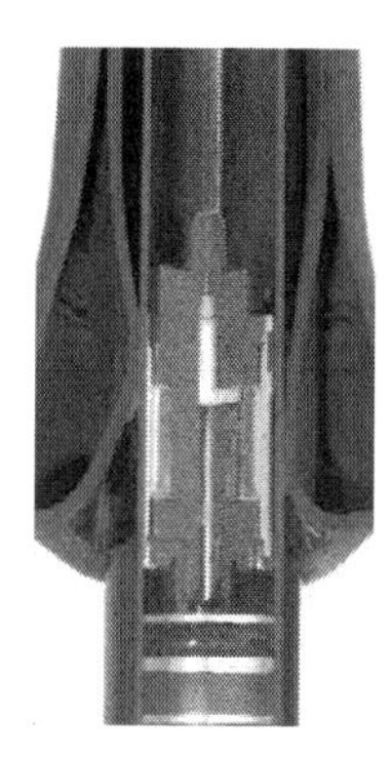

Fig. 5.60 Piston fitted with the exciter-coil winding [General Motors Corp]...

When the strut exciter-coil winding is not energised, the iron particles are dispersed randomly in the MRF. Under this circumstance, the MRF has a mineral oil-like consistency, and this oily fluid flows easily through the strut orifices to provide a soft ride feature [KNOWLES 2002].

If the ECU energises the strut exciter-coil windings, the magnetic field around this winding aligns the iron particles in the MRF into fibrous structures (Fig. 5.61) [KNOWLES 2002].

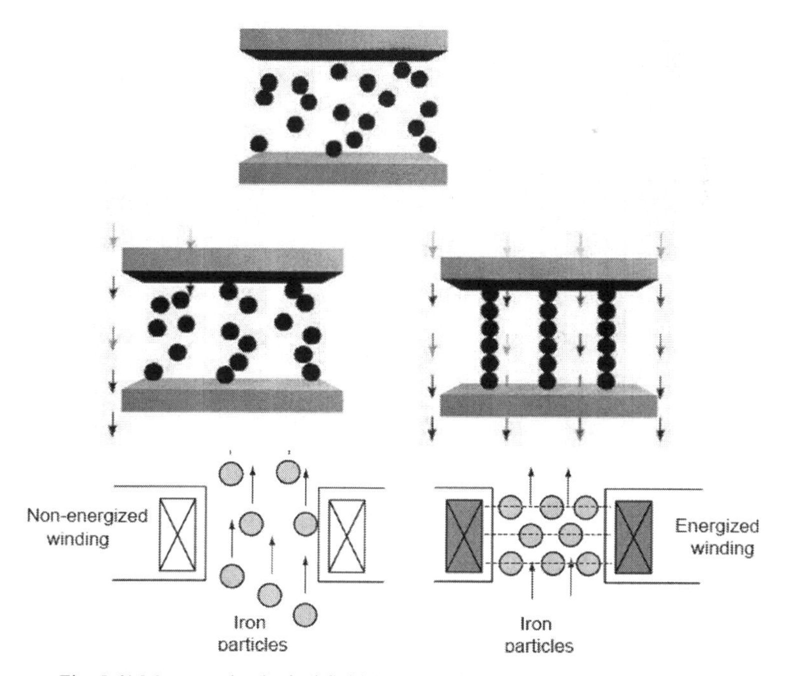

Fig. 5.61 Magneto-rheological fluid (MRF) actions in strut or shock absorber [CARLSON 2001; KNOWLES 2002].

On the other hand, NMRFs are suspension of micro-sized, magnetically polarisable nano-magnets *'embedded'* on micro-particles in oily-fluid or other liquid [FOLONARI 2006].

The flow ratio of NMRFs may be varied by the application of a magnetic field as shown in Figure 5.62 [FOLONARI 2006].

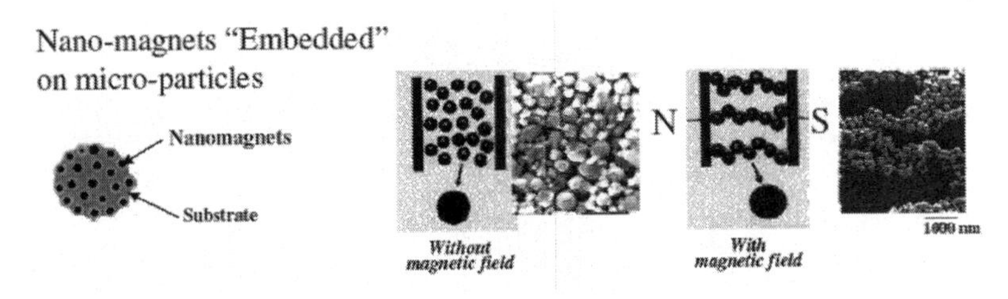

Fig. 5.62 Nano-magneto-rheological fluid (NMRF) action in strut or shock absorber [FOLORANI 2006].

Advantages of working with nano-magnets over micro-particles are as follows:

- Reduced wear;
- Reduced aggregation;
- Reduced power consumption;
- Faster response time.

Applications may be as follows:

- High-speed MR mechanical components;
- Active MR shock absorbers (dampers);
- MR brakes;
- MR clutches.

Success depends on the mechanical design.

Based on the semi-active MR ABW AWA suspension mechatronic control system's inputs from the wheel position sensors and steering-wheel position sensor, the ECU supplies current at rates up to 1 ms to the exciter-coil winding in the appropriate MR shock absorber (damper) or strut.

Therefore, the ECU provides an almost infinite variation in strut damping. The strut may change the damping characteristics of the MRF by 1 ms. The ECU may be located on the bay board in the rear storage compartment. The rotary position sensors may be located on the front and rear suspension components as normal.

The mono-tube MR shock absorber (damper), Figure 5.63 (a), has only one volume of MRF and a fluidical (hydraulical and/or pneumatical) accumulator to accumulate the change in volume due to the additional volume displaced as the piston shaft enters the housing. Mono-tube shock absorbers are the most common configuration.

The twin-tube MR shock absorber, Figure 5.63 (b), has two concentric tubes connected with a pneumatical valve. The pneumatical valve allows for MRF to flow from the inner to outer tube to adjust for piston displacement. The third common configuration is the double-ended design, Figure 5.63 (c).

Because the piston shaft is the same diameter throughout the body, double-ended shock absorbers do not need to account for changes in volume [YANG ET AL. 2002].

Common MR shock absorber (damper) designs also include variations in the number of activation regions, also referred to as the number of stages. Activation regions are the areas where the MRF passes through a magnetic field. Each of the shock absorbers in Figure 5.63 employs one electromagnet coil (exciter coil) winding, and therefore are single-stage designs.

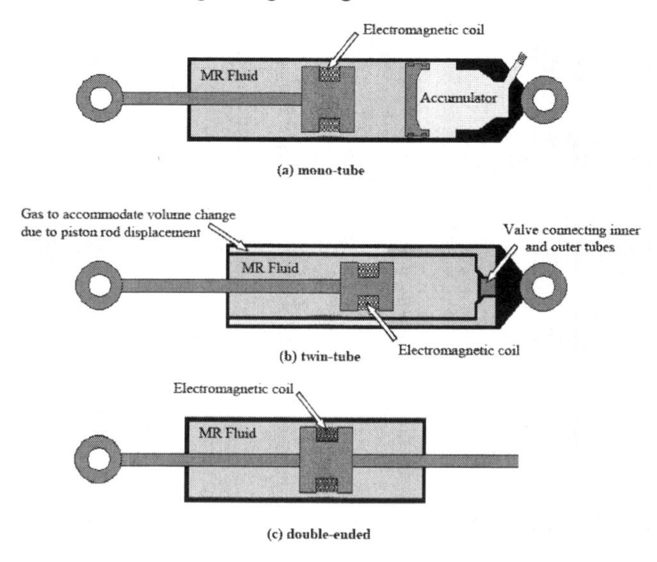

Fig. 5.63 Cross-sections of three common configurations for MR shock absorbers [YANG ET AL. 2002].

The number of stages refers to the number of regions where the MRF is subject to a magnetic field (activation region).

For instance, the MR shock absorber shown in Figure 5.64 is a two-stage design. Both the mono-tube and double-ended MR shock absorbers function in the valve mode [POYNOR 2001].

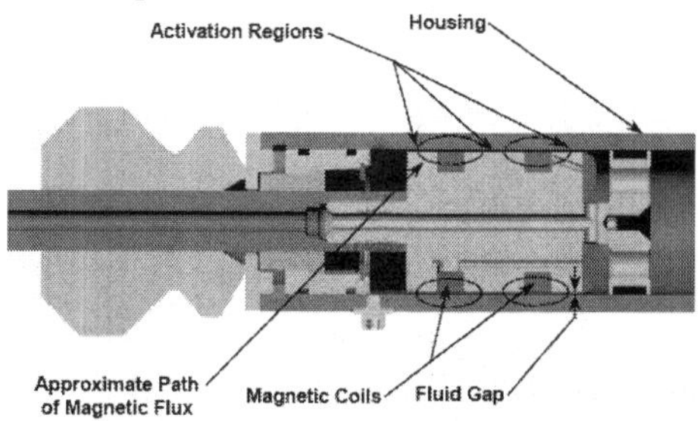

Fig. 5.64 Activation regions in a two-stage design [POYNOR 2001].

Figure 5.65 shows a three-stage, double-ended design that may be used for high-payload automotive vehicles. The magnetic flux lines are sketched in the detail-accompanying Figure 5.65 [YANG ET AL. 2004].

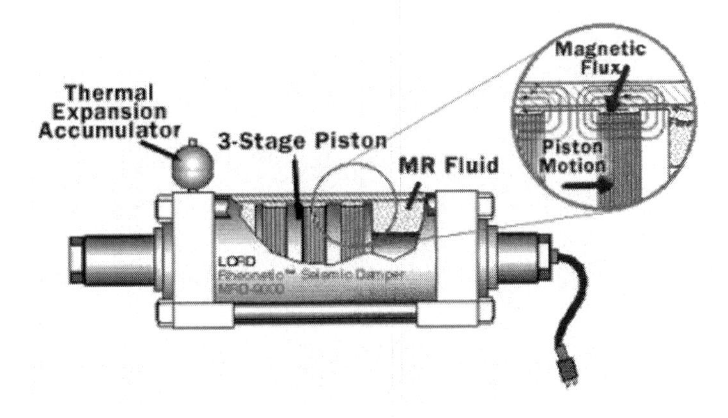

Fig. 5.65 Three-stage, double-ended MR shock absorber for high-payload vehicle damping [YANG ET AL. 2004].

Figure 5.66 shows a cross sectional view of the front two-stage, mono-tube MR shock absorber (damper) that function in the MRF valve mode [POYNOR 2001].

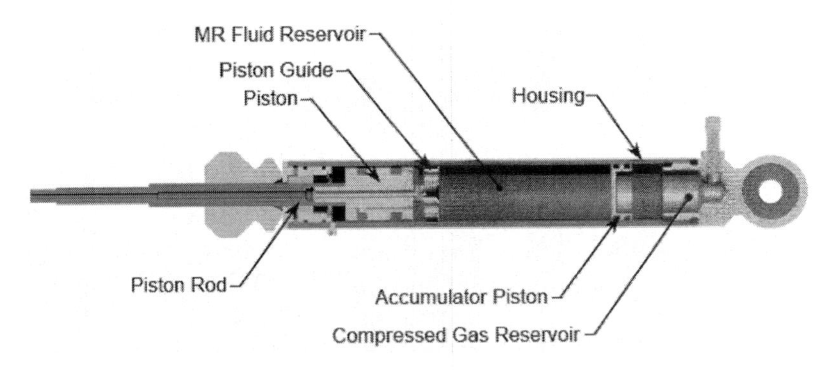

Fig. 5.66 Front two-stage, mono-tube MR shock absorber section view [POYNOR 2001].

The two exciter coils are constructed using *25-gauge* magnet wire wrapped around a paper based phenolic coil form. The mono-tube design includes a nitrogen-gas pneumatical accumulator to reduce the chance of MRF cavitations and accommodate for changes in volume due to the piston rod entering the housing.

In Figure 5.67 and Figure 5.68 are shown cutaways of the two single-stage, mono-tube MR shock absorbers [DALE 2001; GILBERT AND JACKSON 2002; CRAIG 2003].

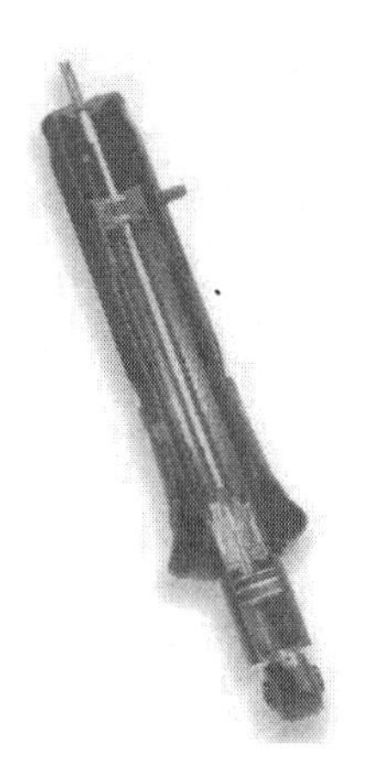

Fig. 5.67 Cutaway of the single-stage, mono-tube MR shock absorber with rear levelling provisions [General Motors Corp.; GILBERT & JACKSON 2002]

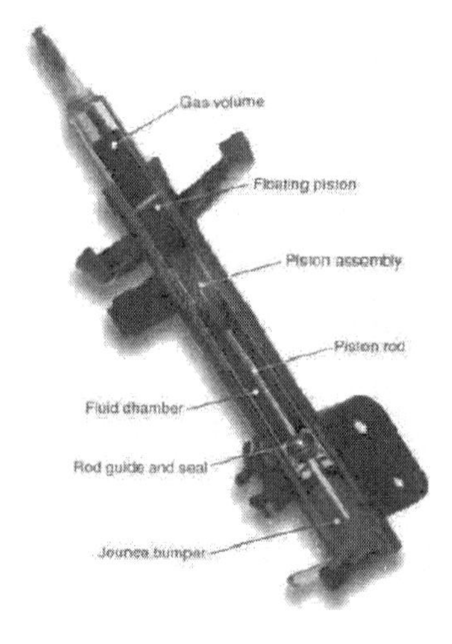

Fig. 5.68 Cutaways of the single-stage, mono-tube Delphi's *MagneRide* MR shock absorber [Delphi Automotive and Lord Corp.; GILBERT AND JACKSON 2002]

Noticeably not present in these cutaways of the MR shock absorber (damper) are the large magnetic or pneumatical solenoids -- normally associated with an active ABW AWA suspension mechatronic control system. The MRF necessitates only a small electromagnetic exciter-coil to control the fluid, and thus the shock's absorbing (damping) rate. As opposed to fluidical (hydraulical and/or pneumatical) valve-based systems, the semi-active MR ABW AWA suspension mechatronic control system has a mono-tube design with no electro-mechanical valves or small moving parts for quieter operation - an industry first, according to the OEMs.

It consists of MRF-based mono-tube struts and MR shock absorbers; a sensor set that consists of a relative position sensor between each control arm and the vehicle body as well as a lateral accelerometer and a steering-wheel angle sensor that are also part of the *Stabilitrak* system (a yaw rate sensor is used indirectly by the semi-active MR ABW AWA suspension mechatronic control system in active-brake-apply events); an on-board controller; and an optional levelling compressor module that has under-body or ECE- or ICE-compartment packaging capabilities and integrates with the existing sensors and controller (Fig. 5.69) [GEHM 2001].

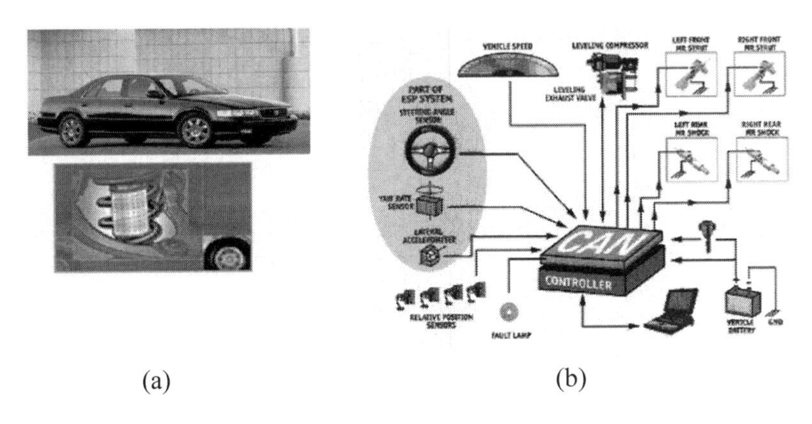

Fig. 5.69 MR DBW AWD suspension in Cadillac *Seville* (a) and the mechatronic control system architecture for *Magne-Ride* (b) [Cadillac – Left photos; GEHM 2001 – Right image].

Based on the wheel inputs from the road-sensing suspension—as well as driver inputs such as steering and braking—the system's on-board computer sends electrical currents up to 1 ms to the electromagnetic exciter coils in each shock absorber (damper) to change the flow properties of the absorbing (damping) fluid. The MRF may change from a mineral-oil-like consistency for low values of the damping force to a jelly-like one for high values of the damping force within 1 ms. the result is continuously variable, real-time damping that provides the following benefits [GEHM 2001]:

- Flatter ride through more controlled vehicle-body motions and reduced wheel bounce that reduces heave, pitch, and roll;
- Greater sense of safety and security due to improved road-holding capabilities on un-even surfaces and during braking;
- Enhanced handling by controlling the lateral and longitudinal load transfer characteristics of the suspension during transient movements;
- Reduction of high-frequency road disturbances transferred through the dampers;
- Greater vehicle-dynamics control when combined with unified chassis control.

The semi-active MR ABW AWA suspension mechatronic control system also contributes to low-frequency vehicle-body control and enhanced roll control during transient steering and evasive manoeuvres.

In addition, this mechatronic control system integrates with other automotive vehicle's mechatronic control systems such as ABSs and TCSs to further improve the automotive vehicle's performance, comfort, and safety characteristics. The operating temperature range for the shock absorbers (dampers) is 233 -- 343 K (-40 -- +70 °C; -40 -- +158 °F) [GEHM 2001].

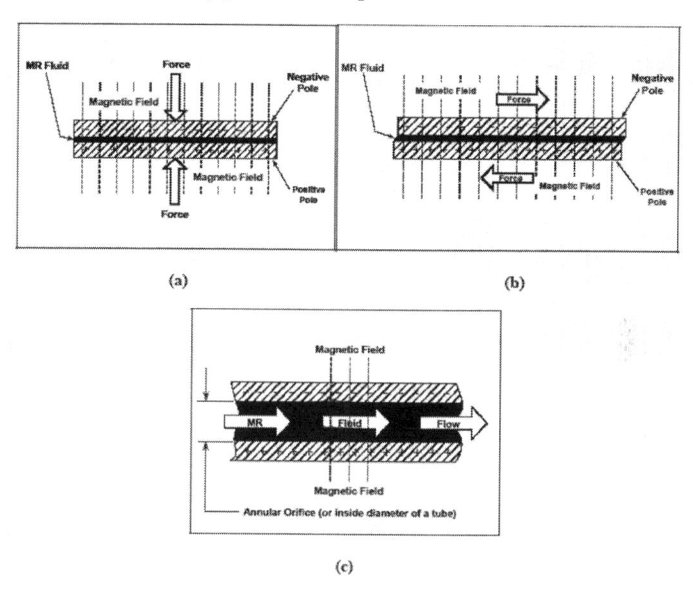

Fig. 5.70 Three common modes of function for MR shock absorbers: (a) squeeze mode; (b) shear mode; and (c) flow mode [POYNOR 2001].

Three common modes of function exist in MR shock absorber designs: squeeze, shear and MRF valve mode, diagrammed in Figure 5.70 [POYNOR 2001].

Figure 5.71 shows an example of the MR shock absorber's relationship between damping force and velocity for a variable applied magnetic field.

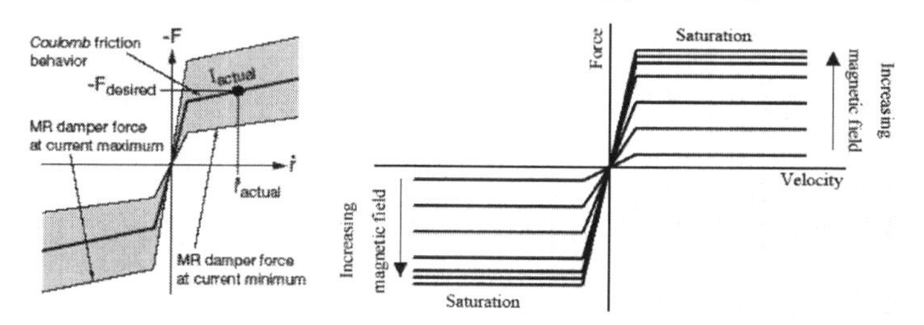

Fig. 5.71 MR shock absorber's relationship between damping force and velocity for variable applied magnetic field [GAVIN ET AL. 2001; WEBER AND FELTRIN 2002].

As the applied magnetic field is increased, the damping force increases until a saturation point is reached. The saturation point is defined by the point where increasing the magnetic field fails to create an increase in damping force.

For a properly designed magnetic circuit, the difference between the off state and saturation damping force may be as high as a factor of 10 or more [GAVIN ET AL. 2001; WEBER AND FELTRIN 2002]. As an example, a commercially available MR shock absorber [LORD 2005] is shown in Figure 5.72 (a), and its force -velocity profile is drawn in Figure 5.72 (b).

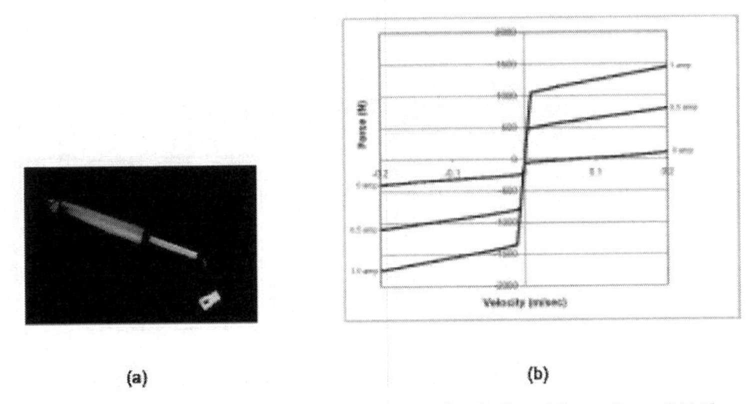

(a) (b)

Fig. 5.72 MR shock absorber (a) and force-velocity profile (b) [Lord Corp.; LORD 2005].

From Figure 5.72 (b) it may be observed, that the force at zero velocity is not equal to zero. This is due to the pneumatical accumulator's floating piston [GONCALVES ET AL. 2003; KOO ET AL. 2005].

Figure 5.73 illustrates the necessity for a pneumatical accumulator in most MR shock absorbers [POYNOR 2001; GONCALVES ET AL 2003].

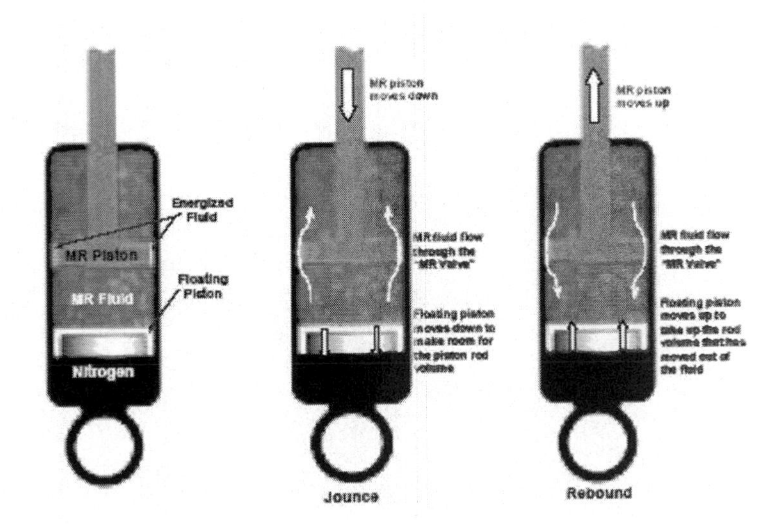

Fig. 5.73 Pneumatical accumulator dynamics in an MR shock absorber [POYNOR 2001; GONCALVES ET AL 2003].

As the MR piston moves into the damper (jounce), the pneumatical accumulator compensates for the increase in volume due to the piston by acting against the charged volume. The pneumatical accumulator floating-piston moves to accommodate the damper rod volume as it enters and leaves the MRF chamber.

In short, the pneumatical accumulator acts as a spring in series with the shock absorber (damper). The spring force comes from the compressibility of the nitrogen gas and the motion of the pneumatical accumulator. With regard to the force-velocity curve, the force observed at zero velocity is proportional to the initial compression of the pneumatical accumulator's floating piston due to the damper being at mid-stroke during testing [KOO ET AL. 2005].

MRFs may be used to construct electrically controllable MR shock absorbers (dampers) for a wide variety of applications. Long-stroke MR shock absorbers (dampers) for high vibration amplitudes usually operate in the flow mode. They contain MR fluidical valves to modify the flow resistance of the fluid and thus the damping force of the actuator. Due to the high vibration amplitudes the MRF works in the post-yield area and its viscous behaviour dominates. At high values of the piston velocity extremely high flow velocities and shear rates can occur that reduce the MR effect. At high shear rates the absorbing (damping) under the influence of a magnetic field is nearly independent of the piston velocity, whereas the zero field absorbing (damping) force still increases. Low-stroke MR shock absorbers (dampers) may be implemented in the squeeze mode. By increasing the magnetic control field a transition from a viscous behaviour to a visco-elastic behaviour may be observed that has a strong influence on the energy dissipated by the MR shock absorber (damper) and on the absorbing (damping) force [BOELTER AND JANOCHA 1998].

In Figure 5.74 is shown cutaways and a physical model of the MR shock absorber [AMESIM 2004].

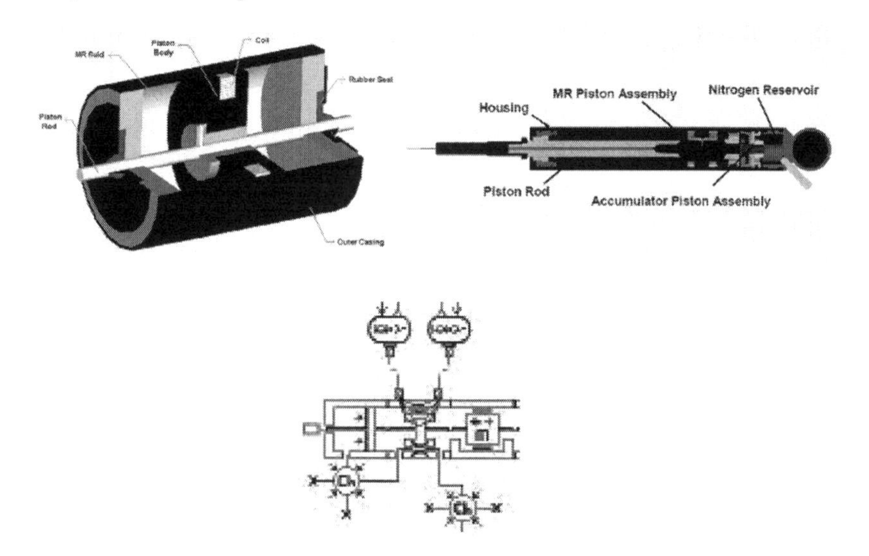

Fig. 5.74 Cutaways and a physical model of the MR shock absorber [AMESIM 2004].

As an example, a heavy-vehicle sized MR shock absorber (damper), shown in Figure 5.75, is intended to provide variable absorbing (damping) for the U.S. Army's *Hummer* that is termed in U.S. military parlance **high mobility multi-purpose wheeled vehicle** (HMMWV) [LIU ET AL. 2004].

Fig. 5.75 HMMWV's MR shock absorber [U.S. Army Research Office; LIU ET AL. 2004].

These automotive vehicles are high-payload and are intended to travel at high rates of velocity across on/off road environments.

The MR shock absorber body is approximately 25 cm in length and 10.8 cm in diameter.

This MR shock absorber uses a single rod design that requires an accumulator to account for the change in volume as the piston moves within the cylinder.

The MR shock absorber is required to dissipate the large amount of energy delivered by a large payload automotive vehicle as it travels in on/off road environments for extended periods of time.

The dynamic range of the MR shock absorber is 0.75 kN at 0.0 A and a velocity of 8 cm/s to 3.2 kN at 2.0 A and 12.5 cm/s velocity.

For instance, the semi-active MR ABW AWA suspension mechatronic control system termed the **magneto-rheological optimised active damper suspension** (MROADS) is designed around MRF that is a water or oil-based fluid whose viscosity, or shear rate, may be changed by applying to it a magnetic field [MILLEN-WORKS 2000B].

When properly engineered, this allows the control of suspension performance in real time with a MR shock absorber (damper) that is inherently less complicated than high performance passive shock absorbers (dampers) with bypass tubes, fluidical valve stacks, or inertial sensing devices.

The MROADS mechatronic control system consists of one MRF computer controlled shock absorber (damper) at each of the vehicle's wheel positions.

The semi-active MR ABW AWA suspension is based on the idea of modulating the forces in a shock absorber (damper) as a function of sensed variables, such as the vehicle velocity (speed), vehicle's body movements, and position of a particular wheel.

This semi-active MR ABW AWA suspension mechatronic control system is lighter, smaller, less expensive, and uses much less power than an active ABW AWA suspension mechatronic control system while providing similar levels of performance.

The MROADS computer-controller damper mechatronic control system automatically optimises the *Hummer* HMMWV's suspension performance based on vehicle loading and operating conditions that yield many benefits [LENOCH 2005; MILLENWORKS 200B]:

- Significant (up to *70 %* on certain terrain) reduction in driver absorbed power;
- Higher mobility speeds over a given terrain;
- Improved wheel-tyre traction;
- Improved wheel-tyre life;
- Reduced fatigue loading of vehicle structure and payload;
- Reduced driver, vehicle and payload damage from terrain impacts at speed;
- Improved vehicle stability and handling;
- Improved accuracy during surveillance, targeting, or weapons firing;
- Suspension system prognostics / diagnostics.

The semi-active MR ABW AWA suspension mechatronic control system was designed to facilitate bolt-on retrofitting.

Only minor modifications to the *Hummer* were required for mounting sensing and control hardware, and the original chassis and lower *A*-arm suspension-mounting holes are used to accommodate the MR shock absorber (damper) and spring assembly.

The gruelling tests underlined the performance advantages of the MROADS mechatronic control system, and demonstrated the high reliability of the system that suffered no failures whatsoever [LENOCH 2005].

As another example, a third-generation, heavy-vehicle sized MR shock absorber (damper), shown in Figure 5.76, is intended to provide variable absorbing (damping) for the U.S. Army's *Stryker* 8-wheeled, 20 Mg **infantry carrier vehicle** (ICV) [RMSV 2004].

Fig. 5.76 *Stryker* 8-wheeled, 20-Mg infantry carrier vehicle's MR shock absorber [RMSV; RMSV 2004].

The third-generation MROADS mechatronic control system was tested head-to -head against a stock suspension for the *Stryker* ICV at the U.S. Army's **Yuma Proving Grounds** (YPG) in Arizona, USA [LENOCH 2005].

The MROADS *Stryker* ICV showed dramatic increases in off-road mobility and on-road handling. The core of the semi-active MR ABW AWA suspension mechatronic control system tested is 8 MR shock absorbers (dampers) and controllers using proprietary algorithms to modulate individual wheel forces within *4 ms* in response to terrain inputs and vehicle-body motion. [ENOCH 2005].

Automotive scientists and engineers retained full functionality of the stock *Stryker* ICV's pressurised gas spring and ride height management system while integrating the electrically controllable MR technology into the physical envelope to the original shock absorber (damper).

The stock ICV use a very capable suspension load-levelling system, and it was a constructive challenge to ensure that important functionality in the process of incorporating controllable MR shock absorbers (dampers) was not lost.

The net result is a high-performance semi-active MR ABW AWA suspension with operation that is completely transparent to the vehicle and its driver.

Mobility gains measured during comparison testing showed great promise for *Stryker* ICVs that could be retrofitted with this semi-active MR ABW AWA suspension mechatronic control system in THE not-too-distant future.

Over a range of off-road bump courses, the MROADS *Stryker* ICV was 40 – 60% faster than the stock ICV at the same level of driver absorbed power, a measure of transmitted vibration [LENOCH 2005].

The MROADS *Stryker* ICV's best performance was a 72% increase in the vehicle's 6 W absorbed power vehicle velocity (speed) from 35 km/h (22 mph) for the stock ICV to 40 km/h (38 mph). Increases in vehicle platform stability were immediately obvious to drivers and bystanders.

The MROADS mechatronic control system showed marked improvements during aggressive on-road manoeuvres like lane changes.

The maximum lane change vehicle velocity increases from 40 km/h (38 mph) for the stock ICV to over 80 km/h (50 mph) with the MROADS mechatronic control system [LENOCH 2005].

The MROADS mechatronic control system has proven to provide many of the benefits of a semi-active MR ABW AWA suspension mechatronic control system to be scalable and adaptable to a number of different automotive vehicles, both wheeled and tracked.

Magneto-rheological elastomers (MRE) are metal particles impregnated rubbers whose mechanical properties may be altered by submission of peripheral magnetic fields.

The most recently existing, but not yet widely known MRE materials, have also led to the discovery of an escalating number of applications in the automotive field, such as vibration absorbing (dampening) shock absorbers (dampers), variable stiffness ECE or ICE mounts and so on [TRIANTAFYLLIDIS AND KANKANALA 2005].

Schematic view of a semi-active MR ABW AWA suspension that displays MRE bushing linking the control arms with vehicle body is shown in Figure 5.77.

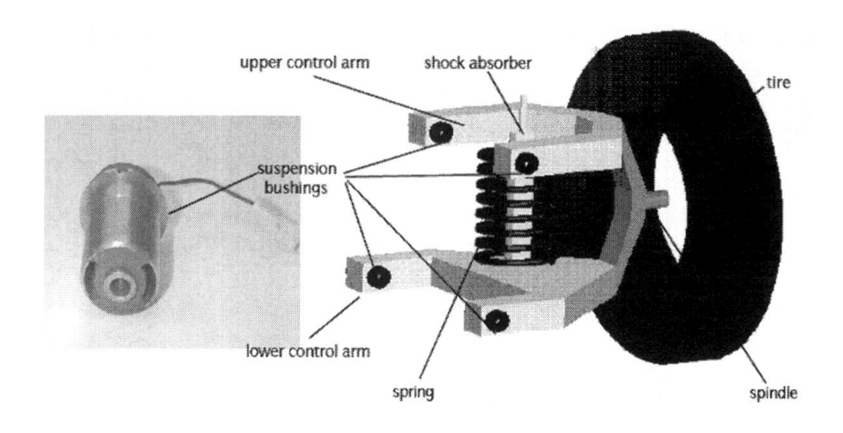

Fig. 5.77 MRE bushing linking the control arms with the vehicle body [*Dr John Ginder* –FORD SRL; TRIANTAFYLLIDIS AND KANKANALA 2005].

MREs may be also used as variable-stiffness springs as shown in Figure 5.78. Iron particles are added to a two-part silicone prior to curing. This composite liquid is then placed in a specially designed mould to cure.

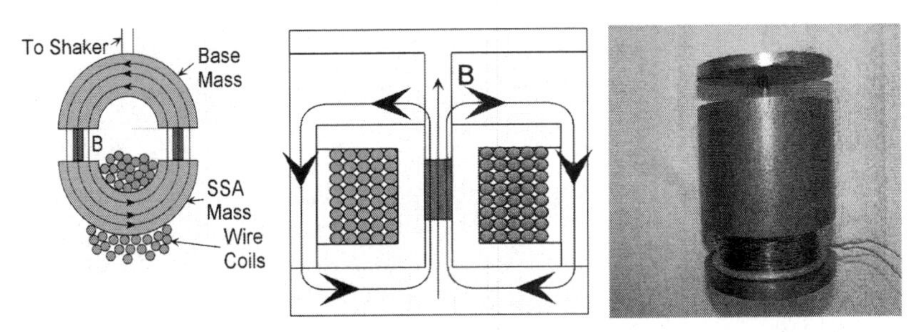

Fig. 5.78 Variable stiffness spring [TRIANTAFYLLIDIS AND KANKANALA 2005].

The purpose of this mould is to generate a strong magnetic flux path that goes through the elastomer during the cure process. This causes the iron particles within the composite to align along the lines of flux and then be solidified into these chains. This enhances the stiffness change effect once cured. The cured silicone composite under the influence of no magnetic flux density is soft, whereas variable amounts of magnetic flux can increase its stiffness.

The **state-switching absorber** (SSA) must be designed such that a magnetic flux path can be generated to run through the elastomer. For this reason, two half-cylinders of low-carbon steel were constructed as base and absorber masses.

For instance, roughly 300 turns of gauge-25 magnet wire are wrapped around the absorber mass. With this set-up, **direct current** (DC) through the magnet wire can generate a magnetic flux path through each silicone half. The semi-active MR absorber, or a SSA, is a hybrid of the reliable **tuned vibration absorber** (TVA) and the more effective active vibration controller.

The SSA is capable of switching one or more of its properties, in an exemplary case its spring stiffness, to fundamentally increase its operational bandwidth. It is much more stable and simple than the **active vibration absorber** (AVA), though, because the control algorithm only allows switching to occur at discrete times and to discrete states.

In this way, the risk of adding energy to the system is eliminated since between states the SSA behaves as a classical, stable TVA.

Some of the benefits of semi-active MR ABW AWA suspension mechatronic control systems include [GILBERT & JACKSON 2002]:

- Higher maximum values of the absorbing (damping) force for superior levels of vehicle-body motion control (better handling);
- Lower minimum values of the absorbing (damping) force for improved isolation, rolling smoothness and impact harshness (better ride);
- Specific amount of damping customised for each set of circumstances to optimise ride comfort (continuously variable real-time damping);
- Compatible with rear levelling capability.

The semi-active MR ABW AWA suspension mechatronic control system provides closer control of pitch and roll body motions that improve on/off road-holding capabilities, steering control, and safety.

These rapid advances in ABW AWA suspension technology emphasise the fact that an automotive technician may require frequent update training to accurately diagnose and service the vehicles of today and tomorrow [KNOWLES 2002].

5.5.10 Semi-Active ER ABW AWA Suspension Solution

Over the past *60* years, it has been known that there exist liquids that respond mechanically to electrical simulation. These liquids that have attracted a great deal of interest from automotive scientists and engineers change their viscosity electro-actively. These **electro-rheological fluids** (ERF) exhibit a rapid, reversible, and tuneable transition from a fluid state to a solid-like upon the application of an external electric field [PHULE AND GINDER 1998].

As an example, corn flour and oil do not conduct electricity - they are non-polar molecules with an even *+Ve/-Ve* charge distributed throughout them. However, corn flour is a *'dielectric'*. That is, when corn flour is placed within an electric field, the molecules of the substance become polarised -- the charges shift so that one part of the molecule is *+Ve,* and another part is *-Ve*. For the oil/oil mixture to flow the oil must flow and the corn flour molecules must flow with the oil and over each other. However by introducing the mixture to an electric field, the corn flour molecules became polarised and started to stick together in long strands (the opposite charges on the molecules becoming attracted to each other). These long strands then constricted the movement of the oil. The whole thing then became more viscous and unable to flow!

In the other words, these materials represent fluid micro-particle suspensions – emulsions wherein the smart material is dispersed, though not dissolved, within a liquid solution. When subjected to an electric field, these micro-particle suspensions experience reversible changes in rheological properties such as viscosity, plasticity, and elasticity. These reversible changes take place because of controllable interactions that occur between various micron-sized *'smart-particles'* suspended within the emulsion. Some of the advantages of ERFs are their high-yield stress, low-current density, and fast response (less than 1 ms). ERFs can apply very high electrically controlled resistive forces while their size (mass and geometric parameters) may be varying small. Their long life and ability to function in a wide temperature range, as much as 233 -- 473 K (--40 -- +200 ^{0}C), allows for the possibility of their use in distant and extreme environments. ERFs are also nonabrasive, non-toxic, and non-polluting, meeting health and safety regulations. One potential use for ERFs is in the creation of **electro-rheological** (ER) shock absorbers (dampers) for automotive vehicles.

Some automotive **original equipment manufacturers** (OEM) recently created an ER shock absorber filled with an ERF that reacts to the on/off road beneath it like no other set of shocks before it. The shocks use a series of sensors to monitor the on/off road surface and the vertical movement of the automotive vehicle's wheels. These data are then used to apply an electric field to an ERF within the shocks.

The ERF then stiffens or softens accordingly, dampening the bounce of the automotive vehicle on its springs. Rather than responding by taking the same action to every bump, these ER shock absorbers may react appropriately to each deviation in the on/off road's surface, providing the smoothest ride ever.

ERFs may be combined with other actuator types such as F-M or P-M or even E-F-M or E-P-M actuators so that novel, **hybrid-mechanical** (HM) actuators may be manufactured with high-power density and low-energy requirements. The controlled rheological properties of ERFs can be beneficial to a wide range of technologies requiring absorbing (damping) or resistive force generation.

Examples of such applications have been explored, mostly in the automotive industry, such as ER shock absorbers (dampers), clutches and brakes as well as ECE or ICE mounts and seat dampers [BULLOUGH ET AL. 1993; SPROSTON ET AL. 1994; STANWAY ET AL. 1996]. Other applications include variable-resistance exercise equipment, earthquake-resistant tall structures, and positioning devices [PHULE AND GINDER 1998]. ERFs are liquids that experience dramatic changes in rheological properties, such as viscosity, in the presence of an electric field. The ER effect was first explained in the 1940s by use oil dispersions of fine powders [WINSLOW 1949].

The ERFs are made from suspension of an insulating base fluid and particles on the order of 0.10 – 100 μm in size. The volume fraction of the particles is between 20 – 60% [FISCH ET AL. 2003; MAVROIDIS ET AL. 2005].

The ER effect, sometimes, termed the *Winslow* effect, is thought to arise from the difference in the dielectric constants of the fluid and particles. In the presence of an electric field, the particles, due to an induced dipole moment, may form chains along the field lines, as shown in Figure 5.79 (b) [AMPTIAC 2001].

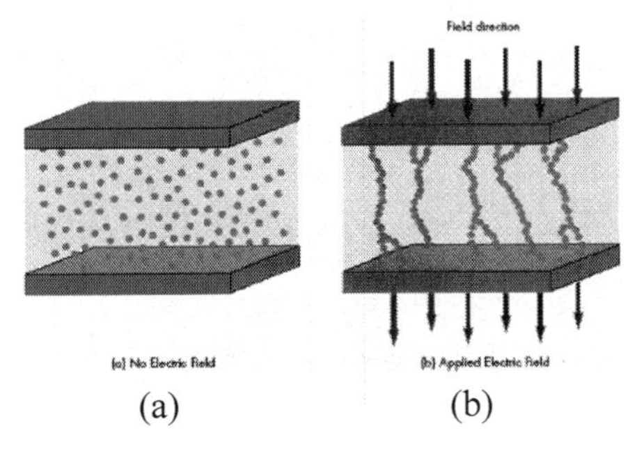

Fig. 5.79 Particle suspension forms chains when an electric field is applied: (a) no electric field – particles in random motion; (b) with electric field – particles forms chains along electric field lines [AMPTIAC 2001].

The structure induces changes to the ERF's viscosity, yield stress, and other properties, allowing the ERF to change consistency from that of a liquid to something that is visco-elastic, such as a gel, with response times to changes in electric fields on the order of milliseconds.

Figure 5.79 (a) shows the fluid state of an ERF without applied electric field and the solid-like state that results when an electric field is applied [ERFD 1998].

The flow ratio of ERFs may be varied by the application of an electric field as shown in Figure 5.80 [ERFD 1998].

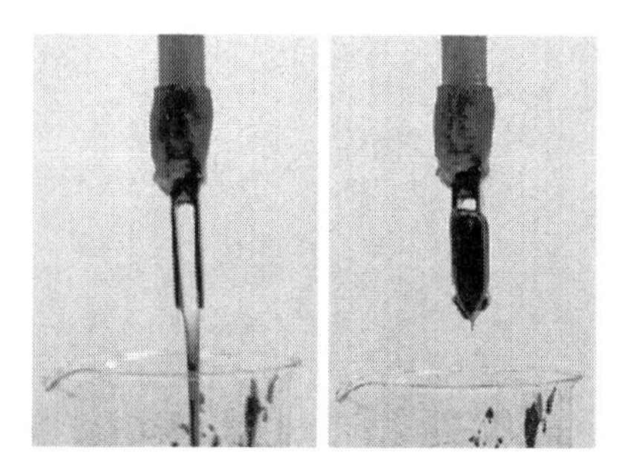

Fig. 5.80 Electro-Rheological Fluid (ERF) at reference (left) and activated states (right) [ER Fluid Developments Ltd, *UK;* ERFD 1998].

Under the influence of an electric field, the ERF alters its state from Newtonian oil to a *non-Newtonian Bingham* plastic. As a *Bingham* plastic, the ERF exhibit a linear relationship between stress and strain rate like *Newtonian* fluid, only after a minimum required yield stress is exceeded. Before that point, it behaves as a solid. At stresses higher than this minimum yield stress, the fluid may flow, and the shear stress may continue to increase proportionally with the shear stress rate [ERFD 1998].

The ER shock absorber (damper) stiffness is modified electrically by controlling the flow of an ERF through slots on the side of a piston (Fig. 5.81).

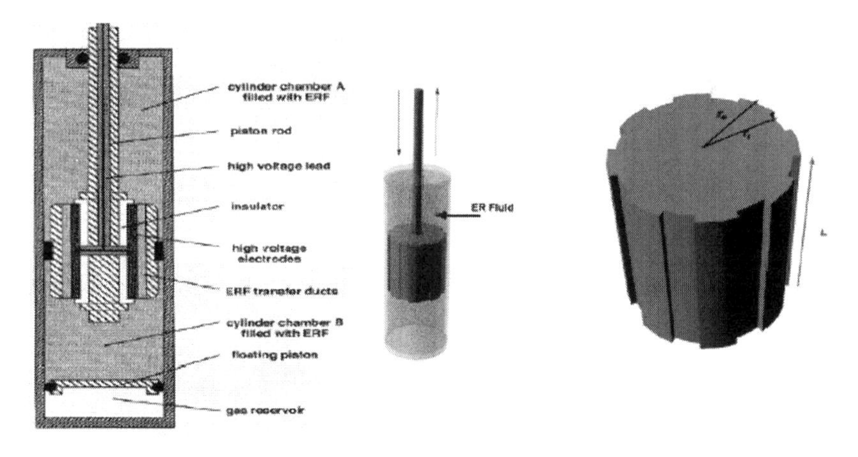

Fig. 5.81 ER shock absorbers with the different pistons and ERF transfer ducts [CFC 2002; MAVROIDIS ET AL. 2005].

The shock absorber (damper) consists of a piston that is designed to move inside a sealed cylinder filled with ERF. Electrodes facing the flowing ERF while inside the channel control the rate of flow electrically.

To control the *'stiffness'* of the ER shock absorber; a voltage is applied between electrodes facing a slot, affecting the ability of the liquid to flow. Thus, the slot serves as an ER fluidical valve since the increased viscosity decreases the flow rate of the ERF and varies the stiffness that it felt.

To increase the stiffness bandwidth from free flow to maximum viscosity, multiple slots may be made along the piston surface. To wire such a piston to an electric power source, the piston and its shaft may be made hollow and electric wires may be connected to electrode plates mounted on the side of the slots.

The inside surface of the ER shock absorber cylinder surrounding the piston may be made of a metallic surface and serve as the ground and opposite polarity. A sleeve covers the piston shaft to protect it from dust, jamming or obstruction.

When a voltage is applied, potential is developed through the ERF along the piston channels, altering viscosity. As a result of the increase in the ERF viscosity, the flow is slowed significantly and resistance to external axial forces increases.

As recently as 2010, scientists have made suspensions of coated nano-sized particles that can become as hard as plastic at the flick of an electrical switch [WEN ET AL. 2003; KALAUGHER 2003; NOVAK 2005].

The suspensions could have applications in *'smart'* ER shock absorbers (dampers), valves, clutches and brakes for RBW or XBW automotive vehicle mechatronic control systems. For the first time one may controllably harness the very large, nano-scale electric field of molecular dipoles to reversibly generate solid-liquid transition within $\geq$ 10 ms. This may be achieved by going to nano-scale systems because only there can one find large electrical forces.

To demonstrate this **giant electro-rheological** (GER) effect [WEN ET AL. 2003], scientists used dielectric nano-sized particles of barium titanyl oxalate coated with a 3 – 10 nm thick layer of urea, a material that has a large molecular dipole moment. The scientists suspended the dielectric nano-sized particles that had an average size of 50 – 70 nm, in silicone oil and applied an electric field. This polarized the dielectric nano-sized particles, causing them to form columns aligned with the direction of the electric field. The presence of the columns meant that the material acted as a solid when shear forces were applied perpendicular to the columns.

The dielectric nano-sized particle suspension's static yield stress increased roughly linearly with increasing electric field. Under an electric field of 5 kV/mm, for example, a dielectric nano-sized particle suspension containing roughly 30% dielectric nano-sized particles by volume had a static yield stress of above 100 kPa — roughly 20 times greater than the theoretical value calculated from the electrical properties of the materials making up the dielectric nano-sized particle suspension. These innovative ERFs can realise many of the applications thought of before but never made practical due to the insufficient yield stress of conventional ERFs.

In the 1980s, detailed engineering studies were carried out to see what kind of yield stress is necessitated for widespread applications.

The consensus at that time was a value of 20 – 50 kPa. The scientists believe that short-range interactions between the urea-coating molecules and the core dielectric nano-sized particle and/or the oil are crucial to the GER effect.

In general, the ERF can serve as an effective interface that translates electrical signals that are easiest to make, into mechanical ones. So when ERFs are coupled with sensors that generate electrical signals according to environmental variations, one may achieve *'smart'* ER shock absorbers (dampers), for example, shown in Figure 5.82 [WEN ET AL. 2003; KALAUGHER 2003].

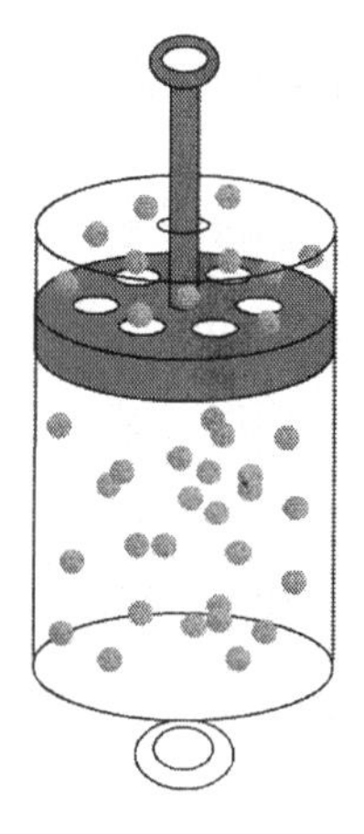

Fig. 5.82 Smart ER shock absorber [WEN ET AL. 2003; KALAUGHER 2003].

The innovative ERF can go from water-like to as hard as plastic. When dielectric nano-sized particles of the ERF are suspended in an insulating liquid, the mixture behaves like a fluid. But apply an electric field and the dielectric nano-sized particles become polarised, lining up in rows, positive end to negative end that stiffens the MRF as shown in Figures 5.83 and Figure 5.84 [WEN ET AL. 2003; NOVAK 2005].

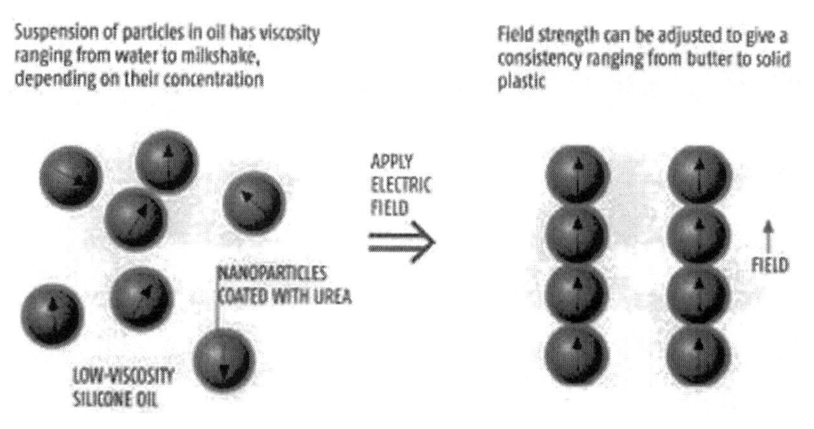

Fig.. 5.83 Turning liquids into solids [WEN ET AL. 2003; NOVAK 2005].

Fig. 5.84 An electric field strings together tiny beads in a liquid solution (inset), forming a solid [WEN ET AL. 2003].

ER shock absorbers thus represent an excellent class of interfaces between ECUs and mechanical components that have gained increased scientific and economic interest in recent years. As a result, a new generation of ERFs with optimised properties are now available. In particular, the difficulties like stability over long time periods and sedimentation of the polarisable nano-sized particles have largely been resolved.

The application areas for ER shock absorbers are numerous. High frequencies and forces may be relatively easily controlled using flexible ECUs. Already many different applications have been reported including a prototype of an adaptively controllable ER shock absorber (see Fig. 5.85) [BUTZ AND STRYK 2001; STRYK 2004

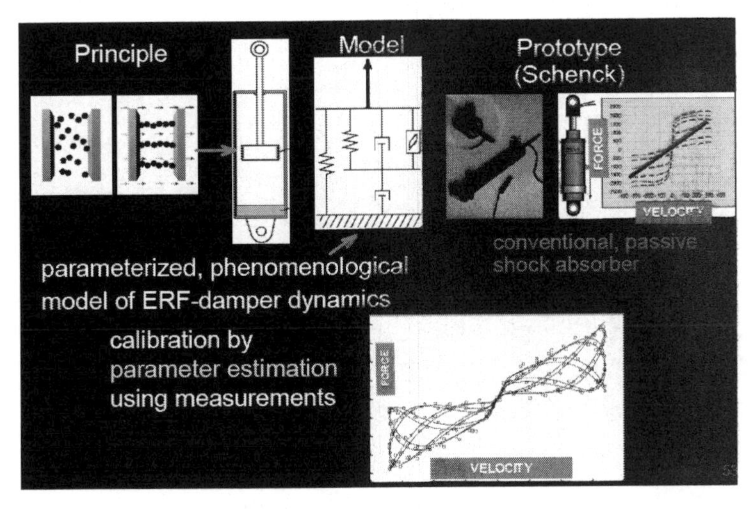

Fig. 5.85 Controllable ER shock absorber (ERF-damper) [STRYK 2004]

There are two major problems that arise when investigating the simulation and optimal control of the continuously controllable ER shock absorbers and their application to semi-active vehicle-body damping [RETTIG AND STRYK 2000, 2005].

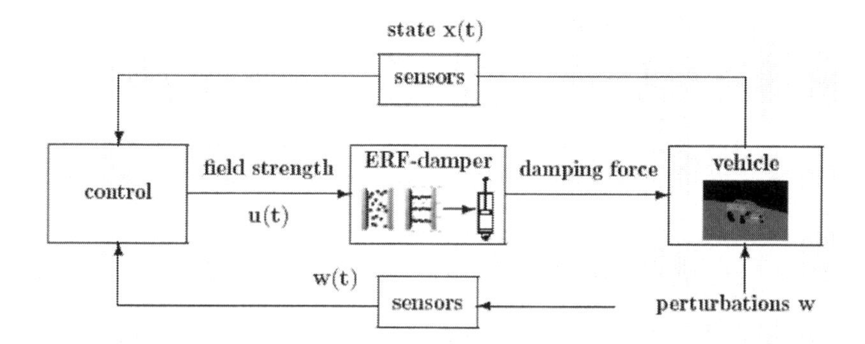

Fig. 5.86 Structural and functional block diagram and flow chart of control, perturbation, and state variables for semi-active ER ABW AWA suspension mechatronic control system using ER shock absorbers [RETTIG AND STRYK 2000, 2005].

First there lies the question of modelling and simulating the dynamic behaviour of ER shock absorbers that depend on the applied electrical field. Second, with regards to the computation of controls that maximise safety and comfort objectives must be developed together with applicable numerical methods (Fig. 5.86) [RETTIG AND STRYK 2000, 2005].

A control framework that serves well to evaluate the potential of semi-active vehicle-body damping with optimally controlled ER shock absorbers is that of deterministic optimal control methods that provide an open-loop trajectory numerically. For its application in a semi-active suspended automotive vehicle, feedback controls must be computed under real-time conditions, thus, the mechatronic control may likely only be sub-optimal [RETTIG AND STRYK 2000, 2005].

Looking to the automotive industry, we see commercial advances in smart fluids, networked automotive vehicle's semi-active ER ABW AWA suspension mechatronic control systems, and smart ER shock absorber where a sensor at the front of the automotive vehicle detects variations in the on/off road surface. That signal is sent to an ECU, that is, a single-chip microprocessor controller that determines whether the ER shock absorber should be more or less stiff. By altering the electrical field in the ER shock absorber, the viscosity of the ERF inside is also changed, tuning the suspension within milliseconds to match on/off road circumstances. The semi-active ER ABW AWA suspension mechatronic control system consists of ER absorbers, a sensor set and on-board ECU. The on-board ECU continually adjusts the absorbing (damping) forces up to once every millisecond. When exposed to an electric field, ER materials change consistency from a fluid state to a near-solid state [FIJALKOWSKI 1997].

In vibration control, the ECU based on the single-chip microcontroller first may obtain five system-state variables from the sensors.

The on/off road surface circumstance, whether smooth or rough, may be judged by statistics operations of relative vertical velocities to select control gains. From these data, the mechatronic control amounts may be calculated, and the control DC voltages of the absorbing (damping) force ERF valves may be decided. Furthermore, floor vibration due to other wheel manoeuvres may be compensated based on the gains selected by its mode and wheel system-state variables of all the wheels [FIJALKOWSKI 1997].

Vibration control may be realised with the ER fluidically suspended automotive vehicle. Position **ultra-sonic** (US) sonars or linear potentiometers or electromagnetic sensors along-side the variable-damping ER shock absorber (damper) filled with the ERF may read the on/off road surfaces [FIJALKOWSKI 1997].

The cornerstone of automotive VAT has been the ER fluidical valve and the resultant ER shock absorber (damper). In the ERF valve the ERF is pumped through a channel capacitor wherein the DC voltage difference between the walls may be varied to regulate the flow resistance up to the point where the fluid solidifies and flow ceases [JORDAN AND SHAW 1989].

A schematic diagram of a force-control-based predictive and adaptive semi-active ER ABW AWA suspension mechatronic control system with the ER shock absorber and ER fluidical valve is shown in Figure 5.87 [FIJALKOWSKI 1997].

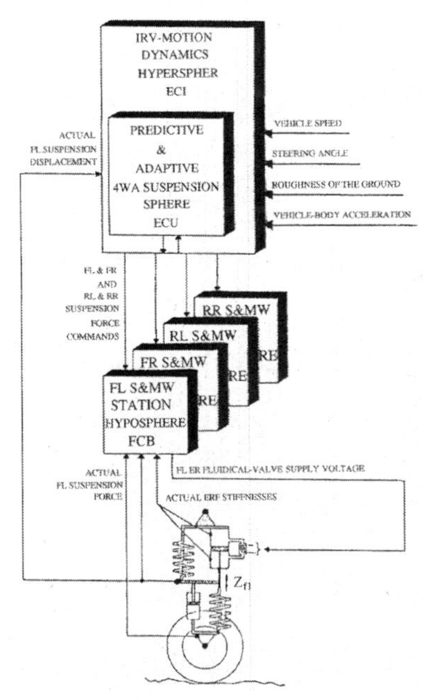

Fig. 5.87 Principle layout of a full-time force control based, predictive and adaptive 4WA ABWs suspension mechatronic control system with ER shock absorber, ER fluidical valve and its hardware (ECI – electronic control instrument; ECU – electronic control unit; ECB – electronic control bit) [FIJALKOWSKI 1997].

Variable-damping ER shock absorbers exploit the property of electric-field-dependent stiffness for a virtually limitless range of damping response. When one considers the milliseconds response time involved, and exciting automotive VAT in position US sonars for anticipating on/off road surface roughness. It is easy to understand what motivates competitors in the automotive industry, to be the first to offer ultra-smooth, single-chip microcomputer- or transputer-controlled predictive and adaptive semi-active ER ABW AWA suspension mechatronic control system for controlling handling stability, ride comfort, attitude and vehicle height.

An ECU based on a single-chip microcontroller may be responsible for achieving command control over a variety of forces despite disturbances and non-linearities. This force -- servo approach may facilitate design of an onboard ECU, but may also motivate systems where ECUs for the wheels may only be used for some local stabilisation, but the onboard ECU may directly inject control signals into the ER fluidical valves.

In general, a large number of sensors, four ER fluidical valves and a large number of mechanical **degree-of-freedoms** (DoF) on an automotive vehicle may create a demanding multivariable control problem. All motions interact. Control designs must also take the non-linearities of fluidics into consideration.

In the attitude control, the reactive real (measured) displacements between the top and bottom of the suspensions may coincide with the set (aimed) displacements by feedback controls.

In addition, for immediate reduction in roll angles caused by steering manoeuvres, feedforward controls based on the estimation of roll angles from steer angles and compensation of roll angles based on changes in semi-active ER ABW AWA suspension mechatronic control system's electric-field-dependent stiffness due to lateral acceleration may be performed. Attitude controls for roll and pitch directions decreases low-frequency vibration.

Acceleration in the horizontal (longitudinal and lateral) directions of a full-time *4WD* × *4WB* × *4WS* × *4WA* automotive vehicle during turning (cornering), acceleration and deceleration may be detected by the **acceleration-and-jerk** (A&J) sensor [YAMAKADA AND KADOMUKAI 1994] or the other acceleration sensor, the so-termed *G* sensor [YOKOYA ET AL. 1989] and the control signal may be given to the linear control ERF valve of each wheel to control the ER shock absorber-cylinder ERF solidification.

As a result, attitude change including roll and pitch may be eliminated and the vehicle attitude may be kept constant. To compensate for the delay a control response to a roll may be controlled by foreseeing the start of the roll with the steer angles and vehicle-velocity sensors at the beginning of turning.

The predictive and adaptive semi-active ER ABW AWA suspension mechatronic control system detects the abruptness of an on/off road bump almost quickly enough to react before the shock of the impact reaches the vehicle spring.

In practice, the predictive and adaptive semi-active ER ABW AWA suspension mechatronic control system swallows high-speed undulations, prevents bottoming and retains a smooth ride on harsh on/off road surfaces.

5.5.11 Semi-Active PF ABW AWA Suspension Solutions

Friction damping has long been used as an effective and simple method to add passive damping to mechanical systems. It requires only the direct contact of two parts moving relative to each other and it may be incorporated into harsh environments and vacuum environments where the use of elastomeric damping treatments and fluid filled shock absorbers (dampers) is limited [LANE ET AL. 1992].

In the 1990s, the idea of varying the normal force in a frictional joint to enhance energy dissipation, for example, from a vibrating vehicle body was first presented [FERRI AND HECK 1992].

A semi-active dry friction shock absorber (damper) feeds back an actuation force to the mechanical system whose dynamics may be altered in this way.

The properties of the system, such as stiffness and damping may be actively changed through the control of this actuation force.

A cross-sectional drawing where the dry friction shock absorber (damper) actively alters the normal force between the vehicle body and the outer housing of the semi-active ABW AWA suspension is seen in Figure 5.88 [UNSAL 2002].

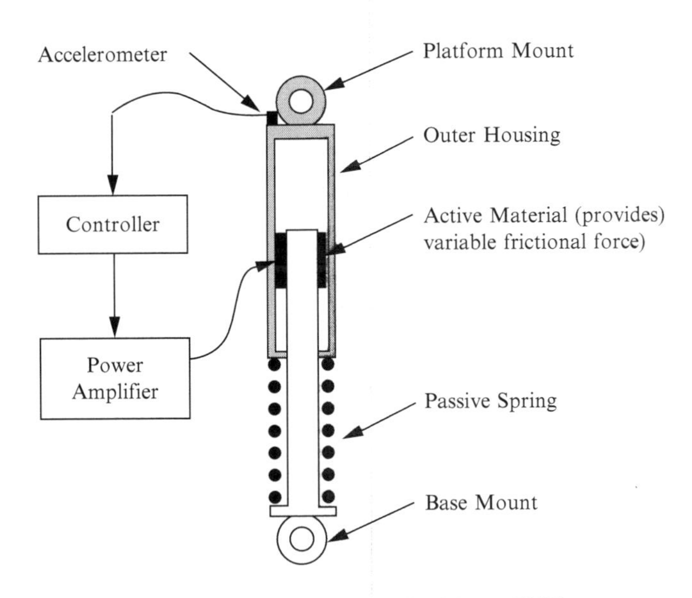

Fig. 5.88 Semi-active PF shock absorber [UNSAL 2002].

In contrast to F-M shock absorbers (dampers), dry friction shock absorbers can provide an excellent mechanism for shock suppression since the friction force transmitted through a dry friction shock absorber is limited. This is the reason why it has been incorporated into automotive vehicle shock absorbers and other types of automotive vehicle shocks.

In theory, the performance of dry friction shock absorbers can rival that of semi-active F-M dampers in every respect. Through the use of a simple low bandwidth feedback mechatronic control system, it may be possible to provide good vibration suppression while retaining the excellent shock suppression characteristics.

A dry friction shock absorber can supply considerable force even for small values of the velocity, something that is not possible for F-M shock absorbers, as they require a relatively large value of the velocity, to transmit the same amount of force [FERRI AND HECK 1992].

The development of dry friction shock absorbers (dampers) to the extent of other semi-active shock absorbers has been impeded due to three primary reasons.

First of all, because of the discontinuity of friction at zero velocity, the differential equation is dependent on a sense of the velocity direction [LANE ET AL. 1992].

Secondly, when in the kinetic coefficient, the *'stick-slip'* phenomenon occurs, this phenomenon is caused by the fact that the friction force does not remain constant as a function of some other variable, such as temperature, displacement, time, or velocity. For the two reasons stated, dry friction shock absorbers (dampers) are non-linear and may require a non-linear controller.

The third and most important reason dry friction shock absorbers (dampers) have not been fully developed is due to the actuator.

In past research, the normal force was altered through the use of fluidics (hydraulics and/or pneumatics) [KANNAN ET AL. 1995].

The main disadvantage of fluidics is the time delay that is required for the actuator to reach the required fluid or air pressure. Rapid modulation of the actuation force is not possible and it could cause a backlash effect when used.

In a variable dry friction shock absorber (damper) system, the speed with which the actuation force may be adjusted is of utmost importance [GARRET ET AL. 2001].

Modern electromagnetic actuators are well suited to provide rational motion (E-M motors); however, their use as linear actuators is limited.

Although they are capable of generating sufficient force and displacement, the large size, mass, electrical demands, and cost of these actuators make them currently impractical.

Recently piezoelectric actuators have been proposed as a method of applying the varying normal force [CHEN AND CHEN 2000; GARRET ET AL. 2001].

Piezoelectrics are materials that exhibit an electrical polarisation with an applied mechanical stress (direct effect), or a dimensional change with an applied electric field (converse effect). They are used for both actuating and sensing devices. Lead zirconate **titanate** ($PbZr_{1-x}Ti_xO_3$) is the premier piezoelectric material as it may be doped to produce an n-type or p-type material with a range of dielectric constants to meet the requirements of numerous applications. Other piezoelectric materials that may be used are **barium titanate** ($BaTiO_3$), **lead meta-niobate** ($PbNb_2O_6$), and PVDF. Polymers are generally favoured for sensing applications while ceramics are favoured for actuating applications.

Figure 5.89 depicts the piezoelectric effect observed in lead zirconate titanate upon the application of compressive forces relative to the crystal structure [AMPTIAC 2001].

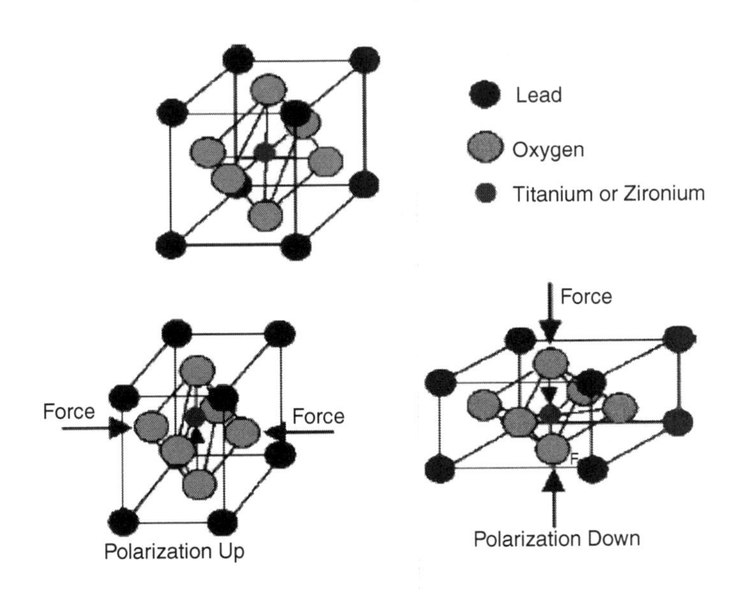

Fig. 5.89 The piezoelectric effect [AMPTIAC 2001].

Piezoelectrics is a mature technology with numerous applications throughout the automotive and aerospace sectors. For instance, automotive vehicle devices using piezoelectrics include actuators, sensors, and wheel-tyre pressure indicators among many others.

Piezoelectricity is the ability of certain crystalline materials to develop an electric charge proportional to a mechanical stress and vice versa.

Piezoelectric materials can generate a significant amount of stress/strain in a constrained condition when exposed to an electric field. This property has been used to suppress excessive vibration of mechanical as well as automotive and aerospace systems and is still an active area of research [GARRET ET AL. 2001].

Due to their high force and bandwidth capability, piezoelectric actuators appear to be a natural candidate for use in **piezoelectric friction** (PF) shock absorbers (dampers). However, until only recently, the maximum (freely loaded) mechanical strain of these devices did not exceed 0.1%. This means that an actuator 25.4 mm (1 inch) long could only deflect 2.54 mm (0.001 ") (significantly less under load). As a result the development of a practical frictional shock absorber (damper) has been hampered [UNSAL 2002; UNSAL ET AL. 2004].

A flex-tensional piezoelectric amplifier (using an ordinary piezoelectric material) has recently been developed (FPA-1700, Dynamic Structures and Materials, LLC) that may generate a 1.6 mm displacement having a load of 44.5 N (10 lbs). This specific actuator was chosen to be implemented in the developed vibration isolator due to its high displacement capability and also due to the inherent characteristics of piezo-electric actuators that make them favourable when compared over a wide frequency range with high-speed actuation, low power consumption, reliability and compactness [UNSAL 2002; UNSAL ET AL. 2004].

A novel semi-active PF shock absorber (damper) has potential application to space environments in which other viscous shock absorbers (dampers) are not suitable. Thus, it may have application to semi-active PF ABW AWA suspension mechatronic control systems. An exemplary PF actuator is shown in Figure 5.90 [UNSAL 2002; UNSAL ET AL. 2004].

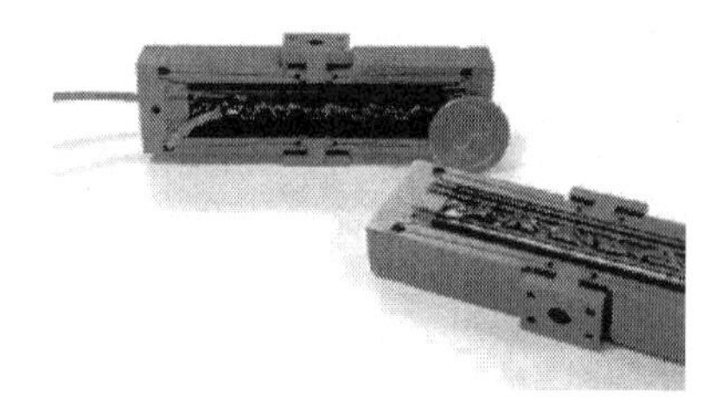

Fig. 5.90 FPA-1700-LV piezoelectric actuator
[Dynamic Structures & Materials LLC; UNSAL 2002; UNSAL ET AL. 2004].

The actuator incorporates a shape memory alloy preload wire for bi-directional motion and a titanium flexure-based amplification mechanism. The peak output stroke of the actuator when it is not loaded is approximately 1.6 mm. The displacement of the mechanism is transmitted through two parallel output plates.

The friction pads are *Kevlar* bike disc brake pads (CODA QPDPAD/BLU) and the housing is stainless steel. The PF shock absorber (damper) consists of several moving and stationary components as shown in Figure 5.91 and Figure 5.92 [UNSAL 2002; UNSAL ET AL. 2004].

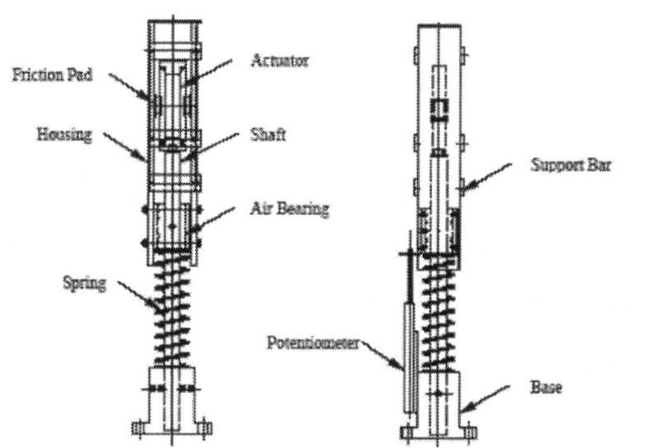

Fig. 5.91 PF shock absorber (front and side view)
[UNSAL 2002]

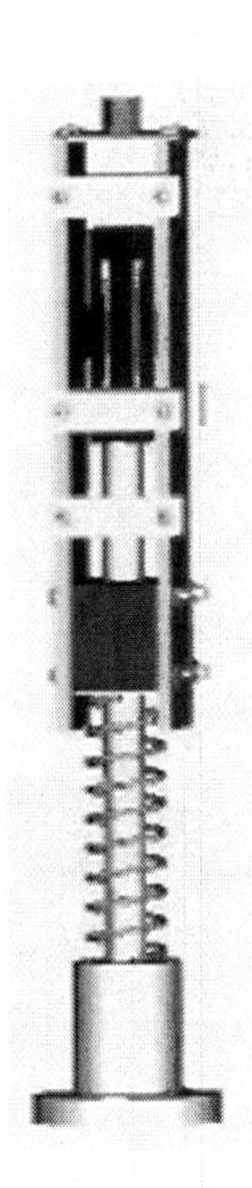

Fig. 5.92 PF shock absorber (3D view) [UNSAL 2002].

A 19 mm (0.75 inch) diameter shaft is fixed to the base of the PF shock absorber. Mounted on the shaft is the flex tensional mechanical amplifier of the piezoelectric actuator. The moving components consist of the outer housing and the air bearing.

The outer housing also comes in contact with the friction pads as it vibrates. The friction pads are fixed to both sides of the actuator so that the normal force that the actuator applies is symmetrical.

The normal force provided between the friction pads and the outer housing induces a frictional load that retards the motion of the outer housing. Within this PF shock absorber, there is also a spring that connects the moving housing to the stationary base.

With the frictional pads not engaged, the air bearing provides a relatively frictionless contact surface. As a result, the PF shock absorber is essentially an ideal **single-degree-of-freedom** (SDoF) spring-mass system [UNSAL 2002; UNSAL ET AL. 2004].

An innovative type of semi-active PF shock absorber (damper) has been developed and has the potential to be used in several active vibration control applications. The heart of the shock absorber is a piezoelectric stack with a mechanical amplifying mechanism.

As shown in Figure 5.93, the friction force increases linearly with the actuator input voltage [UNSAL 2002; UNSAL ET AL. 2004].

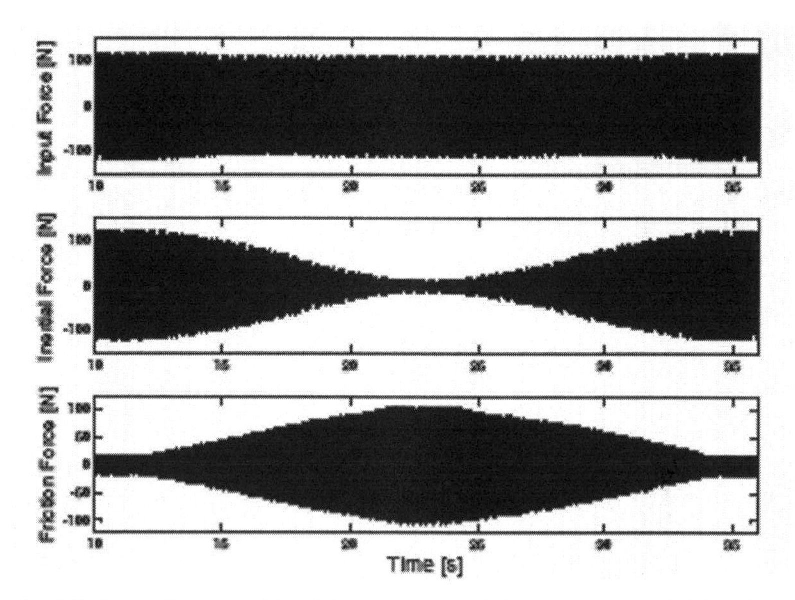

Fig. 5.93 Friction force capability of the actuator when its input voltage is varied linearly [UNSAL 2002].

It was determined that the force generated within the isolator is proportional to the input voltage. Originally developed by the **National Aeronautics and Space Administration** (NASA), **th**in layer **un**imorph ferroelectric **d**riv**er** and sensor (THUNDER™) actuators generate comparatively higher force together with larger displacement compared to conventional piezoelectric actuators [FACE 2004].

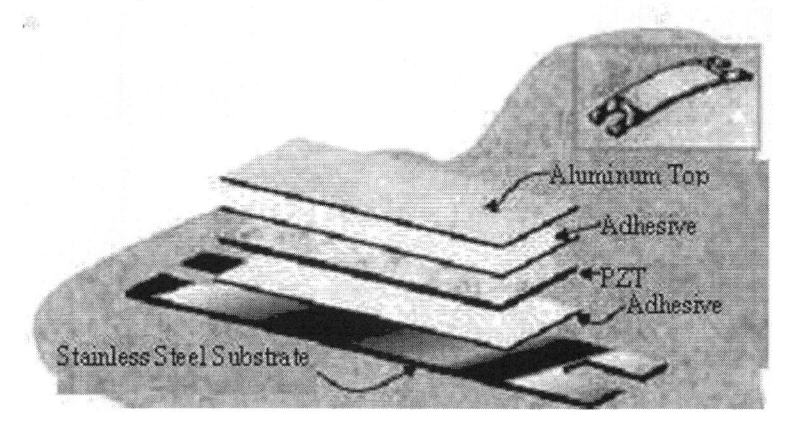

Fig. 5.94 Structure of the classic THUNDER device [FACE 2004].

A classic THUNDER™ device (Fig. 5.94) is composed of a thin wafer of piezoceramic such as PZT bonded to an electrically conductive substrate and a superstrate with high performance epoxy. The epoxy used in the manufacture of a thunder device is LaRC-SI, a high performance bonding material with a very specific cure cycle that was also developed by NASA.

The manufacturing process is comprised of precise pressure and temperature cycles. The resulting piezoceramic composite has a characteristic curvature that is the result of a mismatch in thermal coefficient of expansion and the *Young*'s modulus of elasticity of the composite materials.

The prestress within the actuator is such that the piezoceramic is in a state of compression and the substrate is in a state of tension. This prestress enhances the large deflection producing capability of THUNDER™ elements that is not observed in any other smart material actuators. A normal piezoceramic wafer actuator could not withstand this magnitude of curvature.

The standard THUNDER™ elements that FACE® International Corporation manufactures have a stainless steel substate, a piezoceramic layer and an aluminium top layer. The aluminium layer serves principally as a means for soldering lead wires [FACE 2004].

A very important feature of THUNDER™ actuators is the versatility of operation in addition to the extraordinarily large deflections. THUNDER™ elements can be operated in many different ways pertaining to mounting, stacking configuration and voltage application.

The method of operation may be chosen depending on the type of application with force and displacement being the two major governing physical quantities. The tabs provided with standard THUNDER™ actuators provide a means for mounting.

The method of mounting changes the force and displacement characteristics of these actuators. THUNDER™ actuators produce comparatively higher force together with larger displacement compared to other traditional piezoelectric actuators.

In general, actuators generating high displacements are poor on force generation capability and vice versa. A THUNDER™ actuator may often link these two requirements.

However, if more force or more displacement is required for an application, using multiple THUNDER™ elements in different stacking configurations may enhance these performance characteristics.

The two main configurations commonly implemented are the parallel stack configuration and the clamshell configuration.

The parallel stack configuration increases the stiffness of the actuator assembly and the clamshell reduces it.

The parallel and clamshell stacking are equivalent to parallel and series spring arrangement respectively.

Appropriate measures should be taken to avoid short circuit conditions when stacking THUNDER™ actuators. Thin *TEFLON* sheets act well as electrical insulation layers.

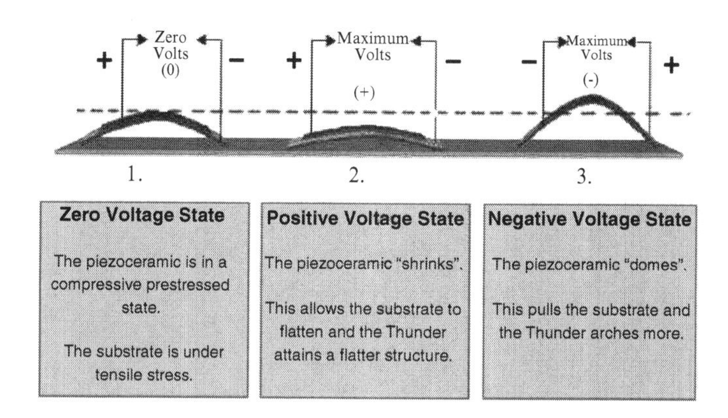

Fig. 5.95 Classic THUNDER™ actuator function [FACE 2004].

In addition to the stacking configurations shown in Figure 5.94, a stacking configuration comprised of a combination of these two configurations (Fig. 5.95) may be used as well. The maximum voltage that may be applied to a THUNDER™ is limited by the thickness of the piezoceramic and the direction of movement is dependant on the polarity of the driving signal. The direction of movement of the THUNDER™ in response to a periodic driving signal is also dependant on the direction in which the THUNDER™ is poled after it is manufactured.

At FACE®, applying a DC voltage with the top layer always positive with respect to the bottom substrate polarizes the THUNDER™ elements. With this kind of polarization, the THUNDER™ curvature decreases (attains a flatter structure) when positive voltage is applied to the top layer and the actuator curvature increases (attains a more domed shape) when the substrate is subjected to positive voltage with respect to the top layer [FACE 2004].

5.6 Active Suspensions

5.6.1 Foreword

Active ABW AWA suspensions (Figure 5.96, 5.97 and 5.98) [DAS 2006] are extremely sophisticated, computer controlled mechatronic control systems that use powered actuators instead of conventional springs and shock absorbers.

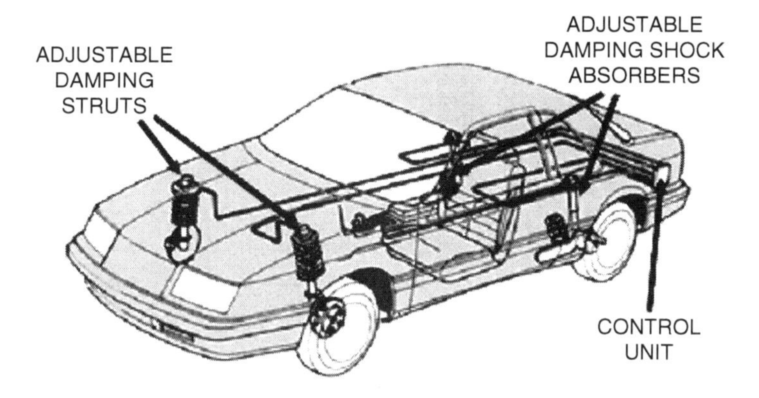

Fig. 5.96 Layout of a typical active ABW AWA suspension [DAS 2006].

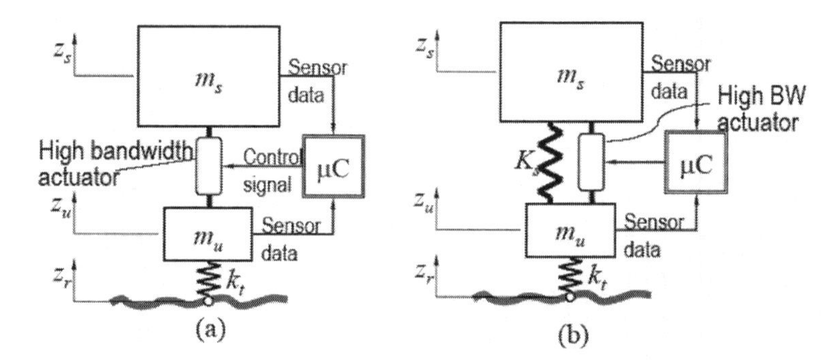

Fig. 5.97 Physical models of fast-active ABW AWA suspensions: actuator provides total suspension force; (b) static load supported by passive spring [DAS 2006].

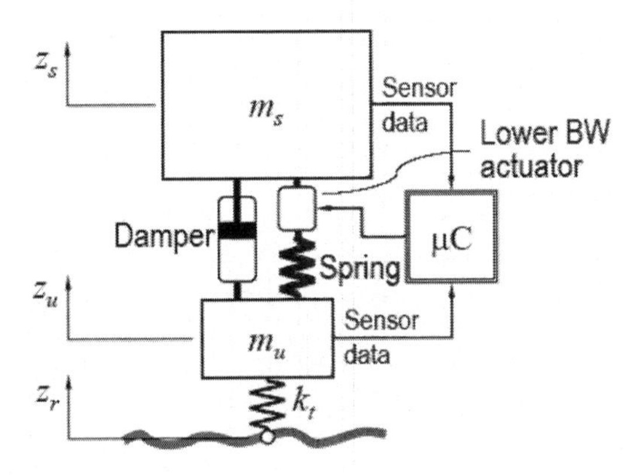

Fig. 5.98 Physical model of a slow-active ABW AWA suspension [DAS 2006].

The actuators position an automotive vehicle's wheels in the best possible manner to deal with on/off road surface disturbances and handling loads.

Active ABW AWA suspension mechatronic control systems move each wheel up and down to control vehicle-body motion in response to on/off road surface abnormalities. The mechatronic system responds to inputs from the on/off road surface and the driver.

With an active ABW AWA suspension mechatronic control system, an automotive vehicle may simultaneously provide the smooth ride of a soft suspension along with superior handling associated with a firm suspension.

Most active ABW AWA suspension mechatronic control systems use a high-pressure M-F pump or M-P compressor with F-M or P-M actuators (cylinders) at each wheel to position the wheels with respect to the automotive vehicle.

Up and down motion of the wheels is actuated by electronically controlled fluidical or pneumatical valves. Other alternatives to power active ABW AWA suspension mechatronic control systems include E-M motors or electromagnets.

In any system, sensors at each wheel determine vertical wheel position and the force of the on/off road surface acting on the wheel.

Some mechatronic control systems use *'road preview'* sensors (microwave radar or laser) to provide data about on/off road surface abnormalities before the front wheels reach them.

Accelerometers tell the computer when the automotive vehicle is accelerating, braking or cornering. The computer uses complex algorithms to continuously process data and decide the position of each wheel. Coil springs may be used at each wheel to avoid *'bottoming out'* of the suspension in case of mechatronic control system failure; they also may reduce the power required to support the sprung mass of the automotive vehicle.

A purely active ABW AWA suspension mechatronic control system has no springs and no shock absorbers (dampers). Instead, an F-M or P-M or even E-M ram positions each suspension corner.

The oily fluid or air or even electrical current is supplied from a high pressure fluid or pneumatical or even electric accumulator, kept charged by an M-F pump or M-P compressor or even M-E generator, respectively, on the ECE or ICE.

At each corner is an accelerometer. When one of these detects upward or downward motion of the automotive vehicle, the computer fluidical or pneumatical or even electrical valves emit oily fluid or air or even electrical current to that corner's RAM to zero out that motion, or to reduce it to a practical minimum. Thus, ECE or ICE power moves the suspension in response to what the automotive vehicle is doing.

Naturally, this requires quite a lot of power - as much as 15% of the ECE/ICE's maximum. Yet the range of effects that may be created is unlimited - for example the automotive vehicle may roll inward as it rounds a turn, rather than outward, or it may squat down during hard acceleration or turning to prevent wheel lifting.

The F-M or P-M or even E-M struts that actuate the suspension are controlled by proportional fluidical or pneumatical or even electrical valves, respectively, originally developed for mechatronic control, and capable of operation at up to 40 cycles per second.

This technology really works but is expensive and complex. It was banned from *Formula 1* for those reasons, but may be used on some high-end production automotive vehicles.

The classic mechatronically controlled active ABW AWA suspension mechatronic control system consists of several components in addition to the normal passive suspension components.

Perhaps the most important component is the computer that interprets input from various sensors that monitor such data as the automotive vehicle's height, pitch, and roll; how fast the wheels are spinning; and how quickly the automobile is turning.

The simplest mechatronically controlled active ABW AWA suspension mechatronic control systems merely maintain a level ride height, counteracting the tendency of the mass of passengers and luggage to lower the rear end.

For instance, mechatronic control systems with four-wheel height adjustment lower the automotive vehicle's ride height to reduce aerodynamic drag and improve fuel economy at highway values of the vehicle velocity.

In off-road vehicles, these mechatronic control systems may raise the automotive vehicle to increase ground clearance over rough terrain. Other mechatronic control systems are adjustable and allow the driver to switch manually between a soft-ride mode and a firm-handling mode. Some mechatronic control systems also offer intermediate choices.

The most advanced systems automatically switch back and forth between soft and firm modes in milliseconds, depending on the condition of the on/off road surface. These mechatronic control systems also work to keep ride height constant and to minimise roll.

Consequently, the automotive vehicle has a better combination of ride comfort and road handling characteristics under various conditions than do vehicles with conventional passive suspension mechatronic control systems.

Mechatronically controlled active ABW AWA suspension mechatronic control systems cost considerably more than conventional passive suspension mechatronic control systems.

They are typically found only on relatively expensive luxury-class automotive vehicles, high-performance SUVs and racing cars.

The latest innovation in the world of automotive vehicle ABW AWA suspension mechatronic control systems is the active ABW AWA suspension mechatronic control.

For instance, the mechatronic control system incorporates a single-chip microprocessor to vary the orifice size of the restrictor fluidical valve in an F-M suspension or shock absorber. This changes the effective spring rate. Control inputs may be vehicle velocity, load, acceleration, lateral force, or a driver preference.

Automotive vehicle drivers are familiar with the fact that an automotive vehicle rolls over during cornering (rolling) and dives to the front during braking (pitching). Also, unpleasant vertical vibrations (bouncing) of the vehicle body may occur while driving over on/off road surface irregularities. These dynamic motions do not only have an adverse effect on comfort but may also be unsafe because the wheel-tyres might loose their grip on the on/off road surface.

The main task of an automotive vehicle suspension is to ensure ride comfort and road holding for a variety of on/off road surface conditions and vehicle manoeuvres. In general, only a compromise between these two conflicting criteria may be obtained if the suspension is built from passive components, such as springs and dampers with fixed rates. This applies for modern wheel suspensions and therefore a breakthrough to build a safer and more comfortable automotive vehicle out of passive components is not to be expected.

The answer to this problem seems to be found in the development of an active ABW AWA suspension mechatronic control system. This type of suspension is characterized by a built-in actuator that may generate control forces (calculated by a computer) to suppress the above-mentioned roll- and pitch-motions.

In addition the road holding has been improved because the dynamic behaviour of the contact forces between wheel-tyres and on/off road surface has been controlled better, due to the absence of the low frequency resonance of the vehicle body (typically 1÷2 Hz).

Furthermore, these mechatronic control systems can improve the comfort considerably by the ability to control the vibrations of the sprung mass (vehicle body) in an active way.

At the moment, F-M driven active ABW AWA suspension mechatronic control systems have been developed. These mechatronic control systems are only available on very few automotive vehicles because they are costly and have a high-energy consumption. The high-energy consumption of these F-M driven active ABW AWA suspension mechatronic control systems finds its origin in the way the control forces are generated in the suspension.

Most of these active ABW AWA suspensions have an F-M actuator built in series with a compression spring (usually an F-P-M spring).

In order to produce the desired control force the F-M actuator has to compress the spring that causes the energy consumption.

These systems are termed slow active ABW AWA suspension mechatronic control systems because of their low bandwidth (typically up to 3 Hz). This control bandwidth must embrace the normal range of sprung mass resonance frequencies in bounce, pitch and roll.

More advanced and therefore also more expensive suspensions are the purely active ABW AWA suspension mechatronic control systems.

The purely active ABW AWA suspension mechatronic control system involves replacing the conventional suspension elements such as the spring and the damper with an F-M actuator that is controlled by a high frequency response servo fluidical valve. In order to obtain good performance it is necessary that the F-M actuator control bandwidth extends substantially beyond the wheel-hop natural frequency (typically 10 -- 15 Hz).

These active ABW AWA suspension mechatronic control systems need such a high bandwidth because the vibrations of the un-sprung mass (initiated by the on/off road surface irregularities) have to be absorbed actively unlike the slow active ABW AWA suspension mechatronic control systems.

Achieving a reasonable bandwidth is not easy within the cost price and packaging limitations and the higher the bandwidth needs to be; the more the power consumption of the system may tend to. Beyond the actuator bandwidth, noise transmission (harshness) is likely to be a problem unless some significant flexibility is added in series with the actuator.

Required force levels, and consequently either the F-M actuator size or the fluidical supply pressure, may be reduced by relieving the strut from the static load of the sprung mass. This may be achieved by incorporating a spring in parallel to the actuator.

To vanquish the problem of the high-energy consumption, an innovative active ABW AWA suspension mechatronic control system has been invented that reduces the required energy level to a minimal amount [VANHOVENS AND KNAPP 1995].

The essence of these inventions is that the effective length of a mechanical lever with which a preloaded spring is connected to the wheel-axle can be adjusted by means of a servo E-M motor.

During this adjustment the length of the spring does not change (the spatial motion of the spring generates an imaginary cone), so the needed actuator force is very small. In this special manner, the desired control forces are generated in the wheel suspension with the major advantage that E-M motors can be used for driving the system. This results in a more cost effective and simpler system than the currently available alternatives.

As already stated the primary goal for developing an active ABW AWA suspension mechatronic control system may be to design a hardware system that is able to control the attitude of the sprung mass by using only a very small amount of energy.

The fact that the attitude aspects (roll and pitch behaviour) is of major importance in influencing the handling behaviour of a vehicle is the reason why most of the attention has been paid to achieving proper roll and pitch control.

Normally, a soft suspension that provides good isolation from the on/off road surface un-evenness input also allows significant vehicle-body roll during cornering that is undesirable for good handling. Having an attitude control system available in an automotive vehicle makes it possible to make a compromise between good handling (high spring stiffness and firm damping) and good ride comfort (low stiffness and soft damping). This feature allows the suspension designer to put more emphasis on the ride comfort factor than the road-holding factor (wheel-tyre load variations).

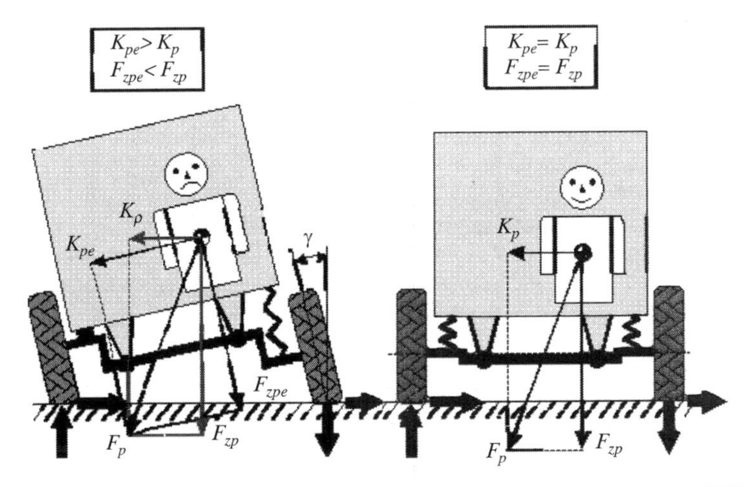

Fig. 5.99 A rolling vehicle versus a non-rolling vehicle [VANHOVENS AND KNAPP 1995].

The suppression of the sprung mass roll and pitch motions as initiated by cornering, braking or accelerating forces has several advantages with respect to ride comfort, handling and active safety of the automotive vehicle [VANHOVENS AND KNAPP 1995]:

- If an automotive vehicle is free from vehicle-body roll during cornering, the driver is not visually impaired with a tilted horizon as would occur with a rolling vehicle;
- The elimination of vehicle-body roll guarantees a more steady seating position of the driver and therefore she/he is able to control the vehicle better (ride comfort/safety); this phenomena is illustrated in Figure 5.99 by the magnitude of the side force K_{pe} and the normal seating force F_{zpe} that the driver experiences; the experienced side force and respectively normal seating force in the rolling vehicle is increased and respectively decreased by a component of the gravitational force F_{zp} acting on the driver's body;
- The roll stabiliser bars may be omitted which yields an additional ride improvement when the vehicle is subjected to asymmetric road excitations;

- The handling and active safety may be refined because the attitude of the wheels with respect to the on/off road surface is close to optimal in a non-rolling/pitching vehicle with relation to the generation of wheel-tyre slip forces (handling); an example of such a roll steer effect that strongly depends on the suspension geometry, is given in Figure 5.99 by the wheel alignment changes such as camber angle of the rolling vehicle;
- The handling aspects may be influenced favourably by the ability to control the ratio of the load transfer between the front and rear axle; this is possible because of the freedom of distributing the four compensating forces necessary to keep the vehicle body on an even keel between the front and rear suspension; the attitude control algorithm may be designed, by using this feature, in such a way that even at limit performance driving good steerability is provided; this makes the manoeuvring operation of the vehicle more convenient for the driver and less unexpected behaviour of the automotive vehicle (braking out, oversteer) may occur in emergency situations;
- In contrast to what most drivers tend to believe, the absence of the roll angle of the vehicle body gives the driver of the vehicle better and more accurate information on the state of the vehicle during cornering; the side forces acting on the wheel-tyres supply a direct indication for the condition of adhesion and thus the limit of cornerability of the vehicle; in a rolling vehicle the perception of wheel-tyre side forces is polluted by a component of the gravitational force (see Fig. 5.99); furthermore, in a non-rolling vehicle the feedback of the side forces through the steering wheel is much better and more direct; in a conventional vehicle first the vehicle-body starts to roll, than the vehicle begins to turn (yawing) and at last the lateral acceleration is built up; this means that the feedback through the steering wheel is not only less (caused by the inclination angle between the track rod and the steering rack) but is also delayed in time; this aspect has a strong impact on the location of the rack-and-pinion in a passive vehicle.

The major contributions to attitude changes of the automotive vehicle are the steering wheel and accelerator/brake input as generated by the driver.

Attitude changes due to, for example, load variations and aerodynamic forces are considered to be quasi-static and may therefore easily be controlled by a rather simple load levelling controller. The attitude controller is discussed in VENHOVENS AND KNAPP [2005] as well as its performance by the demonstration of simulation results and road tests. This controller considers attitude changes due to driver inputs only, but can easily be extended with a load levelling option.

The control software that is implemented in an on-board computer needs three accelerometers (two sensitive in the lateral direction and one sensitive in longitudinal direction) that can be regarded as sensors for measuring the disturbances acting on the vehicle body.

The measured signals are the input for the controller that basically consists of a stable and simple feed forward block for establishing quasi-stationary equilibrium forces in the suspension and a sophisticated PID-feedback control loop on the basis of the 7-DoF internal automotive vehicle physical model.

The PID-action is only activated when the response of the feedforward controller is too sluggish and therefore not able to suppress the roll motion completely in case of rapid braking and steering manoeuvres.

During the progress of the hardware design of an active ABW AWA suspension, interest grew in decreasing ride discomfort due to on/off road surface unevenness.

The first laboratory tests of the actuator system showed that the actuator bandwidth might be sufficiently high enough such that it could be applied to improve the vehicle's ride in addition to all the benefits of controlling the vehicle's attitude.

The minimum energy requirement of attitude control may no longer be maintained because the aim of constant suspension deflections is incompatible with the objectives of vibration control.

From VENHOVENS [1993] it may be concluded that the feedback of the absolute vertical velocity of the sprung mass at each wheel suspension is an extremely simple but also a very effective way to improve the ride in the low frequency range.

A controller based on this strategy is often denoted by *'skyhook control'* because the force actuator that is placed between the vehicle body and the wheel axle, is used to emulate damping forces on the sprung mass as though they were generated by a passive shock absorber (damper) that is connected with one end to a cloud in the sky (i.e. the fixed world) and the other end to the sprung mass. The firmer this shock absorber (damper), the less absolute vertical displacements of the vehicle body may occur and thus the more comfortable the automotive vehicle may be. The sensor hardware of an active ABW AWA suspension mechatronic control system had to be extended with seven additional accelerometers (all sensitive in vertical direction) with three of them mounted on the sprung mass and one on every wheel hub. From the signals of these accelerometers together with the three other accelerometers used for attitude control it is possible to estimate the absolute vertical velocity of the vehicle body at each suspension by using *Kalman* filter techniques [VENHOVENS 1993].

With the hardware of an active ABW AWA suspension the suspension researcher was able to test his developed control algorithm in practice. Many of these test results as well as of those from the developed attitude controller have been published in various papers [VENHOVENS ET AL. 1992, 1993; VENHOVENS 1993; KNAAP ET AL. 1994; VENHOVENS AND KNAAP 1995].

An early attempt, and one that is still being used and researched, to make a smart suspension was to make an active ABW AWA suspension that has a central suspension ECU that monitors the relative position of the wheels to the vehicle's body as well as a velocity (speed) reading and a gyroscope that measures the vehicle body's relative position to the on/off road surface. With this data, the ECU controls something termed an actuator.

For instance, an actuator replaces the shock absorber found in the passive suspension mechatronic control system and is a pressurized piston filled with oily fluid or air.

A reservoir of oil or air and a system of fluidical or pneumatical valves allows the ECU to vary the pressure of oil or air inside the actuator, high pressure to absorb the energy of a large bump, low pressure to absorb a smaller bump.

The downside of an active ABW AWA suspension mechatronic control system is that the oily-fluid or air (gas) reservoir and fluidical or pneumatical valves on the actuator takes up more space than the shock absorber of a passive shock absorber suspension mechatronic control system and is heavier too. Also, being a complicated fluidical or pneumatical system, an active ABW AWA suspension has the potential to develop oil or air pressure problems.

In an active ABW AWA suspension mechatronic control system, the passive shock absorber (damper) or both the passive shock absorber and spring are replaced a force actuator, as illustrated in different suspension physical models shown in Figure 5.100 [SIMON 2001].

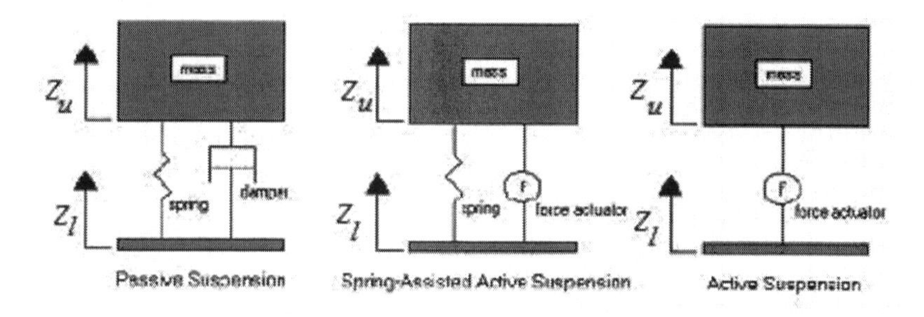

Fig. 5.100 Passive and active suspensions [SIMON 2001].

The force actuator is able to both add and dissipate energy from the mechatronic control system, unlike a passive shock absorber (damper) that may only dissipate energy. With an active ABW AWA suspension, the force actuator may apply force independent of the relative displacement or velocity across the suspension.

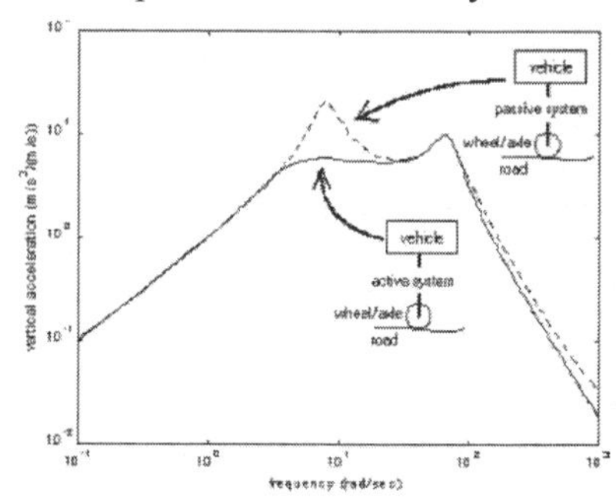

Fig. 5.101 Passive and active suspension comparison [CHALASANI 1986A, SIMON 2001].

Given the correct control strategy, this results in a better compromise between ride comfort and vehicle stability as compared to a passive suspension, as shown in Figure 5.101 for a quarter-vehicle physical model [SIMON 2001].

A quarter-vehicle physical model shown in Figure 5.102 is a 2-DoF physical model that emulates the vehicle body and axle dynamics with a single mass, that is, one-quarter of an automotive vehicle [SIMON 2001].

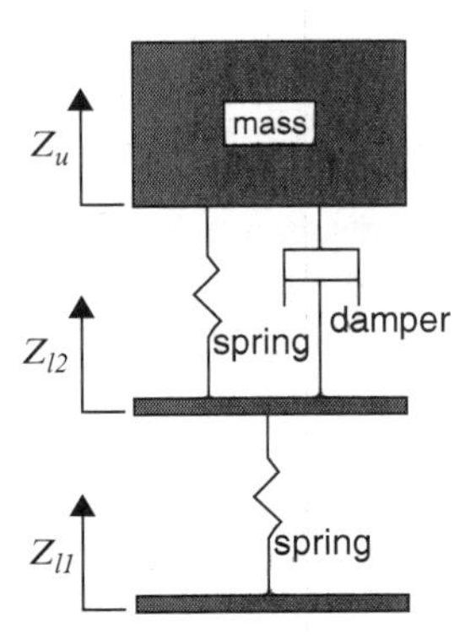

Fig. 5.102 A quarter-vehicle physical model [SIMON 2001].

In CHALASANI [1986A], a quarter-vehicle physical model was used to investigate the performance gains possible with an active ABW AWA suspension mechatronic control system. In the study, the on/off road surface input was modelled as white-noise velocity input. The study found that within practical design limitations, an active ABW AWA suspension might reduce the **root mean square** (RMS) value of the sprung-mass acceleration by 20%. This suspension configuration exhibited approximately the same level of suspension travel and wheel-hoop damping ratio as a lightly damped, soft passive suspension.

In a further study [CHALASANI 1986B], similar simulations and analysis were performed for half-vehicle physical model. This study estimated that active ABW AWA suspensions could reduce the RMS value of the sprung-mass acceleration by 15 %.

Active ABW AWA suspension mechatronic control systems have the added advantage of controlling the attitude of an automotive vehicle. They may reduce the effects of braking that causes a vehicle to nose-dive, or acceleration that causes a vehicle squat. They also reduce the vehicle roll during cornering manoeuvres.

Active ABW AWA suspension mechatronic control systems, though shown to be capable of improving both ride and stability, do have disadvantages. The force actuators necessary in an active ABW AWA suspension mechatronic control system typically have large power requirements, typically circa 3 – 4 kW (4 -- 5 HP). The power requirements decrease the overall performance of the automotive vehicle, and are therefore often unacceptable.

A detraction of active ABW AWA suspension mechatronic control systems is that they may have unacceptable failure modes. In the case of actuator failure the automotive vehicle would be left undamped, and possibly unsprung. This is a potentially dangerous situation for both the automotive vehicle and driver.

The shock absorber suspension mechatronic control system described in Chapter 5.3 is a passive one for the reason that only the absorbing (damping) force is adjusted. Through active adjustment of the ABW AWA suspension mechatronic control system or because of the action of supplementary F-M or P-M or even E-M forces it turns out to be realisable to minimise unnecessary vehicle roll ensuing from dynamic lateral loads, as verified in Figure 5.103 [SEIFFERT AND WALZER 1991].

Fig. 5.103 *VW Jetta* slalom driving: left -- conventional suspension system; right -- active suspension system with roll suppression [Volkswagen; SEIFFERT AND WALZER 1991].

Purely active suspensions add a fourth dimension, using sophisticated components to minimise dive during braking and roll during cornering. They are a mixture of high-tech mechatronic add-ons and fast-acting M-F pumps or M-P compressors or even M-E generators. They tend to be unpopular because of their complexity, cost and questionable reliability. So far, none of these mechatronic control systems has proven to be a measurable improvement over a well-designed conventional suspension. Active ABW AWA suspension mechatronic control systems are similar to that of semi-active ABW AWA suspension mechatronic control systems accept that the spring and actuator units are completely replaced by a single F-M or P-M or even E-M actuator.

An active F-M ABW AWA suspension mechatronic control system of similar arrangement is being used in the *'Formula 1'* racing car [DOMINY AND BULMAN 1985]. It was first developed by Lotus during 1977 to 1982 [FORTUNECITY 2005]. The main reason for this development was due to the aerodynamic changes that took place during that time in *Formula 1*. With the improvements to front wing, rear wing, and skirt design, a significant increase in down force was created. This contributed to a massive increase in load on the suspension by up to three times the normal amount. Although active ABW AWA suspension solved their original problem, another one was created by the *F1A* (*Formula 1* governing body) who banned the use of it due to the enhancements of the vehicle handling. Active ABW AWA suspension mechatronic control systems are usually all high bandwidth within racing applications.*)

*) NB : - *Formula 1* racing cars must have conventional sprung suspension up to now. Any system, such as an active ABW AWA suspension, that can alter the suspension or its geometry while the racing car is moving is forbidden.

However, for conventional use it is possible to benefit more from a control low bandwidth system due to its cheapness.

Active ABW AWA suspension mechatronic control systems work by altering the vehicle's ride height, in turn influencing the vehicle's behaviour. This is done automatically by an ECU that monitors the vehicle velocity, steering wheel position, acceleration, and vehicle body movement. The ECU then controls an M-F pump or M-P compressor or even M-E generator to apply or reduce oil or air pressure or even voltage at the relative F-M or P-M or even E-M actuators, respectively. The ECU has to be able to measure and interpret data at a very fast rate in order to calculate the required ride heights. This is due to the rate at which the on/off road surface can change even at low values of the vehicle velocity, as well as the velocity that vehicle-body roll, anti-dive, and anti-squat can effect. A spring is sometimes fitted in parallel with the F-M, P-M or E-M actuator, but not to add to the performance of the system. The purpose of the spring is in case a failure occurs, in which case a passive suspension system is created with the actuator acting like a conventional shock absorber (damper). This is very important as, if the suspension system was to fail during a vehicle manoeuvre, then it would be possible to regain control of the vehicle. One of the reasons why active ABW AWA suspension mechatronic control systems are so expensive is due to the amount of power that is required to run the M-F pump or M-P compressor or even M-E generator. In prototype versions approximately 3 kW (4 hp) is required for effective performance. Realistically, active ABW AWA suspension mechatronic control systems can only be expected on top-end luxury vehicles with ECEs or ICEs large enough to afford the loss of 4-5% of its performance. As with semi-active ABW AWA suspension mechatronic control systems, the initial stages of design are undertaken using the *'skyhook theory'* [FORTUNECITY 2005].

Automotive vehicle's active ABW AWA suspension mechatronic control systems in which absorbing (damping) forces are generated in response to feedback signals by active elements obviously offer increased design flexibility compared to the conventional suspensions using passive elements such as springs and shock absorbers (dampers). The attention is concentrated on several aspects of such an unconventional actuator controlled to obtain a variable mechanical force for the automotive suspension. The main advantage of such a solution is the possibility to generate desired absorbing (damping) forces acting between the unsprung (wheels) and sprung (vehicle body) masses.

An active ABW AWA suspension mechatronic control system optimises the automotive vehicle attitude and the wheel-tyres' contact with on/off road surfaces under any dynamic and static load and surface conditions.

Usage of an automotive vehicle's active ABW AWA suspension mechatronic control system has two main reasons -- to increase ride comfort and to enhance handling performance with respect to the reduction forces.

Both these requirements are contradictory and it is impossible to satisfy them simultaneously with a P-M suspension system.

In its principle arrangement such an active ABW AWA suspension mechatronic control system would contain F-M or P-M or even E-M components, for example, F-M or P-M or even E-M actuators, M-F pumps or M-P compressors or even M-E generators and fluidical (hydraulical and/or pneumatical) or even electrical accumulators as well as electromagnetic servo-fluidical (hydraulical and/or pneumatical) or even electrical valves, respectively, and the control mechatronics.

Active ABW AWA suspensions fall into two broad categories, high and low bandwidth. High bandwidth systems are what might be described as *'purely active'*, providing complete control through all frequency ranges, from gentle wallowing over undulating on/off road surfaces to high velocity pattering over small irregularities. Low bandwidth systems use conventional springs to handle the low frequency work and support the automotive vehicle's mass. In the sense that springs, shock absorbers (dampers) and anti-roll bars are actually hard to beat given the right level of expertise [LOTUS 2005].

Attempts to introduce active ABW AWA suspension mechatronic control systems into mainstream manufacturing have been going on for over *25* years but have been frustrated in the past mainly through the unavailability of single-chip microprocessors capable of reacting fast enough, cost and the energy required to drive them. Since the super-fast single-chip microprocessor has come of age only some automotive vehicle manufacturers have taken the plunge so far but other concepts are emerging. Some automotive tyre manufacturers have a working concept for an active F-M suspension unit comprising a spring and F-M actuator.

In the 2010s, one of them revealed its **active wheel** (AW) concept that as the name suggests, packages active suspension, disc brake unit, wheel hub and even E-M motor, inside the envelope of the wheel rim. Are we ever likely to see fully active, spring free suspension in manufacture? It's unlikely, simply because supporting the automotive vehicle's mass consumes a huge amount of energy, while supporting it with a spring is free. The real advantages of active suspension components of all types arguably lie in their potential not for enhancing the driving experience, but in providing an added safety margin for the driver.

Integration with stability, braking and steering systems already make it possible for the automotive vehicle to respond to a dynamic crisis even if the driver does not.

Active front steering systems that can influence the steering angle independently of driver input, are already in the marketplace and tentatively being integrated with stability and active ABW AWA suspension mechatronic control systems to complement the effect of both if the automotive vehicle becomes unstable [LOTUS 2005]. All of that implies a substantial role in the use of electrical chassis components that in itself brings manufacturing benefits.

Electro-mechanical brakes (EMB) already being successfully tracked by *Tier 1* suppliers, not only promise stick integration with stability mechatronic control systems and active F-M suspension but in manufacturing the prospect of *'dry'* assembly lines of brake fluid is an attractive one.

In the 2010s, yet further advanced technology automotive vehicle manufacturers revealed another possibility when they pulled the wraps off a semi-active ABW AWA suspension that swaps an F-M or a P-M actuator for a linear or rotary EM motor.

Research and development (R&D) work, for that reason, moves in the trend of F-M, P-M or E-M support parallel to a passive ABW AWA suspension mechatronic control system. Since even these mechatronic control systems are still very complicated and luxurious, supplementary innovations must be carried out before mechatronic control systems of this arrangement can be accepted for mass-manufactured vehicles. ABW AWA suspension mechatronic control systems for passenger vehicles have improved in many steps over the years. R&D in the line of work originated with the P-M suspension for controlling vehicle height and then progressed to the F-P-M suspension and different suspensions with absorbing (damping) force and spring rate control.

At the present R&D works are in progress to improve an active ABW AWA suspension mechatronic control system. It is characterised as one that has the subsequent features [AKATSU, 1994]:

- Energy is continuously fed to the suspension system and the force created by that energy is constantly controlled;
- The suspension system contains different kinds of sensors and an ECU for processing their signals that initiates forces that are a function of the signal outputs.

In contrast to semi-active ABW AWA suspension mechatronic control systems that do not require particular energy from outside, active ABW AWA suspension mechatronic control systems help reduce vehicle vibration and improve vehicle cornering as the latter is actively stretched by the high oily-fluid or air (gas) pressure or even electrical current fed by the M-F pump or M-P compressor or even M-E generator, respectively, in the ECE or ICE drive.

This also allows an automotive vehicle to maintain its roll angle, as the vehicle tilts while cornering, at almost zero in $5\ m/s^2$ of lateral acceleration equivalent to a sudden lane change on highways. This feature enables an automotive vehicle active ABW AWA suspension mechatronic control system to perform far better than a semi-active ABW AWA suspension mechatronic control system designed only to control the absorbing (damping) force [UEKI ET AL. 2004].

For instance, the automotive vehicle's active ABW AWA suspension mechatronic control system has continuously adjustable shock absorbers (dampers) and accelerometers in the front right damper dome and also in the left rear. Other sensors detect vertical vehicle-body movements, steering angle, vehicle velocity, brake oily fluid or air pressure or even electrical current and ECE or ICE torque. With these data, the suspension controller automatically provides optimum shock absorber (damper) control for each wheel. If the system should fail, a bypass fluidical or pneumatical or even electrical valve automatically closes and selects, respectively, the hardest driving position to ensure safe driving.

The active ABW AWA suspension may also have five software modules to provide optimum settings for lane change, vertical control, lateral acceleration, brake and load change.

For example, during a sudden lane change manoeuvre, the suspension increases absorbing (damping) forces on both axles to eliminate swaying and rocking [SCHULDINGER 2005].

The vertical control module helps smooth out driving on bumpy on/off road surfaces. In normal mode, damping forces increase when vertical vehicle-body movement exceeds a threshold. The result is less of a tendency for the body to rock. Sport mode also reduces the damping effect to maintain wheel-tyre contact with on/off road surface. The lateral acceleration module damping adjusts with vehicle velocity and lateral acceleration. In sudden hard braking, the suspension firms up its damping action to minimise vehicle-body dive and allows quicker transference of brake forces to the on/off road surface. A millisecond later, the suspension switches to softer damping and applies different absorbing (damping) forces to the front and rear ends of the vehicle. This helps shorten stopping distance, even on bumpy on/off road surfaces. When a driver's foot is lifted off the accelerator, for example, while shifting gears, the load change module uses harder damping in normal setting to prevent squat. In sports mode, softer damping is used to increase traction [SCHULDINGER 2005].

Summing up, all current suspensions are reactive. When a vehicle's wheel rolls over a bump or dip in the pavement, the change in wheel position causes the suspension to compress or extend in response. Cornering, braking, and acceleration similarly cause the suspension to move and let the vehicle-body roll, squat, or dive. That's been the case since springs were added to horse-drawn carriages in the middle of the XVIIth century [CSERE 2004].

Such suspensions have evolved to a high level and deliver a very good combination of ride and handling. But in the end, they are followers, not leaders. Their springs and shocks only let the wheels move in response to some force that has acted on them, and even with modern adjustability, their actions may be limited by the fact that they are always playing defence-acting only after something has happened.

An active ABW AWA suspension, however, is playing offence. It has a computer that tells a powerful actuator at each wheel exactly when, which way, how far, and how fast to move. The wheel motions are no longer subject to the random interactions between the on/off road surface and the various springs, shocks, and anti-roll bars.

Just as a human being may bend her/his knees and suddenly extend them to jump up, an active ABW AWA suspension automotive vehicle (programmed to perform this parking-lot trick) may leap over a *'two-by-six'*, or *'proactively'* counteract the forces acting on an automotive vehicle. The latter may be shown actually jumping a two-by-six as grace-fully as a horse leaps over a steeplechase fence. This technical trick has no application, but it does demonstrate the profound differences between an active ABW AWA suspension mechatronic control and even the most sophisticated, computer-controlled semi-active ABW AWA suspension one.

The computer making these decisions uses a network of sensors to measure, for example, the vehicle velocity, longitudinal and lateral accelerations, and forces and accelerations acting at each wheel. This then commands the wheel to move in the ideal way for the existing circumstances. No more compromise between ride comfort and road handling, rough on/off road surfaces and smooth on/off road surfaces, high- and/or low-values of the vehicle velocity [CSERE 2004].

The purpose of an automotive vehicle's active ABW AWA suspension mechatronic system is to isolate the shock and vibration due to road/terrain irregularities from the main body of a vehicle, yet still provide the necessary ground contact to allow the vehicle to stay in control.

Components in such suspensions have traditionally been passive springs and shock absorbers (dampers).

ABW AWA suspensions, made of components that may change their properties or add energy to the suspension, may be developed to increase performance by:

- Improving rider comfort;
- Increasing handling;
- Allowing faster speeds over more irregular terrain;
- Increasing durability of components.

Recent testing has also confirmed that radar vibration sensing technology is well suited for non-contact measurement of road/terrain surfaces immediately ahead of a moving vehicle. This microwave terrain sensor's measurement data (Fig. 5.104) have excellent utility as a control input for a modern active ABW AWA suspension mechatronic control system [RADATEC 2002].

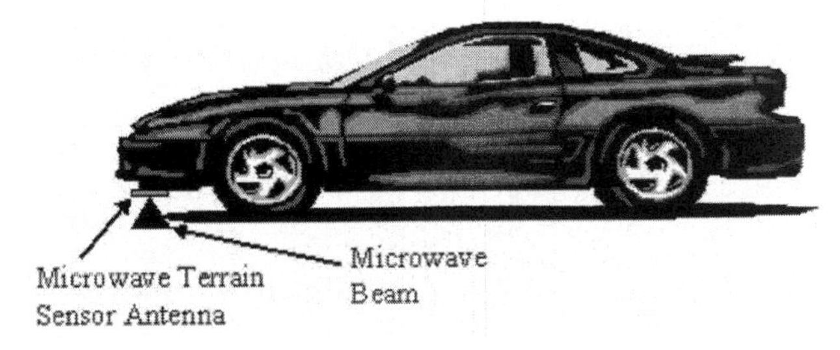

Fig. 5.104 Microwave terrain sensor [Radatec Inc.; RADATEC 2002].

Although this microwave terrain sensor has wide potential applications, a case study proposing its use as a control input to an active ABW AWA suspension mechatronic control system is proposed.

This microwave radar-based sensor has significant advantages over other means of road sensing [RADATEC 2002]:

- The sensor is constructed from simple, commonly available, inexpensive microwave components, making it cost effective and reliable;
- The method of sensing is unaffected by dust, debris, rain, or other obscurants, making it robust to typical road conditions;
- It may be mounted on a vehicle behind a radio-transparent shield, out of harm's way.

Other means of non-contact road sensing are largely based on expensive optical systems that are impractical for widespread use.

Automotive vehicles with active ABW AWA suspension can take large bumps in the on/ off road surface with little difficulty, do not lean in sharp turns, provide a smooth ride and overall makes the vehicle drivable by even the most inexperienced of drivers.

Summing up, active F-M or F-P-M or P-M or even E-M ABW AWA suspension mechatronic control systems that use conventional passive shock absorbers (dampers) and springs made active by an outside energy source, such as a M-F pump or M-P compressor or even M-E generator, respectively, and one or more accumulators adding or removing energy from the units.

The low-frequency active F-M or F-P-M or P-M or even E-M ABW AWA suspension mechatronic control systems being developed are currently using a single central M-E pump or M-P compressor or even M-E generator, respectively, and one or more accumulators to provide and store fluidical or even electrical energy at a constant oily fluid or air pressure or even electrical current, respectively, for actuating the suspension units.

The low-frequency active F-P-M ABW AWA suspension mechatronic control system (sometimes referred to as low-band active or soft active) uses a passive spring element, usually in the form of an F-P-M spring/shock absorber that isolates the high frequency road vibrations from the body.

Each F-P-M suspension unit is made active by adding or removing oily fluid. In effect, the predeflection is actively changed for each passive F-P-M spring.

Each F-P-M spring is actuated to minimise the low-frequency vehicle body motions of pitch, roll, and heave that range from 1 to 3 Hz.

Most of the benefits of an active F-M or F-M-P or P-M or even E-M ABW AWA suspension mechatronic control system may be obtained with a low-frequency active ABW AWA suspension mechatronic control system, but with much less complexity, lower cost, and reduced power requirements.

An ABW AWA suspension may be considered active when an outside energy source is used to change its characteristics, and these mechatronic control systems may be placed in one of three different categories: semi-active damping, active, and low-frequency active.

A regenerative M-F pump or M-P compressor or even M-E generator concept may minimise the energy requirement for the low-frequency active F-M or F-P-M or P-M or even E-M ABW AWA suspension mechatronic control system.

It uses four independent variable displacement M-F/F-M pump/motor or M-P/P-M compressor/motor or even M-E/E-M generator/motor combinations, respectively, on a common shaft to actuate each individual suspension unit.

5.6.2 Active F-M ABW AWA Suspension Mechatronic Control Systems

Vehicular suspension is a compromise between comfort and control. ABW AWA suspension mechatronic control systems are comprised of springs and dampers that store and release energy.

Spring rates and damping coefficients are adjusted to set the balance between these two opposing criteria.

Using conventional means, this compromise is set once and for all in the developmental stage.

Whether it is a vehicular suspension for a Cadillac limousine or a Ferrari sport car, some balance between comfort and control is chosen.

Active suspension is the name given to an approach that allows this compromise to vary based on driving circumstances [TREVETT 2002].

The first level of advanced vehicular suspension is semi-active suspension. In this vehicular suspension one or more parameters of the ABW AWA suspension are automatically varied based on sensor inputs.

The adjustments capable of on-the-fly management are the ride height of the vehicle, the stiffness of the shock absorber (damper), and the spring rate.

The adjustments are realised by altering either the oily-fluid pressure in the components or some internal geometry.

Quick operating fluidical valves are used to alter orifice size and control the movement of oily-fluid.

The shock absorbers (dampers) may be regulated by varying the volume of the hole between the two compartments. Spring rate alterations are based on the available volume the oily-fluid may fill.

Overall ride height may be controlled by the oily-fluid pressure in the chambers of the shock absorber unit.

Using M-F pumps, reservoirs, and fluidical valves the parameters of the vehicular suspension may be modified [GORDON 1999].

Fully active suspension is quite simple in concept but challenging in design. It consists of four F-M actuators, one at each wheel, and sensors and software to control them.

Fully active suspension detects bumps and compensates for them before the passenger compartment responds. It is capable of reducing roll, pitch, and yaw. Automotive vehicles with full-active suspension may *'lean'* into turns to enhance handling [GORDON 1999].

Fully active suspension is an objective of suspension designers; however, the power requirements and cost considerations make it impractical at the present.

Combinations between semi and fully active suspension also exist. These are known as series active and parallel active suspensions.

They incorporate springs in either series or parallel, as their names indicate, with F-M actuators.

Series active suspensions reduce the necessary bandwidth from that of fully active suspensions, while parallel active ones reduce the load that the F-M actuators must carry [LEIGHTON AND PULLEN 1994].

Diagrams explaining the configuration for the four categories of suspension mentioned are shown in Figure 5.105 [LEIGHTON AND PULLEN 1994; TREVETT 2002].

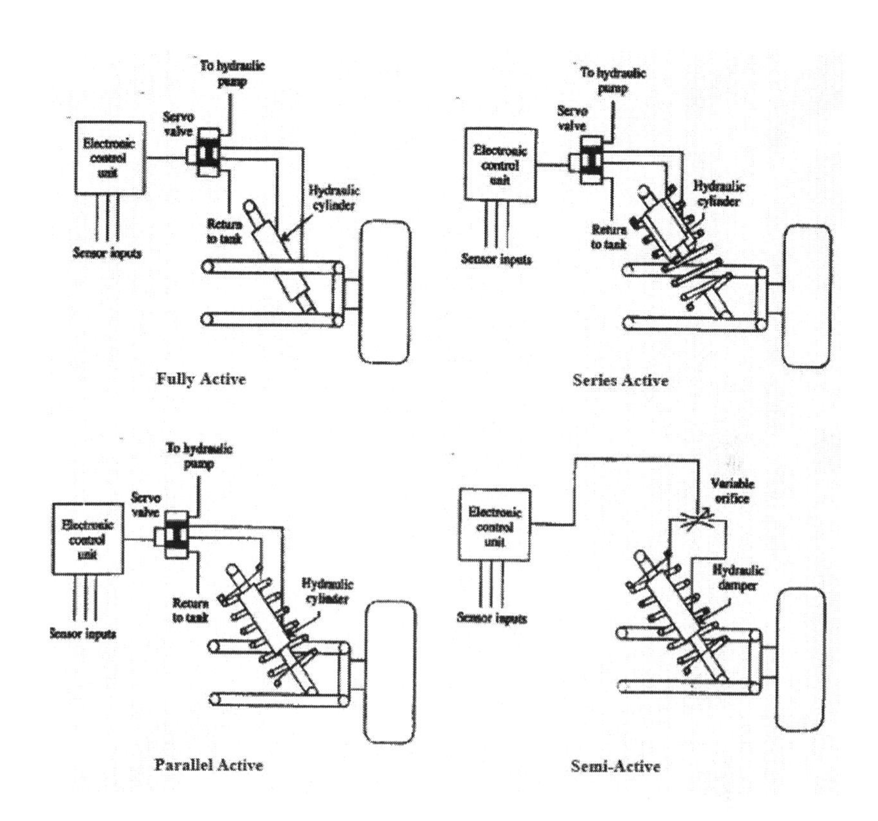

Fig. 5.105 Active suspension configurations [LEIGHTON AND PULLEN 1994].

The price of enhanced comfort and control may be foreseen in the increased mass, power consumption, and cost of active suspensions.

Sensors are indispensable to monitor functions such as, on/off road surface modulations, driving, braking, steering and vehicle velocity and position as well as acceleration in **three dimensions** (3D). Sophisticated logic is necessary to account for all possible driving circumstances.

Fully active suspensions necessitate F-M actuators capable of large displacements in microsecond time intervals [KNUTSEN 2002] under loads of several thousand newtons [N].

The wheel-tyres are capable of absorbing frequencies above 20 -- 40 Hz, but the F-M actuators must eradicate the rest [LEIGHTON AND PULLEN 1994].

Semi-active suspensions are becoming a reality in manufacture while fully active ones remain distant. Increased competition on the automotive market has forced vehicle manufacturers to research alternative strategies to classical passive suspension mechatronic control systems.

In order to improve ride comfort and road handling performance, instead of a conventional static spring and shock absorber (damper) mechatronic control system, semi-active and active F-M ABW AWA suspension mechatronic control systems are being developed.

A semi-active F-M ABW AWA suspension mechatronic control system involves the use of a shock absorber (damper) or spring with variable gain. Such mechatronic control systems may only operate on three fixed positions: soft, medium and firm absorbing (damping) or stiffness. Additionally, a semi-active F-M ABW AWA suspension mechatronic control system may only absorb the energy from the motion of the vehicle body.

On the other hand, an active F-M ABW AWA suspension mechatronic control system possesses the ability to reduce acceleration of sprung mass continuously as well as to minimise suspension deflection that results in improvement of wheel-tyre grip with the on/off road surface, thus, brake, traction control and vehicle manoeuvrability may be considerably improved.

A computerized active F-M ABW AWA suspension mechatronic control system uses F-M actuators instead of conventional springs and shock absorbers to support the vehicle's mass.

An on-board computer monitors ride vehicle–body height, wheel deflection, vehicle-body roll and acceleration to control ride and vehicle-body attitude. Bumps are sensed as they are encountered, causing the computer to vent oil pressure from the wheel actuator as the wheel floats over the bump. Once the bump has passed, the computer opens a vent that allows oil pressure to extend the actuator back to its original length.

Beginning in 1996, some vehicle manufacturers began offering active F-M ABW AWA suspension mechatronic control system on its automotive vehicles.

Known as **continuously variable road-sensing suspension** (CVRSS) mechatronic control system (as shown in Figure 5.106), the mechatronic control system uses a series of sensors to actuate F-M shock absorbers at all four corners, improving road feel and dampening.

Fig. 5.106 Active F-M ABW AWA suspension known as *Cadillac*'s continuously variable road-sensing suspension (CVRSS) [Ford Motor Company; MEMMER 1999].

This mechatronic control system adjusts in a fraction of a second -- the amount of time it takes an automotive vehicle going circa 100 km/h (65 mph) to travel 38 cm (15 inch) [MEMMER 1999].

Starting with the 1999 model year, another automotive vehicle manufacturer has installed an **active cornering enhancement** (ACE) shown in Figure 5.107 [BRAUER 1999; MEMMER 1999].

Fig. 5.107 Active cornering enhancement (ACE) [Ford Motor Company; BRAUER 1999; MEMMER 1999].

The ACE system is a first for SUVs. It uses an F-M system that replaces the more conventional front and rear anti-roll bars, applying torque to the vehicle-body via two piston/lever configurations. The ACE system has the capability to counteract up to 1.0 g lateral acceleration in 250 ms [MEMMER 1999].

The most significant improvement to the vehicle is the availability of ACE. The conventional automotive vehicle regularly threatened to turn turtle in tight corners and rocked on its moorings in stiff crosswinds. The amount of vehicle body roll was ridiculous. Combined with numb, slow steering, the conventional automotive vehicle simply made no sense as an urban commuter. But ACE, an F-M system designed to counteract lateral vehicle-body movement, solves this problem completely. ACE measures lateral acceleration during cornering and applies torque to the vehicle body to prevent roll. The result is a perfectly flat cornering stance up to 0.4 g of grip, and then the system dials in progressively higher amounts of roll as the conventional vehicle nears its limits. Combine a wider front and rear track (for example, thanks to new solid axles similar to those on the more upscale *Range Rover*), permanent 4WD, an updated and more on/off road-worthy suspension, and the 18 inch wheels and wheel-tyres that come with the performance package, and drivers can have one heavy SUV that handles more like a sports automotive vehicle than a sport-commute around town.

Predictably, since this is an off-road automotive vehicle first and foremost, the steering is still slow with little feedback about what's happening under the wheel-tyres, but feel is improved over the conventional truck and the turning circle has been tightened up. ACE really acts, and transforms the vehicle from klutz to capable in one fell swoop [BRAUER 1999].

For many years automotive scientists and engineers have struggled to achieve a compromise between vehicle handling, ride comfort and stability. The results of this are clear in the automotive vehicle people see today. In general, at one extreme are large passenger and luxury vehicles with excellent ride qualities but only adequate handling behaviour. At the other end of the spectrum are SUVs with very good handling but very firm ride quality. In between are any number of variations dictated by the vehicle manufacturer and target customer requirements [DEMEIS 2005; TRW 2005].

Active F-M ABW AWA suspension mechatronic control systems (Fig. 5.108) help to resolve the rigid drop links of one stabilizer bar end against an F-M actuator (active stabiliser bar) [DEMEIS 2005; TRW 2005].

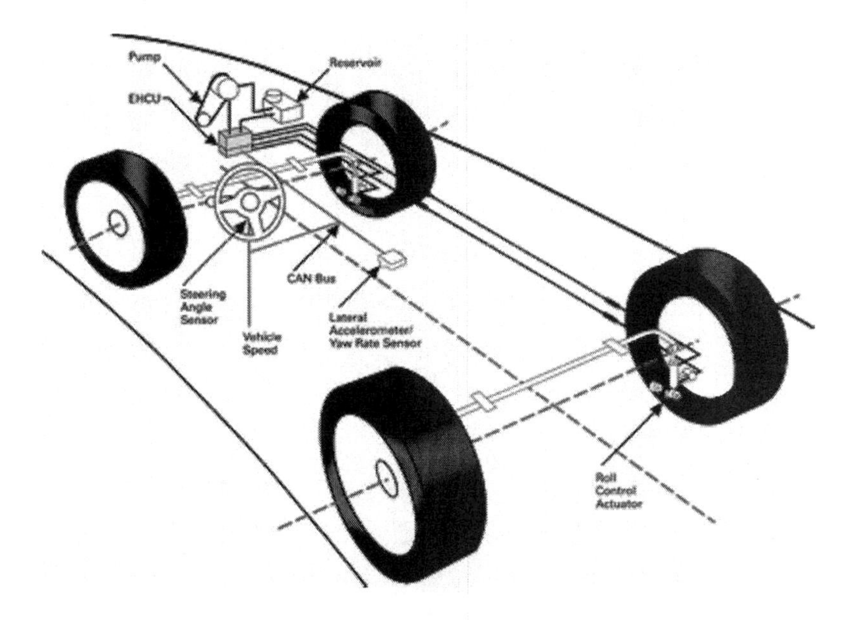

Fig. 5.108 Principal layout of the active F-M ABW AWA suspension [TRW Automotive; TRW 2005].

Depending on the sensed driving circumstances, an **electro-fluidical control unit** (EFCU) controls these actuators. That gives automotive scientists and engineers innovative possibilities for suspension tuning.

The existing manufacture active F-M ABW AWA suspension mechatronic control system architecture may be divided into three different hyposystems [DEMEIS 2005; TRW 2005]:

- **Active dynamic control 1** (ADC1):
  - *Set-up*: active stabiliser bar at front or rear axle;
  - *Sensor signals*: controlled by sensor signals lateral acceleration, steering angle, vehicle velocity and yaw rate;
  - *Control goal*: agile if possible, stable if necessary;
  - *Benefits*: improved, controlled neutral handling and agile steer response, improved roll comfort.
- **Active roll control** (ARC):
  - *Set-up*: active stabiliser bar at front and rear axle, controlled by a common fluidical circuit;
  - *Sensor signals*: lateral acceleration, steering angle and vehicle velocity;
  - *Control Goal*: vehicle-body roll compensation;
  - *Benefits*: flat vehicle-body, while cornering improved roll comfort and roll over mitigation.
- **Active dynamic control 2** (ADC2):
  - *Set-up*: active stabiliser bar at front and rear axle, controlled by two separate fluidical circuits;
  - *Sensor signals*: lateral acceleration, steering angle, vehicle velocity and yaw rate;
  - *Control goal*: agile if possible, stable if necessary and vehicle-body roll compensation;
  - *Benefits*: ADC2 combines ADC1 and ARC benefits in one system.

Active ABW AWA suspension mechatronic control systems in which changes in the attitude or height of the vehicle body relative to the on/off road surface are detected by means specifically provided for the purpose, other than the driver, the detection means providing an input to the mechatronic control system itself to cause the latter to compensate for the change in attitude or height of the automotive vehicle [MOTTA ET AL. 2000].

Typical examples of detection means are accelerometers, pendulum devices, electrical valves (mercury switches), weighted fluidical valves, devices connected to the steering mechanism, for instance, levers, rheostats, or simple levers attached to an electrical valve (switch) and/or fluidical valve.

Active body control (ABC) is a semi active F-M ABW AWA suspension mechatronic control system, by Mercedes-Benz, that virtually eliminates vehicle-body roll in many driving situations including cornering, accelerating, and braking. The mechatronic control system's computer detects vehicle-body movement from sensors located throughout the vehicle, and controls the action of the active F-M ABW AWA suspension with the use of F-M servos. A total of thirteen sensors continually vehicle-monitor body movement and vehicle level and supply the ABC computer with new data every ten milliseconds. Two sensors at each end of the automotive vehicle measure ride level, while nine other sensors monitor vertical and transverse vehicle-body movement [WIKIPEDIA 2005].

As the ABC computer receives and processes data, it operates four F-M servos, each mounted on a spring strut, one beside each wheel.

Almost instantaneously, the servo-regulated suspension generates counter forces to vehicle-body roll, dive, and squat during various driving manoeuvres. A suspension strut, consisting of a steel coil spring and a shock absorber are connected in parallel, as well as a fluidically (hydraulically) controlled adjusting cylinder, are located between the vehicle body and wheel. These components adjust the cylinder in the direction of the suspension strut, and change the suspension length. This creates a force that acts on the suspension and dampening of the automotive vehicle in the frequency range up to 5 Hz. The mechatronic control system also lowers the automotive vehicle up to 11 mm between the speeds of 60 km/h (37 mph) and 160 km/h (99 mph) for better aerodynamics, fuel consumption, and handling [WIKIPEDIA 2005].

The ABC active F-M ABW AWA suspension mechatronic control system also acts as an automatic levelling control that raises or lowers the automotive vehicle in response to changing load (i.e. the loading or unloading of passengers or cargo).

Each automotive vehicle equipped with ABC has an “ABC Sport” button that allows the driver to adjust the suspension range for different driving style preferences. This feature allows the driver to adjust the suspension to maintain a more level ride in more demanding driving conditions [WIKIPEDIA 2005].

Mercedes-Benz comes with ABC, shown in Figure 5.109, that uses thirteen different on-board sensors, and that feed into four F-M servos positioned atop each coil spring. The ABC computer adjusts the ride every 10 ms [MEAD 1999; MEMMER 1999; SCHÖNER 1999].

ABC system is an active F-M suspension mechatronic control system that uses a coil spring and a mechatronically controlled F-M actuator (cylinder) in series plus a separate gas-pressurised shock absorber (damper) at each wheel.

Using a total system pressure of up to 20.38 MPa (2,900 psi), ABC continually adjusts each wheel's suspension to counteract vibration, pitch, dive, squat and roll. ABC also provides automatic *4*-wheel level control driver-selectable ride height and automatic lowering at higher values of the vehicle velocity (speed).

ABC system is an active F-M ABW AWA suspension mechatronic control system that keeps the automotive vehicle level under acceleration, braking and cornering. F-M actuators (cylinders) at each corner support a conventional spring and shock absorber (damper), providing adjustments to ride height without intruding on the automotive vehicle's comfortable ride. Reconciling the conflicting objectives of dynamic road handling and ride comfort, ABC active F-M ABW AWA suspension uses high-pressure fluidics (hydraulics), sophisticated sensors, and powerful single-chip microprocessors to adjust suspension response to current driving conditions with split-second accuracy. ABC active F-M ABW AWA suspension virtually eliminates vehicle-body roll in cornering, squat under acceleration, and dive during braking. While active suspension technology has the potential to actually bank a vehicle into every turn, automotive scientists and engineers use its interplay of fluidical (hydraulical), electronical and mechanical parts to reduce vehicle-body roll by 68%, ensuring that the suspension still provides the driver with feedback through the vehicle chassis, and a switch on the console permits the driver to further reduce vehicle-body roll by 95%.

Mercedes-Benz has now developed this active F-M ABW AWA suspension mechatronic control system even further and added supplementary functions that further improve ride comfort and road handling dynamics.

With the new feature of load adaptation, ABC now also registers the actual vehicle load and takes this value into account when calculating the active F-M ABW AWA suspension mechatronic control parameters.

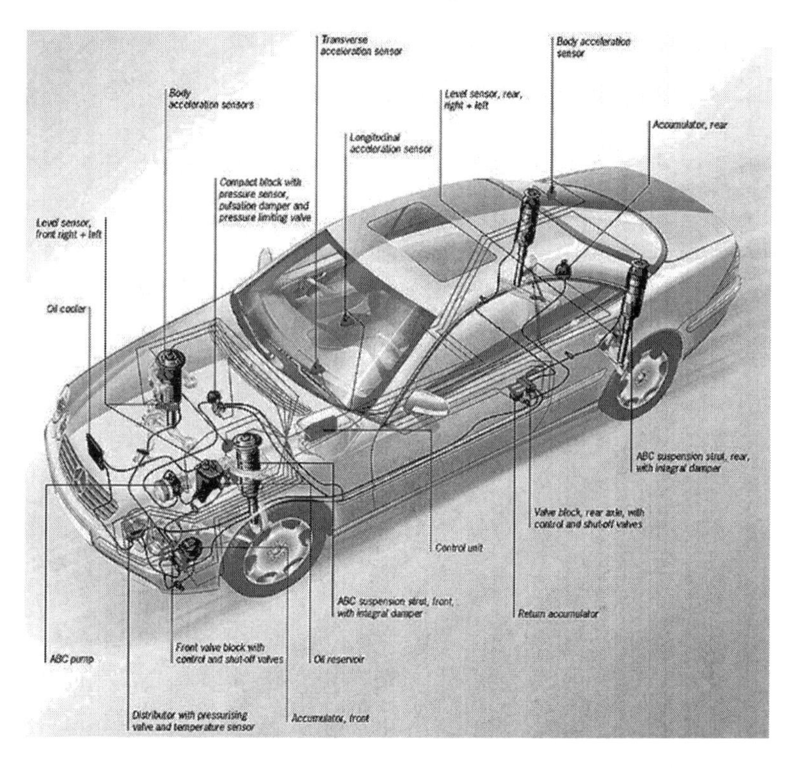

Fig. 5.109 Active F-M ABW AWA suspension known as Mercedes-Benz's active body control (ABC) – Top image; ABC in Mercedes-Benz *CL* – Bottom image [Mercedes-Benz; MEAD 1999; MEMMER 1999; SCHÖNER 1999].

Using the spring travel measured by sensors during the journey, the ABC computer constantly monitors the current vehicle mass and activates appropriate control algorithms that compensate major diving and squatting movements when the vehicle is heavily loaded.

If the vehicle mass changes, for example because luggage has been removed from the boot or a passenger has left the automotive vehicle, the mechatronic control system automatically performs a new online calculation and ABC adapts the suspension characteristics to the new load conditions. To provide for this precise load control every automotive vehicle is weighed when it leaves the assembly line at the Mercedes plant. The actual kerb mass established in this way is stored in the ABC unit and forms the basis for all subsequent calculations. Load adaptation significantly reduces the angle of vehicle-body roll when the vehicle is loaded and ensures that despite carrying an additional load, the vehicle attains the same handling performance values as with an empty boot. Remember, in the case of active suspension, today's innovation is tomorrow's standard feature. Expect to see active ABW AWA suspension mechatronic control systems in the coming years on much lower price point vehicles [MEAD 1999; MEMMER 1999]. Currently, the ABC is the nearest thing to an active F-M ABW AWA suspension on the market. Although faster acting than the other ones, ABC still operates only one way and is too slow to deal with individual bumps. For now, the progress of the active F-M ABW AWA suspension mechatronic control system appears to be stalled [CSERE 2004]. The active F-M ABW AWA suspension mechatronic control system helps to resolve the conflict between vehicle handling and ride comfort by replacing the rigid drop links of one stabilizer bar end against an F-M actuator (active stabilizer bar). These F-M actuators (Fig. 5.110) are controlled by an EFCU depending on the sensed driving circumstances. [TRW 2005].

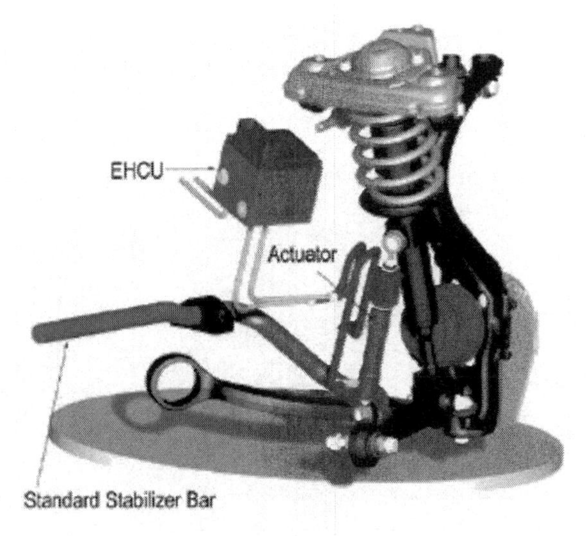

Fig. 5.110 F-M actuator controlled by electro-fluidical control unit (EFCU/EHCU) [TRW Automotive; TRW 2005].

As the effects undoubtedly verified, the active F-M ABW AWA suspension mechatronic control system, because of advanced roll control and bounce control, affords a ride and a level of mechatronic control far advanced to that of other suspension mechatronic control systems. The active F-M ABW AWA suspension mechatronic control system significantly improves ride and handling qualities. As an example, four F-M struts powered by an on-board E-M-F pump replace conventional shock absorbers (dampers). The computer mechatronic control system inside the automotive vehicle controls the movement of each strut in response to the bumps on the on/off road surfaces and vehicle-body motions.

Computer controlled M-F struts (Fig. 5.111) pulls wheels up over bumps and levels the chassis around corners [DRDC 2005].

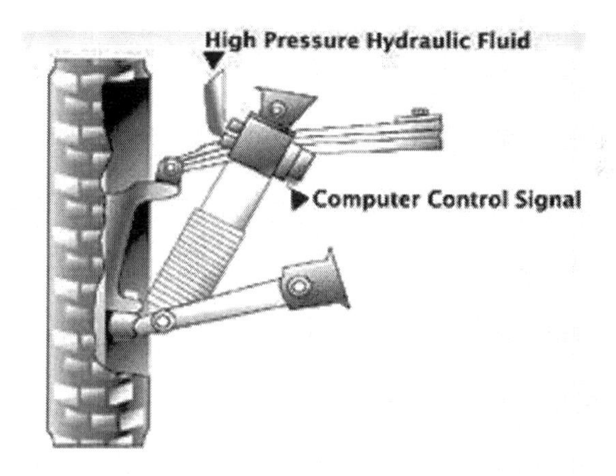

Fig. 5.111 A computer controlled F-M strut of the *Iltis* truck [DRDC 2005].

As the automotive vehicle encounters a bump, force sensors in the struts relay information to the computer mechatronic control system. The computer in turn moves F-M actuators to lift the wheel over the bump. It results in a smoother ride both on- or off-road surfaces.

Servo-controlled F-M actuators working in pairs control vehicle-body sway and improve stability during cornering. Vertical acceleration of the driver may be reduced approximately 10% over discrete bumps. Slalom vehicle velocity may have increased control, with reduced steering effort. Safety may also be increased because rollover is less likely [DRDC 2005].

The active F-M ABW AWA suspension mechatronic control system resolves the well-known conflict between active safety, responsive handling and ride comfort and is thus an imperative milestone in passenger automotive vehicle design.

Using high-pressure F-M servos, a smart sensor system and high-performance microprocessors, a mechatronic control system adapts the suspension and absorbing (damping) to different driving situations.

The computer-controlled F-M servos or *'plungers'* that are mounted in the spring struts between the coil springs and the vehicle body, develop additional absorbing (damping) forces that act on the suspension and damping to control vehicle-body motion.

The active F-M ABW AWA suspension mechatronic control system is designed to control vehicle-body vibrations in the frequency range up to 5 Hz - the kind of vibrations typically caused by uneven on/off road surfaces or by braking and cornering. To control the higher-frequency wheel vibrations, passive gas-pressure shock absorbers and coil springs are used that may be tuned for high ride quality.

The active F-M ABW AWA suspension mechatronic control system virtually eliminates vehicle-body movements when moving off from rest, when cornering and when braking.

Cornering roll on automotive vehicles equipped with the active F-M ABW AWA suspension mechatronic control system is significantly reduced and there are also safety advantages in high-speed evasive manoeuvres compared with automotive vehicles with conventional suspension systems. A press of a button on the centre console allows drivers to choose between comfortable or sporty suspension settings.

An active F-M ABW AWA suspension mechatronic control system may be detached into two unique systems: the F-M (hydraulic) system and the mechatronic control system, as exposed in Figure 5.112 [AKATSU, 1994].

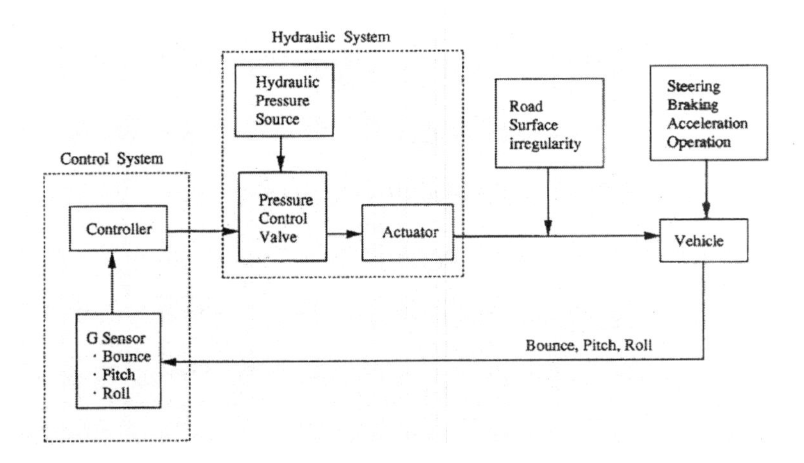

Fig. 5.112 F-M (hydraulic) and control systems for an F-M ABW AWA suspension mechatronic control system [AKATSU, 1994]

The oily fluid pressure of each of the F-M actuators positioned at each wheel is controlled in compliance with the output fluidical valves from the *G* sensors to eliminate alterations in vehicle-body position (bounce, pitch, roll) and trim down vibration from the on/off road surface.

A principal layout of the active F-M ABW AWA suspension mechatronic control system is illustrated in Figure 5.113 [AKATSU, 1994].

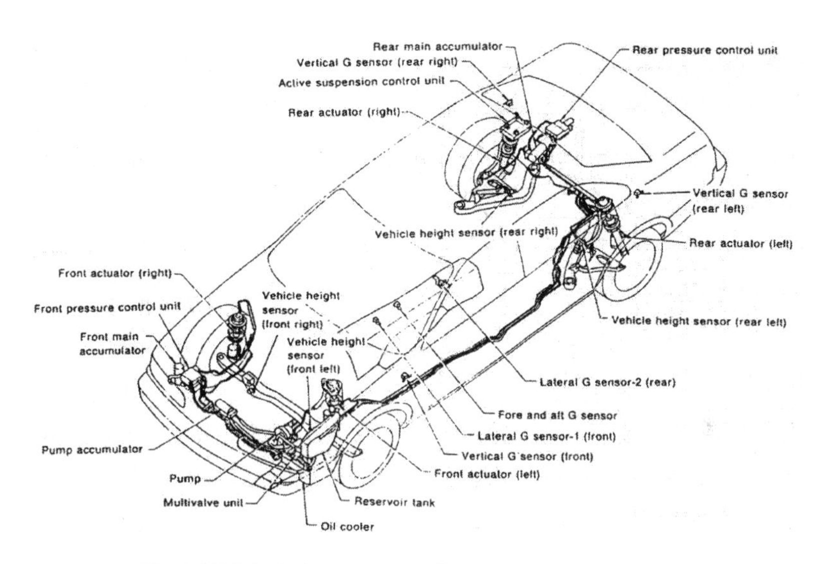

Fig. 5.113 Principal components of an active F-M ABW AWA suspension mechatronic control system [AKATSU, 1994]

As shown in Figure 5.114, the mechatronic control system contains the controller and every one of the sensors implicating the vertical *G* sensors, lateral *G* sensors, fore and aft *G* sensors, and vehicle-height sensors [DOMINY AND BULMAN 1985; AKATSU 1994].

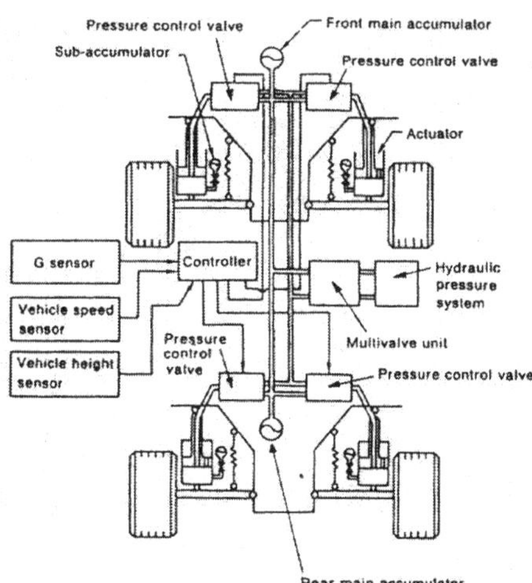

Fig. 5.114 Mechatronic controls for an active F-M ABW AWA suspension mechatronic control systems [DOMINY AND BULMAN 1985; AKATSU, 1994].

Active F-M ABW AWA Suspension Mechatronic Control System Components - The ***M-F pump*** has seven cylinders sited around the circumference and may ascertain, for example, a maximum oily fluid flow of 200 ml/s. The M-F pump is coupled as a pair with the power-steering vane M-F pump. The M-F pump supplies the necessary oily fluid for system function (power supply). To dampen pulsating oil pressure generated by the M-F pump, ***pump fluidical accumulators*** are bedded in the fluidical supply unit, plus one on the side of the M-F pump. To dampen the high-frequency pulsations, a metal-bellows-type fluidical accumulator is used. The pump fluidical accumulator removes the pulsating action from pressurised oil supplied by the M-F pump. As shown in Figure 5.115, the ***fluidical multi-valve unit*** is the fundamental control of oily fluid pressure for the overall system. The fluidical multi-valve unit controls the supply of pressurised oily fluid, fail-safe function and so on. The function of the fluidical multi-valve unit is shown in Table 5.7 [AKATSU 1994].

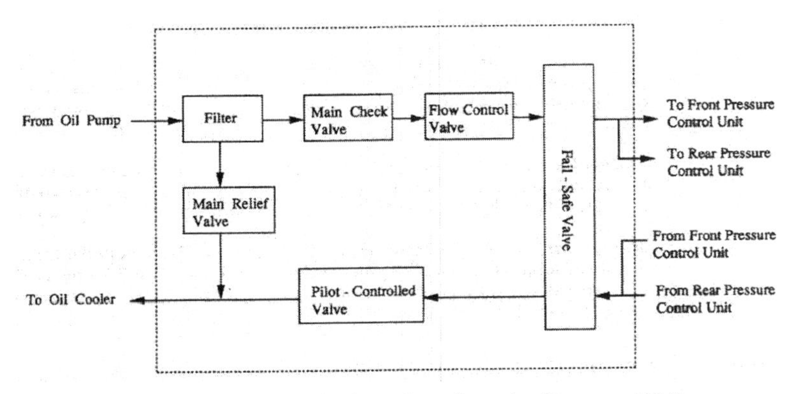

Fig. 5.115 Fluidical multi-valve unit configuration [AKATSU, 1994].

The ***main fluidical accumulator*** is located at both the front and rear axles. The main fluidical accumulator has two main functions: it accumulates oily fluid from the fluidical multi-valve unit and affords supplementary flow to the F-M actuators when indispensable and it conserves vehicle height when the ECE or ICE is immobilized. The main fluidical accumulator maintains oily fluid pressure, compensates when large amount of flow is required and preserves vehicle body height.

The ***oily-fluid pressure control unit*** controls the oily-fluid movement of the F-M actuator of each wheel in compliance with commands received from the control unit. The oily-fluid pressure control unit controls the fluidics for the actuators on each wheel according to signals received from the control unit.

Table 5.8 illustrates the fluidical valve's construction and operating principle [AKATSU, 1994].

The **electro-fluidical** (E-F) pressure control system contains an oil pressure F-M actuator and an oily-fluid pressure control fluidical valve. The F-M actuator is of the single acting type and is endowed with a damping fluidical valve and a fluidical accumulator underneath the F-M cylinder.

Table 5.7 Fluidical Multi-Valve Unit Functions [AKATSU 1994].

Function	Fluidical valve	Outline
Pressure supply management function	Main relief Fluidical valve	When the oily-fluid pressure exceeds a constant value, the main relief fluidical valve may return some of the oil flow. This keeps the supply oil pressure at a constant pressure.
Vehicle height maintenance function	Main check fluidical valve Pilot-controlled check fluidical valve	The main check fluidical valve is non-return fluidical valve that controls the flow from the line filter and directs it to the flow control fluidical valve. The pilot controlled check fluidical valve is a supply pressure-reaction-type open/closed fluidical valve. When the oil pressure exceeds a constant value, the fluidical valve opens and when the oil pressure falls below that level, it closes. In addition, it maintains the oily-fluid pressure at a constant level when the ECE or ICE is turned off.
Vehicle height control function	Flow control fluidical valve	When the ECE or ICE is turned off, the flow control fluidical valve closes the main passage and directs the flow through the bypass passage orifice, slowly increasing the oily-fluid pressure, after that the main passage is opened. This prevents any sudden changes in vehicle height when starting the ECE or ICE.
Failsafe function	Failsafe fluidical valve	When any irregularities occur to the mechatronic control system, it changes the oily-fluid passage, preventing any sudden changes in vehicle height.

Table 5.8 Pressure Control Fluidical Valve Functions [AKATSU 1994].

Active control function	The pressure in the control port (actuator) is controlled in response to the electrical current applied to the electromagnet solenoid, thus controlling the vehicle body attitude.
Passive damping function	When various pressure levels are caused in the interior of the actuator by on/off road surface forces, this pressure passes through the control port, causing feedback on the spool and the generation of appropriate absorbing (damping) force.

The oily-fluid pressure control fluidical valve is constructed with three ports and uses a pilot-type proportional electromagnetically controlled fluidical valve.
The oily-fluid pressure has two principal functions [AKATSU, 1994]:

- ❖ It controls the oily-fluid pressure of the F-M actuator in relation to the control input; this is realised by driving the solenoid so that it regulates the pilot fluidical valve, activating the spool to move;
- ❖ Feedback control is relevant to move the spool in response to oscillations in F-M actuator pressure generated by on/off road surface inputs, the performance of the spool functions to maintain the actuator pressure at a particular permanent level.

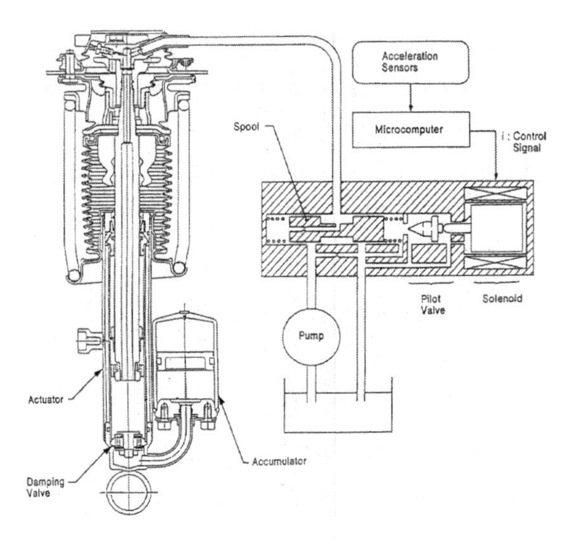

Fig. 5.116 Schematic diagram of E-M-F oily-fluid pressure control system [AKATSU, 1994].

As shown in Figure 5.116 the ***F-M actuator*** contains the power F-M cylinder, fluidical subaccumulator, damping fluidical valve, and so on [AKATSU, 1994].

Supplementary coil springs are also engaged to trim down the oily-fluid pressure indispensable for the whole system and to decrease the amount of power consumed.

The F-F sub-accumulator and damping fluidical valve at the base of the power F-M cylinder absorb and/or damp the high-frequency vibration from the on/off road surface.

The F-M actuator controls vehicle attitude and absorbs external forces from the on/off road surface.

As shown in Figure 5.117, the ***controller*** may be built employing two 16- or 32-bit semi-custom **microcontroller units** (MCU) that incorporates normal MCU elements plus application-specified peripheral devices such as high-power port outputs, special timer units, and so on. [AKATSU 1994].

MCU1 processes signals from the *g* sensors and then transmits attitude control signals to the oily fluid pressure fluidical valve solenoid drive circuit. MCU2 processes signals from the vehicle-height sensors and then transmits attitude control signals to the solenoid drive circuit.

MCU1 and MCU2 as a rule carry out mutual transmission, but should a malfunction emerge, the signal may be dispatched to the failsafe circuit, activating the failsafe fluidical valve to function and as a result ensure a high degree of safety.

The ***G sensors*** are ball position sensitive-type sensors. They sense alterations in the magnetic field resulting in the position alteration of a steel ball as the outcome of acceleration.

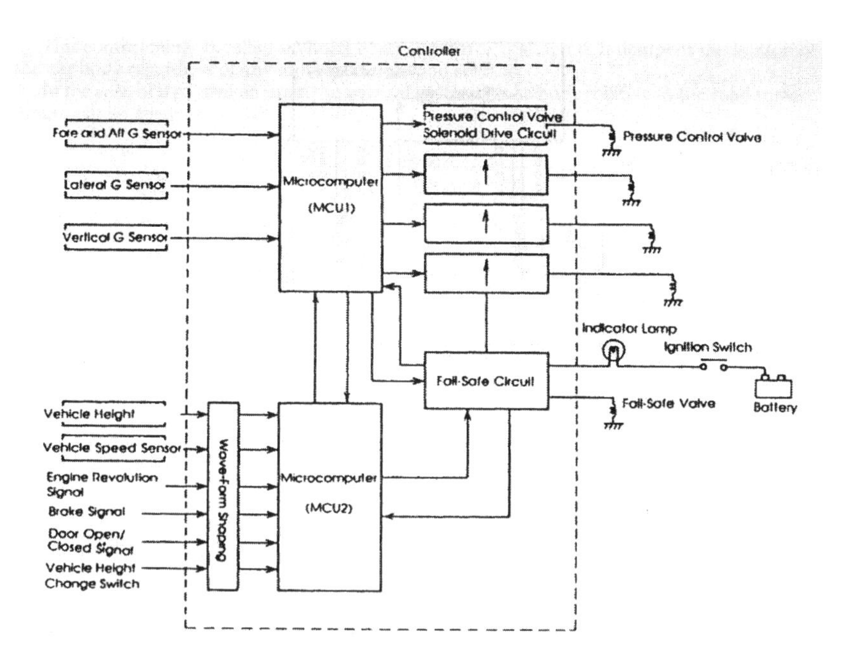

Fig. 5.117 Interior construction of the E-M-F oily fluid pressure controller [AKATSU, 1994].

Active F-M ABW AWA Suspension Mechatronic Control System Arrangement --The inertia force that initiates the automotive vehicle to roll is sensed by the lateral *G* sensor. ***Roll control*** is caused by increasing the control pressure on the wheels on the outside of the turn and by lessening the control pressure for the wheels on the inside of the turn. Figure 5.118 illustrates this functioning rule.

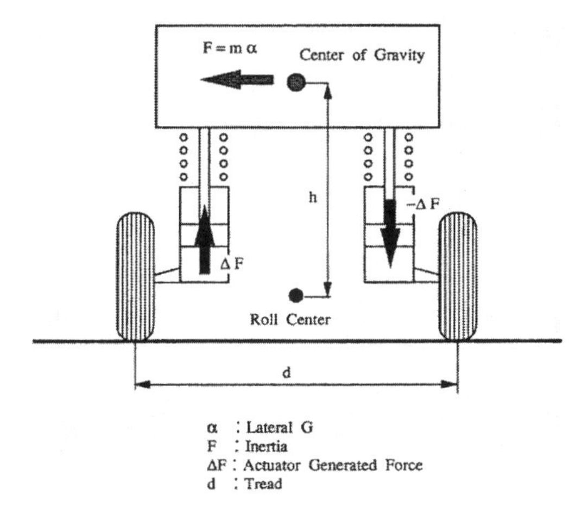

Fig. 5.118 Functioning rule for roll control [AKATSU 1994].

The particular relationship between the lateral *G* and the force caused by the F-M actuator can be given as [AKATSU 1994]:

$$F = m\,\alpha \tag{5.1}$$

$$F\,h = \Delta F\,d$$

$$\Delta F = \frac{F\,h}{d} = \frac{m\alpha h}{d}$$

where: F – inertia force [N];
m – vehicle body mass [kg];
α – lateral G [ms^{-2}];
ΔF – F-M actuator generated force [N];
d – tread [m].

During dispulsion (braking), inertia is created at the automotive vehicle's **centre of gravity** (CoG) and initiates pitching. The longitudinal G sensor senses this inertia and annuls it to weaken nose dive by raising the control pressure to the front and lessening control pressure to the rear, as shown in Figure 5.119 [AKATSU 1994].

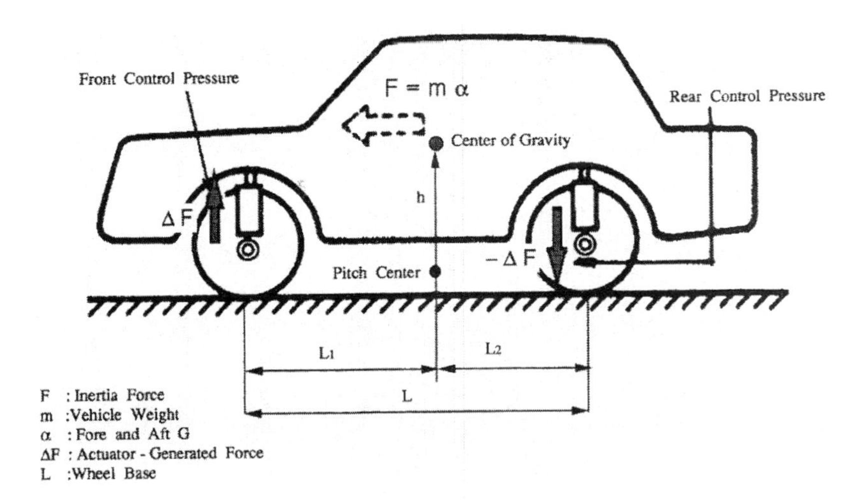

Fig. 5.119 Pitch control parameters [AKATSU 1994].

The relationship between longitudinal *G* and the F-M actuator initiated force is [AKATSU 1994]:

$$F\,h = \Delta F\,(l_f + l_r)$$

$$\Delta F = \frac{F\,h}{l_f + l_r} = \frac{m a h}{l} \tag{5.2}$$

where: F – inertia force [N];
m – vehicle body mass [kg];
h – vehicle height [m];
α – fore and aft G [ms^{-2}]'
ΔF – F-M actuator generated force [N];
l – wheel base [m].

The vertical G sensor partial to the vehicle body senses the value for vehicle-body acceleration.

By integration of the acceleration, the absolute value of the vehicle-body velocity is computed. The pressure control fluidical valve creates a force proportional to the absolute value of the vehicle-body velocity. This *'bounce control'* mode, termed skyhook damper control, is able to be adapted. It dampens the motion of the vehicle body irrespective of any inputs from the on/off road surface.

$$\frac{z_l}{z_u} = \frac{2\,j\,\omega_s\,\xi_a\,\omega + \omega_s^2}{2\,j\,\omega_s\,\xi_a\,\omega + \omega_s^2 - \omega^2} \tag{5.3}$$

where: $\omega_u = \sqrt{\dfrac{K_T}{m_u}}$ - natural frequency of unsprung mass [rad s^{-1}];

$\omega_s = \sqrt{\dfrac{K_S}{m_s}}$ - natural frequency of sprung mass [rad s^{-1}];

$\xi_a = \frac{1}{2} C_p \sqrt{\dfrac{m_s}{K_S}}$ - active damping ratio;

$\xi_p = \frac{1}{2} C_a \sqrt{\dfrac{m_s}{K_S}}$ - passive damping ratio;

C_p – passive damping coefficient;
C_a – active damping coefficient;
K_T – tyre stiffness;
K_S – spring stiffness;
m_u – unsprung mass [kg];
m_s – sprung mass [kg].

The vibration transmission ratio at the resonant point is [AKATSU 1994]:

$$\left|\frac{z_l}{z_u}\right|_{\omega=\omega_s} = \sqrt{1+\frac{1}{\xi_a^2}} \tag{5.4}$$

and always has a value greater than one.

When a contrast is made, the vibration characteristics of the skyhook shock absorber (damper) are prearranged as [AKATSU 1994]:

$$\frac{z_l}{z_u} = \frac{2\,j\,\omega_s\,\xi_a\,\omega + \omega_s^2}{2\,j\,\omega_s\left(\xi_a + \xi_p\right)\omega + \omega_s^2 - \omega^2} \tag{5.5}$$

The vibration transmission ratio at the resonant point is [Akatsu 1994]:

$$\left|\frac{z_l}{z_u}\right|_{\omega=\omega_s} = \frac{\sqrt{4\xi_a^2+1}}{2\left(\xi_a+\xi_p\right)} \tag{5.6}$$

hence,

$$\xi_p \geq \sqrt{\xi_a^2 + \tfrac{1}{4}} - \xi_a \tag{5.7}$$

and it is possible to reduce the ratio to less than one.

Table 5.9 Effect of Active Suspension Control [AKATSU 1994].

Control	Automotive vehicle related effects
Roll control	During transient control of wheel loading, as when changing lanes, the steering characteristics of the automotive vehicle can be optimally controlled. The wheel-tyres are used to their utmost performance ability because there is minimal roll, minimal change in the camber of the wheel-tyres to the ground, and because the wheel-tyres are continually kept in square contact with the road.
Pitch control	Nosedive and tail lift are minimised during braking. Squats are minimised during stars and rapid acceleration.
Bounce control	Vertical vibration of the automotive vehicle is reduced and continuity is improved. Vertical load fluctuations are minimal, and the contact of the wheel-tyres with the on/off road surface is greatly improved.

The effects of the F-M active ABW AWA suspension mechatronic control system are stated in Table 5.9 compatible with those related to the automotive vehicle [AKATSU 1994].

Active F-M ABW AWA Suspension Mechatronic Control System Effectiveness -
- Figure 5.120 shows the lateral *G* and angle of roll during cornering. Figure 5.121 shows the fore and aft *g* force, angle of nosedive, and squat angle during starts and stops.

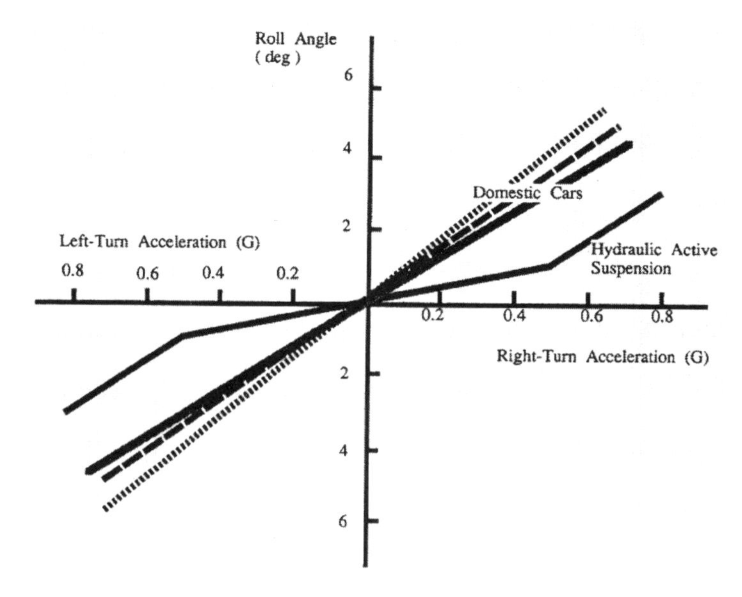

Fig. 5.120 Comparison of roll angles during cornering [AKATSU 1994].

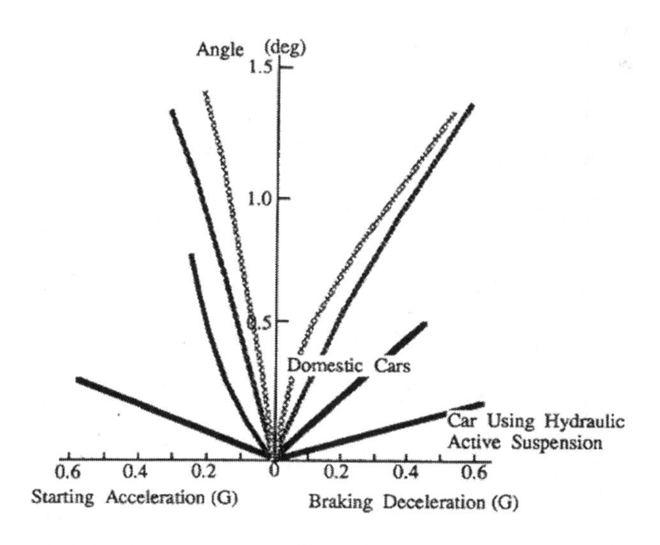

Fig. 5.121 Comparison of nosedive and squat angles [AKATSU 1994].

In either instance, the automotive vehicle with an active F-M ABW AWA suspension out-performed the other vehicles with conventional suspensions.

Figure 5.122 shows the effects of bounce control for the skyhook shock absorber (damper).

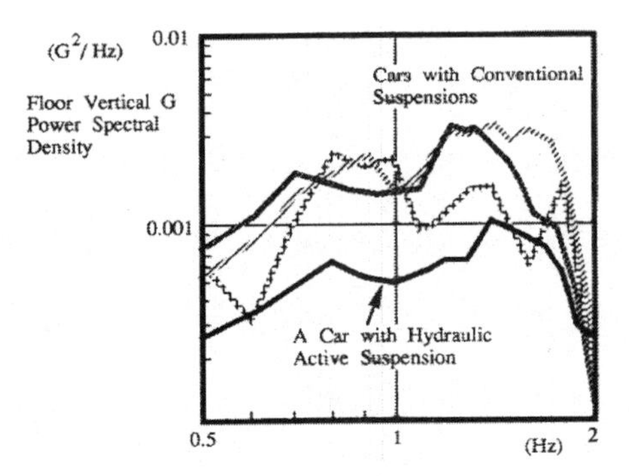

Fig. 5.122 Comparison of vertical vibrations [AKATSU 1994].

Relative to automotive vehicles using conventional suspension systems, the automotive vehicle with the active F-M ABW AWA suspension mechatronic control system, shows evidence of superior performance and low vibration levels.

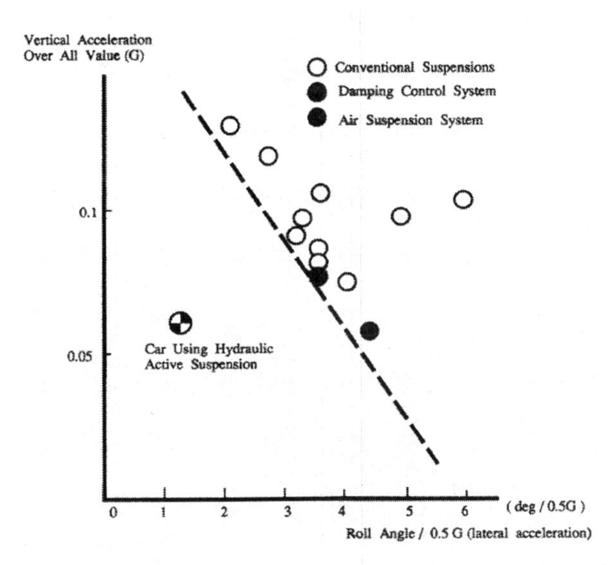

Fig. 5.123 Ride characteristics and roll rate for various automotive vehicles with different suspension systems [AKATSU 1994].

Figure 5.123 shows the ride characteristics and roll rate for different automotive vehicles with various ABW AWA suspension mechatronic control systems.

5.6.3 Active F-P-M ABW AWA Suspension Mechatronic Control Systems

An active **fluido-pneumo-mechanical** (F-P-M) ABW AWA suspension mechatronic control system is a combination of oily fluid and air that has been developed , in which the elastic medium is a sealed-in, fixed mass of air, and no M-P compressor is necessary.

The fluidical (hydraulical) part of each spring is an F-M cylinder set up on the vehicle-body sill and fitted with a plunger that is pivotally attached to the wheel linkage to form an F-M strut. Each spring cylinder has a spherical air chamber attached to its outer end. The sphere is separated into two chambers by an adaptable diaphragm; the upper engaged by air and the lower by oily fluid that is in interaction with the F-M cylinder through a two-way restrictor fluidical valve. This fluidical valve restricts the rate of movement of the plunger in the cylinder, since oily fluid must be pushed into the sphere when the vehicle's body moves down and moves back when it moves up. This damping action thus controls the motion of the wheel relating to the sprung mass of the automotive vehicle held up by the spring. Sometimes, they are also termed as active ABW AWA P-M suspension mechatronic control systems.

An active F-P-M ABW AWA suspension mechatronic control system, despite a reputation for complexity, complication and unreliability is actually very simple in operation and very reliable and furthermore, offers an unrivalled level of comfort coupled with excellent handling and grip [CITROËN 2001].

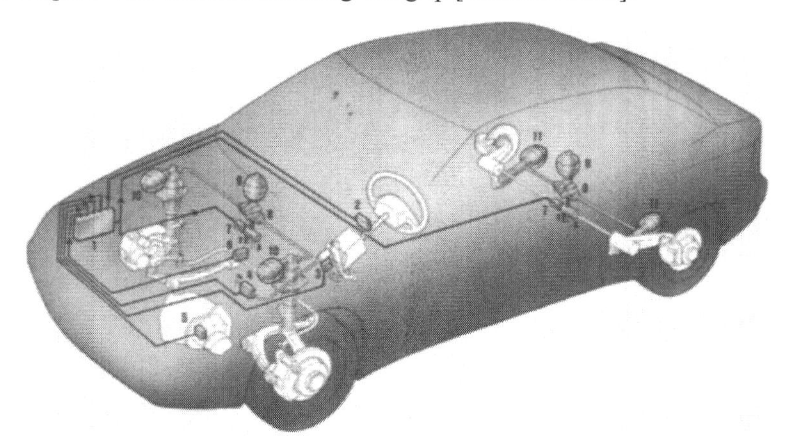

Fig. 5.124 Layout of the *'Hydractive 2'* F-P-M 4WA ABW suspension mechatronic control system: (1)- computer; (2) - steering wheel movement and speed of movement sensor, (3) accelerator sensor - reads accelerator movement and rate of movement;(4) - brake sensor; (5) – vehicle velocity (speed) sensor; (6) – vehicle's body movement sensor; (7) – electro-valve; (8) - stiffness regulator; (9) - extra sphere; (10) - front suspension sphere; (11) - rear suspension sphere [Citroën; CITROËN 2001].

The active F-P-M ABW AWA suspension mechatronic control system takes advantage of the fact that the driver can compress a gas but cannot compress an oily fluid. Thus gas acts as the springing medium while the oily fluid does all the clever things such as providing absorbing (damping) and self-levelling. The mechatronic control system switches from *'soft'* to *'firm'* modes according to a number of parameters programmed into the on-board computer. Sensors provide the necessary input to the computer to determine which mode is appropriate. A central sphere on each axle is switched in and out of circuit to alter the amount of suspension travel and damping.

Unlike the fluido-elastic system fitted to certain automotive vehicles, this mechatronic control system relies on an ECE- or ICE-driven M-F pump to pressurise the fluidical (hydraulic) system and it is this power source that enables self-levelling, variable ride height, assisted jacking and zero roll, and also allows for fully powered braking systems and power steering too. The high-pressure fluidical (hydraulical) system may also operate the clutch and gear change. For instance, the so called *'Hydractive'* F-P-M ABW AWA suspension mechatronic control system allows for variable damping and automatic switching between soft and firm modes that results in an unparalleled combination of ride comfort and good handling. The *'Hydractive 2'* F-P-M 4WA ABW suspension mechatronic control system (Fig. 5.124) is fitted to high specification variants of the Citroën *Xantia* and *XM* [CITROËN 2001].

For example, the Citroën *Xantia* front suspension (Fig. 5.125) comprises *MacPherson* type struts with the F-P-M suspension spheres mounted on top [CITROËN 2001].

Fig. 5.125 Front suspension [Citroën; CITROËN 2001].

Each sphere contains a diaphragm behind which a quantity of nitrogen is trapped. A height corrector is attached to the anti-roll bar and when a load is placed in the vehicle, the vehicle body sinks. The height corrector that opens a fluidical valve to admit oily fluid under pressure to *'lengthen'* the F-P-M strut and thereby re-establish the correct ride height registers this movement. A control inside the vehicle allows the ride height to be varied by the driver, thereby aiding wheel changes.

The rear suspension of the Citroën *Xantia* (Fig. 5.126) comprises trailing arms connected to an anti-roll bar with height corrector attached [CITROËN 2001].

Fig. 5.126 Rear suspension [Citroën; CITROËN 2001].

The rear spheres lie flat which permits an unobstructed boot floor with no suspension unit intrusions.

In Figure 5.127 is shown a schematic diagram of the active F-P-M 4WA ABW suspension mechatronic control system when its suspension is in *'soft'* state [CITROËN 2001].

When an active F-P-M 4WA ABW suspension is in *'soft'* state; the electromagnetic solenoid fluidical valve (1) is energised, the slide fluidical valve opens allowing oily fluid to flow between the suspension cylinders and the spheres (4) and (3) via the shock absorbers (dampers) (6). All six spheres are in use.

In Figure 5.128 is shown a schematic diagram of the active F-P-M 4WA ABW suspension mechatronic control system when its suspension is in *'firm'* state [CITROËN 2001].

When active F-P-M 4WA ABW suspension is in *'firm'* state; the electromagnetic solenoid fluidical valve (1) is not energised, the slide fluidical valve takes up a position that blocks the movement of oily fluid between the two main spheres (4) on each axle and isolates the additional spheres (3) from them both.

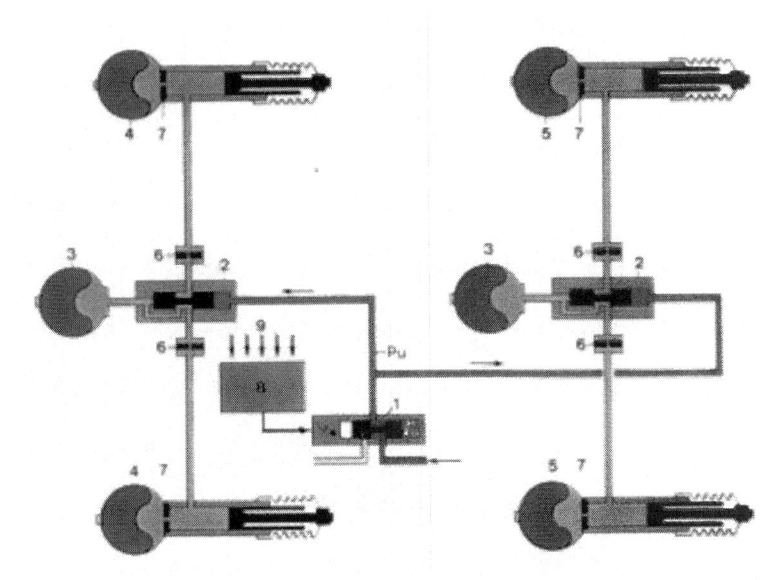

Fig. 5.127 Schematic diagram of the active F-P-M 4WA ABW suspension mechatronic control system when its suspension is in *'soft'* state: (1) electromagnetic fluidical valve; (2) stiffness regulator; (3) additional spheres; (4). front spheres; (5) rear spheres: (6) additional shock absorbers (dampers); (7) shock absorbers (dampers); (8) central computer; (9) sensors [Citroën; CITROËN 2001].

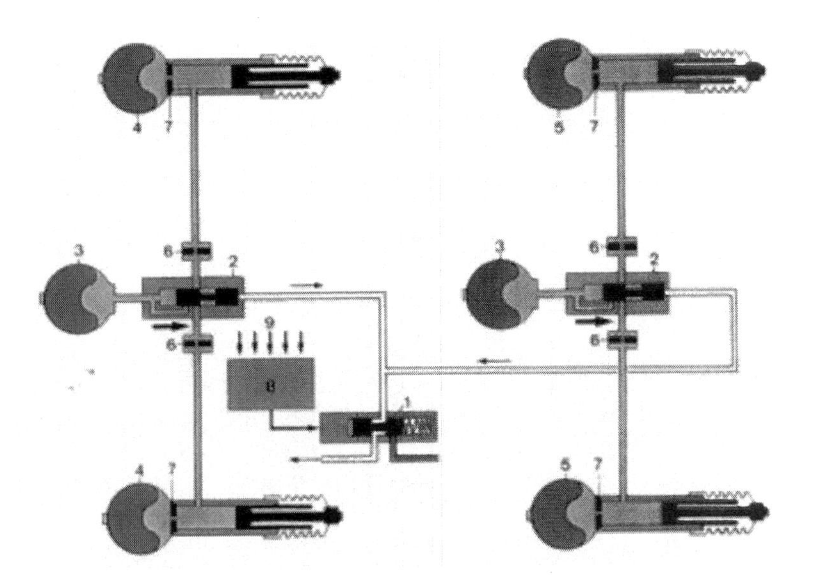

Fig. 5.128 Schematic diagram of the active F-P-M 4WA ABW suspension mechatronic control system when its suspension is in *'firm'* state: (1) electromagnetic fluidical valve; (2) stiffness regulator; (3) additional spheres; (4). front spheres; (5) rear spheres: (6) additional shock absorbers (dampers); (7) shock absorbers (dampers); (8) central computer; (9) sensors [Citroën; CITROËN 2001].

The active F-P-M ABW AWA suspension mechatronic control system offers an unparalleled level of comfort and control although it must be admitted that it is not perfect. No *'Hydractive'* or *'Activa'* automotive vehicles suffer from considerable vehicle-body roll (the basic system is poor in roll stiffness), the ride tends to be harsh at low values of the vehicle velocity, the fluidical (hydraulical) system is very good at transmitting road noise into the cabin, some people suffer from motion sickness in an F-P-M *Citroën*, there is a tendency for pitching if one is heavy footed when driving a *Citroën* and some drivers dislike the feeling of being divorced from the on/off road surface.

The system may be caught out by hump backed bridges - the vehicle body rises from the on/off road surface as the apex of the bridge is reached, the height corrector then decides the vehicle body is raised too high and reduces ride height as the vehicle descends the other side of the bridge -- the vehicle body then crashes into the suspension bump stops [CITROËN 2001].

On the plus side, the driver may drive much faster on indifferent on/off road surfaces than might be believed possible and handling is almost unaffected by a change in ride height caused by a full load.

Additional benefits include an absence of diagonal pitching, the constant ride height ensures headlamp beams are always properly aligned, irrespective of load and aerodynamics remain unaffected by load thanks to a constant angle of attack.

More routine but nevertheless valuable advantages include the ability to drive on terrain that is normally restricted to off-road vehicles, the ability to raise or lower the suspension to assist in loading or unloading the vehicle, to assist in changing a wheel, to facilitate towing and to make washing of the vehicle easier.

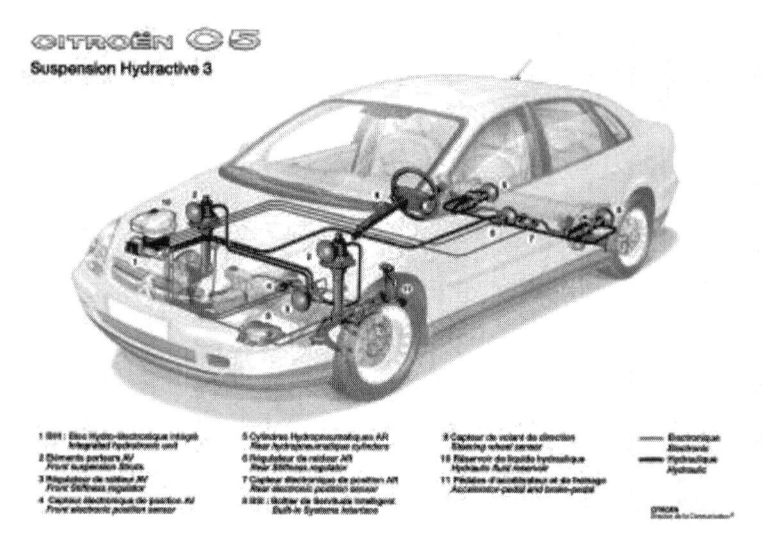

Fig. 5.129 Layout of the *'Hydractive 3'* F-P-M 4WA ABW suspension [Citroën; CITROËN 2001].

The Citroën *C5* (Fig. 5.129) features a decentralised system employing separate systems for power steering and braking while future developments may feature a E-M-F pump for each road wheel and a purely active F-P-M 4WA ABW suspension mechatronic control system (the Citroën *Activa* is, in truth, *reactive*, albeit its reactions are so quick as to almost deserve the appellation) that reads the on/off road surface ahead of the automotive vehicle and provides precisely the correct amount of suspension travel [CITROËN 2001].

Citroen's *'Hydractive 3'* F-P-M ABW AWA suspension mechatronic control system (Fig. 5.130) allows the height of an automotive vehicle to vary according to vehicle velocity (speed).

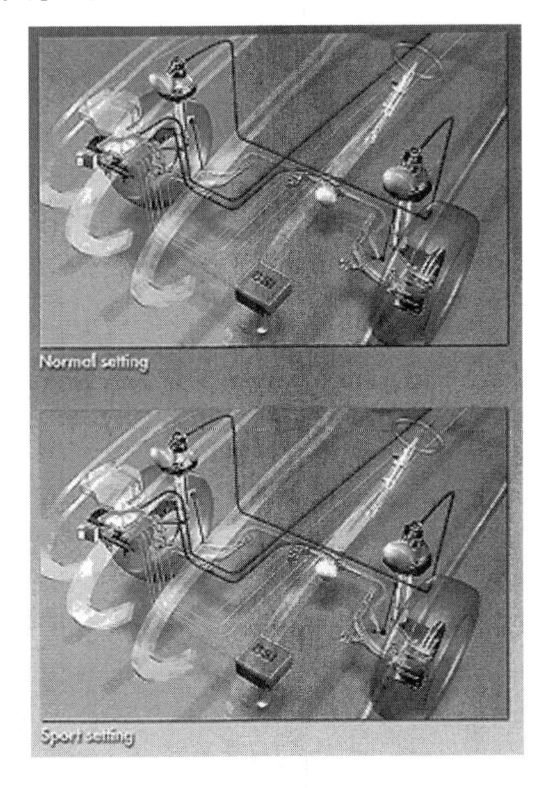

Fig. 5.130 Layout of the *'Hydractive 3'* active F-P-M ABW AWA suspension [Citroën; CITROËN 2001].

The CoG shifts by about 15 mm (0.6 inch) for improvements in drag. Over very poor surfaces ride height may increase by 20 mm (0.8 inch). The active F-P-M ABW AWA suspension system offers normal or sport settings.

In Figure 5.131 is shown a normally suspended automotive vehicle undertaking the slalom test in comparison with the Citroën *Xantia* Activa undertaking the same slalom test at the same vehicle velocity (speed) - note the total absence of vehicle-body roll [CITROËN 2001].

Fig. 5.131 Comparison of a conventionally passive suspended automotive vehicle with the active F-P-M 4WA ABW suspended automotive vehicle [Citroën; CITROËN 2001].

One of the most significant innovations in the passenger automotive vehicle BMW *E65 7* Series is an active P-F-M suspension - and at the same time it was one of the greatest challenges in terms of both its technology and its acoustic integration within the vehicle.

An active P-F-M suspension makes utmost demands of the automotive vehicle's **fluidical-and-electronic control units** (FECU). Oily fluids are pumped up to a pressure of up to 20 MPa (200 bar) within fractions of a second, flow through tiny pipes to rotary F-M actuators with utmost precision down to a few millilitres, expand and give up their pressure, cool down to operating temperature in a multi-stage cooling process, and flow back to their fluidical reservoir.

Fluidical valves open instantaneously, excess flow ducts are opened and rotary F-M motors generate enormous power for a brief moment in order to keep the vehicle body in a stable, horizontal position, particularly when the automotive vehicle is meandering through a succession of *S*-bends. But whenever incompressible media are in rapid motion the enormous pressure peaks generated in this way not only cause a substantial load for the components directly involved, but also lead to the generation and transmission of solid-borne sound within the entire automotive vehicle. When driving on a winding road this whole process is constantly repeated in hundreds of cycles [BMW 2001].

Through its driving characteristics and handling, the passenger automotive vehicles BMW *E65 7* Series clearly confirm BMW's leadership in the design, construction, set-up and manufacture of the suspension and chassis. The foundation for this superiority is provided by the all-aluminium suspension with unsprung masses consistently reduced to a minimum for excellent driving comfort on less fuel than ever before. Aluminium wheels and extra-large aluminium brake callipers round off this light-mass construction concept also offering optimum safety on the road [BMW 2001].

An active P-F-M ABW AWA suspension mechatronic control system filters out bumps and suppresses virtually all vehicle-body roll on cornering through the presence of two active anti-roll bars physically integrated into the front and rear axles. This further enhances agility and stability. Up to 80% of an automotive vehicle's normal vehicle-body roll in sharp bends is removed by two separated mechatronic control systems, which also has active safety implications since directional stability is enhanced when taking emergency evasive actions.

Active P-F-M ABW AWA suspension and two separated active mechatronic control systems serve to further enhance ride comfort and safety:

A **dynamic drive** (DD) active P-M ABW AWA suspension (Fig. 5.132) compensates vehicle-body sway in bends, an **electronic damper control** (EDC) adjusts the absorbing (damping) effect permanently and infinitely to on/off road surface conditions and the automotive vehicle's dynamic driving behaviour [BMW 2001].

Fig. 5.132 Dynamic drive (DD) active P-M ABW AWA suspension [BMW AG; BMW 2001].

The rear axle is a further development of BMW's proven integral design concept offering unique benefits in terms of track control and motoring comfort that are the main functions of the suspension. The wheel mounts to the rear axle subframe by means of four-track control arms that transmit forces flowing into the wheel suspension at the wheel/road contact point.

The front swinging mount on the rear axle subframe provides the longitudinal wheel guidance and suspension so important to enhanced roll comfort. The mounts on the rear axle subframe provide additional, longitudinal suspension of the entire rear axle system, making a significant contribution to the automotive vehicle's roll and noise comfort.

The active P-F-M ABW AWA suspension on the rear axle offers two essential advantages: The level and ground clearance of the automotive vehicle remains virtually unchanged regardless of the load it is carrying, and at the same time suspension comfort is absolutely excellent at all times. The simple and straightforward principle applied for this purpose is that the mechatronic control system automatically increases gas pressure in the spring bellows with an increasing load on the rear axle. This keeps the suspension virtually unchanged in the ratio between load and spring hardness, such a consistent balance helping to improve ride comfort most significantly.

On the active P-F-M ABW AWA suspension, pneumatical spring struts replace the conventional elements on the two rear wheels, representing a combination of a twin-sleeve gas pressure shock absorber (damper) and a pneumatical spring.

An E-P-M compressor driven by an E-M motor supplies the mechatronic control system with compressed air. To recognise the level of the vehicle body, two sensors measure the angle of the two rear axle swinging arms, the level then being adjusted individually for each wheel whenever it deviates from its target value. Through its inherent logic, the FECUs even considers the driving condition of the vehicle in this process, for example avoiding any control or adjustment function when the vehicle is in a bend or when a wheel is being changed.

In all, this concept offers a number of decisive advantages [BMW 2001]:

- The mechatronic control function operates independently of the vehicle's ECE or ICE;
- The mechatronic control function operates individually for each wheel;
- The mechatronic control system is able to compensate any uneven load in the vehicle.

A DD active P-F-M ABW AWA suspension mechatronic control system for the first time combining the superior ride comfort of a passenger vehicle BMW *E65 7* Series with the equally superior, dynamic performance of the passenger vehicle BMW *Z8*: While both driver and passengers may hardly notice any bumps on a straight stretch of road, the mechatronic control system suppresses vehicle-body sway in bends, thus ensuring supreme agility and stability under all driving conditions. The improvement in driving safety ensured in this way, the nimble response of the vehicle and the accurate steering even in extreme manoeuvres set new standards in suspension technology.

Passengers travelling at the rear also experience this unprecedented smoothness and tranquillity referred to by the expert as stabilisation of the vehicle in vertical-dynamic terms as a substantial improvement enhancing ride comfort to a level never seen before. Clearly, this makes the process of reading or working on the rear seats much easier and more pleasant. A DD P-F-M ABW AWA suspension mechatronic control system largely eliminates the contradiction in terms between handling, on the one hand, and smoothness and comfort, on the other, since with the help of DD the pneumatical springs and F-M shock absorbers (dampers) may be tuned mainly for enhanced ride comfort under all on/off road surface conditions. DD consists of two active anti-roll bars, a fluidical valve unit with integral sensors, a tandem M-F pump, a lateral acceleration sensor, a FECU and other supply components. The core elements are the two active anti-roll bars integrated in the front and rear axles instead of conventional mechanical anti-roll bars. An active anti-roll bar or actuator consists of a fluidically operated sway F-M motor with its sway F-M motor shaft and housing being connected in each case to one half of the anti-roll bar. The active anti-roll bar converts oily fluid pressure into torsional momentum and, via its connection, into the stability force required to keep the automotive vehicle stable.

Pressure generated by an M-F pump is regulated by two electronically controlled pressure fluidical valves to ensure that [BMW 2001]:

- ❖ Any sway movement of the body in bends is minimised or eliminated completely;
- ❖ The automotive vehicle maintains a high standard of agility and directional stability throughout its entire speed range;
- ❖ The automotive vehicle own steering behaviour is optimised, as is the automotive vehicle's response to any change in load.

When travelling straight ahead or under very low lateral acceleration, in turn, the actuators are not subjected to any pressure, thus ensuring that [BMW 2001]:

- ❖ The torsional spring rate of the anti-roll bars cannot cause any undue hardness in the basic suspension set-up;
- ❖ Yaw motion of the vehicle body is reduced to a minimum.

The main signal used by the mechatronic control system in order to perform its control function is lateral acceleration measured by a lateral acceleration sensor. Further signals on the automotive vehicle's lateral and longitudinal dynamics are obtained from the CAN bus for evaluation, providing even better and more *'robust'* data on the lateral dynamic behaviour of the new passenger vehicle BMW *E65 7* Series. The lateral dynamic signal calculated in this way serves to determine the exact fluidical valve flow functions in the fluidical valve unit. Through its configuration DD active P-M ABW AWA suspension substantially reduces the sway angle under driving conditions relevant to the customer: Under lateral acceleration from 0 -- 0.3 G sway angles are relatively small and are eliminated 100%. Up to 0.6 G DD active P-M ABW AWA suspension generates quasi-stationary sway behaviour well known from vehicles with passive suspension up to a maximum of 0.1 G.

The reduction of sway under 0.6 G lateral acceleration, in turn, is more than 80 %. A further advantage is that DD active P-M ABW AWA suspension reduces the steering angle required as compared with a vehicle running on a conventional passive suspension. The response and intervention of DD active P-M ABW AWA suspension under high lateral acceleration above 0.6 G is intended to warn the driver that she/he is approaching the extreme limit. Accordingly, the sway angle gradient increases consistently and quite perceptibly as of this crucial point. This indication that driver is reaching the extreme limit -- that does not, however, make the driver feel insecure or alarmed in any way -- reminds the driver that even DD active P-M ABW AWA suspension mechatronic control system cannot override the laws of physics [BMW 2001].

In all, the new passenger automotive vehicles BMW *E65 7* Series with DD active P-M ABW AWA suspension under low lateral acceleration offers neutral driving behaviour. Then, with lateral acceleration increasing, the vehicle may build up slightly growing understeer.

While the previous system automatically switched from one to the other of three shock absorber (damper) control curves while driving, infinitely variable EDC offers any random number of shock absorber (damper) curves to provide an optimum balance of ride comfort and safety on the road. This infinite variability is ensured by a continuously adjustable pneumatical valve inside the shock absorber (damper) itself.

Setting the **controller and control display** (CCD), the driver is able to choose in the menu between a more comfortable and a more sporting, dynamic damper setting. In the sports mode, EDC responds to bumps and vibrations on the on/off road surface by exerting higher damper forces than in the ride comfort mode.

Reducing vehicle-body sway in the process, the system gives the vehicle much firmer road holding, the entire automotive vehicle literally *'sticking'* to the road beneath. In bends, when braking and accelerating, driving safety is enhanced in both modes by a higher level of damper forces, improving the vehicle's sway and yaw behaviour in the process. A further advantage of EDC is the maintenance of consistently good vibration behaviour regardless of the load the vehicle is carrying and regardless of the vehicle's service life thanks to automatic re-adjustment of the shock absorbers (dampers).

Four gas-pressure twin-sleeve shock absorbers (dampers) with continuously adjustable pneumatical valves form the core elements of EDC. The control pneumatical valve is integrated in each case within the damper piston, electrical energy being supplied through the hollow piston rod. The four shock absorbers (dampers) are masterminded by a single-chip microprocessor receiving data from three vertical acceleration sensors near the front axle and, respectively, rear axle pneumatical spring struts. Signals from the wheel angular-velocity sensors and the CAN system are also used to calculate current driving and vibration conditions. This data are used to determine the pneumatical valve control current required in order providing an appropriate damping effect. Switching over from the conventional increment pneumatical valves used on EDC so far to a continuously variable pneumatical valve configuration with EDC called for the development of an all-innovative, complex control strategy.

To distinguish between low frequency excitation of the vehicle's body and high-frequency bumps on the on/off road surface, the appropriate signal components are filtered out of the input signals and provided as virtual sensor signals for calculating the pneumatical valve control current.

Wherever high- and low-frequency signals are generated at the same time, for example on bad country roads, the control strategy chosen by the system enhances ride comfort most significantly [BMW 2001].

Originally conceived through an internally funded research project, the **Southwest Research Institute** (SwRI) regenerative active F-P-M ABW AWA suspension mechatronic control system offers both increased ride quality and better handling.

Conventional passive suspension mechatronic control systems cannot offer both because ride quality is associated with spring stiffness - the softer the spring, the better the ride - and soft springs contribute to more vehicle-body roll and pitch, thus sacrificing good handling [SwRI 1997].

The SwRI-developed regenerative active F-P-M ABW AWA suspension mechatronic control system counteracts compression and extension of vehicle F-P-M struts during turns by recovering fluidical (hydraulic) energy from the inside struts and transferring fluidical energy to the outside struts.

This energy transfer occurs through a combination of computer-controlled M-F/F-M pumps/motors, one per strut on the automotive vehicle, all ECE or ICE driven and mechanically coupled to each other.

By recovering most of the actuation energy, a design engineer may improve ride quality by softening the suspension spring and damping rates while maintaining control of the vehicle chassis during turns, braking events, and operation over undulating on/off road surfaces.

Currently, no other active ABW AWA suspension mechatronic control system permits actuation energy to be recovered during operation.

For large automotive vehicles such as buses, conventional active F-P-M AWA suspension mechatronic control systems may require a total of 22.4 kW (30 hp) for activation.

Tests show the SwRI regenerative active F-P-M suspension mechatronic control system requires only 5.2 kW (7 hp) to maintain chassis control while recovering more than 75% of actuation energy during demanding manoeuvres [SwRI 1997].

The improvement in ride that results from reduced spring and damping accompanied by improved control makes this mechatronic control system ideally suited for large passenger automotive vehicles such as buses and recreational vehicles.

This system has been installed and demonstrated on a 13.6 Mg (30,000 lb) bus (Fig. 5.133) with exceptional results [SwRI 1997].

Ride was improved by reducing the spring rate by 50%, and chassis motions during steering and braking events were reduced by 50% when compared to conventional bus passive suspension mechatronic control systems.

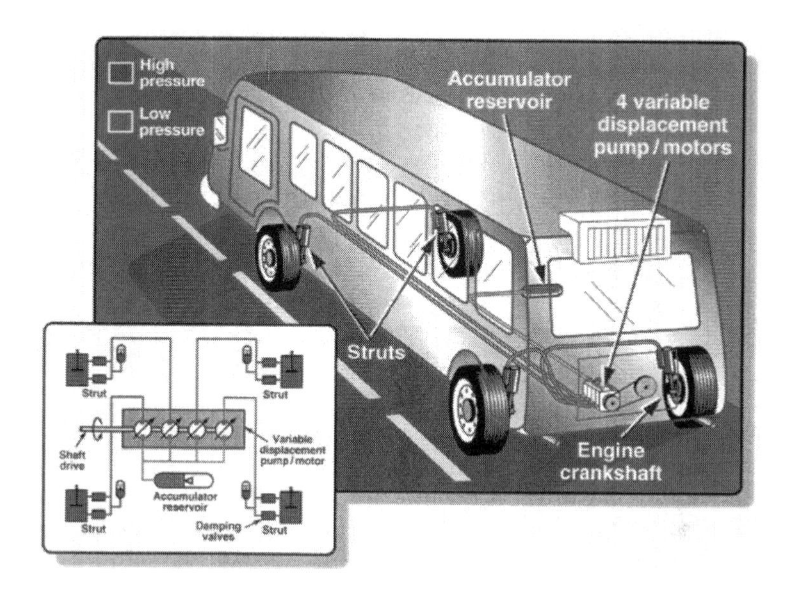

Fig. 5.133 One application of the recuperative active F-P-M ABW AWA suspension for a passenger bus [SwRI®; SwRI 1997].

Other applications for the regenerative active F-P-M ABW AWA suspension mechatronic control system include emergency automotive vehicles such as ambulances, where a smooth ride may be critical to the care of the injured, and military vehicles that must travel at high speeds over rough terrain [SwRI 1997].

5.6.4 Active P-M ABW AWA Suspension Mechatronic Control Systems

An active P-M ABW AWA suspension mechatronic control system replaces a passive shock absorber by generating forces opposing uneven on/off road surfaces in the roadway or terrain. The active P-M ABW AWA suspension mechatronic control system can provide a smoother ride allowing for better vehicle control, and less automotive vehicle abuse.

For instance, the active P-M ABW AWA suspension mechatronic control system may consist of springs, a P-M actuator, electromagnetic solenoid pneumatical valves, vehicle-body position and vehicle velocity sensors. The on/off road disturbance input may be simulated using a DC-AC macrocommutator IPM magneto-electrically excited motor. Signals from the position and velocity sensors may be sent to a microprocessor that controls the P-M actuator. Some of the control algorithms that may be implemented include proportional and linear **optimal quadratic control** (OQC) law. Presently, a proportional control law attenuates on/off road disturbance input effect by 40% [Green and Bush 2003].

Adaptive active P-M ABW AWA suspension (sport) that is offered as an option for the automotive vehicles, for example, Audi *A8* (Fig. 5.134), is the sport's suspension to be based on the principle of P-M suspension [AUDI 2005B].

Fig. 5.134 Adaptive active P-M ABW AWA suspension schematic
Audi AG; AUDI 2005B].

The P-M shock absorber (damper) and suspension characteristics are noticeably firmer than in the dynamic mode of the vehicle's standard adaptive P-M suspension.

For instance, in the automatic and comfort modes, the vehicle ride height is always 2.0 cm lower than is the case with the standard suspension [AUDI 2005A].

Like the standard adaptive P-M suspension, the adaptive active P-M ABW AWA suspension - sport has four driving modes. The comfort mode offers a level of ride comfort that cannot be achieved with conventional sport's suspension.

Adaptive active P-M ABW AWA suspension is a mechatronically controlled P-M suspension system (Figure 5.135) at all four wheels with a continuously **adaptive damping system** (ADS) that unites sporty handling and a high level of ride comfort [AUDI 2005A; ANNIS 2006].

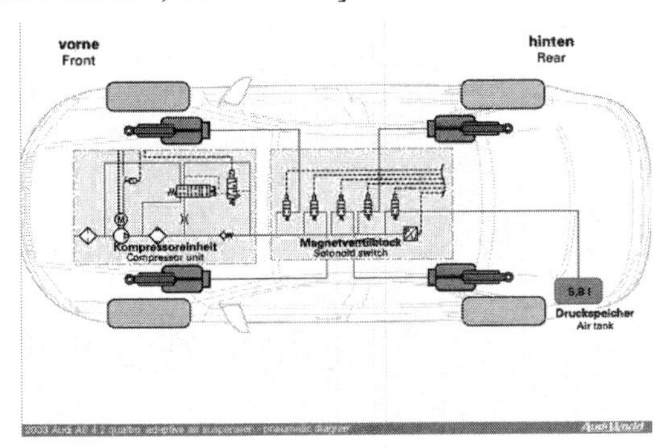

Fig. 5.135 Variable P-M shock absorber (damper) – Left image; pneumatical circuit diagram of the active P-M ABW AWA suspension mechatronic control system – Right image [Audi AG; AUDI 2002, 2005A; ANNIS 2006].

In addition, the active ABW AWA suspension allows the speed-dependent lowering of the vehicle body - this change in ride height means a low CoG and significantly increased directional stability as a result. The vehicle's aerodynamics is improved at the same time.

The vehicle has P-M suspension struts on all four wheels. The data from sensors on the axles and acceleration sensors on the vehicle body is evaluated in the adaptive active ABW AWA suspension's central control unit. This computer controls the adjustment of the individual P-M shock absorbers (dampers) in milliseconds depending on driving situation.

Provided no higher absorbing (damping) forces are required – for instance when driving straight ahead on good on/off road surfaces -- the shock absorber (damper) settings remain comfortably soft.

Controlled changes to the absorbing (damping) force at individual wheels help to eliminate vehicle-body movements at any time that could reduce occupant comfort.

The ADS automatically reduces rolling or pitching movements when cornering, braking or driving off. Adaptive active P-M ABW AWA suspension moreover offers the advantages of a traditional self-levelling suspension.

The vehicle's suspension height remains constant irrespective of the load it is carrying.

An exemplary active ABW AWA suspension mechatronic control system known as standard-specification *AIRMATIC* suspension combines P-M suspension with an ADS that adjusts the front and rear P-M shock absorber forces to the current payload, the condition of the on/off road surface and driving style [CARLIST 2005C].

A steering angle sensor, three acceleration sensors on the vehicle body, the ABS velocity sensor and the brake pedal sensor constantly measure the lateral and longitudinal acceleration of the vehicle body.

From this data, the ADS ECU calculates the optimal P-M shock absorber (damper) setting for each individual wheel and transmits the relevant signals with split-second velocity to special pneumatical valves on the gas-pressure P-M shock absorbers. These pneumatical valves are able to switch between four different damping characteristics. Using a selector on the instrument panel, the driver may also switch between a standard mode and a tauter, sportier mode. This adjusts the thresholds at which the different absorbing (damping) characteristics may be activated. In sporty mode, the firmest characteristic may be selected earlier, while in comfort mode the softest characteristic remains activated for longer.

The adaptive active P-M ABW AWA suspension - a four-corner active P-M ABW AWA suspension mechatronic control system with continuous damping control (Fig. 5.136) -- is the P-M suspension concept to enter the luxury class vehicles that resolves the classic conflict of aims between good handling and a level of ride comfort befitting a vehicle of this class.

At the same time the lowering of the suspension at high values of the vehicle velocity enhances the aerodynamic characteristics of the vehicle and therefore also improves its economy [AUDI 2002; ANNIS 2006].

Fig. 5.136 Adaptive active P-M ABW AWA suspension design – Left image; conventional passive suspension design – Right image [ANNIS 2006].

Adaptive Active P-M ABW AWA Suspension with Continuously Variable Damping - resolves the classic conflict of objectives between the handling qualities of a SUV and the suspension comfort of a luxury class automotive vehicle. In addition, the active P-M ABW AWA suspension allows the vehicle velocity-dependent lowering of the vehicle body - this change in ride height means a low CoG and significantly increased directional stability as a result. At motorway values of the vehicle velocity the lowering of the vehicle body also optimises the automotive vehicle's aerodynamic properties and thus reduces **specific fuel consumption** (SFC). At low values of the vehicle velocity the adaptive active P-M ABW AWA suspension establishes greater ground clearance if the driver so wishes. This is an additional bonus on uneven terrain. As an example, for innovative Audi *A8 Quattro,* four different defined ground clearance stages are therefore available [AUDI 2005B]:

- The raised level with a ground clearance of 14.5 cm;
- The standard level with a ground clearance of 12.0 cm;
- The sport level with a ground clearance of 10.0 cm;
- The motorway level with a ground clearance of 9.5 cm.

The Components and Their Functions - This is what the adaptive active P-M ABW AWA suspension, that entirely replaces the conventional steel suspension of the previous automotive vehicle, consists of [AUDI 2002]:

- At all four wheels, the innovative automotive vehicle has P-M-suspension struts on which there are P-M suspension bellows arranged concentrically around the continuously variable twin-tube shock absorbers; the P-M suspension bellows consist of a special, multi-layered elastomer material with polyamide cord inserts to increase strength; this layer absorbs the forces produced in the P-M spring;
- The advantage of this design is that the combination of individual layers enables the innovative vehicle to achieve its excellent running characteristics and the sensitive response of the suspension in the event of minor impacts;

- The P-M suspension struts are pressurised by an E-M-P compressor in the ECE or ICE compartment and e.g. a circa 6 l pressure reservoir located at the rear of the vehicle with a maximum pressure of 1.6 MPa (16 bar); plastic air lines connect the individual components; electromagnetic solenoid pneumatical valves regulate the air flows;
- The data from four sensors on the axles and three acceleration sensors on the body is evaluated in the adaptive active P-M ABW AWA suspension's central ECU; this computer prompts the adjustment of the individual shock absorbers in a matter of milliseconds, based on the driving situation identified, thereby always assuring optimum driving dynamics and ride comfort;
- Adaptive active P-M ABW AWA suspension moreover offers the advantages of a conventional self-levelling suspension mechatronic control system; the automotive vehicle's suspension height remains constant irrespective of the load it is carrying; this ensures that the vehicle is consistently presented to its best advantage.

The active P-M ABW AWA suspension also has a system-specific quality: the occupants additionally benefit from a constantly agreeable vibration behaviour, again irrespective of the automotive vehicle's load; this is because extra air is compressed into the P-M springs at high gross masses, and discharged again when the load is removed; the firmness of the springs thus adapts to the load situation at all times.

Vehicle-Body Movements Effectively Eliminated - The active P-M ABW AWA suspension in conjunction with the mechatronically controlled, continuously ADS provides perfect driving comfort because the system responds to numerous different sensor signals: the acceleration of the vehicle body, the vertical movements of the four wheels and many other parameters too, such as the current steering angle and the position of the accelerator and brake pedals, are used to determine the optimum damping force for every driving situation [AUDI 2002; 2005B].

Provided no higher damping forces are required - for instance when driving straight ahead on good roads - the damper settings remain comfortably soft. Specific adjustments to the damping force at individual wheels permanently eliminate body movements that could be detrimental to occupant comfort. This configuration is beneficial both to occupant comfort and dynamic handling properties. When cornering, braking or moving away, the innovative vehicle's adaptive damping automatically reduces rolling or pitching movements.

Dynamic Response To Order - The adaptive active P-M ABW AWA suspension also allows the driver to influence the suspension characteristic -- and thus the operating dynamics -- as individually preferred. The damping characteristics and ride height can be adjusted in just a single process via the **"CAR"** menu of the **multi-media interface** (MMI) terminal (Fig. 5.137) [AUDI 2005C].

The MMI is an integral operating concept for the logically simple and intuitive operation of fitted automotive vehicle and infotainment components. The MMI consists essentially of two elements: the MMI terminal in the centre console and the MMI display, a monochrome or colour monitor in the dashboard.

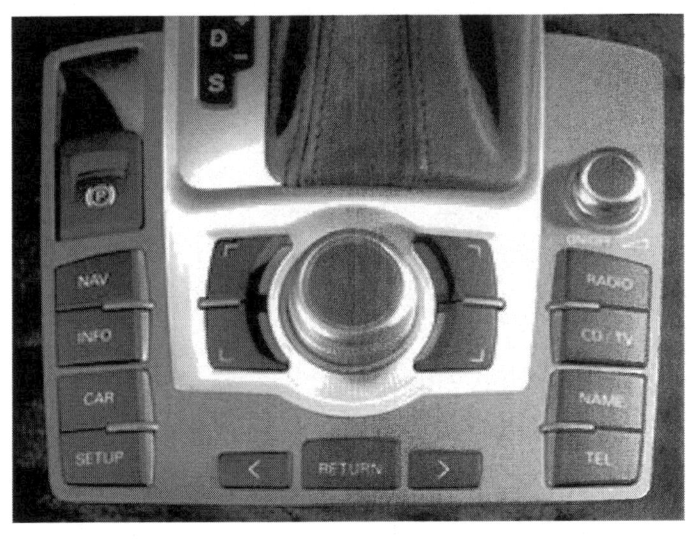

Fig. 5.137 Multi-media interface (MMI) CAR menu system [Audi AG; AUDI 2005C].

The central element of the MMI terminal is a control button that can be turned and pressed, with four control keys grouped around it. On either side of the terminal there are a total of eight function keys with which the main menus can be called up directly.

The main functions of the system - **entertainment, communication, information and control** (ECI&C) of automotive vehicle systems -- are accessed using the eight permanently assigned function keys. Within the individual menus, the driver activates the required functions by turning/pressing the control button (Fig. 5.138) [AUDI 2005C].

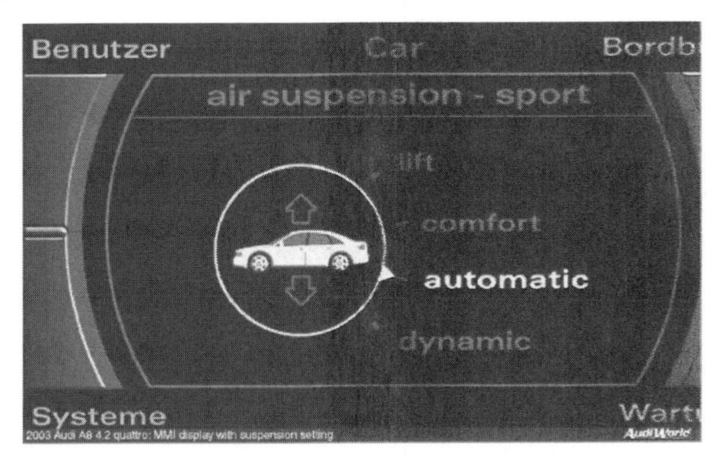

Fig. 5.138 Multi-media interface (MMI) display with suspension setting [Audi AG; AUDI 2005C].

It is intentional that not all settings may be altered in the MMI. For instance, the driver can choose between various driving modes, namely [AUDI 2002]:

- In the balanced automatic mode, the vehicle body is at the standard height when stationary and at medium value of the vehicle velocity, with 12.0 cm of ground clearance; if the innovative vehicle is driven at more than 120 km/h for over 30 s, the body is lowered by 2.5 cm; in the reverse order, the vehicle body is raised if the automotive vehicle is driven more slowly than 70 km/h for more than two minutes or as soon as it drops below a speed of 35 km/h.;
- The dynamic mode lowers the automotive vehicle by 2.0 cm to a ground clearance of 10.0 cm even before it sets off; its **centre-of-gravity** (CoG) is lower and the active P-M ABW AWA suspension operates with firmer springs and a harder damping characteristic; the suspension tuning is particularly agile as a result; once again the same threshold values of the vehicle velocity apply for automatic changes in ride height: if the vehicle is driven faster than 120 km/h for more than 30 s, the vehicle body is lowered by a further 0.5 cm; and the body is raised again if the vehicle is driven more slowly than 70 km/h for more than two minutes or as soon as it falls below a speed of 35 km/h.
- The comfort mode enables the vehicle to glide extra-smoothly over all kinds of on/off road surface bumps because the system increases damping forces less frequently at lower speeds; the automatic lowering at motorway speeds is suppressed in order to obtain maximum compression travel;
- The lift mode may be activated in the vehicle-velocity range below 80 km/h as required; it raises ground clearance constantly by 2.5 cm and offers the balanced damping characteristics of the automatic mode. It is therefore particularly suitable for driving on uneven terrain; at values of the vehicle velocity of above 100 km/h, active P-M ABW AWA suspension automatically cancels the lift mode and restores whichever mode was previously selected.

Two other, less frequently required modes may be activated via a sub-menu of the MMI control system [AUDI 2002, 2005A; 2005B]:

- In the trailer-towing mode, regardless of the mode previously selected, the vehicle body is not lowered further so as to avoid altering the nose mass; this mode is automatically activated when a trailer is hitched to the innovative vehicles;
- Finally, the jack mode prevents all control processes so that when changing a tyre or wheel, for example, work is not impeded by an automatic change in the vehicle height.

Distinctly Dynamic, Consistently Versatile: Adaptive Active P-M ABW AWA Suspension – Sport - In addition to the standard options, the *'adaptive active P-M ABW AWA suspension-sport'* sport's suspension may be available for the vehicle; the control modes available have the same name as those offered with conventional adaptive active P-M ABW AWA suspension, but have their own characteristics.

Here, the P-M shock absorber (damper) and suspension characteristics are noticeably firmer than in the dynamic mode of the vehicle's standard adaptive active P-M ABW AWA suspension. The vehicle level is always 2.0 cm lower than the standard suspension.

Like the standard suspension, adaptive active P-M ABW AWA suspension - sport has four driving modes.

In the comfort mode, the P-M shock absorbers are actuated in such a way that damping forces are not increased as often in the lower vehicle-velocity range. This adds up to an impression of comfort that conventional sports suspension could never offer.

In order to be able to exploit the full spring travel when driving at high motorway values of the vehicle velocity, the vehicle-body level is not lowered further even above 120 km/h.

In the dynamic mode, the spring and damping rates are higher than in the automatic mode; the vehicle's ride height does not alter. Finally, in the lift mode the vehicle body is raised by 2.5 cm and, at the same time, the spring/shock absorber characteristics are adjusted to the same level as in the automatic mode.

5.6.5 Active E-M ABW AWA Suspension Mechatronic Control Systems

Active E-M ABW AWA suspension mechatronic control systems keep passengers on an even keel by sensing accelerations and controlling actuators at several points on the chassis.

The active E-M ABW AWA suspension mechatronic control system demands high peak power -- about 12 kW. Although its average power consumption is a modest 200 – 400 W, the energy storage components and **electrical energy distribution** (EED) network must be able to handle the high peak load.

E-M actuators are devices intended to provide the force input called for by automotive vehicle active E-M ABW AWA suspension mechatronic control systems. Experimental active E-M suspensions have been devised, in which F-M, P-M or P-F-M actuators provide the force input.

Such mechatronic control systems leave much to be desired regarding safety of equipment and personnel.

In addition, F-M, P-M or P-F-M actuators require an electromechanical or fluidical valve to interface the electrical and fluidical (hydraulical and or pneumatical) aspects of the overall mechatronic control system.

Accordingly, E-M actuators are being studied as an alternative approach to active E-M ABW AWA suspension design.

Such E-M actuators (Fig. 5.139) would provide a direct interface between mechatronic controls and the suspension dynamics, eliminating the need for F-M ABW AWA suspension mechatronic control systems and their attendant problems of leakage and vulnerability to encounter damage [Kowalick 2007].

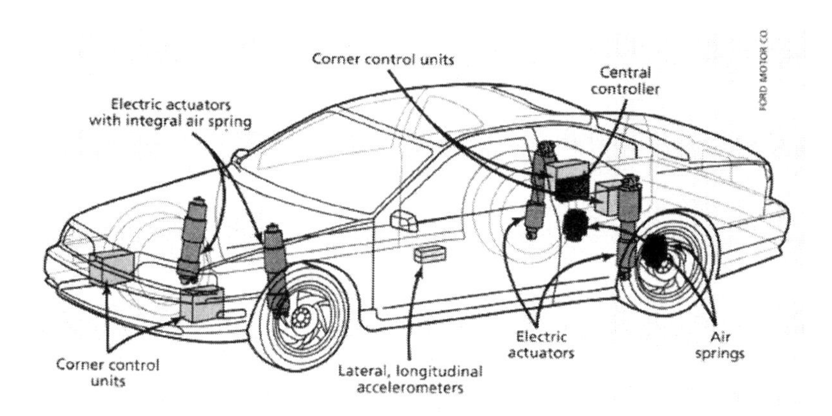

Fig. 5.139 Principal layout of an active ABW AWA suspension [Ford Motor Co.; KOWALICK 2007].

The heart of the active E-M ABW AWA suspension mechatronic control system (Fig. 5.140) is a linear or rotary E-M motor installed at each wheel in a modified *MacPherson* strut arrangement [MEILLAUD 2005].

When electricity is fed to armature wire coils inside, the E-M motor expands and contracts so quickly and forcefully that it prevents pitch and roll during times when the automotive vehicle is driven hard, all the while maintaining passenger vehicle isolation from the sort of wheel impacts that would typically slam driver or passenger head to the ceiling. Anti-roll bars are no longer necessary. The bobbing-head dog for the parcel shelf is obsolete.

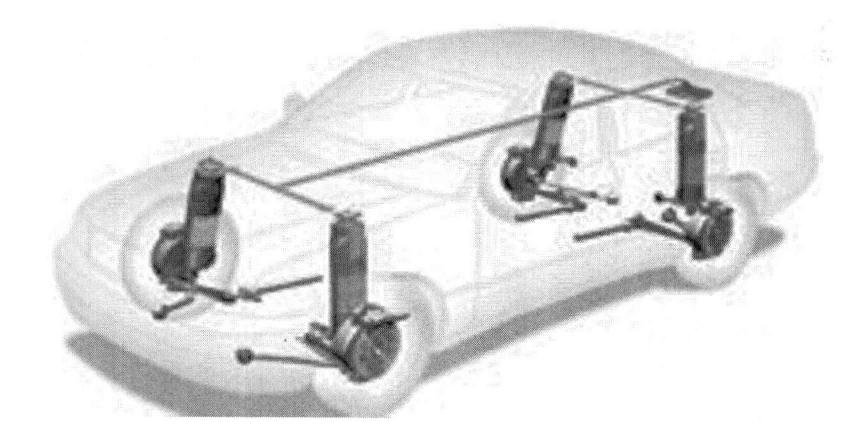

Fig. 5.140 Bose® **active** E-M ABW AWA suspension [MEILLAUD 2005].

The innovative design is modular, allowing automotive scientists and engineers to easily retrofit existing automotive vehicles with four independent modules (fronts incorporate a two-piece lower control arm, wheel shock absorber (damper) mounted inside the wheel, and torsion bar; rears include wheel shock absorbers (dampers) and suspension links, with E-M motors laid out more in double-wishbone fashion) mounted in aluminium cradles directly to original suspension hyposystem attachment points [JENNINGS 2005].

Some automotive OEMs are currently working to develop active E-M ABW AWA suspension mechatronic control systems using electric machines that became available as linear and/or rotary brushless AC-AC or DC-AC macrocommutator electromagnetically- or magnetoelectrically-excited motor and macro- and microelectronics advanced technology developed.

Active E-M ABW AWA suspension mechatronic control systems are highly responsive and are expected to offer control at higher frequencies that are beyond the control of existing active F-M or P-M ABW AWA suspension mechatronic control systems.

For instance, the active E-M ABW AWA propulsion mechatronic control systems currently in development use a tubular linear brushless DC-AC macrocommutator **interior-permanent-magnet** (IPM) magnetoelectrically excited motor without a reduction gear mechanism as shown in Figures 5.141 and 5.142. This makes it easy to absorb (dampen) the impact from the on/off road surfaces, and this enhances ride feature [BREVER ET AL. 2006; GILSDORF ET AL. 2006].

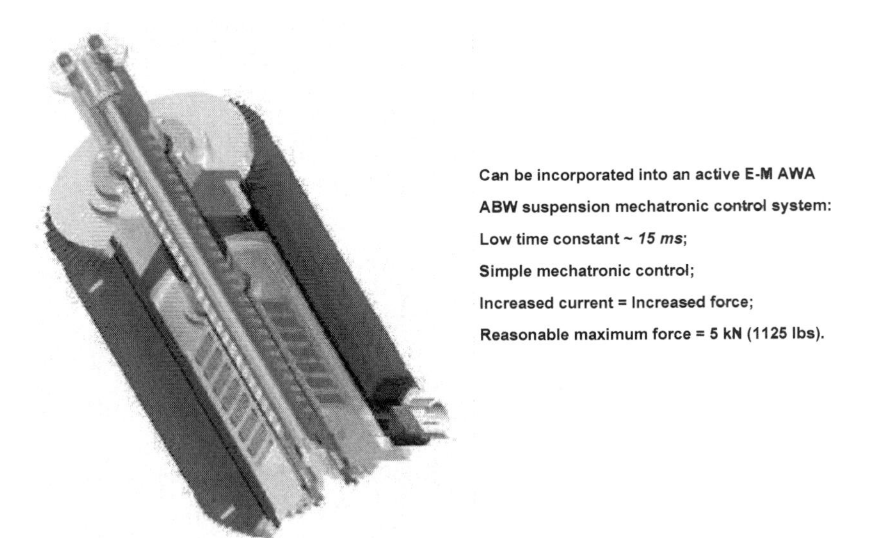

Fig. 5.141 Tubular linear DC-AC macrocommutator IPM magnetoelectrically excited motor [California Linear Devices; BREVER ET AL. 2006].

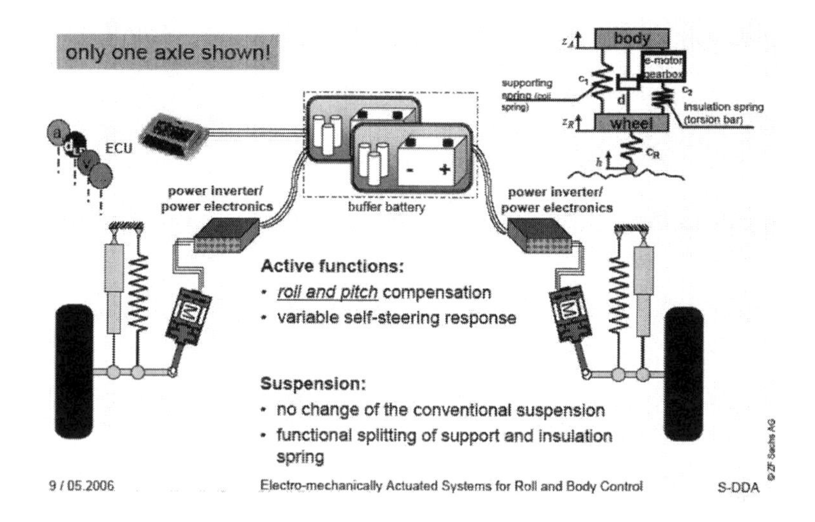

Fig. 5.142 Active E-M ABW AWA suspension mechatronic control system (only one axle shown) [ZF Sachs AG; GILSDORF ET AL. 2006].

Usage of an automotive active E-M ABW AWA suspension mechatronic control system has two main reasons -- to make a vehicle more comfortable for drivers and passengers and to enhance a vehicle handling.
Ride comfort can be classified as:

- ❖ Vehicle body acceleration attenuation during driving when an on/off road surfaces causes jumping on the vehicle's wheel-tyres;
- ❖ Sprung mass vertical displacement peaks reduction.

In addition, it seems to be better to take into consideration that human beings are most sensitive in a frequency range from 4 -- 8 Hz, therefore an add-on requirement for ride comfort is primarily to attenuate vehicle-body acceleration on these values of the frequency.
An automotive vehicle handling may be characterised:

- ❖ As vehicle stability and controllability, thus as rolling and pitching reduction during accelerating, decelerating (braking) and driving through curves;
- ❖ As wheel-tyre jumping, thus as acceleration of unsprung masses.

For conventional passive suspension systems both performance requirements are contradictory. That means if the comfort level is improved then handling stability deteriorates.

This contradiction is the reason why an active ABW AWA suspension mechatronic control system has been used. It should be noticed that it is possible to improve performance in the full frequency bandwidth, but the comfort frequency range for most sensitive humans is about 4 – 8 Hz.

The main problem in the control design is robustness, because some parameters in the automotive vehicle's active ABW AWA suspension mechatronic control system may vary, especially the vehicle-body mass including vehicle load. From this viewpoint, for instance, using H_∞ control may be advisable. All these aspects ought to be taken into consideration and an H_∞ controller satisfying the mentioned requirements may be designed, with tiny trade-offs of course [KRUCZEK ET AL. 2003].

To satisfy all of these requirements, it is necessary to change the damping characteristic dynamically with respect to on/off road surfaces.

The E-M actuator has to be added into the ABW AWA suspension mechatronic control system, by what the active or semi-active E-M ABW AWA suspension mechatronic control system may be attained.

An idea of using a tubular linear E-M motor as an E-M actuator for an active ABW AWA suspension mechatronic control system may be currently possible. The force/velocity profile is shown in Figure 5.143 [KRUCZEK ET AL. 2003].

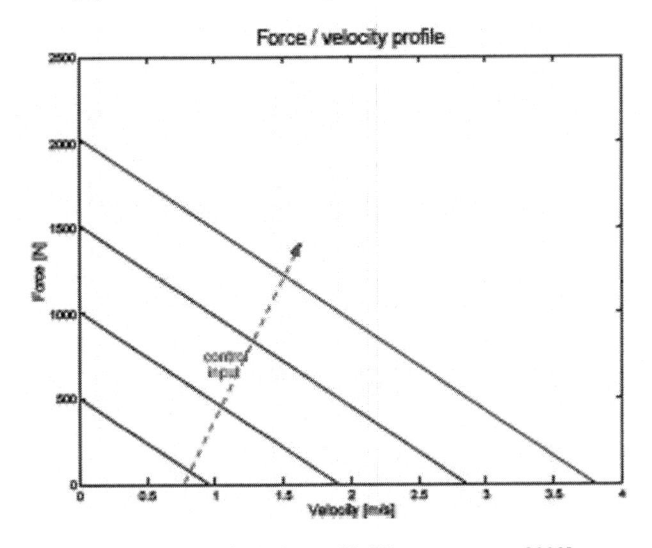

Fig. 5.143 Force/velocity profile [KRUCZEK ET AL. 2003].

This profile assumes the continuous armature voltage is available across the tubular linear E-M motor. The beauty of tubular linear E-M motors is that they directly convert electrical energy into mechanical energy, that is, a usable mechanical force and motion, and vice versa.

Compared to conventional rotary E-M motors, the armature stator and the electromagnetically- or magnetoelectrically-excited mover of direct-drive tubular, linear, brushless E-M motors are linear-shaped. One may imagine such an E-M motor with infinite stator diameter. Tubular, linear, brushless E-M motor mover movements take place with high values of velocity (up to approximately 3.33 m/s), large accelerations (up to g multiples), and forces (up to 1 kN).

As mentioned above, the electromagnetic force can be applied directly to the payload without the intervention of a mechanical transmission, which results in high rigidity of the whole system, its higher reliability and longer lifetime.

In practice, the most often used type is the tubular, linear, brushless DC-AC/AC-DC macrocommutator IPM magnetoelectrically excited motor/generator. For the most part, tubular, linear, brushless E-M motors function within an active ABW AWA suspension mechatronic control system in the same manner as rotary E-M motors. The major difference lies in macroelectronic commutation.

Tubular, linear, brushless E-M motors commutate based on linear position; rotary E-M motors – on angular position. For a tubular, linear brushless DC-AC macrocommutator IPM magnetoelectrically excited motor to produce electromagnetic force, the armature phase-windings must be switched in polarity and amplitude relative to the permanent magnetic field.

In the case of brushed DC-AC mechanocommutator electromagnetically excited motors, the electromagnets are usually stationary (stator). In brushless DC-AC macrocommutator IPM magnetoelectrically-excited motors, the permanent magnets are typically on the mover or rotor, either sliding or rotating, respectively.

Commutation is cyclical in nature, and is based on a fixed ratio of magnetic poles to electric phase windings. With rotary DC-AC mechanocommutator electromagnetically excited motors, the cycles repeat over a fixed distance.

While active E-M ABW AWA suspension mechatronic control systems (Fig. 5.144) have been around for a while, most have used variations on conventional spring/strut, fluidical or pneumatical spring designs [UT-CEM 2004].

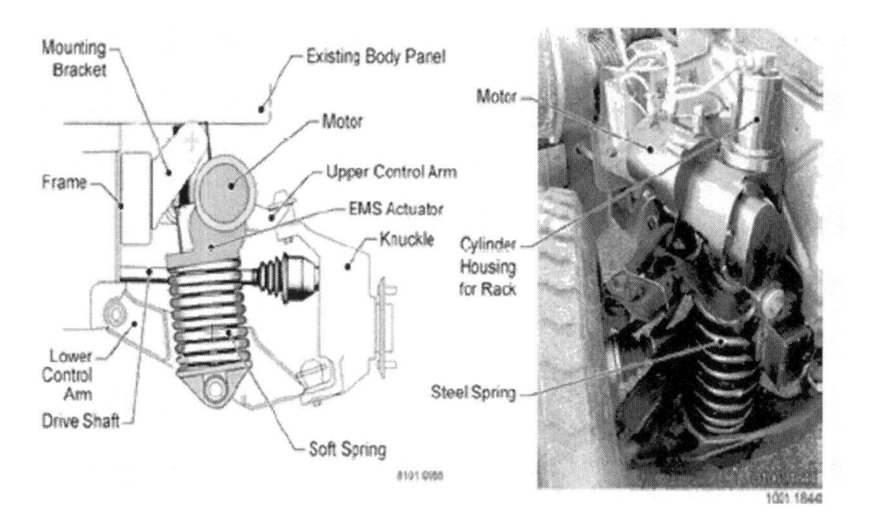

Fig. 5.144 The active E-M ABW AWA suspension (with the rotary E-M motors) that is termed *'electronically active controlled suspension system' (ECASS)'* [The University of Texas at Austin – Center for Electro-Magnetics; UT-CEM 2004].

Some automotive scientists and engineers have ditched all these in favour of a design that uses extremely fast acting tubular linear brushless DC-AC/AC-DC macrocommutator IPM magnetoelectrically excited motors/generators of their own design [McCormick 2004].

Coupled with wheel shock absorbers (dampers), these E-M/M-E motors/ generators can effectively isolate the up and down motions of a vehicle's wheels from the vehicle body (Fig. 5.145) [McCormick 2004; Kirsner 2004; Longoria 2004; Lotus 2005].

Fig. 5.145 The A E-M ABW AWA suspension (with the tubular linear E-M motors) that is termed *'Bose® active suspension'* [Longoria 2004; McCormick 2004].

The result is a ride quality over rough on/off road surfaces so smooth that it is hard to believe as shown in Figure 5.146 [Jennings 2005].

Fig. 5.146 The active E-M ABW AWA suspension-equipped automotive vehicle (below) has far better vehicle-body control than the standard automotive vehicle (above) [AKA Bose Corporation Headquarters, Framingham, Massachusetts, USA; Jennings 2005].

One of the keys to this design is that its DC-AC/AC-DC macrocommutators have a regenerative function that allows power to be returned to the automotive vehicle's **electrical energy-distribution** (EED) system. The bottom line is that the ABW AWA suspension on average uses only a third the power taken by an air conditioning system. This overcomes a conventional drawback to very advanced active E-M ABW AWA suspension mechatronic control systems, which is that they draw so much energy as to be impractical. That's the good news. Still in the debit column are two major issues -- mass and cost. As it stands, the active E-M ABW AWA suspension mechatronic control system adds a great deal of mass when compared to a conventional suspension. Even if may be achieved its goal of minimising the net mass increase to less than 100 kg, that figure is bound to meet resistance from vehicle manufacturers' engineering teams. As for cost, there's no question that the active E-M ABW AWA suspension mechatronic control system may be expensive, destined primarily for luxury vehicles.

The automotive scientists and engineers claim these tubular, linear E-M motors have significant advantages in response time and power consumption over the F-M motors (cylinders) they have seen on all previous active ABW AWA suspensions. According to them, the system may have a total stroke of circa 20 cm (8.5 inch), may respond faster than 10 ms, and consumes about 1.5 kW (2 hp) as a complete system [CSERE 2004].

The active E-M ABW AWA suspension mechatronic control system may be installed in front of the vehicle as shown in Figure 5.147 [LONGORIA 2004; LOTUS 2005].

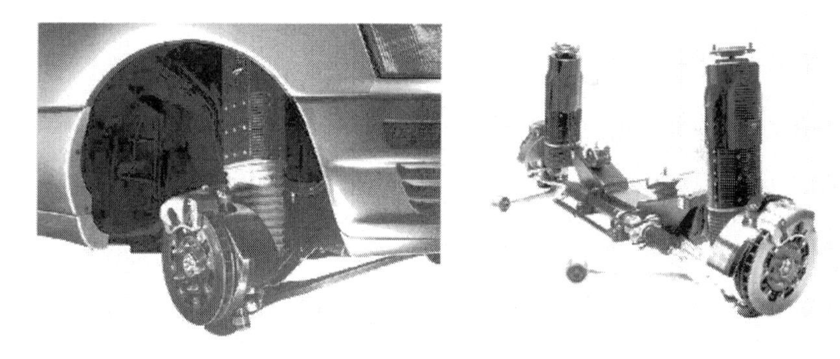

Fig. 5.147 The active E-M ABW AWA suspension (with the tubular linear E-M motors) that is termed *'Bose® active suspension'* [LONGORIA 2004; LOTUS 2005].

In Figure 5.147 is also shown an active E-M ABW AWA suspension design that uses tubular, linear E-M motors in a modified *MacPherson* strut type layout as exposed to the right.

The OEMs plan is to start working with an interested vehicle manufacturer in the 2010's and with five years have the active E-M ABW AWA suspension mechatronic control system appear in a manufacture vehicle [MCCORMICK 2004].

Actually, automotive scientists and engineers reveal that tubular, linear, brushless DC-AC/ AC-DC macrommutator IPM magnetoelectrically excited motors are scaleable and could be used in other applications, specifically for ICE's intake and/or exhaust valve actuation.

That could be a very promising avenue to explore, because gaining complete control over operation of intake and/or exhaust valves, with less complex systems than those currently available, is one of the automotive industry's holy grails [FIJALKOWSKI 1998; MCCORMICK 2004].

A novel concept of the IU-shaped E-M actuator [LEBEDEV ET AL. 2004] is a part of an innovative linear slider shock absorber (damper) system (Fig. 5.148) for **six degrees-of-freedom** (6-DoF) suspension and propulsion.

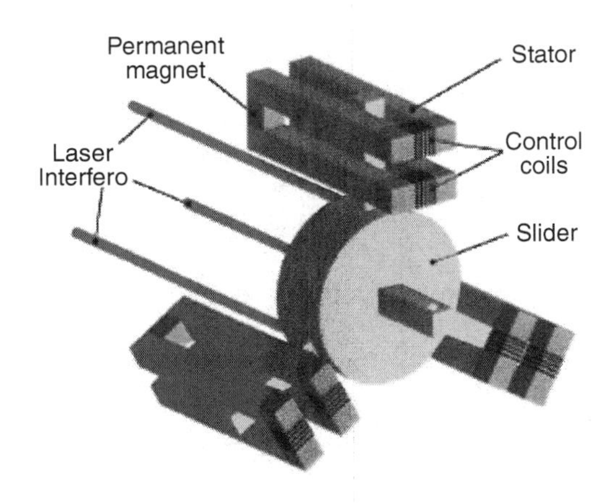

Fig. 5.148 Six degrees-of-freedom (DOF) contactless slide system [LEBEDEV ET AL. 2004].

Such a synergic system is reached by integration of electromagnetic and mechanical structures of magnetic bearing and linear E-M actuator.

Decoupled forces, originated from IPMs and suspension windings (coils) are functions of rotor position and current values in the control windings (coils).

An IU-shaped E-M actuator may be a part of an active E-M ABW AWA suspension mechatronic control system.

This mechatronic control system may be tailored for suspensions mastering applications and should provide the movement along one long stroke sliding direction while being stabilised in other directions.

From E-M and mechatronic control viewpoints such an apparatus, that combines the linear E-M actuator providing a bi-sense directional movement along x-axis and IPM biased active magnetic bearing, belongs to the class of non-linear dynamic systems of considerably complexity, it is amenable to a formal mathematical analysis.

However, it is not a trivial matter to comprehend the principles of its operation, *e.g.* under static and transient conditions, in an imaginative way.

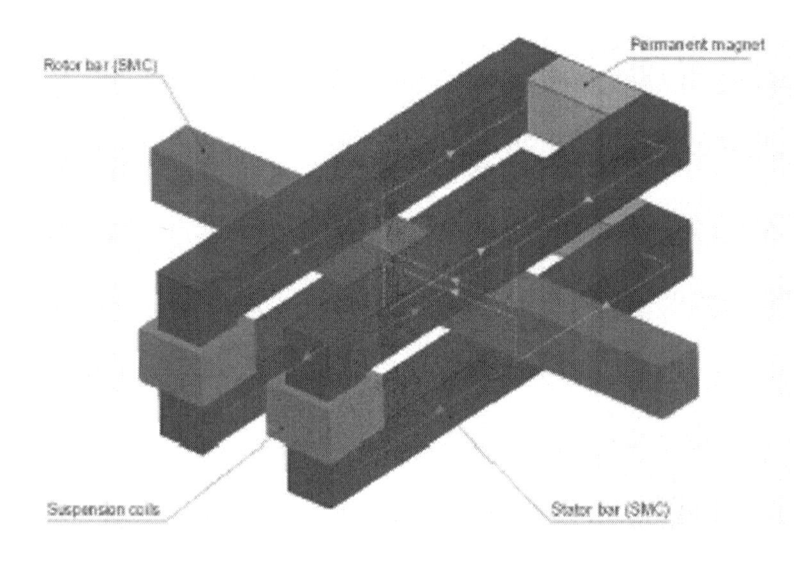

Fig. 5.149 IU-shaped actuator topology with indicated main magnetic-flux paths [LEBEDEV ET AL. 2004].

The general view of the IU-actuator is shown in Figure 5.149 [LEBEDEV ET AL. 2004]. A stator consists of two identical U-shaped iron parts spaced from each other and magnetically coupled by means of two IPM bars (for the biasing purposes) with the same sense of magnetisation direction (along the y-axis) in upper and lower branches.

The IPMs' driven magnetic fluxes (Fig. 5.149) travel along the stator core legs (the upper and lower ones), cross the air gaps, both enter and pass through the rotor length, return across the two next air gaps and stator yokes to the IPMs. Both U-shaped parts are supported (connected) by vertical iron legs, suspension windings (coils), controlling the suspension of the I-mover (I-rotor), and are wounded around each of them.

The magnetic fluxes caused by two suspension windings (coils) travel along the upper (right and left ones) stator cores, cross the air gaps and pass through the mover and return across the two next air gaps and lower stator yokes.

Four propulsion windings (coils) are placed consequently around each long iron bar (two upper and two lower mover bars). A mover (rotor) is inserted in the clearance between halves of the U-shaped stator parts. The mover includes no field sources. All resulting forces acting on the mover are reluctance forces, caused by the IPMs and the suspension windings (coils), and the *Lorenz* forces caused by the propulsion windings (coils). Usually, in magnetically levitated systems the role of the biasing is to improve the linearity and dynamic performance.

Practical application of the biasing scheme based on incorporating IPMs into the magnetic structure may be used. The primary advantage of this scheme is a reduction in electrical energy consumption.

The magnetic flux created by the upper IPMs crosses two upper air gaps that cause reluctance forces in both air gaps between two upper stator bars and mover [LEBEDEV ET AL. 2004].

The magnetic flux from the lower IPMs causes reluctance forces in both air gaps between two lower stator bars and mover. These magnetic fluxes are termed *bias* magnetic fluxes. To be able to control the reluctance forces on the mover, two additional windings (coils) are placed at the ends of stator legs and extra control forces are generated. These control forces are dependent on the applied current and mover position. The total suspension force acting on a mover is the sum of the reluctance forces and the control forces. It has a small x-component in the sense of motion (propulsion) direction and a vertical significant suspension force (z-component). The horizontal position of the mover bar is controlled by the propulsion windings (coils).

The propulsion principle is based on the interaction between the magnetic fluxes crossing the air gaps and the currents in the propulsion windings (coils) that cause the *Lorenz* forces to act on the mover along the x-direction. Obviously, the electromagnetic field generated by the propulsion windings (coils) distorts the distribution of the magnetic flux density in the stator bars and generates additional reluctance forces. The y-component of the forces can neglected, as it remains very weak with respect to the other x- and z-force components. As a result, the field distribution possesses a three-dimensional nature. Besides, the large air gaps cause significant values of leakage magnetic fluxes. Thus, the analysis of the actuator is becoming very complex [LEBEDEV ET AL. 2004].

E-M shock absorbers play a key role in active security, smoothing out road imperfections and adapting suspension firmness. Mechatronics allow even more progress to be made as regards comfort and safety.

In recent years, the **active wheel** (AW) technology makes possible an application of an innovative active E-M ABW AWA suspension with a dynamic -absorbing **steered, motorised and/or generatorised wheel** (SM&GW) that integrates both an E-M shock absorber and an in-wheel-hub motor/generator for traction and braking (Fig. 5.150) [MEILLAUD 2005].

Fig. 5.150 The *Michelin* Concept vehicle can lean thanks to its active AW ABW AWA suspension [Challenge Bibendum in Shanghai; MEILLAUD 2005].

A SM&GW equipped with the AW technology appears from the outside to be a standard wheel and tyre package. But on the inside, there are springs and in-wheel-hub motors/generators and other E-M components and connections, all packaged within the circumference of the wheel and tyre [FIJALKOWSKI 1991, 1997, 1999, 2000; EDSALL 2004].

Each of the SM&GWs is piloted independently. Also, there is no longer any mechanical link with the ECE or ICE that has allowed the suppression of some mechanical parts such as the clutch, gearbox, drive shaft, anti-roll bar or else conventional E-M shock absorbers.

In practice, the automotive vehicle may roll in bends like a motorbike. This suspension may stabilise the chassis so as to avoid plunging forward when braking abruptly.

The road clearance may also be altered to clear an obstacle in cross-country mode, from a minimum of 10 cm to a maximum of 35 cm.

According to Michelin, this type of suspension opens the way to the design of innovative automotive vehicles, giving free rein to the creativity of automotive vehicle designers [MEILLAUD 2005].

In-wheel-hub motors/generators not only turn the SM&GWs, but may be used to slow and stop them, so conventional disc, ring or drum brakes might eventually be eliminated, or at least reduced to smaller, redundant systems.

And by using in-wheel-hub motors to turn the SM&MWs, large and heavy transmissions and mechanical differentials become obsolete [FIJALKOWSKI 1991, 1997, 1999, 2000; EDSALL 2004].

With the AW technology (Fig. 5.151), automotive vehicle designers could become more creative because suspension components that now intrude on interior space would be re-packaged within each of the vehicle's four wheels, creating more room for people and cargo [EDSALL 2004].

Fig. 5.151 The active wheel (AW) technology uses a traction E-M motor to turn the wheel and all active ABW AWA suspension components within the wheel and tyre [*Larry Edsall*/Special to the Detroit News; EDSALL 2004].

Anything that makes anything smaller is good for customers. Automotive designers would appreciate a suspension system with *'nothing sticking up or sticking out'*.

However, there are likely to be issues regarding the additional mass the E-M components add within the wheel-known as *'unsprung'* mass.

An electrically controlled ABW AWA suspension may adjust instantly from a silky smooth ride that soaks up bumps to a firm setting for the kind of evasive manoeuvres needed to avoid a collision.

The ABW AWA suspension may even make a vehicle roll into a turn, or hop up and down like a low rider.

The AW system (Fig. 5.152) may be designed to work with the **all-electric vehicles** (AEV), that is, the **battery-electric vehicles (BEV)** and/or **fuel-cell-electric vehicles** (FCEV) as well as **hybrid-electric vehicles** (HEV) that many experts see in the world's automotive future.

Fig. 5.152 The *Michelin* concept vehicle is equipped with the tyre maker's active wheel (AW) technology that packages the active ABW AWA suspension mechatronic control system inside the wheels [*Larry Edsall*/Special to the Detroit News; EDSALL 2004].

A dynamic-absorbing in-wheel-hub E-M motor DBW AWD propulsion mechatronic control system uses the in-wheel-hub E-M motors themselves to absorb vibration, which enhances handling, safety and comfort in AEVs and/or HEVs. This dynamic-absorbing SM&GW overcomes disadvantages of existing in-wheel-hub motor DBW AWD propulsion mechatronic control systems that have downgraded the applicability of those mechatronic control systems.

Installing the E-M/M-E motors/generators inside the wheel-hubs supports controlling each wheel independently. That provides for first-rate road-handling performance.

It also eliminates the necessity for the M-M differential and driveshaft (or chains and sprockets) and therefore permits exceptional freedom in designing vehicles. Automotive designers may provide more space to the driver and passengers without enhancing the overall mass and size of the vehicle.

An intractable disadvantage of in-wheel-hub E-M/M-E motors/generators has been the mass that they add to each wheel. That influences comfort and road-holding performance unfavourably, and it has restricted the applicability of in-wheel-hub motors/generators in AEVs and/or HEV. New technology overcomes this disadvantage of in-wheel-hub E-M motor DBW AWD propulsion mechatronic control systems by using the in-wheel-hub E-M/M-E motors/generators to absorb vibration.

The in-wheel-hub motors/generators themselves operate as vibration shock absorbers (dampers). Their own vibration absorb the vibration from the on/off road and wheel-tyres. That permits enhanced traction and a more-comfortable ride than is practicable with other in-wheel-hub E-M motor DBW AWD propulsion mechatronic control systems or with other modes of AE or HE DBW AWD propulsion.

In dynamic-absorbing SM&GWs (Fig. 5.153) dynamic shock absorbers (dampers) suspend the shaftless in-wheel-hub motors/generators to insulate them from the unsprung mass [BRIDGESTONE 2003].

Fig. 5.153 Dynamic-absorbing steered, motorized and/or generatorised wheel (SM&GW) [Bridgestone Corporation; BRIDGESTONE 2003].

The in-wheel-hub E-M motor's vibration as well as the vibration from the on/off road and wheel-tyres dampen each other, which enhance road-handling performance. The flexible coupling (Fig. 5.154) consists of four cross guides that transfer the drive power from each in-wheel-hub E-M/M-E motor/generator smoothly to its wheel.

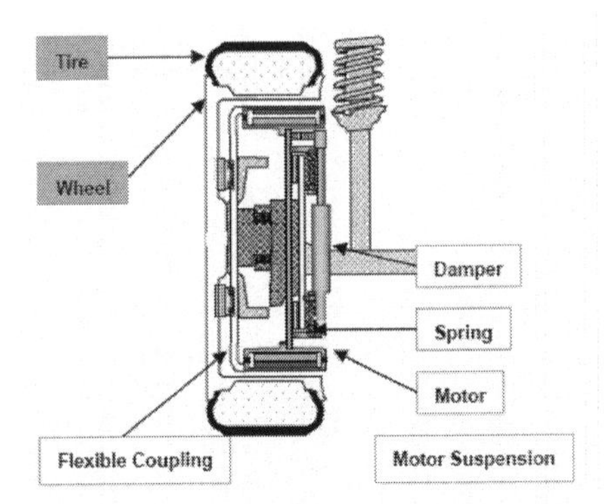

Fig. 5.154 Layout of the dynamic-absorbing steered, motorized and/or generatorised wheel (SM&GW) [Bridgestone Corporation; BRIDGESTONE 2003].

The cross guides provide balance for continuous, satisfactory shifting in the rotational positioning of the in-wheel-hub motor/generator and wheel.

Analytical comparisons of the performance of a conventional AEV or HEV with a single E-M/M-E motor/generator, an AEV or HEV equipped with conventional SM&GWs with in-wheel-hub motors and an AEV or HEV equipped with dynamic-absorbing SM&GWs, emphasize E-M differentials in road-handling performance (contact force fluctuation) and right quality (vertical acceleration frequency) on a rough on/off road surface. The dynamic-absorbing in-wheel-hub motor DBW AWD propulsion mechatronic control system results in enhanced road-holding performance and a more comfortable ride than are practicable with conventional in-wheel-hub motor DBW AWD propulsion mechatronic control systems. It also submits advantages over conventional, single E-M motor AEVs and/or HEVs in safety and comfort [BRIDGESTONE 2003].

5.6.6 Active E-P-M ABW AWA Suspension Mechatronic Control Systems

It was only the last few years that active F-M or F-P-M or P-M ABW AWA suspension mechatronic control systems were developed. These were designed to eradicate the unavoidable pitch and roll performance of a spring suspension by fast alteration of the length of the firm strut between each axle and the vehicle body. Active ABW AWA suspension of these kinds might be envisaged to elevate each wheel over a bump and to push it down into a hole as necessitated. But it was indispensable for exactly controlled physical movement to occur repetitively – and as each movement acquired time to perform, its aptitude to cope with diminutive disturbances at high values of the vehicle velocity was restrained. A guaranteed amount of springing had to be engaged to deal with this. The peak energy requirement troubles and the intricacy of the fluidics (hydraulics) also made equipment costly and unfeasible for high-volume application [AMT 2005B].

A highly developed design for an active E-P-M ABW AWA suspension mechatronic control system is currently accessible, using an advanced technology of mechatronic force control. There is currently an entirely novel E-P-M suspension module that may take action in a fraction of a millisecond. Wheel forces may currently be selectively separated from or attached to the automotive vehicle, to any necessary degree, instant by instant, for every centimetre of the automotive vehicle motion, even at full vehicle velocity [AMT 2005A, 2005B].

The E-P-M suspension module is a very quick force E-M modulator, different from any other kind of E-M actuator. Nothing has to move for it to control the forces transferred to the automotive vehicle by the wheels. Really it transmutes a passenger automotive vehicle, a truck, an off-road vehicle or a military wheeled or tracked vehicle into an exactly-stabilised platform, moving under real-time mechatronic control. The mechatronically-controlled E-P-M suspension module is electrically and mechanically uncomplicated and principally reliable.

It is appropriately designed so that it doubles as an air or a gas spring that, without driver intervention, adjusts to on/off road surface circumstances, temperature deviations, vehicle mass distribution and so on. In active mode it attracts mechanical energy from the automotive vehicle to control the vertical forces at each wheel corner, for a safe and even ride to be feasible. In semi-passive mode it receives mechanical energy from the vertical motions of the wheels and uses this to control the attitude and position of the automotive vehicle, restoring any excessive energy to a central energy source.

The E-P-M suspension module thus is a development for an active E-P-M ABW AWA suspension mechatronic control system design. Under accurate mechatronic control it functions as [AMT 2005B]:

- ❖ A force generator that compensates instantaneously for the disturbing effects of the on/off road surface;
- ❖ An air or a gas spring that is repeatedly adjusted to optimise vehicle attitude and ride height;
- ❖ An energy-recovering E-P-M shock absorber (damper) that reacts instantaneously to vehicle accelerations;
- ❖ A vehicle attitude control element, whether the automotive vehicle is in motion or stationary;
- ❖ A vehicle height control element, whether the vehicle is in motion or stationary;
- ❖ A *'fail-soft'* subsystem, preserving function through several grades of degradation.

This novel active E-P-M ABW AWA suspension's E-P-M suspension module may impressively affect the near-term design of vehicle chassis, wheels and tyres, SBW AWS conversion mechatronic control system and electrical/electronic equipment. In view of the fact that it also gives better fuel economy and a magnitude of sequential improvements in ride quality, it may enhance significantly the value of any kind of automotive vehicle [AMT 2005B].

An interesting invention may be the dual-action magneto-electrical E-P-M shock absorber (damper) shown in Figure 5.155 [DENNE 2001].

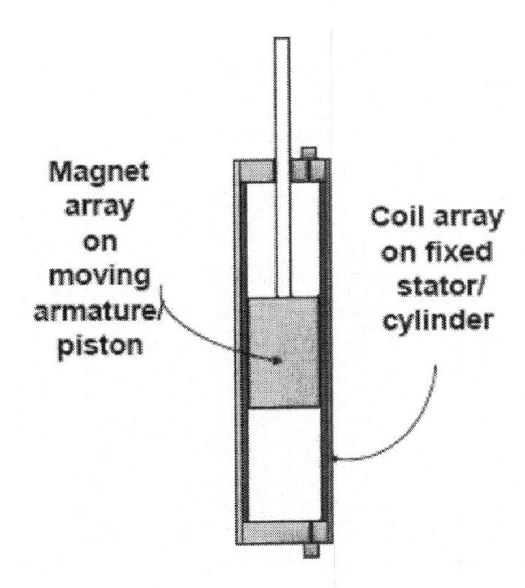

Fig. 5.155 Dual-action magneto-electric E-P-M shock absorber [DENNE 2001].

The first step may be to suspend the automotive vehicle on air (gas) springs. Having arranged for it to float in this way, there may be no method of moving it along except by electro-magnetic forces.

So the inventor put **interior permanent magnets** (IPM) on the pneumatical piston and windings (coils) inside the cylinder, to produce a dual-action machine with only one moving part, looking like the F-M shock absorber (damper) that it replaced [DENNE 2001].

It appears that the magneto-electric E-P-M shock absorber (damper) may also revolutionise the technology of vehicular suspensions. In the past, the wheels have transmitted their forces to the vehicle body by means of springs.

The compliance of a spring permits the wheel to move some distance vertically – but the force on the vehicle body increases in proportion to that movement. Since mechanical energy is then stored in the spring, shock absorbers (dampers) must be used to dissipate the mechanical energy so as to prevent spurious oscillation or *'bounce'*.

If a magneto-electric shock absorber is used to replace the spring and damper combination, it is possible for the M-E generator action of the shock absorber to recycle the spring energy back into the E&IN -- which is especially useful for the economy of AEVs or HEVs.

In another mode of mechatronic control of a magnetoelectrical shock absorber as a suspension component, it is possible to arrange for the force on the body of the automotive vehicle to be maintained at an almost constant value, so that the ride is akin to that of flying.

That is possible because the magneto-electric forces can be made equal and opposite to the mechanical forces and with a very fast reaction time, on the order of a millisecond. It is well known that 1 ms is approximately equivalent to circa 25 mm (1 inch) of travel at circa 100 km/h (60 mph) [DENNE 2001].

A simplified structural and functional schematic diagram of the ServoRam™ E-P-M actuator is shown in Figure 5.156 [AMT 2005B].

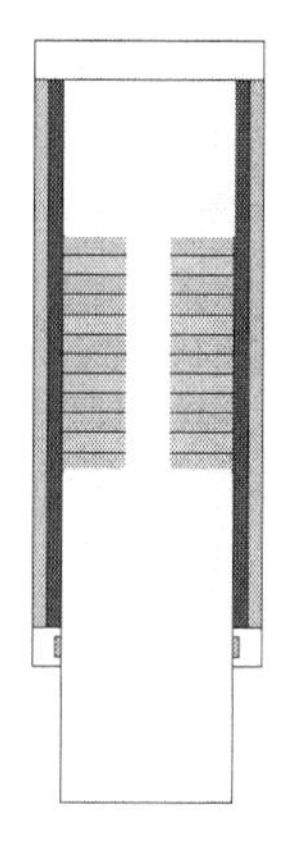

Fig. 5.156 Structural and functional schematic diagram of the ServoRam™ E-P-M actuator [AMT® AMT 2005B].

It comprises an outer mild steel case lined with electrical windings and fitted with aluminium end pieces. The case and winding system is the stator element of an electrical machine. A piston element or armature moves along the axis of the stator bearing on the inner surface of a thin dielectric tube that lines the winding assembly. This liner tube is not shown in Figure 5.155.

The piston comprises a stack of planar rings forming magnetic force elements. Each magnetic element is made from a magnetic ring and two mild steel pole-pieces. The stack of piston elements drives a thrust tube that transpires from one end of the E-P-M actuator. The thrust tube is fitted with a sliding seal that makes the complete assembly airtight.

The tube is hollow and the space within it is in contact with the reminder of the volume of the E-P-M actuator through a hole in the centre of the *'piston'*.

In action the E-P-M actuator is filled with air (gas) that develops a spring whose rate may be selected in relation to the ratio of the separated volumes of the device when contracted or extended. The P-M force may be adjusted by altering the air (gas) pressure at any time. The E-P-M actuator is exclusive in that it creates an electromagnetic force and pneumatical force acting concurrently on the same output element.

A linear transducer (not shown in Figure 5.156) evaluates the position of the piston, so that the unit may be specifically controlled by any standard ECU drive of the kind used for rotary three-phase servomotors.

The E-P-M actuator has zero mechanical hysteresis, in view of the fact that the force is used directly to the output element. It has zero electrical hysteresis – a micro-amp in one sense of direction creates a positive force, an electrical current in the opposite sense of current-flow direction generates a negative force, so that the force output is an accurately linear function of the electrical current input.

There is zero transport lag – nothing moves except the armature itself, and as soon as the electrical current is in the windings, the output force is present. The small control time constant permits the force to be altered at a rate of thousands of N/ms. ***Mechatronic Control of a ServoRam™ E-P-M Actuator*** -- The windings (coil system) in the stator of the electrical machine are designed to be energised by a 3-phase power driver stage for a rotary 3-phase servomotor (Fig. 5.157) [AMT 2005B].

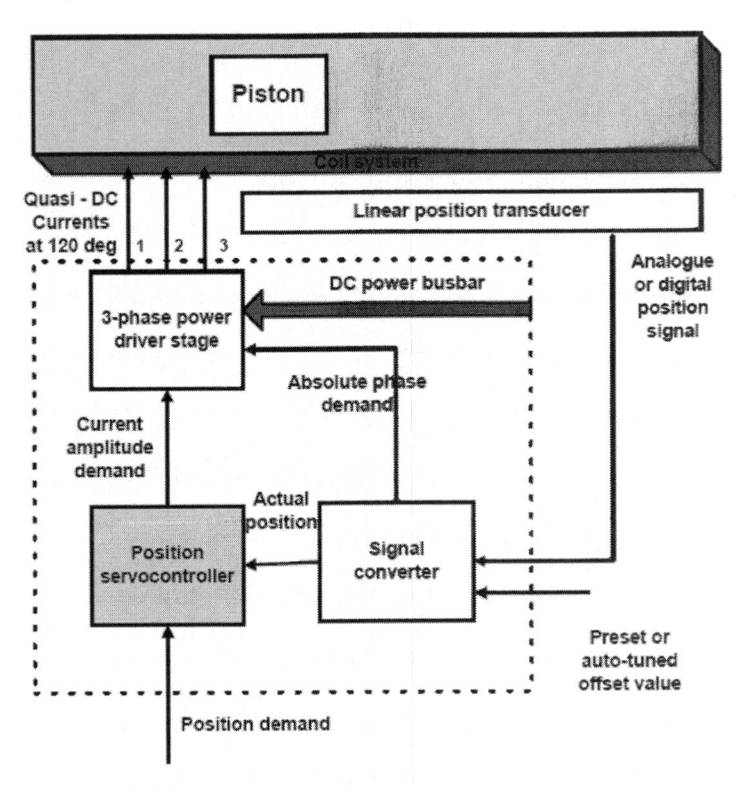

Fig. 5.157 Structural and functional block diagram of the ServoRam™ E-P-M actuator piston's mechatronic control [AMT®; AMT 2005B].

These function by converting the incoming electrical energy into a DC rail voltage (if it is not DC already) altered as though $2\pi/3$ rad out of phase with one another. Thus three quasi-DC currents energise the windings of the electrical machine. The phase of these electrical currents is locked to the position of the piston, so that the thrust is always optimised.

So that the phase of the stator currents may be locked to the armature position, it is necessary for there to be a position transducer -- usually along the central axis of the E-P-M actuator. The magnitude of the electrical current (the amplitude of the sinusoidal function or the peak value of the trapezoid) determines the value of the thrust.

It is set by the parameters of a servo control loop around a controlled position of the armature. An ancillary circuit processes an output signal that specifies the magnitude and sign of the drive current. This controls the pressure in the air (gas) spring with which the E-P-M actuator is coupled, so as to minimise the mechatronic control system power consumption. One of the primary aims of an active E-P-M ABW AWA suspension mechatronic control system is to design the passenger compartment insensitive to the vertical motions of the wheels, whilst permitting the wheels to continue firmly in contact with the on/off road surface. Explicitly, within sensible limits, the vertical position of the passenger compartment should be retained stable – and when it has to move (to follow a large vertical excursion), it should move as smoothly and quietly as feasible.

Newton's Fist Law of Motion tells us that there cannot be any vertical acceleration unless the vertical force functioning on the automotive vehicle is altered. Explicitly, if the ride is to be a smooth one, the vertical forces at the wheel corners must continue constant.

Automotive suspension designers begin by abandoning the conventional concept of a suspension and presume that the automotive vehicle is supported on very soft undraped air (gas) springs at the four wheel-corners (Fig. 5.158) [AMT 2005B]

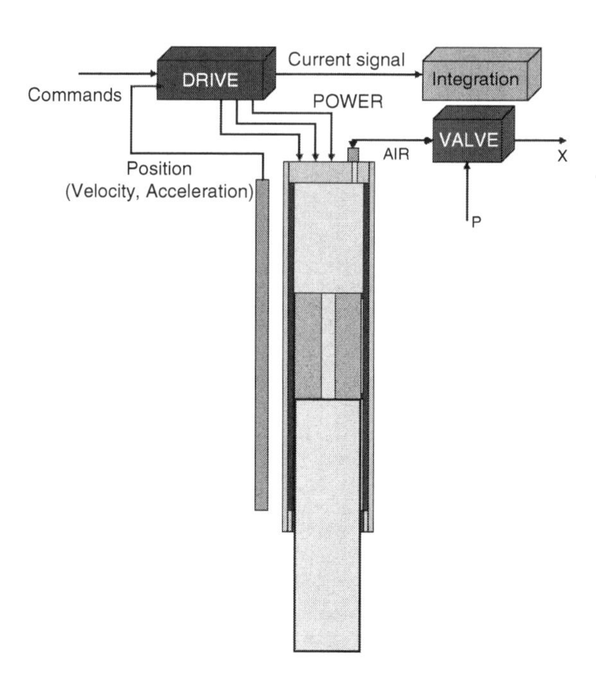

Fig. 5.158 Structural and functional schematic diagram of the Servo Ram™ E-P-M actuator's mechatronic control system [AMT®; AMT 2005B].

The passenger compartment may therefore continue to move at the same height even as the wheels move chaotically (within the sensible limits of their movement) across uneven on/off road surface. It may move on an air cushion. It may move smoothly even if it is moving on a flat on/off road surface. Evidently, this ideal concept disregards the unavoidable seal and bearing friction of the pneumatical elements and their thermodynamic losses, both of which may proceed to degrade the separation of the air (gas) springs and so transfer a quantity of the wheel forces to the automotive vehicle.

The air (gas) springs themselves also use a restoring force that fluctuates with vertical wheel movement. On the other hand imagine that the suspension elements are ServoRam™ E-P-M actuators described above.

Electromagnetic forces than may be used to compensate, over a wide bandwidth, for all the redundant forces in a real active E-P-M ABW AWA suspension mechatronic control system.

As the extremes of the wheel move are emerged the E-P-M actuators are designed to transmit force smoothly to the vehicle body so as to offer a comforting ride.

It should be particularly well-known that, since the ServoRam™ E-P-M actuator is capable of reacting in real time – in less than a millisecond, or about 2.54 cm (1 inch) of move at 96 km/h (60 mph) – there is indeed no necessity for any look-forward optical systems [AMT 2005b].

Since the CoG of an automotive vehicle is normally above the level surface of the wheel corners, an automotive vehicle has an inclination to roll when turning and to pitch fore and aft when altering vehicle velocity. Conventional passive suspension mechatronic control systems are set up to have non-zero spring rates to oppose this motion.

On the contrary the E-P-M actuator may generate very influential quick-acting electro-magnetic forces to suspend the level surface of the automotive vehicle stable, instant by instant as the motion continues.

Accelerometers at each separate wheel corner may be set up to command alterations to the instant values of the E-P-M actuator's forces, generating a very intense anti-roll and anti-dive action.

Since the full up-ward force generated by the four wheel-corners continues the same, the vehicle body remains at the same height. It should be well known that the E-P-M actuator does not generate a constant electro-magnetic force.

The latter is modulated instant by instant so as to preserve the net force at the wheel corner at such value that the vehicle-body height and attitude does not alter.

That desired value of the net force alters only gradually in relation to the electromagnetic signal -- in hundreds of milliseconds rather than in fractions of milliseconds.

Despite the fact that the automotive vehicle occupant may have the impression that the suspension unit reacts only slowly, that is not the truth of the matter.

The air (gas) spring pressure at each wheel corner is recurrently optimised, averaged over a period of several seconds by a simple algorithm, so as to trim down the electrical energy demand.

In automotive vehicle application the algorithm has the consequence of, without driver intervention, reducing the ride height if it has been adjusted too low for the unevenness of the on/off road surface.

Since the ServoRam™ E-P-M actuator does not necessitate moving itself or any other item in order to alter the state of the mechatronic control system, there is no need for discussion of normal bandwidth restrictions of actuators that do.

Explicitly, there is no 3 db per octave roll-off as an effect of velocity margins, nor any 6 db per octave roll-off initiated by inertial effects [AMT 2005B].

The identical E-P-M actuator (Fig. 5.159) may be used in a semi-passive or first stage fallback mode as an extremely controllable shock absorber (damper).

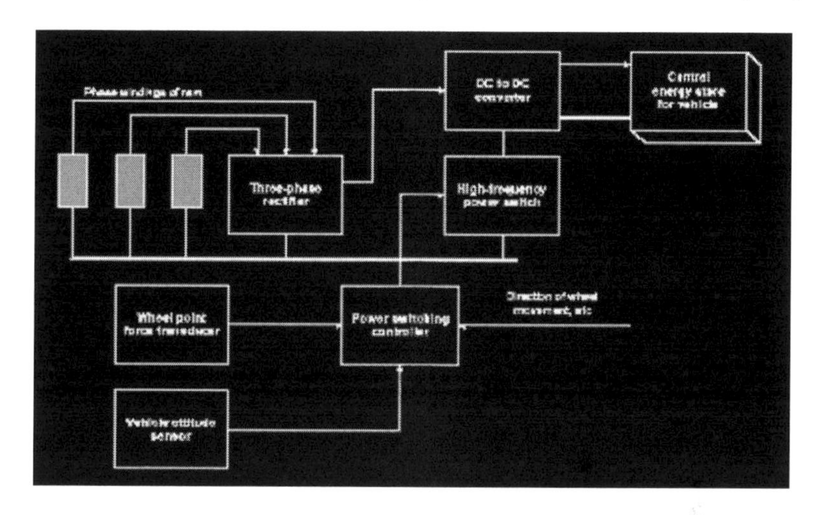

Fig. 5.159 Structural and functional block diagram of damping control for the semi-passive E-P-M ABW AWA suspension mechatronic control system [AMT®; AMT 2005B].

In this arrangement it is the velocity-correlated coupling between the two parts of the E-P-M actuator that is modulated to continue the passenger compartment at the same height and attitude. As the E-P-M actuator piston moves, it generates electrical energy that is AC voltages in the control windings. If the windings are done in an open circuit, no current may flow and there is no reaction force on the piston, but if the windings are done in a short circuit, the reaction force may be very great, reliant on the piston velocity. Thus by controlling the impedance of a load accessible to the windings it is doable to control the velocity of the ServoRam™ E-P-M actuator -- to control the damping of its motion [AMT 2005B].

It should be taken into account that the outer part of the E-P-M actuator is not fixed -- it is coupled to the vehicle body. Consequently when automotive suspension designers refer to *'damping'* they essentially insinuate the degree of correlation between the vertical motion of the wheel and that of the automotive vehicle itself.

The damping coefficient of an E-P-M actuator may be altered in a fraction of a millisecond by changing the mark/space ratio of a high-frequency electrical valve (e.g. switching transistor) that is actually connected across the phase windings. It is sensible to generate slow alterations in the mean height of the automotive vehicle by altering the air (gas) spring pressure at each wheel corner. Although it may be evident from the previous paragraph that the vehicle attitude may also be controlled quickly and exactly by altering the damping coefficient of each E-P-M actuator suspension module. For instance, the proper suspension modules may be stiffened against upward motions of the wheels to oppose a redundant pitch or roll motion but undisturbed on every wheel downward motion so that the wheel may continue its grip on the on/off road surface. The predisposition to control vertical-damping forces immediately and asymmetrically is an inimitable attribute of the E-P-M actuator. When moving over a hypothetically level on/off road surface, the automotive vehicle moves on its air (gas) spring suspension. For diminutive amplitude motions that do not disturb the mean attitude of the automotive vehicle, the damping coefficient is kept low, in an attempt to create a smooth ride. On the contrary, as the automotive vehicle starts to alter its mean height or attitude outside preset restrictions, the damping is escalated asymmetrically, in an attempt to stiffen the suspension against the unfavourable motion.

The same as the air (gas) spring pressure is controlled by the mean direction of current in the windings of the E-P-M actuator in active mode, the mean direction of the damping current may also be sensed and used similarly to reduce the average air (gas) spring settings for each wheel.

An easy algorithm that controls the position of the automotive vehicle in height and attitude thus harnesses in an inimitable way the forces functional to the wheels by the on/off road surface. It may be recognized that the suspension module does incorporate a compliant element at the lower extreme of its move, in an attempt to create a residual restoring force if the active control system, shock absorber (damper) control system and the air (gas) spring control system should all break down concurrently.

Conventional automotive vehicle shock absorbers (dampers) convert motion mechanical energy into thermal energy (heat) that is wasted into the slipstream. On the contrary, the output of the E-P-M shock absorber (damper) is in the form of electrical energy, a large fraction of which may be supplied back in to a central store (accumulator) -- e.g. the automotive vehicle's CH-E/E-CH storage battery – via a DC-DC macrocommutator (converter) and stored.

By using a combination of the semi-passive and active E-P-M ABW AWA suspension mechatronic control technologies, it is also feasible to set up for the energy accumulated by the damping system to be used to drive the wheel corners in active mode during the return stroke. This escalates the force suspending the wheels in contact with the on/off road surface and creates a smoother ride.

Vehicle Attitude Control for an Active E-P-M ABW AWA Suspension – An automotive vehicle's active E-P-M ABW AWA suspension mechatronic control system may simply be modified to roll into a bend or to recompense for reverse road camber, since the mechatronic control system may identify the side forces and alter the suspension height on separate wheels as relevant.

The active E-P-M ABW AWA suspension mechatronic control system also allows the ride height to be trimmed down (compatible with the smoothness of the on/off road surface) in an attempt to reduce vehicle drag as vehicle velocity escalates. Further, the automotive vehicle may, in a simple manner, be set up to *'kneel'* or drop to a supplementary appropriate height when stationary for the loading and unloading of goods or for the entry and exit of disabled persons [AMT 2005b].

Advantages of the Active E-P-M ABW AWA Suspension Mechatronic Control System - The E-P-M absorber combines in one single and robust mechanical element [AMT 2005b]:

- An active E-P-M ABW AWA suspension mechatronic control system using a minimum energy principle based on *Newton*'s First Law of Motion;
- A regenerative shock absorber (damper) with very fast and functioning control of coupling between wheels and chassis;
- Almost instant action in response to external forces that evaluate to adjust vehicle ride height and attitude;
- A constantly adjusted and self-levelling air (gas) spring suspension;
- An adaptable mechatronic control system for ride height and attitude control in response to altering vehicle velocity and on/off road surface circumstances;
- A multi-layer fail-safe mechatronic control system.

E-P-M Actuator Control for the Active E-P-M ABW AWA Suspension Unit – The physical model of the ServoRam™ E-P-M actuator is shown in Figure 5.160 [AMT 2005B].

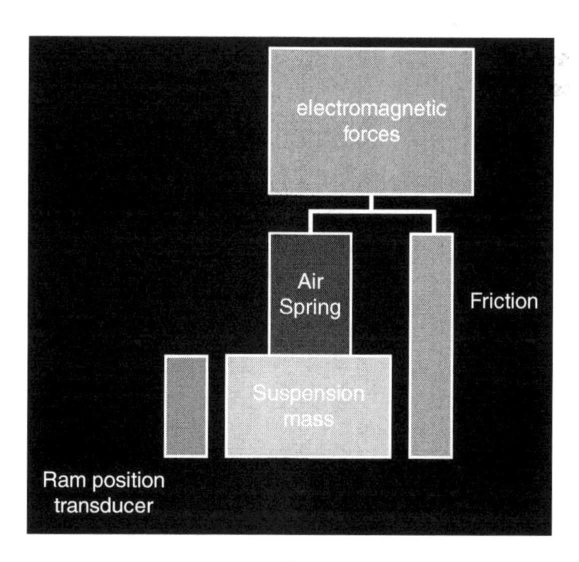

Fig. 5.160 Physical model of the ServoRam™ E-P-M actuator [AMT®; AMT 2005B].

The suspension mass (piston and thrust tube, plus couplings or fittings) is separated from the stator body by an air (gas) spring, but this separation is made worse by the friction in the piston bearings and seals.
The electromagnetic forces function directly between the piston and stator and make ineffective the other forces.

In use, the ServoRam™ E-P-M actuator (Fig. 5.161) is located between the wheel corner (on the vehicle chassis) and the wheel stub axle, in an attempt to transmit all the vertical forces, the lateral forces are transmitted by the wishbones [AMT 2005B].

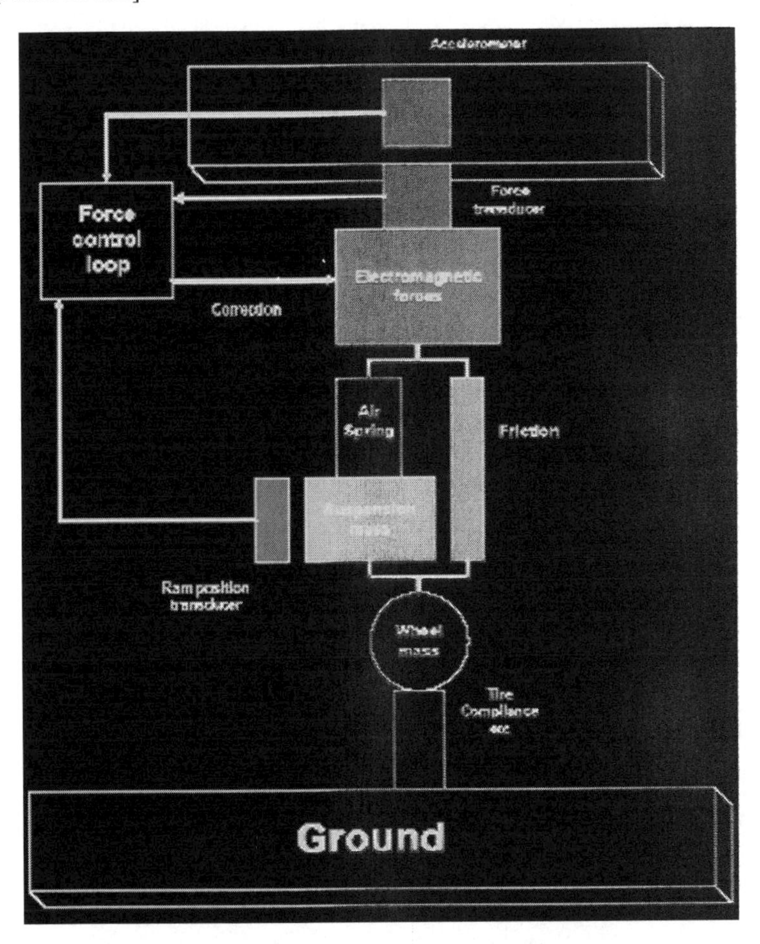

Fig. 5.161 A quarter-vehicle physical model of the active E-P-M ABW AWA suspension [AMT®; AMT 2005B].

It should be recognized that, within restrictions of the E-P-M actuator move, electromagnetic forces can exactly control the forces transmitted from the wheel to the chassis.

A force-evaluating transducer may be used to control the electrical current to the winding system, in an attempt to continue the total upward force at a constant value, regardless of the wheel vertical motion. The desired value of this *'constant'* force may be set up consecutively by the output from a wheel-corner accelerometer, in an attempt to suspend the automotive vehicle's stability against pitch and roll motions, for instance.

Air (Gas) Spring Pressure Control for an Active E-P-M ABW AWA Suspension – The force control loop also creates a signal that is altered to control the pressure in the air (gas) spring system. If an especially downward-acting force is a requirement of the electromagnetic system, the pressure becomes more intense, and vice-versa [AMT 2005b].

Attitude Clearance Control for Active E-P-M ABW AWA Suspension – The ride height of the automotive vehicle at each wheel corner may be processed from the average extension of the ServoRam™ E-P-M actuator. It is therefore feasible to alter the attitude or ground clearance of the automotive vehicle (Fig. 5.162) by an external input to each E-P-M actuator controller [AMT 2005B].

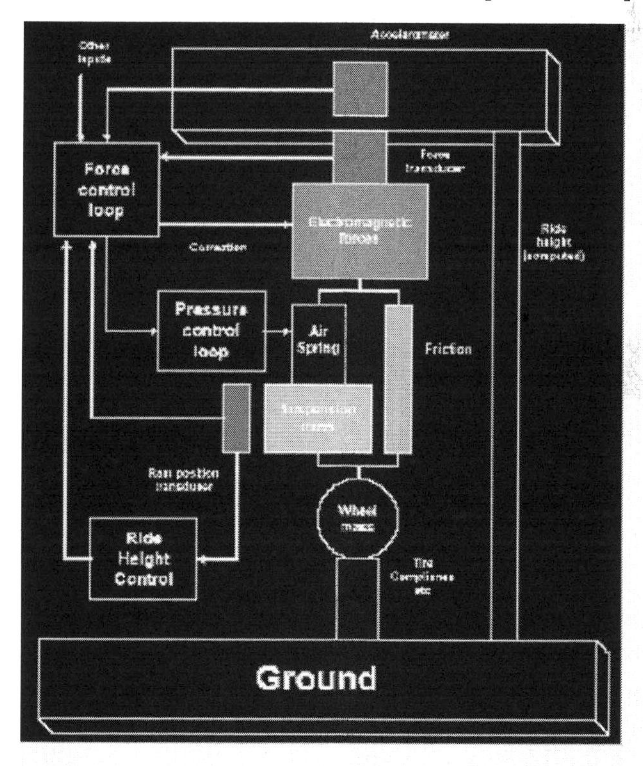

Fig. 5.162 A quarter-vehicle physical model of the active E-P-M ABW AWA suspension [AMT®; AMT 2005B].

The suspension E-P-M actuator moves to its average position and the air spring pressure is, without driver intervention, attuned to set-up. In an analogous mode, a series of strong down-ward-acting (vehicle-body lifting) electro-magnetic forces on all wheels, initiated by the reaction of the automotive vehicle to uneven on/off road surface, may without driver intervention cause the air (gas) spring pressures to intensify and the ride height to intensify.

A programmed bias arrangement may be set up to permit the automotive vehicle to move without driver intervention to low attitude (ground) clearance on an even on/off road surface.

Air (Gas) Spring Pressure Control for a Semi-Passive E-P-M ABW AWA Suspension -- It may be evident that the same air (gas) spring pressure control techniques may be used for the *ServoRam*™ E-P-M actuator when no external energy is fed and the module is a controlled shock absorber (damper). The energy for the attitude and ride height mechatronic control system then arrives from the vertical forces applied to the wheels by the on/off road surface -- explicitly from the energy of the vehicle motion. To lift the automotive vehicle the damping is intensified when the wheel moves upwards and it is reduced when the wheel moves downwards. The mean magnitude and sense of flow direction of the electrical currents drawn from the E-P-M actuator may therefore be controlled in the same mode as for an active E-P-M ABW AWA suspension mechatronic control system and the air (gas) spring pressure may be attuned in the proper sense, consistent with an analogous algorithm [AMT 2005B].

5.6.7 Active E-M-M ABW AWA Suspension Mechatronic Control Systems

Only a few years ago, active **electro-magneto-mechanical** (E-M-M) ABW AWA suspension mechatronic control systems were developed. These were designed to eliminate the compulsory pitch and roll performance of a spring suspension by quick changing the length of the firm strut between each axle and the vehicle body.

Active E-M-M ABW AWA suspension of these kinds might be predicted to raise each wheel over a bump and to push it down into a hole as required. But it was necessary for exactly controlled physical movement to occur repetitively – and as each movement reached its time to perform, its ability to deal with diminutive disturbances at high values of the vehicle velocity was held down.

An extremely developed design of an active E-M-M ABW AWA suspension mechatronic control system may be soon available, using an advanced technology of mechatronic force control. There is an entirely novel E-M-M suspension module that may take action in a fraction of a millisecond.

A linear E-M-M actuator consists of a rod of magnetostrictive material surrounded by an exciter (electrical coil). Energising the exciter, with electrical current causes the magnetic material to elongate in relation to the current magnitude.

Precision positioning may be achieved by accurately controlling the current. Such a simple E-M-M actuator may be used for precise mechanical positioning, vibration control, or switch and valve operation.

Magnetostriction is a physical phenomenon that may be described as the deformation of a body in response to an alteration in its magnetization (for considered materials due to an alteration of external magnetic field). This phenomenon is the most common magnetostrictive mechanism employed in magnetostrictive actuators.

It is well known that deformation λ depends on such parameters as temperature T, applied external stress σ_0 (prestress), external magnetic field intensity H and mechanical and magnetic load spectrum. In conventional materials, as iron, nickel or cobalt, λ amount to ca. 0.005%.

The reciprocal phenomenon consists in a altering of magnetization due to applying a stress to the material. It is known as reverse magnetostriction or *Villary*'s effect. This phenomenon is commonly used in magnetostrictive sensors [BOMBA AND KALETA 2003].

In 1965 in the Naval Ordnance Lab and the Ames Laboratory it was discovered [JOSHI AND BOBROV 2000], that some rare-earth elements and alloys, for example. **terbium** (Tb), **dysprosium** (Dy) and **samarium** (Sm), exhibit much higher magnetostriction in cryogenic temperatures than nickel ca. 0,2%.

A few years later an alloy of rare-earth elements and iron was developed that exhibits giant magnetostriction at ambient temperature. This group of materials is termed **giant magnetostrictive materials** (GMM). GMM belong to the group of **smart magnetic materials** (SMM).

Currently the most widely known is intermetallic alloy $Tb_xDy_{1x}Fe_y$ where $x = 0.27 \div 0.30$ and $y = 1.92 \div 2.00$ [DOPINO ET AL. 1999] – termed *TERFENOL-D*. They are manufactured in a form near to uniform crystal or as powders by the *Bridgman* or *Czochralski* method and as thin films by magnetron sputtering [BOMBA AND KALETA 2003]. They permit interchange of mechanical and magnetical energies (Fig. 5.163).

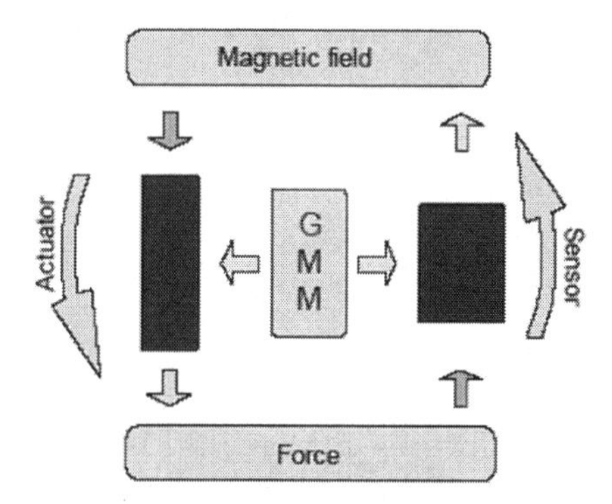

Fig. 5.163 Physical model of energy transformation in magnetostrictive materials [BOMBA AND KALETA 2002].

Magnetostriction and reverse magnetostriction have been widely applied since the 1950s to transducer construction. In many devices the conversion between magnetical and mechanical energies facilitates their construction. Normally to generate force or displacement in a magnetostrictive actuator one needs just to send a current through an exciter (electrical coil) to generate a magnetic field. To sense a magnetic field strength one needs to measure current induced by a magnetic field. Because of this, magnetostrictive devices are in fact **electro-magneto-mechanical** (E-M-M) transducers. Thus, magnetostriction is the SMM property that causes a material to change its length when subjected to a magnetic field. Magnetostriction properties also cause materials to generate magnetic fields when they are deformed by an external force. E-M-M actuators are one of the most exciting innovative actuator technologies available today. The recent GMM advances, mainly *TERFENOL-D*, have created new design options for automotive scientists and engineers alike [HATHAWAY AND CLARK 1993; CLARK 1994; JOSHI 2000; BOMBA AND KALETA 2003].

As the trend towards an E-M area continues, more and more magnetostrictive material such as the rare-earth lanthanides, or GMMs, may be of great benefit to automotive vehicle design engineers. GMMs are ones that exhibit a strain when exposed to a magnetic field. In other words, GMM undergo a deformation when a magnetic field is present. These GMM are referred to as the rare-earths. Rare-earth GMMs normally consist of the lanthanides group in the transition metals on the periodic table. The difference between conventional magnetostrictive materials as opposed to GMMs is the amount of strain per unit volume of the material. GMMs are ones that undergo large amounts of strain (i.e., large deformations) for a given applied magnetic field. One such material in the giant magnetostriction category is *TERFENOL-D* that is a proprietary alloy consisting of terbium, dysprosium, and iron in varying compositions. The use of GMMs such as *TERFENOL-D* may advance the development of E-M devices in automotive engineering. Meaning, conventional mechanical devices may be replaced by E-M devices that benefit from GMMs. This may usher in a new era in vehicular suspension actuator and sensor technology.

Some of the recent developments surrounding magnetostrictive applications include solid state speakers, vibration *'shake'* tables, transducers and various types of sensors, as well as vehicular suspension actuators.

Magnetostrictive devices show potential for replacing conventional piezoelectric devices as well. This last claim comes from the fact that magnetostriction is a material property that does not decay over time—experience less hysteresis. Furthermore, a magnetostrictive device is fairly robust as far as wear and tear is concerned.

HATHAWAY and CLARK [1993; LACHEISSERIE DE [1993]; CLARK [1994]; JOSHI [2000] AND BOMBA AND KALETA [2003] suggest that *TERFENOL-D* based actuators and sensors exposed to overheating may experience problems, however after the material is brought back to reasonable temperatures it may be functional once more. This is in contrast to some conventional materials that are ruined after an overheating event. GMMs may thus be used for both sensing and actuation.

TERFENOL-D is a commercially available magnetostrictive material that has *'giant'* magnetostriction at ambient temperatures. Where other actuation materials, such as piezoelectric crystals, require high voltages (200 - 300 V_{DC}) to create desired mechanical deformations, GMMs readily respond to significantly lower voltages. This is an attractive design characteristic in that a GMM may be excited with automotive available low-voltage power supplies. The advantages of GMM E-M-M actuators are high force and frequency, as well as simple construction. The disadvantage is small displacement. An important role in construction of E-M-M actuator is played by applied external stress (prestress) and magnetic field (bias-magnets). Figure 5.164 presents the influence of prestress on magnetostrictive characteristics in GMM samples.

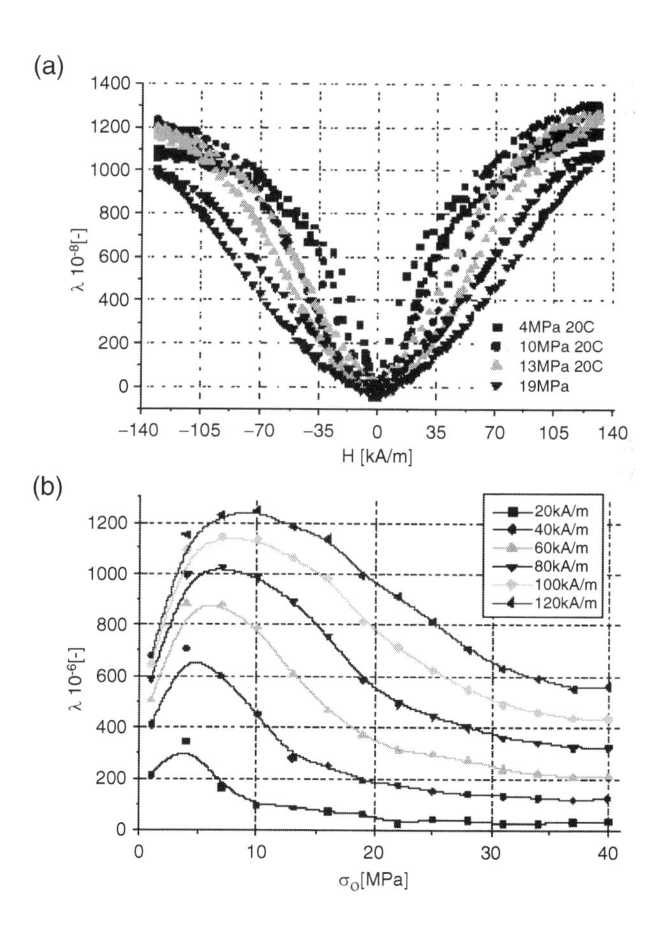

Fig. 5.164 Influence of prestress on magnetostriction.(a); prestress vs.magnetostriction for different magnetic field intensity (b) [BOMBA ET AL. 2003A, 2003B].

Figure 5.165 depicts a conventional construction of magnetostrictive E-M-M actuator. The GMM is in the form of bulk material. Of importance is applied external stress realised by pre-load spring what enhances magnetomechanical coupling (Figures 5.163 (a) and 5.164).

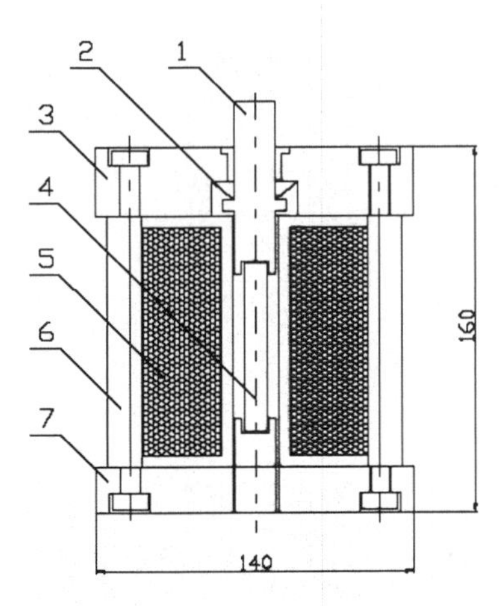

Fig. 5.165 Physical model of magnetostrictive E-M-M actuator;
1 – inserted flex pivot, 2 – preload washer, 3 – upper housing, 4 – magnetostrictive material, 5 – field coils, 6 – housing pivot, 7 – bottom housing [BOMBA AND KALETA 2003].

In some construction bias magnets are used in order to shift from a nonlinear range of magneto-mechanical coupling to a linear one.

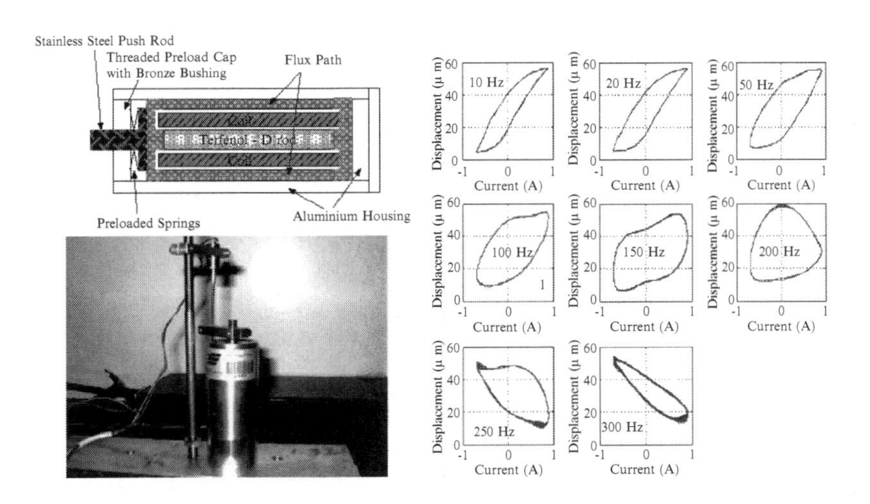

Fig. 5.166 An electro-magneto-mechanical (E-M-M) actuator and its rate-dependent hysteresis behaviour [TAN 2004].

Applications for magnetostrictive E-M-M actuators include high-force linear E-M-M motors, ultrasonic and seismic sources, accurate positioners, active vibration or noise mechatronic control systems, pumps, sonar and so on. There are a lot of magnetostrictive fuel injection patents which means that much attention is paid to application of GMM in the automotive industry [SENSORMAG 2001].

In Figure 5.166 is shown a magnetostrictive E-M-M actuator and its rate-dependent hysteresis behaviour for various frequencies [TAN 2004].

In a conventional application, an exciter (coil) would be placed around the material and the material is actuated by driving the exciter (coil).

The basic concept of a linear E-M-M actuator to create a linear stepper E-M motor is shown in Figure 5.167 [JOSHI 2000].

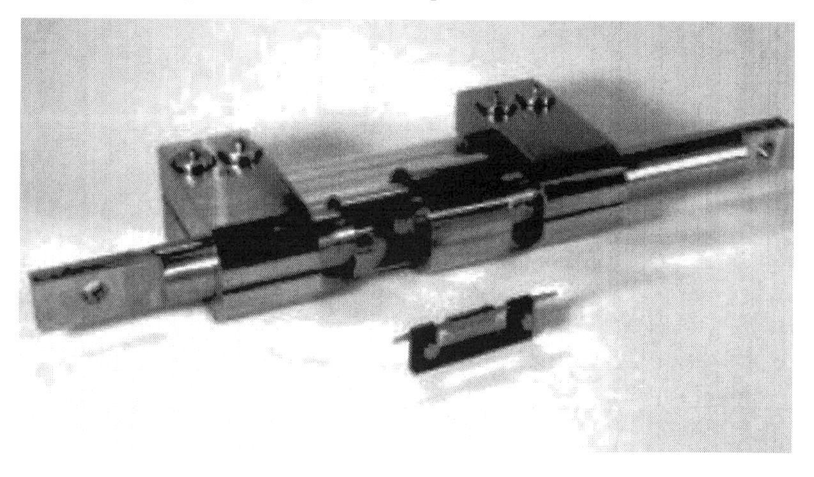

Fig. 5.167 Outlook of three E-M-M actuators to create a linear stepper E-M-M motor [Energen Inc.; JOSHI 2000].

The linear stepper E-M-M motor uses, for example, a set of three E-M-M actuators that enable it to move a rod up (forward) or down (backward) in a stepwise fashion. This E-M-M motor may provide a large stroke of several millimetres. Variations on this design may yield an MS-E actuator with a stroke limited only by the length of the translating rod.

Figure 5.167 shows a photograph of two linear stepper E-M-M motors. The translating rod contains a magnetostrictor element with connecting rods on each end [JOSHI 2000].

The centre section is a magnetostrictive rod surrounded by a superconducting exciter (electrical coil). Clamps on each side of the actuator grab onto connecting rods. These clamps contain a magnetostrictor actuator that when energised may cause the clamp to release its hold on the rod.

In the not-too-distant future, E-M-M actuators and linear stepper E-M-M motors may be also utilized to active E-M-M ABW AWA suspension mechatronic control systems.

5.7 Hybrid ABW AWA Suspension Mechatronic Control Systems

The suspension of most automotive vehicles is purely passive, that is, may be schematised as composed of passive elements, for instance, shock absorbers (dampers) and springs.

In an active ABW AWA suspension mechatronic control system an actuator of variable length regulates the interaction between vehicle's body and wheel. The active ABW AWA suspension mechatronic control systems [THOMPSON 1976; HAC 1985; CORRIGA ET AL. 1991] have better performance than passive suspensions [CORRIGA ET AL. 1996 HROVAT 1997].

However, the associated power that must be provided by the automotive vehicle's ECE or ICE may reach the order of several 10 kW depending on the required performance. As a viable alternative to a purely active ABW AWA suspension mechatronic control system, the use of a hybrid ABW AWA suspension mechatronic control system (an actuator in parallel with a passive suspension) has been considered [CROSBY AND KARNOPP 1973; KRASNICKI 1980; CHEOK ET AL. 1985; GÖRING ET AL. 1993; GRIMA AND RENOU 1993; ROBERTI ET AL. 1993; GIUA ET AL. 2000].

Such an active ABW AWA suspension mechatronic control system requires a lower power controller. Furthermore, even in case of malfunctioning of the active subsystem the automotive vehicle need not halt because the passive suspension can still function. It may also be required that the total absorbing (damping) forces between vehicle body and wheel and the fraction of the absorbing (damping) forces generated by the actuators never exceed given bounds.

The constraint on the actuator forces can be used to measure the total absorbing (damping) forces by the actuators, so that it is required to provide only a fraction of the total absorbing (damping) forces generated by the hybrid ABW AWA suspension mechatronic control systems.The constraint on the absorbing (damping) forces bounds the acceleration on the sprung masses so as to ensure the comfort of drivers and passengers and to reduce the risk of loss of contact between wheel-tyres and on/off road surfaces. Simulations showed that these constraints are active only when the hybrid ABW AWA suspension mechatronic system's state is far from the origin [GIUA ET AL. 2000].

It is well known that passenger automotive vehicle ABW AWA suspension mechatronic control systems provide comfort by isolating passengers from on/off road surface disturbances and improve handling by regulating the contact forces between the vehicle-chassis, wheel-tyres and the on/off road surface. It is also well known, that these requirements are mutually conflicting. Softer ABW AWA suspensions generally offer more ride comfort at the cost of degraded handling [KARNOPP AND HEESS 1991].

Automotive vehicle suspension design often trades these requirements off by grouping them into a weighted performance function for optimisation [HROVAT 1993; SMITH AND WALKER 2000].

An ABW AWA suspension's optimal performance depends on whether it is active or passive. Active ABW AWA suspensions use external energy sources (e.g. F-M or P-M or even E-M actuators), whereas passive suspensions consist solely of energy storage and dissipation components (e.g. springs and viscous shock absorbers) [KARNOPP AND HEESS 1991].

Passive components may only impart forces that depend on relative chassis/wheel-tyre motion, whereas active (sky-hook) elements may generate forces that depend on absolute chassis motion.

Consequently, active ABW AWA suspension may outperform their passive counterparts significantly, at the expense of using external energy [KARNOPP AND HEESS 1991; HROVAT 1993; SMITH AND WALKER 2000; SHARP AND HASSAN 1986A].

When a hybrid ABW AWA suspension comprises both passive and active components, these components *"compete, rather than help each other"* [SMITH ET AL. 1992], thus failing to reach the suspension's full synergistic performance potential [ULSOY ET AL. 1994].

This competition exemplifies the well-known coupling between the plant and controller optimisation problems [SMITH ET AL. 1992; BRUSHER ET AL. 1997B; FATHY ET AL. 2001].

Sequentially optimising a suspension's plant (*i.e.* passive components) and controller (i.e. active components) does not account for this coupling and hence fails to guarantee system optimality. To prevent sub-optimality and to design synergistic suspensions, one must optimise passive and active constituents simultaneously.

Such a simultaneous optimisation approach is given in FATHY ET AL. [2003] that contributes an integrated passive/active suspension optimisation problem and its solution using a nested strategy. The resulting system-optimal hybrid ABW AWA suspension outperforms its passive, active and sequentially optimised passive/active counterparts in both the time and frequency domains [FATHY ET AL. 2003].

5.8 Discussion and Conclusions

What is active ABW AWA suspension, and how does it differ from conventional vehicular suspension? When automotive suspension designers refer to a conventional vehicular suspension, they mean a system that comes *'as is'*. In other words, a conventional system is a passive suspension. Once it has been installed in the automotive vehicle, its character modifies very little. This has certain advantages and disadvantages. On the plus side, the vehicular suspension is very predictable. Over time, vehicular suspension designers may develop a familiarity with their automotive vehicle's suspensions. They may understand their capabilities and theirs limitations. On the down side, once the suspension has reached these restrictions, it has no way of compensating for circumstances beyond its design parameters. Thus shock absorbers (dampers) bottom out, struts overextend, springs respond sluggishly, and torsion bars get tweaked.

An active ABW AWA suspension mechatronic control system, on the other hand, has the capability to adjust itself continuously to altering on/off road surface circumstances. It *'artificially'* extends the design parameters of the mechatronic control system by constantly monitoring and adjusting itself, thereby altering its character on an unchanging foundation. It is schizophrenic, as people may be, but with a purpose.

With advanced sensors and microprocessors feeding it data all the time, its identity remains fluid, contextually amorphous. By altering its character to respond to varying on/off road surface circumstances, active ABW AWA suspension offers superior handling, on/off road feel, responsiveness and safety. Before people dive into active ABW AWA suspension mechatronic control systems, though, a word about why they should care. True, this type of technology typically appears on very expensive automotive vehicles. But, as with any new technology, a *'trickle-down'* effect takes place.

Rapid advances in single-chip microprocessor science may soon bring these features to a whole innovative range of automotive vehicles, including family sedans, minivans, trucks, SUVs, even compact automotive vehicles.

An active ABW AWA suspension mechatronic control system uses conventional passive shock absorbers and springs made active by an outside power source, such as a M-F pump or M-P compressor or even M-E generator and one or more fluidical (hydraulical and/or pneumatical) or even electrical accumulators adding or removing oily fluid or air (gas) or even electric current from the units, respectively.

The low frequency active ABW AWA suspension mechatronic control systems being developed are currently using a single central M-F pump or M-P compressor or even M-E generator and one or more fluidical (hydraulical and/or pneumatical) or even electrical accumulators to provide and store fluidical (hydraulical and/or pneumatical) or even electrical energy at a constant oily-fluid or air (gas) pressure or even electric current, respectively, for actuating the suspension units.

For instance, the low frequency active ABW AWA suspension mechatronic control system (sometimes referred to as low band active or soft active) uses a pneumatical spring element, usually in the form of an F-P-M spring/shock absorber that isolates the high frequency on/off road surface vibrations from the vehicle body. Each F-P-M suspension unit is made active by adding or removing oily fluid. In effect, the pre-deflection is actively changed for each pneumatical F-P-M spring. Each F-P-M spring is actuated to minimise the low frequency vehicle-body motions of pitch, roll, and heave that range from 1 to 3 Hz. Most of the benefits of an active ABW AWA suspension mechatronic control system may be achieved with a low frequency active system, but with much less complexity, lower cost, and reduced power requirements.

A suspension may be considered active when an outside power source is used to alter its characteristics, and these systems can be placed in one of three different categories: semi-active damping, fully active, and low frequency active.

A regenerative M-F pump or M-P compressor or even M-E generator concept may minimise the power requirement for the low frequency active system.

For example, the regenerative M-F pump concept uses four independent variable displacement M-F/E-M pump/motor combinations on a common shaft to actuate each separate suspension unit.

Active ABW AWA suspension mechatronic control systems, also known as **computerised ride control** (CRC), consist of the following components: a computer or two, sometimes referred to as an **electronic control unit** (ECU), adjustable shocks and springs, a series of sensors at each wheel and throughout the automotive vehicle, and an actuator or servo atop each shock and spring. The components may vary slightly from OEM to OEM, but these are the basic parts that make up an active suspension system. As mentioned above, active suspension operates by constantly sensing alterations in the on/off road surface and feeding that data, by means of the ECU, to the outlying components. These components then act upon the mechatronic control system to modify its character, adjusting shock stiffness, spring rate and the like, to enhance ride performance, drivability, responsiveness, and so on. In a common sense, active suspension mimics the functions of the human body.

Consider:

- The sensors are nerve ends -- seeing, feeling, hearing, even tasting the on/off road surface and delivering that data back to the ECU;
- The ECU mimics a human mind, defective though it may be (residual damage from the 1960s) -- collecting, classifying, interpreting and analysing sensory input; once it has interpreted this data, it makes decisions and sends *'marching orders'* to the outposts in the hinterlands;
- The wires connecting the whole thing are the central nervous system, stout deliverers of commands and directives; they dispense decrees, elucidate edicts, and issue instructions.

A driver cruises down the on/off road in an automotive vehicle with an active ABW AWA suspension mechatronic control system, rounds the first turn, and hits a series of potholes, each one larger than the next.

Now, in an automotive vehicle with conventional suspension, these potholes might present a serious challenge to the suspension. Their ever-increasing size could even max out the system, setting up an oscillation loop -- a situation wherein the automotive vehicle begins to bob up and down, higher and higher, and gets a little out of control. But the driver is not worried, because the automotive vehicle has active ABW AWA suspension.

The sensors on the right front of the automotive vehicle (assuming the driver is turning left, by the way) begin to monitor the situation. They pick up yaw and transverse vehicle-body motion, and send this data back to the ECU. They also sense excessive vertical travel, particularly in the right-front region of the automotive vehicle. This data is likewise forwarded to the ECU.

Rotary-position wheel sensors and a steering angle sensor confirm the data coming off the other *'nerve endings'* on the automotive vehicle. The ECU collects, analyses, interprets the data in approximately 10 ms, and immediately sends an urgent message to the servo atop the right-front coil spring to *'stiffen up'*. To accomplish this, an ECE- or ICE-driven oil M-F pump operating at nearly 21.3 MPa (3,000 lb. per sq. in.) sends additional fluid to the servo that increases spring tension, thereby reducing vehicle-body roll, yaw, and spring oscillation. A similar message, but with a slightly less intense nature, is sent to the servo atop the right-rear coil spring, with similar results.

At the same time, another set of actuators kicks in to temporarily increase the rigidity of the suspension shock absorbers (dampers) on the right-front and -rear corners of the automotive vehicle.

The automotive vehicle glides through the turn, not even breathing hard. And neither is the driver. The ride potential for such a mechatronic control system is truly spectacular. As we mentioned at the outset, several state-of-the-art producers of passenger automotive vehicles offer active ABW AWA suspension mechatronic control systems as standard or optional equipment. Here's a brief rundown.

Cadillac - Beginning in 1996, Cadillac began offering active suspension on its *Eldorado Touring Coupe, DeVille Concours* and *Seville STS* models. Known as **continuously variable road-sensing suspension** (CVRSS), this mechatronic control system uses a series of sensors to actuate F-M shock absorbers at all four corners, improving road feel and dampening. The mechatronic control system adjusts in a fraction of a second -- the amount of time it takes an automotive vehicle going circa 100 km/h (65 mph) to travel 38 mm (15 in).

Land Rover - Starting with the 1999 model year, the Land Rover Discovery Series II sport Ute comes with **active cornering enhancement** (ACE). The Land Rover system is a first for SUVs. It uses an active F-M ABW AWA suspension that replaces the more conventional front and rear anti-roll bars, applying torque to the vehicle body via two piston/lever configurations. The mechatronic control system has the capability to counteract up to 1.0 g lateral acceleration in 250 ms. Take a look at these driving impressions of the Land Rover:

Mercedes-Benz - The new *2000 CL500* boasts what MB modestly terms *'the world's only active suspension system'*. Anyway, the Mercedes system, known as **Active Body Control** (ABC), uses 13 different on-board sensors, which feed into four F-M servos positioned atop each coil spring.

The ABC computer adjusts the ride every 10 ms. This system is the result of over 20 years of research. Remember, in the case of active suspension, today's innovation is tomorrow's standard feature. Expect to see active ABW AWA suspension mechatronic control systems in the coming years on much lower price point automotive vehicles.

Passive suspensions are designed to dispel the energy otherwise transmitted to an automotive vehicle's body because of interactions with an on/off road surface. The dispelled energy related with vertical and transverse coordinates as a rule creates from the longitudinal motion of the automotive vehicle and is transmitted because of the wheel-tyre-ground contact patch. Besides, since the longitudinal energy is created by the automotive vehicle's ECE or ICE, the energy dispelled by means of the suspension shock absorber (damper) as well as other components (for example, mechanical joints, and so on) in effect dispel some ECE or ICE energy.

The theory has been confirmed that, when driving on a rough on/off road surface, the automotive vehicle's longitudinal velocity diminishes when vertical vibrations escalate. Results show that an automotive vehicle set up with a passive suspension is subjected to a greater velocity reduction contrasted to one with an active suspension negotiating the same rough on/off road surface.

Computer simulations confirm that corresponding to passive suspensions, not only do active suspensions yield substantial enhancement in ride quality, they can also result in considerable energy savings.

It may be concluded that if F-M, P-M or E-M actuators are augmented by passive springs to support the vehicle static mass, the amount of energy necessary for function of F-M, P-M or E-M actuators is considerably less than the amount dispelled by conventional shock absorbers (dampers). The active ABW AWA suspension mechatronic control system affords exceptional levels of performance that are inaccessible with other suspension mechatronic control systems and conventional passive suspensions.

Estimations done with actual automotive vehicles verified the effectiveness of the active ABW AWA suspension mechatronic control system in getting better ride comfort and handling stability predispositions.

On the topic of ABW AWA suspension mechatronic control systems, semi-active and active F-M, P-M and/or E-M ABW AWA suspension mechatronic control systems have previously been used in practical applications during the early development of enhancements to driving stability and ride feature.

Eventually, automotive scientists and engineers focused their centre of attention on developing vehicle roll control by the E-M FA ABW AWA suspension mechatronic control systems -- again to enhance vehicle dynamics and driving safety.

The proprietary active E-M ABW AWA suspension mechatronic control system couples tubular, linear, brushless DC-AC/AC-DC macrocommutator IPM magnetoelectrically excited motors/generators and their DC-AC/AC-DC macrocommutators with a set of unique mechatronic control algorithms. For the first time, it is possible to have, in the same automotive vehicle, a much smoother ride than in any luxury automotive vehicle and less roll and pitch than in any SUV.

The research automotive vehicle may provide superior comfort by gliding smoothly over bumpy on/off road surfaces and superior control by keeping the vehicle-body level during aggressive manoeuvres.

E-M actuators have the necessary force and displacement capabilities for automotive applications. However, their implementation is often not practical because of their large mass, electrical demands, and limited bandwidth.

An E-M actuator required to induce necessary force and displacement would be extremely heavy and demand a significant electrical power. This makes E-M actuators poorly suited to directly control vibrations of large amplitude and high bandwidth. Besides requiring a significant external power supply for E-M actuators, active control has the inherent danger of becoming unstable through the injection of mechanical energy into the active ABW AWA suspension mechatronic control system.

The semi-active MR or ER AWA ABE suspension features MR or ER shock absorbers (dampers) able to detect on/off road surfaces and adjust the absorbing (damping) rates to those surfaces almost instantly for optimal ride and vehicle-body control. It is the fastest reacting suspension, replacing mechanical valves with nearly instantaneous reactions of MRF or ERF or even MRE.

MRFs and/or ERFs or even MREs have the very interesting aptitude to experience an exciting change in their viscosity, and thus their physical and mechanical properties, in less than a millisecond through the function of an electric or magnetic field, respectively. This incorporation of a fast response time with a considerable adjustment of properties translates into a gear aptitude for absorbing (damping) and other applications. These smart fluids or elastomers may be used in vibration control and shock absorbing (dampening) systems as well as in exercise equipment, valve, clutch and brake systems. The ground rules of conventional shock absorbers (dampers) have not transformed considerably since they were first introduced into automotive vehicles, but there is an innovative system employment. MRFs and/or ERFs or even MREs are a major advancement in this technology. MR or ER shock absorbers, embedded in some automotive vehicles, afford a more stable and smooth ride by continually varying their absorbing (dampening) rate. The innovative system is about *5* times faster responding to on/off road bumps than conventional shock absorbing (dampening) systems.

The semi-active MR or ER ABW AWA suspension mechatronic control system is a vehicle suspension mechatronic control system that uses a revolutionary shock absorber (damper) design to control wheel and body motion with innovative MRF- or ERF-based actuators. By controlling the current to an electromagnetic exciter-coil or voltage to an electrostatic exciter-condenser inside the piston of the damper, the fluid's consistency can be changed, resulting in real-time, continuous variable control of vehicle damping. As a result, drivers feel a greater sense of security, with smooth, well-controlled ride motions, and more precise and responsive handling, particularly on uneven surfaces and during challenging manoeuvres.

The semi-active MR or ER ABW AWA suspension mechatronic control system helps maintain the maximum amount of wheel-tyre patch in contact with the on/off road surface.

The system isolates and smoothes the action of each wheel-tyre, resulting in less bouncing, vibration and noise. At all values of the vehicle velocity, on all road surfaces and twisting variations, the system integrates with ABS and TCS to keep the vehicle balanced and poised.

The semi-active MR or ER ABW AWA suspension mechatronic control system is the fastest reacting vehicle suspension mechatronic control system, responding in one millisecond. Using a simple combination of sensors, as well as steering wheel and braking inputs from the driver, the system's onboard ECU reacts to wheel inputs by sending an electrical signal to exciter coils or exciter-condenser in each shock absorber (damper), changing the absorbing (damping) fluid's flow properties. This MRF or ERF or even MRE contains randomly dispersed ferrous micro-particles or dielectric nano-particles that, in the presence of a magnetic field or an electric field, respectively, align themselves into structures adopting a near-plastic state. This action regulates the damping properties of the mono-tube struts, capable of changing up to 1,000 times per second.

A semi-active MR ABW AWA suspension mechatronic control system, designed for bolt-on retrofitting, consists of one MRF computer controlled shock absorber (damper) at each of the vehicle's wheel positions. The active shock absorber (damper) suspension modulates the forces in a shock absorber (damper) as a function of sensed variables, such as the vehicle velocity (speed), body movements, and position of a particular wheel. The innovative active shock absorber (damper) suspension mechatronic control system is lighter, smaller, less expensive, and uses much less power than a fully active system while providing similar levels of performance. Tests showed significant (70% on certain terrain) reduction in driver absorbed power, excellent reliability and no failures.
Other advantages of the MRF system include:

- Higher values of the mobility velocity (speed) over a given terrain;
- Improved wheel-tyre traction;
- Improved wheel-tyre life;
- Reduced fatigue loading of vehicle structure and payload;
- Reduced driver, vehicle and payload damage from terrain impacts at speed - Improved vehicle stability and handling;
- Improved accuracy during surveillance, targeting, or weapons firing;
- Suspension system prognostics / diagnostics.

MagneRide™ currently is the automotive industry's only real-time control system without electromechanical valves or small moving parts. The system consists of MR fluid-based mono-tube shock absorbers, a sensor set and on-board controller. The onboard controller continually adjusts the damping forces up to once every millisecond based on input from four suspension displacement sensors, a two-position driver input switch on the Chevrolet *Corvette*, a lateral accelerometer and a steering wheel angle sensor. By controlling the current to an electromagnetic coil inside the piston of the damper, the MR fluid's resistance to flow can be changed. This characteristic, and other features of the *MagneRide*™ system, provides smooth, continuously variable damping in a cost effective and reliable package that reduces body motion and increases tire road contact on all types of surfaces.

MR and ER flow mode shock absorbers (dampers) may exhibit bi-viscous absorbing (damping) behaviour. Such behaviour is characterized by a high damping pre-yield region for low velocities, with a transition to a relatively lower post-yield damping, once the MR or ER damper force exceeds the static yield force of the damper. The bi-viscous damping behaviour is typically the result of leakage, that is, a second path of *Newtonian* flow in addition to the *Bingham* plastic flow through the ER or MR fluidical valve.

Chassis Continuously Variable Real-Time Damping (CCVRTD) suspension system with a semi-active MR ABW AWA suspension mechatronic control system. The resulting uncompromised comfort and handling with redesigned jounce bumper for superior wheel control and large impact feel, and automatic and independent adjustment of rebound and compression damping loads are a perfect upgrade to MR shock absorbers that use state of the art MRF, which is a complex, synthetic -- hydrocarbon -- based liquid having somewhere between 20% and 40% (by volume) Carbony iron particles (97 -- 99% iron) and offers excellent electromagnetic properties. The flow characteristics (rheology) change when subjected to direct current -- magnetic field strength. The MR controller is a powerful dual-processor device that is capable of making 1000 adjustments per second in each of the automotive vehicle's four shocks. The controller receives each wheel movement in the automotive vehicle via position sensors and according to the wheel movement data; the controller may determine the vehicle body motions (pitch, roll, and lift). Utilizing the two-mode centre console switch, in (tour) mode, the controller emphasizes driver comfort, while in (sport) mode the road feel and steering response may be maximised. This upgrade kit uses the existing wheel-position sensors, driver select switch, and main suspension system wiring harness.

Glossary

Absorb-by-wire (ABW) represents the replacement of conventional components such as the shock absorbers (dampers) with mechatronic adjustable electrorheological or magneto-rheological shock absorbers (dampers) or linear tubular E-M motors as well as sensors and actuators; a ABW suspension mechatronic control system, by nature, is a safety critical system and therefore fault tolerance is a vitally important characteristic of this system; as a result, an ABW suspension mechatronic control system is designed in such way that much of its essential information would be derived from a variety of sources (sensors) and be handled by more than the bare necessity hardware.

Absorbing (damping) is the mechatronic control of motion or oscillation, as seen with the use of fluidical (hydraulical or pneumatical) gates and valves in an automotive vehicles' shock absorber; this may also vary, intentionally or unintentionally; like spring rate, the optimal absorbing (damping) for comfort may be less than for control; absorbing (damping) controls the travel velocity (speed) and resistance of the vehicular suspension; An undamped automotive vehicle may oscillate up and down; with proper damping levels, the automotive vehicle may settle back to a normal state in a minimal amount of time; most damping in modern automotive vehicles may be controlled by increasing or decreasing the resistance to oily-fluid or air (gas) flow in the shock absorber.

AC-DC commutator - The commutator is a mechanical AC-DC rectifier; for a rotary DC-AC commutator generators, the commutator mechanically switches the armature windings so that the resultant induced source AC armature voltages always act with the same sense of voltage polarisation; this requires a reversal of the armature winding connection every π rad; the induced source AC armature voltages are mechanically rectified to induced source DC armature voltage via commutator segments that contact the carbon brushes.

AC-DC macrocommutator - The macrocommutator is an ASIM AC-DC rectifier; for a rotary DC-AC commutator generators, the macrocommutator electronically switches the armature windings so that the resultant induced source AC armature voltages always act with the same sense of voltage polarisation; this requires a reversal of the armature winding connection every π rad; the induced source AC armature voltages are electronically rectified to induced source DC armature voltage via inputs of the ASIM that contact via bipolar electrical valves an output of the ASIM.

Active rollover protection (ARP) is a system that recognizes impending rollover and selectively applies brakes to resist; ARP builds on **electronic stability control** (ESC) and its three chassis control systems already on the vehicle -- **anti-lock braking system** (ABS), traction control and yaw control; ARP adds another function: detection of an impending tip or rollover; excessive lateral force, which is generated by driving too fast in a corner or turn, may result in a rollover because of the high vehicle **centre of gravity** (CoG); ARP automatically responds whenever it detects an unstable condition leading to a potential rollover; ARP rapidly applies the brakes with a high burst of pressure to the appropriate wheels to interrupt the rollover before it occurs.

Active steering is an automotive technology that varies the degree that the wheels turn in response to the steering hand wheel; at lower values of the vehicle velocity, this technology reduces the amount that the steering hand wheel must be turned -- enhancing performance in circumstances such as parking and other conurbation area traffic manoeuvres; at higher speeds, the performance is such that steering becomes more responsive and provides improved directional stability.

Active suspension allows mechatronic control of the vehicle body motions and therefore virtually eliminates body roll in many driving circumstances including cornering (steering), accelerating (driving), and decelerations (braking); this technology allows automotive vehicle manufacturers to achieve a higher degree of both ride quality and vehicle and keeps the wheel-tyres perpendicular to the on/off road surface in corners, allowing for much higher levels of grip and control; not possible before the advent of modern computing technology, an on-board computer detects body movement from sensors located throughout the vehicle, sensors continually monitor vehicle-body movement and vehicle level and supply the computer with new data constantly.

Actuator - The component of an open-loop or closed-loop mechatronic control system that connects the **electronic control unit** (ECU) with the process; the actuator consists of a commutator and a final-control element; positioning electrical signals are converted to mechanical output.

Analog signal **-** A signal in which the information of interest is communicated in the form of a continuous signal; the magnitude of this signal is proportional (or analogous) to the actual quantity of interest.

Analog input -- Sensors usually generate electrical signals that are directly proportional to the mechanism being sensed; the signal is, therefore analogue or may vary from a minimum limit to a maximum limit.

Analog-to-digital (A/D) converter - An electronic device that produces a digital result that is proportional to the analog input voltage.

Algorithm - A set of software instructions causing a computer to go through a prescribed routine; because embedded computer ECE or ICE controls have become so common, algorithm has become essentially synonymous with control law for automotive scientists and engineers.

Analog input - Sensors usually generate electrical signals that are directly proportional to the mechanism being sensed; the signal is, therefore analogue or may vary from a minimum limit to a maximum limit.

Absorbing (damping) - Any effect, either deliberately engendered or inherent to a system that tends to reduce the amplitude of oscillations of an oscillatory system.

ASIC - Application-specific integrated circuit, an IC designed for a custom requirement, frequently a gate array, single-chip microprocessor or programmable logic device.

ASIM - Application-specific integrated matrixer, an IM designed for a custom requirement, frequently a gate array or single-chip macrocommutator.

Automotive vehicle - A self propelled vehicle or machine for land transportation of people or commodities or for moving materials, such as a passenger car, bus, truck, motor-cycle, tractor, or earthmover.

Brake is a device for slowing or stopping the motion of a machine or vehicle, and to keep it from starting to move again; the kinetic energy lost by the moving part is usually translated to heat by friction.

Brake-by wire (BBW) - represents the replacement of conventional components such as the pumps, hoses, fluids, belts and brake boosters and master cylinders with electronic sensors and actuators; a BBW dispulsion mechatronic control system, by nature, is a safety critical system and therefore fault tolerance is a vitally important characteristic of this system; as a result, a BBW dispulsion mechatronic control system is designed in such way that many of its essential information would be derived from a variety of sources (sensors) and be handled by more than the bare necessity hardware.

Braking is the term applied to the collection of components, linkages, etc. which allows a driver to use a brake foot pedal; alternatively, in regenerative braking, much of the energy is recovered and stored in a flywheel, capacitor, inductor or turned into alternating current by an alternator, then rectified and stored in a battery for later use.

Closed-loop mechatronic control - A process by which a variable is continuously measured, compared with a reference variable, and changes as a result of this comparison in such a manner that the deviation from the reference variable is reduced; the purpose of closed-loop mechatronic control is to bring the value of the output variable as close as possible to the value specified by the reference variable in spite of disturbances; in contrast to open-loop mechatronic control, a closed-loop mechatronic control system acts to offset the effect of all disturbances.

Central processing unit (CPU) - The portion of a computer system or microcontroller that controls the interpretation and execution of instructions and includes arithmetic capability.

DC-AC commutator - The commutator is a mechanical DC-AC inverter; for a rotary DC-AC commutator motors or actuators, the commutator mechanically switches the armature windings so that the resultant force always acts in the same sense of rotary direction; this requires a reversal of the armature winding connection every π rad; the DC supply to the armature is via carbon brushes that contact the commutator segments.

DC-AC macrocommutator - The macrocommutator is an ASIM DC-AC inverter; for a rotary DC-AC commutator motors or actuators, the macrocommutator electronically switches the armature windings so that the resultant force always acts in the same sense of rotary direction; this requires a reversal of the armature winding connection every π rad; the DC supply to the armature is via an input of the ASIM that contact via bipolar electrical valves outputs of the ASIM.

Defuzzification - The process of translating output grades to analogue output values.

Dependent suspension - normally has a live axle (a simple beam or *'cart'* axle) that holds wheels parallel to each other and perpendicular to the axle; when the camber of one wheel changes, the camber of the opposite wheel changes in the same way (by convention on one side this is a positive change in camber and on the other side this a negative change).

Digital signal - A signal in which the information of interest is communicated in the form of a number; the magnitude of this number is proportional to (within the limitations of the resolution of the number) the actual quantity of interest.

Digital signal processor (DSP) - A monolithic integrated circuit (IC) optimised for digital signal-processing applications; portions of device are similar to a conventional microprocessor; the architecture is highly optimised for the rapid, repeated additions and multiplications required for digital signal processing; digital signal processors may be implemented as programmable devices or may be realised as dedicated high-speed logic.

Dispulsion - The act of dispelling or the state of being dispelled; negative propulsion, the act of reducing the effect of propulsion, the act of decelerating; a braking system is used for the **dispulsion** of an automotive vehicle.

Conversion - The act of turning or the state of being converting; a steering system is used for the conversion of an automotive vehicle.

Drive - The means by which a machine is given motion or power (as in steam engine, diesel-electric drive), or by which power is transferred from one part of a machine to another (as in gear drive, belt drive).

Drive-by-wire (DBW) - technology in automotive industry replaces the conventional mechano-mechanical and fluido-mechanical mechatronic control systems with electro-mechanical mechatronic control systems using electro-mechanical actuators and human-machine interfaces such as accelerator foot pedal and steering feel emulators.

Driver - A solid state device used to transfer electrical energy to the next stage that may be another driver, an electrical load (power driver), a wire or cable (line driver), a display (display driver), etc.

Driving wheel - A wheel that supplies driving power.

Electro-Mechanical (E-M) suspension - A suspension system using electrical energy to support the vehicle body.

Final-control element - The second or last stage of an actuator to control mechanical output.

Fluido-Mechanical (F-M) suspension - A suspension system using oily fluid to support the vehicle body.

Fluido-Pneumo-Mechanical (F-P-M) suspension - A suspension system using oily fluid and air to support the vehicle body.

Fuzzification - The process of translating analog input variables to input memberships or labels.

Fuzzy logic (FL) - Software design based upon a reasoning model rather than fixed mathematical algorithms; a FL design allows the automotive system engineer to participate in the software design because the fuzzy language is linguistic and built upon easy-to-comprehend fundamentals.

Gear - An adjustment device of the transmission in an automotive vehicle that determines mechanical advantage, relative velocity and sense of travel direction.

Groundhook shock absorber (damper) control - The control law applied to control the automotive vehicle as if it were fixed within absolute space suspended from the ground; groundhook control is based on the idea that a fictitious shock absorber (damper) is attached between the unsprung mass and an immovable point in the ground; the groundhook control differs from the skyhook control in that the shock absorber (damper) is now connected to the unsprung mass rather than the sprung mass.

Hybrid skyhook-groundhook shock absorber (damper) control - From the simulation and experimental results, the hybrid skyhook-groundhook shock absorber (damper) control shows significant improvement in vehicle-body acceleration, vehicle-body displacement, suspension displacement without allowing excessive wheel acceleration magnitude; the hybrid skyhook-groundhook controller is also superior to the counterparts in overcoming the shock absorber (damper) constraint by generating the target forces consistently in the same sign with the damper velocity.

Independent suspension allows wheels to rise and fall on their own without affecting the opposite wheel; suspensions with other devices, such as anti-roll bars that link the wheels in some way are still classed as independent.

Input memberships - The input signal or sensor range is divided into degrees of membership, i.e., low, medium, high or cold, cool, comfortable, warm, hot; each of these membership levels is assigned numerical values or grades.

Microcontroller unit (MCU) - A semiconductor device that has a CPU, memory, and I/O capability on the same chip.

Open-loop mechatronic control - A process within a mechatronic control system in which one or more input variables act on output variables based on the inherent characteristics of the mechatronic control system; an open loop is a series of elements that act on one another as links in a chain; in an open loop, only disturbances that are measured by the control unit can be addressed; the open loop has no effect on other disturbances.

Output memberships - The output signal is divided into grades such as off, slow, medium, fast, and full-on; numerical values are assigned to each grade; grades can be either singleton (one value) or *Mandani* (a range of values per grade).

Passive suspension - Conventional springs and shock absorbers (dampers) are referred to as passive suspensions; if the suspension is externally controlled then it is a semi-active or active suspension.

Pneumo-Mechanical (P-M) suspension - A suspension system using air (gas) to support the vehicle body.

Prime mover - The component of a power plant that transforms energy from the thermal or pressure form to the mechanical form.

Powertrain - The part of an automotive vehicle connecting the engine to propeller or driven axle may include drive shaft, clutch, transmission, and differential gear.

Propulsion - The process of causing a body to move by exerting a force against it; a force causing movement.

Protocol - The rules governing the exchange of information (data) between networked elements.

Semi-active suspension includes devices such as air springs and switchable shock absorbers, various self-levelling solutions, as well as systems like hydropneumatical, hydroelastic, and hydragas suspensions; currently semi-active suspension includes shock absorbers filled with a magneto-rheological fluid, whose viscosity may be changed electro-magnetically, thereby giving variable control without switching valves, which is faster and thus more effective; for example, a hydropneumatical Citroën may *'know'* how far off the ground the automotive vehicle is supposed to be and constantly reset to achieve that level, regardless of load.

Semi-dependent suspension - In this case, jointed axles are used, on drive wheels, but the wheels are connected with a solid member, most often a *de Dion axle*; this differs from *'dependent'* mainly in unsprung mass.

Shock absorber (damper) is a mechanical device designed to smooth out or damp a sudden shock impulse and dissipate kinetic mechanical energy.

Skyhook shock absorber (damper) control - The control law applied to control the automotive vehicle as if it were fixed within absolute space suspended from the sky; skyhook control is based on the idea that a fictitious shock absorber (damper) is attached between the sprung mass and an immovable point in the sky; the skyhook is designed to insure control of the whole automotive vehicle with anti-dive and anti-squat features; available as an option, skyhook uses six sensors, one at each wheel and two on the vehicle body, to provide real-time ABW AWA suspension absorbing (damping) forty times per second; additionally, the **human driver** (driver) can select one of two settings: *'Normal'* or *'Performance'*.

Steer or a Steering mechanism used to turn while controlling the operation of a vehicle.

Steer-by-wire (SBW) - The aim of steer-by-wire technology is to completely do away with as many mechanical components (steering shaft, column, gear reduction mechanism, etc.) as possible.

Steering is the term applied to the collection of components, linkages, etc. which allow for a car or other vehicle to follow a course determined by its driver, except in the case of rail transport by which rail tracks combined together with railroad switches provide the steering function.

Steering hand wheel is a type of steering control used in most modern land vehicles, including all mass-production automobiles; the steering hand wheel is the part of the steering system that is manipulated by the driver; the rest of the steering system responds to the movements of the steering hand wheel; this may be through direct mechanical contact in rack and pinion steering, with the assistance of fluidics (hydraulics) in power steering, or in some concept cars and modern production automotive vehicles entirely through computer mechatronic control.

Supersonic sensor - A sensor used to measure the distance between a vehicle body and the on/off road surface using supersonic waves.

Transmission **-** The gearing system by which power is transmitted from the engine to the live axle in an automotive vehicle; also known as gearbox.

Vehicle propulsion refers to the act of moving an artificial carrier of people or goods over a distance; the power plant used to drive the automotive vehicles may vary widely; in modern times, most automotive vehicles use some form of **external-combustion engine** (ECE) or **internal-combustion engine** (ICE), with electro-mechanical (E-M) motors supplementing them.

Vehicular suspension is the term given to the system of springs, shock absorbers (dampers) and linkages that connects a vehicle to its wheels; suspension systems serve a dual purpose – contributing to the automotive vehicle's handling and braking for good active safety and driving pleasure, and keeping vehicle occupants comfortable and reasonably well isolated from road noise, bumps, and vibrations; these goals are generally at odds, so the tuning of suspensions involves finding the right compromise; the suspension also protects the automotive vehicle itself and any cargo or luggage from damage and wear; the design of front and rear suspension of an automotive vehicle may be different.

References and Bibliography

1. ABDEL-HADY MBA AND DA CROLLA (1989): Theoretical Analysis of Active Suspension Performance using a Four-Wheel Vehicle Model. *Proc. ImechE,* Vol. D 203, 1989, pp. 125-135.
2. ABDEL-HADY MBA AND DA CROLLA (1992): Active Suspension Control Algorithms for a Four Wheel Vehicle Model. *Journal of Vehicle Design*, Vol. 13, No. 2, 1992, pp. 142-158.
3. ADAMS (2002): *ADAMS 12.0 User Manual*, ADAMS/Tire Documentation. Creating user tire and road models. Mechanical Dynamics Inc. Ann Arbor, Michigan, USA, 2002, 94 p.
4. ACKER B, W DARENBURG AND H GALL (1991): Active suspension for passenger cars. Dynamics of road vehicles. *Proc. 11th IAVSD Symposium,* 1991.
5. AHMADIAN M AND RH MARJORAM (1989A): Effects of passive and semi-active suspensions on body and wheelhoop control. *Journal of Commercial Vehicles*, Vol. 98, 1989, pp. 596-604.
6. AHMADIAN M AND MARJORAM (1989B): On the development of a simulation model for tractor semitrailer systems with semiactive suspensions. *Proceedings of the Special Joint Symposium on Advanced Technologies*, 1989 ASME Winter Annual Meeting, San Francisco, CA, DSC13, 1989.
7. AHMADIAN M (1993): Ride Evaluation of a Class 8 Truck with Semi-Active Suspensions. *Advanced Automotive Technologies, ASME Dynamic Systems and Control Division, DSC*, Vol. 52, 1993, New York, NY, pp. 21-26. AHMADIAN M (1997): Semiactive control of multiple degree of freedom systems. *Proceedings of the Sixteenth Biennial Conference on Mechanical Vibration, ASME Design Technical Conference,* CA, 1997.
8. AHMADIAN M (1999): On the Isolation Properties of Semiactive Dampers. *Journal of Vibration Control*, Vol. 5, No. 2, 1999, pp. 217-232.
9. AHMADIAN M AND CA PARE (2000): A Quarter-car Experimental Analysis of Alternative Semiactive Control Methods. *Journal of Intelligent Material System and Structures*, Vol. 11, No. 8, 2000, pp. 604-612.
10. AHMADIAN M AND DE SIMON (2002): Improving Roll Stability of Vehicles with High Center of Gravity by using Magneto-Rheological Dampers. *Proceedings of ASME IMECE 2002: 2002 International Mechanical Engineers Conference,* November 17th – 22nd 2002, New Orleans, Lousiana, Paper No. IMECE2002-32950, 7 p.
11. AHMADIAN M AND F GONCLAVES (2004): A frequency analysis of semi-active control methods for vehicle application. *Proceedings of the 2004 SAE Automotive Dynamics, Stability and Controls Conference, SAE Technical Paper Series*, 2004, Paper No. 2004-01-2098. pp. 231–240.

12. AIROCK (2006): *AiRock air suspension systems – What is AiROck?* Off Road Only (ORO), LLC, 1999-2006. Available online at: http://www.offroadonly.com/products/suspension/airock;http://www.jkonoir.com/AiRock.html
13. AKATSU Y, N FUKUSHIMA, K TAKAHASHI, M SATOH AND Y KAWARAZAKI (1990): An active suspension employing an electrohydraulic pressure control system. *XXIII FISITA Congress*, Torino, Italy, 1990, Paper 905123.
14. AKATSU Y (1994): Chapter 17 - Suspension Control: 17.1-17.19. *Automotive Electronics Handbook*, (R. Jurgen, Ed.), McGraw-Hill Inc., New York / London, 1994.
15. AL-HOLOU N, J SUNG AND A SHAOUT (1994): The development of fuzzy-logic based controller for semi-active suspension systems. *Proceedings of the IEEE 37th Midwest Symposium on Circuits and Systems*, Vol. 2, 1994, Paper No. 94CH35731, pp. 1373-1376.
16. ALKHATIB R, NG JAZAR AND MF GOLNARAGHI (2004): Optimal Design of Passive Linear Suspension using Genetic Algorithm. *Journal of Sound and Vibration,* Vol. 275 No. 3-5, 23 August 2004, pp. 665-691.
17. ALLEYNE A AND JK HEDRICK (1992): Nonlinear Control of a Quarter Car Active Suspension. *Proceedings of the 1992 American Controls Conference*, Chicago, IL, 21-25 June 1992.
18. ALLEYNE A, PD NEUHAUS AND JK HEDRICK (1992): Application of Nonlinear Control Theory to Electrical Control Suspensions. *Proceedings of the International Symposium on Advanced Vehicle Control*, Yokohama, Japan, September 1992.
19. ALLEYNE A AND JK HEDRICK (1993): Adaptive Control for Active Suspension. *Advanced Automotive Technologies*, Vol. 52 1993, pp. 7-13.
20. ALLEYNE A, PD NEUHAUS AND JK HEDRICK (1993): Application of nonlinear control theory to electronically controlled suspensions. *Vehicle System Dynamics*, Vol. 22, 1993, pp. 309-320.
21. ALLEYNE A (1994): *Nonlinear and Adaptive Control with Applications to Active Suspension.* Ph.D. Dissertation, The University of California at Berkeley, Berkeley, CA, 1994.
22. ALLEYNE A AND JK HEDRICK (1995): Nonlinear adaptive control of active suspension. *IEEE Transactions on Control System Technology*, Vol. 3, No. 1, 1995, pp. 94-101.
23. ALLEYNE A (1995): Improved Vehicle Performance Using Combined Suspension and Braking Forces. *American Control Conference*, Vol. 3, 1995, pp. 1672-1676.
24. ALLEYNE A AND R LIU (1999): On the limitations of force tracking control for hydraulic servo systems. *Journal of Dynamic Systems, Measurement and Control, Trans. Of the ASME*, Vol. 121, No. 2, 1999, pp. 184-190.
25. AMESIM (2004): AMESIm Applications – Suspension. *Modeling & Simulation Platform for Systems Engineering*, IMAGINE SA, 2004. Available online at: http://www.amesim.com .
26. AMPTIAC (2001): Materials that Sense and Respond: An Introduction to Smart Materials. *The AMPTIAC Quarterly*, Vol. 2, 2001, pp. 9-14.

27. AMT (2005A): Applications – Vehicle Suspension. *AMT Advanced Motion Technologies®*, 2005, Available online at http://www.advancedmotion.net/applications-vehicle-suspension.asp .
28. AMT (2005B): A revolutionary advance in the design of vehicle suspension systems – Real time electronic control of wheel point vertical forces: A quantum leap in the design of vehicle suspension systems - *ServoRam*™. *AMT Advanced Motion Technologies®*, 2005. Available online at http://www.advancedmotion.net/pdf/veh_sus.pdf
29. ANNIS ND (2006): Development of a Visual Demonstration Platform for Parallel Evaluation of Active Suspension Systems. MSc Thesis, Virginia Polytechnic Institute and State University, Blacksburg, Virginia, USA, November 29, 2006, pp. 1-76.
30. APPLEYARD M AND PE WELLSTEAD (1994): ACTIVE SUSPENSION: Some background. *Control Systems Centre Reports*, No. 816, December 1994, University of Manchester, 14 p. Available online at http://www.csc.umist.ac.uk/CSCReports/816.htm .
31. AUDI (2002): The New Audi A8: A New Sporting Dimensions in the Luxury Segment. *AudiWorld News*, September 8, 2002. Available online at http://www.audiworld.com/ news/02/a8launch/content3.shtml .
32. AUDI (2005A): Adaptive Air Suspension, *Audi Glossary*, 2005. Available online at http:// www.audi.com/en1/glossary/adaptive_air_suspension.htm .
33. AUDI (2005B): Adaptive Air Suspension -- Sport, *Audi Glossary*, 2005. Available online at http://www.audi.com/en1/glossary/adaptive_air_suspension_sport_.htm .
34. AUDI (2005C): MMI – Multi-Media Interface, *Audi Glossary*, 2005. Available online at http://www.audi.com/en1/glossary/mmi_multi_media_interface.htm .
35. AUTO PRO (2000): Suspension Systems. *Auto Pro Automotive Enginering Related Site,* 1998-2000. Available online at http://autopro.8k.com/suspension.html .
36. BASTOW D (1990): *Car Suspension and Handling*. Second Edition, Pentech Press, London, 1990, 300p.
37. BASTOW D, G HOWARD AND JP WHITEHEAD (2005): *Car Suspension and Handling*. Fourth Edition, SAE International, Warrendale, PA, 2005.
38. BAJKOWSKI J (1996): Modeling of shock absorbers with dry friction subjected to impact. *Machine Dynamics Problems*, Vol. 16, 1996, pp. 7-19.
39. BARAK P (1989): Design and Evaluation of an Automobile Suspension. *Autotechnologies Conference and Exposition*, SAE Paper #890089, Vol.98, Section 6, January 1989.
40. BARR AJ AND JI RAY (1996): Control of an active suspension using fuzzy logic. *Proceedings of the Fifth IEEE International Conference on Fuzzy Systems*. Vol. 1, 1996, pp. 42-48.
41. BEKKER MG (1969): *Introduction to Terrain-Vehicle Systems*. The University of Michigan Press, 1969.

42. BENDER EK (1968): Optimum linear preview control with application to vehicle suspension. *Journal of Basic Engineering*, Series D, Vol. 90, pp. 213-221.
43. BENO JH, A BRESIE, AM GUENIN AND DA WEEKS (1999): The Design of an Electromagnetic Linear Actuator for an Active Suspension. *SAE International Congress and Exposition.* Detroit, Michigan, March 1-4. 1999, Publication 1999-01-0730.
44. BLOCK H AND JP KELLY (1988): Electro-Rheology. *Journal of Physics,* D: Applied Physics, Vol. 21, 1988, p. 1661.
45. BLUNDELL MV (1999A): The Modelling and Simulation of Vehicle Handling. Part 2: Tire Modelling. *Proc. ImechE – Part K. Journal of Multi-body Dynamics.* Vol. 214, Issue 1, 1999, pp. 119-134.
46. BLUNDELL MV (1999B): The Modelling and Simulation of Vehicle Handling. Part 3: Tire Modelling. *Proceedings of the ImechE – Part K. Journal of Multi-body Dynamics.* Vol. 214, Issue 1, 1999, pp. 1-32.
47. BMW (2001): Dynamics – experiencing what modern suspension technology can do. *E65: A Closer Look*, Part 3, July 20, 2001. Available online at http://www.bmwnation. com/articles/e65_new7_2_03.html .
48. BOELTER R AND H JANOCHA (1998): Performance of long-stroke and low-stroke MR dampers. *SPIE Proceedings on Smart Structures and Materials 1998 – Passive Damping and Isolation*, 03/01 – 03/05/98. San Diego, CA, USA, Vol. 3327 (1998), Paper # 3327-26, pp. 303-313.
49. BOMBA J AND J KALETA (2002): Giant Magnetostrictive Materials (GMM) As A Functional Material For Construction Of Sensors And Actuators. *19th DANUBIA-ADRIA Symposium on Experimental Methods in Solid Mechanics*, 25-28 September 2002, Polanica Zdroj, Poland.
50. BOMBA J AND J KALETA (2003): Giant Magnetostrictive Materials (GMM): Facilitate of sensor and actuator constructions. *AMAS Workshop on Smart Materials and Structures – Smart'03.* Jadwisin, Poland, September 2-5, 2003, pp. 337-342.
51. BOMBA J, J KALETA AND P SAWA (2003A): Wplyw naprezen wstepnych na zjawiska magnetomechaniczne w materialach o gigantycznej magnetostrykcji, *II Symp. Mechaniki Zniszczenia Materialow i Konstrukcji,* Augustow, 4-7 June 2003, Poland (In Polish).
52. BOMBA J, J KALETA AND P SAWA (2003B): The influence of prestress on magneto-mechanical damping in Giant Magnetostrictive Materials. *20th DANUBIA-ADRIA Symposium on Experimental Methods in Solid Mechanics*, 24-27 September 2003, Gyor, Hungary.
53. BRAUER K (1999): Follow-Up Test: 2000 Land Rover Discovery Series II. *Edmunds Reviews*, Road Test: Follow-Up Tests, Date posted: 01-01-1999. Available online at: http://www.edmunds.com/reviews/roadtest/ spin44416/ article.html .
54. BREVER TH, D SLOOPE AND D ROCHELEAU (2006): Modeling Mechanical Systems using Virtual Test Bed. *Presentation to Department of Mechanical Engineering*, Advanced Actuators Research Group, 2006, ss. 1-18.

55. BRIDGESTONE (2003): Bridgestone introduces Revolutionary Dynamic-Damping In-Wheel Motor Drive System – New technology improves performance of electric vehicles. *Bridgestone News Release,* 2003. Available online at: http://www.bridgestone.co.jp/english/news/030904.htm .
56. BROCATO M (1999): Control Theory and Electro-elasticity: Realisation of an Active Damper. RESEARCH AND DEVLOPMENT. ERCIM News, No. 38, July 1999. Available online at http://www.ercim.org/publication/Ercim_News/enw38/brocato.html .
57. BROGE JL (2000): Passive, reactive suspension systems. *Automotive Engineering International Online (AEI)*, Tech Briefs (March 2000), SAE International.
58. BRUNEAU H, R LE SETTY, F CLAEYSSEN, F BARILOT AND N LHERNET (1997): Application of a new amplified piezoelectric actuator to semi-active control of vibrations. *Proceeding of the MV2 Convention on Active Control in Mechanical Engineering,* Lyon, France, (1997), 33 p.
59. BRUSHER GA, PT KABAMBA AND AG ULSOY (1997A): Coupling between the Modeling and Controller Design Problems -- Part 1: Analysis. *Journal of Dynamic Systems, Measurement and Control*, Vol. 119, No, 3, September 1997.
60. BRUSHER GA, PT KABAMBA AND AG ULSOY (1997B): Coupling between the Modeling and Controller Design Problems -- Part 2: Design. *Journal of Dynamic Systems, Measurement and Control*, Vol. 119, No, 3, September 1997.
61. BURTON AW, AJ TRUSCOTT AND PE WELLSTEAD (1995): Analysis, Modelling and Control of an Advanced Automotive Self-levelling Suspension System. *IEE Proc. – Control Theory*, A, Vol. 142, No. 2, March 1995, pp. 129-139.
62. BUTZ T AND O VON STRYK (2001): Modeling and simulation of electro- and magneto-rheological fluid dampers. *Z. Angew. Math. Mech.*, 2001.
63. BULLOUGH WA, AR JOHNSON, A HOSSEINI-SIANAKI, J MAKIN AND R FIROOZIAN (1993): Electro-rheological clutch design, performance characteristics and operation. *Proceedings of the IMechE, Part I – Journal of Systems and Control Engineering,* Vol. 207, No.2. 1993, pp. 87-95.
64. CADILLAC (2003): Cadillac www-pages. 22.07.2003, Available online at: http://www.cadilac.com/cadillacjsp/models/seville/redecontrol.html#more .
65. CAI B AND D KONIK (1993): Intelligent Vehicle Active Suspension Using Fuzzy Logic. *IFAC World Congress*, Sydney, Australia, Vol. 2, 1993, pp. 231-236.
66. CAMBIAGHI D, M GADOLA AND D VETTURI (1994): The effects of nonlinearity on a quarter car model. *SITEV - Automotive Industry International Week*, Torino, Novembre 1994. In attesa di pubblicazione sulla Rivista dell'ATA.
67. CAMBIAGHI D AND M GADOLA (1994): Computer-aided racing car research and development at the University of Brescia, Italy. *Motorsports Engineering Conference and Exposition*, Dearborn (USA), December 1994, *SAE Technical Paper Series*, Paper No. 942507.

68. CAMERON DS, NA KHEIR AND M SHILLOR (2000): Low energy active platform stabilizing suspension system. *Proceedings of the World Automation Congress – ISIAC*, Maui, Hawaii, June 11-16, 2000.
69. CAMINO JF, DE ZAMPIERI AND PL PERES (1999): Design of a Vehicular Suspension Controller by Static Output Feedback. *Proc. of the American Control Conference*, 1999, pp. 3168-3172.
70. CAPONETTO R, G FARGIONE, A RISITANO AND D TRINGALI (2001): Soft computing for the design and optimisation of a fuzzy Sky-Hook controller for semiactive suspensions. *XXX Convegno Nazionale AIAS*, Alghero, Italy, September 2001 (In Italian).
71. CAPONETTO R, O DIAMANTE, G FARGIONE, A RISITANO AND D TRINGALI (2003): A soft computing approach to fuzzy sky-hook control of semiactive suspension. *IEEE Transactions on Control Systems Technology*, Vol.11, Nov.2003.
72. CARBONARO O (1990): Hydractive suspension electronic control system. *XXIII FISITA Congress*, Torino, Italy, 1990, Paper 905101.
73. CARLIST (2005A): Active Tilt Control. *Automotive Glossary*, Carlist.com, October 10, 2005, Available online at http://www.carlist.com/autoglossary/autoglossary_23.html .
74. CARLIST (2005B): Active Suspension System. *Automotive Glossary*, Carlist.com, October 10, 2005, Available online at http://www.carlist.com/autoglossary/autoglossary_27.html
75. CARLIST (2005C): Adaptive Damping System (ADS). *Automotive Glossary*, Carlist.com, October 10, 2005, Available online at http://www.carlist.com/autoglossary/autoglossary_30.html .
76. CARLSON JD, DN CATANZARITE AND KA STCLAIR (1996): Commercial Magneto-Rheological Fluid Devices. *Proceedings 5th Int. Conf. on ER Fluids, MR Suspensions and Associated Technology*, W. Bullough, Ed., World Scientific, Singapore (1996) 20-28. Available online at http://www.lordcorp.com .
77. CARLSON JD AND MR JOLLY (2000): MR fluid, foam and elastomer devices. *Mechatronics*, Vol. 10, June 1, 2000, Issue 4-5, pp. 555-569.
78. CARLSON JD (2001): Introduction to Magnetorheological Fluids. *Advanced Course on Structural Control and Health Monitoring: SMART –2001*, May 22—25, 2001, Staszic Palace, Warsaw, Poland.
79. CARLSON JD (2002): Semi-active control using MR fluids. *2002 ISMA - International Conference on Noise and Vibration Engineering*, III Active suspensions, September 16 – 18, 2002, Katholieke Universiteit Leuven, Belgium.
80. CARLSON JD (2004): MR Fluids and Devices in the Real World. *ELECTRORHEO-LOGICAL FLUIDS AND MAGNETO-RHEOLOGICAL SUSPENSIONS (ERMR 2004) Proceedings of the Ninth International Conference*, Beijing, China, 29 August – 3 September 2004. Beijing, China, 29 August – 3 September 2004, p. 531.

81. CASTILLO JMD, P PINTADO AND FG BENITEZ (1990): Optimization for Vehicle Suspension II: Frequency Domain. *Vehicle System Dynamics*, Vol. 19, 1990, pp. 331-352.
82. CECH I (1994): A Pitch Plane Model of a Vehicle with Controlled Suspension. *Vehicle System Dynamics*, Vol. 23, 1994, pp. 133-148.
83. ČECH I (2000): Anti-roll and active roll suspensions. *Vehicle System Dynamics*, Vol. 33, Issue 2, February 2000, pp. 91-106.
84. CFC (2002): '*Smart Material*' – Electrorheological (ER) Fluids. Sponsor: USDOT/FHWS, Grant No DT-FH61-01C00002, US Department of Transportation, Federal Highway Administration, West Virginia University.
85. CHALASANI RM (1986A): Ride Performance Potential of Active Suspension Systems – Part I: Simplified Analysis Based on a Quarter-Car Model. *ASME Symposium on Simulation and Control of Ground Vehicles and Transportation Systems*. ASME Monograph AMD-80, DSC-1, 1986, American Society of Automotive Engineers, pp. 187-204.
86. CHALASANI RM (1986B): Ride Performance Potential of Active Suspension Systems – Part II: Comprehensive Analysis Based on a Full-Car Model. *ASME Symposium on Simulation and Control of Ground Vehicles and Transportation Systems*. AMD-Vol. 80, DSC-Vol. 2, American Society of Automotive Engineers, pp. 205-226.
87. CHALASANI RM (1987): Ride performance potential of active suspension systems – Part I: Simplified analysis based on the quarter-car model. *ASME Journal of Applied Mechanics*, Vol. 80, 1987, pp. 187-204.
88. CHAN BJ. AND C SANDU (2003): A Ray-tracing Approach to Simulation and Evaluation of a Real-time Quarter Car Model with Semi-active Suspension System using Matlab. Paper No. DETC2003/VIB-48559, *Proceedings of ASME International, 19th Biennial Conference on Mechanical Vibration and Noise (VIB), Active and Hybrid Vibration Control Symposium*, September 2-6, 2003, Chicago, Illinois.
89. CHANTRANUWATHANA S AND H PENG (1999A): Adaptive Robust Control for Active Suspensions. *Proceedings of the 1999 American Control Conference*, June 2-4, 1999, San Diego, CA, USA, pp. 1702-1706.
90. CHANTRANUWATHANA S AND H PENG (1999B): Force tracking control for active suspensions – theory and experiments. *IEEE Conference on Control Applications – Proceedings*, Vol. 1, 1999, pp. 442-447.
91. CHANTRANUWATHANA S AND H PENG (2000): Practical adaptive robust controllers for active suspensions. *Proceedings of the 2000 ASME International Congress and Exposition (IMECE)*, Orlando, Florida, 2000, 9 p.
92. CHANTRANUWATHANA S (2001): *Adaptive Robust Force Control for Vehicle Active Suspensions*. Ph.D. Dissertation, University of Michigan, 2001.
93. CHANTRANUWATHANA S AND H PENG (2004): Adaptive robust force control for vehicle active suspensions. *International Journal of Adaptive Control and Signal Processing*, Vol. 18, 2004, Paper ACS: 783, 20 p.

94. CHAUDHURI A, NM WERELEY, S KOTHA, R RADHAKRISHNAN AND TS SUDARSHAN (0000): Visco-metric characterization of cobalt nanoparticle-based magnetorheological fluids using genetic algorithms. *Journal of Magnetism and Magnetic Materials*, 0 (0000) p.1-9; Article in Press, Available online at http://www.sciencedirect.com and http://www.elsevier.com/locate/jmmm .
95. CHEN G AND C CHEN (2000): Behavior of piezoelectric friction dampers under dynamic loading. *Proceedings of the SPIE – Smart Structures and Materials*, New Port Beach, CA, Vol. 3988, 2000, pp. 54-63.
96. CHEN W, JK MILLS AND L WU (2000): Neurofuzzy Adaptive Control for Semi-Active Vehicle Suspension. *Proc. IASTED Int. Conf. On Control and Applications*, Cancun, Mexico , May 24-26, 2000.
97. CHEOK KC, NK LOH, DRIVER MCGREE AND TF PETIT (1985): Optimal model following suspension with microcomputerized damping. *IEEE Trans. on Industrial Electronics*, Vol. 32, No. 4, November 1985.
98. CHOI SB (1999): Vibration control of a flexible structure using ER dampers. *Transactions of the ASME, Journal of Dynamic Systems, Measurement and Control*, 1999, Vol. 121, pp. 134-138.
99. CHOI SB, MH NAM AND BK LEE (2000A): Vibration Control of a MR Seat Damper for Commercial Vehicles. *Journal of Intelligent Material System, Structures*, Vol. 11, No. 12, 2000, pp. 936-944.
100. CHOI SB, WK KIM (2000B): VIBRATION Control of a Semi-active Suspension Featuring Electrorheolo*gical Fluid Dampers. Journal of Sound and Vibration*, 234 (3), 2000, pp. 537-546.
101. CHOI SB, MS SUH, DW PARK AND MJ SHIN (2001): Neuro-fuzzy control of a tracked vehicle featuring semi-active electro-rheolo*gical suspension units. Vehicle System Dynamics*, Vol. 35, No. 3, 2001, pp. 141-162.
102. CHU SY, TT SOONG AND AM REIKORN (2005): *Active, Hybrid, and Semi-active Structural Control* – A Design and Implementation Handbook. John Wiley & Sons, Ltd, 2005.
103. CITROËN (2001): *Hydropneumatique* Citroën hydraulic system. *Citroënët*, 2001. Avail-able on line at http://www.citroen.mb.ca/citroenet/index.html .
104. CITROËN (2003A): Citroën's www-pages, 22.07.2003. Available online at http://www.citroen.com/site/htm/en/technologies/yesteryear/hydractive .
105. CITROËN (2003B): Citroën's www-pages, 22.07.2003. Available online at http://www.citroen.com/site/htm/en/technologies/yesteryear/activa .
106. CLARK AE (1993): High Power Magnetostrictive Materials from Cryogenic Temperatures to 250 C. *Materials Research Society Fall Meeting*, Boston, MA, November 28-30, 1994.
107. COLE DJ, D CEBON AND FH BESINGER (1994): Optimization of passive and semi-active heavy vehicle suspension. *Proceedings of International Truck & Bus Meeting & Exposition*, Heavy vehicle dynamics and simulation in braking, steering and suspension systems, November 7-9, 1994, SAE International, Warrendale, USA, 1994, Paper No. 942309, pp. 105-117.
108. COLE DJ (2001): Fundamental issues in suspension design for heavy road vehicles. *Vehicle System Dynamics*, Vol. 35, No. 4-5, 2001, pp. 319-360.

109. CONRAD H (1998): Properties and design of electrorheological suspensions. *MRS Bulletin*, August 1998, Vol. 23, pp. 35-42.
110. D.CORONA D, A GIUA AND C SEATZU (2004): Optimal control of hybrid automata: design of a semiactive suspension. *Control Engineering Practice*, Vol. 12, 2004, pp.1305–1318.
111. CORRIGA G, S SANNA AND G USAI (1991): An optimal tandem active-passive suspension for road vehicles with minimum power consumption. *IEEE Trans. on Industrial Electronics*, Vol. 38, No. 3, 1991, pp. 210-216.
112. CORRIGA G, A GIUA AND G USAI (1996): An H_2 formulation for the design of a passive vibration isolation system for cars. *Vehicle System Dynamics*, Vol. 26, 1996, pp. 381-393.
113. CRAIG B (2003): One Fluid, Multiple Viscosities – Numerous Applications: Active Fluids are Changing Damping Technology Forever. *The AMPTIAC Quartely*, Vol., 7, No, 2, 2003, pp. 15-19.
114. CROLLA DA AND AMA ABOUL NOUR (1988): Theoretical comparison of various active suspension systems in terms of performance and power requirements. *Proceedings of ImechE*, 1988, Paper No. C420/88, pp. 1-9.
115. CROLLA DA AND MBA ABDEL-HADY (1991A): Semi-Active Suspension Control for a Full Vehicle Model. *SAE (Society of Automotive Engineers) Transactions*, Vol. 100, Sect. 6, 1991, Paper No. 911904, pp. 1660-1666.
116. CROLLA DA AND MBA ABDEL-HADY (1991B): Active Suspension Control: Performance Comparisons using Control Laws Applied to a Full Vehicle Model. *Vehicle System Dynamics,* Vol. 20, 1991, pp. 107-120.
117. CROSBY MJ AND DC KARNOPP (1973): The active damper: a new concept in shock and vibration control. *43rd Shock and Vibration Bulletin,* June 1973.
118. CSERE C (2004): A surprising new active suspension. The Steering Column. *CAR and DRIVER – C/D*, October 2004, CARandDRIVER®.com.
119. DAE SJ AND N AL-HOLOU (1995): Development and evaluation of fuzzy logic controller for vehicle suspension systems, *27th Southeastern Symposium on System Theory (SSST'95)*, 1995, p. 295.
120. DALE M (2001): Eye On Electronics. *Motor*, November 2001, pp. 20-23.
121. Dan Cho D (1993): Experimental Results on Sliding Mode Control of an Electro-magnetic Suspension. *Mechanical Systems and Signal Processing*, Vol.7, No. 4, 1993, pp. 283-292.
122. DAPINO MJ, FT CALKINS, AB FLATAU (1999): Magnetostrictive devices. *22nd Encyclopedia of Electrical and Electronics Engineering*, Vol. 12, JG Webster (Ed.),John Wiley & Sons, Inc., New York, 1999, pp. 278-305.
123. DARLING J AND LR HICKSON (1998): An experimental study of a prototype active anti-roll suspension system. *Vehicle System Dynamics*, Vol. 29, 1998, pp. 309-329.
124. DAS (2006): Truck suspensions. *Presentation to DAS*, 2006. ss. 1-35.
125. DASSAULT (2003): Dassault systemes www-pages, 22.07.2003. Available online at http://www.3ds.com/en/brands/catia_ipf.asp .

126. DAVIS LI, GV FELDKAMP, GV PUSKORIUS AND F YUAN (1993): A Benchmark Problem for Neural Control: Quarter-Car Active Suspension. *Intelligent Engineering Systems through Artificial Neural Networks*, Vol. 3, 1993, pp. 581-586.
127. DECKER H, W SCRAMM AND R KALLENBACH (1988): A practical approach toward advanced suspension systems. *Proc. IMechE International Conference on Advanced Suspensions,* London, 1988. pp. 93-99.
128. DECKER H, W SCRAMM AND R KALLENBACH (1990): A modular concept for suspension control. *XXIII FISITA Congress*, Torino, Italy, 1990, Paper 905124.
129. DECKER H AND W SCRAMM (1990): An optimised approach to suspension control. *SAE Technical Paper Series,* 1990, Paper 900661.
130. DELPHI (2003A): Delphi Automotive System's www-pages. 27.05.2003, Available online at http://www.delphi.com/pdf/chassispdfs/MSR.pdf .
131. DELPHI (2003B): Delphi Automotive System's www-pages. 27.05.2003, Available online at http://www.delphi.com/pdf/chassispdfs/BSRTDM.pdf .
132. DEMEIS R (2005): Active Suspension Includes Roll Control. *Automotive Design Line*, CMP Limited Business Media, 03/08/2005. Available online at http://www.automotive designline.com/60407794.htm .
133. DEMIC M (1992): Optimization of Characteristics of Elasto-Damping Elements of Cars from the Aspect of Comfort and Handling. *International Journal of Vehicle Design*, Vol. 13, 1992, pp. 29-46.
134. DENNE PH (2001): *The Children of Necessity* Friday evening Discourse to the Royal Institution, February 23rd 2001.
135. DIXON JC (2000A): *The shock absorber handbook.* SAE International, Warrendale, PA, USA, 2000, 495 p.
136. DIXON JC (2000B): *Tires, Suspension and Handling.* Second Edition. SAE International, Warrendale, PA, USA, 2000.
137. DOGRUOZ MB, EL WANG, F GORDANINEJAD AND AJ STIPANOVICH (2002): Heat Transfer from Fail-Safe Magnetorheological Fluid Dampers. *Journal of Intelligent Material System and Structures*, 2002.
138. DOI S, E YASUDA AND Y HAYASHI (1988): An experimental study of optimal vibration adjustment using adaptive control methods. *Proc. IMechE*, London, Oct. 24-25, 1988.
139. DOMINY J AND DN BULMAN (1985): An active suspension for a Formula One Grand Prix racing car. *Transactions ASME, Journal Dynamic Systems Measurement Control*, Vol. 107, 1, 1985, pp 73-78.
140. DONAHUE MD (2001): *Implementation of an Active Suspension, Preview Controller for Improved Ride Comfort.* M.Sc. Thesis, The University of California at Berkeley, Berkeley, CA, 2001.
141. DONG XM, M YU, SL HUANG, Z LI AND WM CHEN (2004): Half Car Magnetorheological Suspension System Accounting for Nonlinearity and Time Delay. *ELECTRORHEOLOGICAL FLUIDS AND MAGNETORHEOLOGICAL SUSPENSIONS (ERMR 2004) Proceedings of the Ninth International Conference*, Beijing, China, 29 August – 3 September 2004, p.398.

142. DRDC (2005): Active Suspension Iltis. *Defence R&D Canada*, Suffield, Canada, Date Published: 2005 02 21. Available online at http://www.suffield.drdc-rddc.gc.ca .
143. DUNCAN AE (1982): Application of modal modelling and mount system optimisation to light duty truck ride analysis. *SAE Technical Paper Series*, 1982, Paper No. 811313, pp. 4075-4089.
144. DUYM S AND K REYBROUCK (1998): Physical Characterization of Nonlinear Shock Absorber Dynamics. *European Journal Mech. Eng.,* Vol. 43, No. 4, 1998, pp. 181-188.
145. DYKE SJ, BF SPENCER JR, MK SAIN AND JD CARLSON (1996): Modeling and Control of Magnetorheological Dampers for Seismic Response Reduction. *Smart Materials and Structures*, **5**, 1996, pp. 567–57.
146. EBAU M, A GIUA, C SEATZU AND G USAI (2001): Semiactive suspension design taking into account the actuator delay. *Proceedings of the 10th IEEE Conference on Decision and Control*, Orlando, Florida, December 2001, pp. 93-98.
147. EDSALL L (2004): Michelin pushes tech envelope. *The Detroit News – AUTOS Insider*, October 19, 2004. Available online at http://www.detnews.com/2004/autoinsider/0419/ 21/index.htm
148. EFATPENAH K (1999): *Manual and automatic control of an active suspension for high-speed off-road vehicles*. University of Texas at Austin, TX. 1999.
149. EFATPENAH K, JH BENO AND SP NICHOLS (2000): Energy Requirements of a Passive and an Electromechanical Active Suspension System. *Vehicle System Dynamics*, Vol. 34, No. 6, December 2000, pp. 437-458.
150. EHRGOTT RC, AND SF MASRI (1992): Modelling the Oscillatory Dynamic Behaviour of Electrorheological Materials in Shear. *Smart Materials and Structures*, **1**(4)**,** 1992, pp. 275–285.
151. EILER MK AND FB HOOGTERP (1994): Analysis of Active Suspension Controllers. *1994 Summer Computer Conference*, San Diego, California, July 1994.
152. EISENSTEIN PA (2005): Bose Unveils – Project Sound – Top secret program has nothing to do with audio. *THE CARCONNECTION.COM*™, *The Web's Automotive Authority*, September 28, 2005.
153. EL BEHEIRY EM, DC KARNOPP. ME ELARABY AND AM ABDELRAAOUF (1995): Advanced Ground Vehicle Suspension System – A Classified Bibliography. *Vehicle System Dynamics*, Vol. 24, 1995, pp. 231-258.
154. EL BEHEIRY EM AND DC KARNOPP (1996): Optimal Control of Vehicle Random Vibration with Constrained Suspension Deflection. *Journal of Sound and Vibration*, Vol. 189, No. 5, 1996, pp. 547-564.
155. EL BEHEIRY EM, DC KARNOPP, ME ELARABY AND AM ABDELRAAOUF (1996): Sub-optimal Control Design of Active and Passive Suspension Based on a Full Car Model. *Vehicle System Dynamics*, Vol. 26, 1996, pp. 197-222.
156. ELMADANY MM (1987): A Procedure for Optimization of Truck Suspensions. *Vehicle System Dynamics*, Vol. 16, 1987, pp. 297-312.

157. ELMADANY MM AND A EL-TAMIMI (1990): On a Subclass of Nonlinear Passive and Semi-Active Damping for Vibration Isolation. *Computer and Structures*, Vol. 36, No. 5, 1990, pp. 921-931.
158. ELMADANY MM (1990): Optimal linear active suspensions with multivariable integral control. *Vehicle System Dynamics*, Vol. 19, 1990, pp. 313-329.
159. ELMADANY MM AND Z ABDULJABBAR (1991): Alternative control laws for active and semi-active automotive suspensions – A comparative study. *Computers & Structures*, Vol. 39, No. 6, 1991, pp. 623-629.
160. EL MONGI BEN GAID M, A ÇELA AND R KOCIK (2004): Distributed control of a car suspension system. *Eurosim 2004*, COSI – ESEE, France, 2004, 6 p.
161. ENGELMAN GH AND G RIZZONI (1993): Including for the force generation process in active suspension control formulation. *Proceedings of the 1993 American Control Conference*, 1993, pp. 701-705.
162. ERFD (1998): Electro-rheological fluid LID 3354. *Technical Information Sheet*, ER Fluids Developments Ltd., UK, 1998.
163. FACE (2004): Brief Introduction to Thunder Technology. *Thunder Technology, FACE® International Corporation*, 2004. Available online at http://www.prestostore.com/cgl-bin/pro23.pl?ref+thunderonline&pg+17412 .
164. FANG X, W CHEN, L WU, Q WANG, D FAN AND Z LI (1999): Fuzzy Control Technology and the Application to Vehicle Semi-Active Suspension, *Chi Hsieh Kung Ch'Eng Hsueh Pao/Chinese Journal of Mechanical Engineering*, Vol. 35, No. 3, 1999, pp. 98-100.
165. FATHY HK, JA REYER, PY PAPALAMBROS AND AG ULSOY (2001): On the Coupling between the Plant and Controller Optimization Problems. *Proceeding of the 2001 American Control Conference*, Arlington, Virginia, USA, June 22-27, 2001, pp. 1864-1869.
166. FATHY HK, PY PAPALAMBROS, AG ULSOY AND D HROVAT (2003): Nested Plant/Controller Optimization with Application to Combined Passive/Active Automotive Suspensions, *Proceedings of the American Control Conference*, Denver, Colorado, June 4-6, 2003, pp. 3375-3380.
167. FERRI AA AND BS HECK (1992): Semi-active suspension using dry friction energy dissipation. *Proceedings of the American Control Conference*, Green Valley, AZ, Vol. 1, 1992, pp. 31-35.
168. FIALHO I AND G BALAS (2000): Design of nonlinear controllers for active vehicle suspensions using parameter-varying control synthesis. *Vehicle Ssystem Dynamics*, Vol.33, 2000, pp.351–370.
169. FIALHO I AND G BALAS (2002): Road adaptive active suspension design using linear parameter-varying gain-scheduling. *IEEE Transactions on Control Systems Technology*, Vol. 10, January 2002.
170. FIJALKOWSKI B (1985): On the new concept hybrid and bi-modal vehicles for the 1980s and 1990s. *Proc. DRIVE ELECTRIC Italy '85*, Sorrento, Italy, October 1985, pp. 4.04.1-4.04.8.
171. FIJALKOWSKI B (1986): Future hybrid electromechanical very advanced propulsion systems for civilian wheeled and tracked vehicles with extremely high mobility. *Proc. EVS-8: The 8th International Electric Vehicle Symposium*, Washington, DC, USA, October 1986, pp. 426-442.

172. FIJALKOWSKI B (1987): *Modele matematyczne wybranych lotniczych i motoryzacyjnych mechano-elektro-termicznych dyskretnych nadsystemow dynamicznych (Mathematical models of selected aerospace and automotive mechano-electro-thermal discrete dynamic hypersystems)*, Monografia 53, Politechnika Krakowska im. Tadeusza Kosciuszki, Krakow, 1987, 274 p. (In Polish).
173. FIJALKOWSKI B (1991): Mechatronically fuzzy-logic controlled full-time 4WA - 4WC -4WD - 4WS intelligent motor vehicle. *Proc. ISATA 91: International Dedicated Conference and Exhibition on MECHATRONICS, USE OF ELECTRONICS FOR PRODUCT DESIGN, TESTING, ENGINEERING AND RELIABILITY in conjunction with the 24th International Symposium on Automotive Technology and Automation (ISATA)*, Florence, Italy, 20-24 May 1991, Paper No. 911238, pp. 101-108.
174. FIJALKOWSKI B AND J KROSNICKI (1993): Smart electro-mechanical conversion actuators for intelligent road vehicles: Application to front- and/or rear-wheel rack-and-pinion steering gears, pp. 16-25. Chapter in the book: *Road Vehicle Automation* (Christopher O. Nwagboso, Ed.), PENTECH PRESS Publishers: London, 1993, p. 30.
175. FIJALKOWSKI B AND J KROSNICKI (1994): Concepts of electronically controlled electro-mechanical/mechano-electrical steer-, autodrive- and autoabsorbable wheels for environ-mentally-friendly tri-mode supercars. *Journal of Circuits, Systems and Computers*, Vol. 4, No. 4, 1995, pp. 501 -516.
176. FIJALKOWSKI B (1995): The concept of a high performance all-round energy efficient mechatronically-controlled tri-mode supercar. *Journal of Circuits, Systems and Computers*, Vol. 5, No. 1, 1995, pp. 93-107.
177. FIJALKOWSKI B (1997): Intelligent automotive systems: Development in full-time chassis motion spheres for intelligent vehicles, pp. 125-142. Chapter 5 in the book: *Advanced Vehicle and Infrastructure Systems: Computer Applications, Control, and Automation* (Christopher O. Nwagboso, Ed.), John Wiley & Sons, New York / Weinheim / Brisbane / Singapore / Toronto (1997), 502 p.
178. FIJALKOWSKI B (1998): Mechatronics electromechanic-activated inlet/outlet fluidical valve control - The ancillary benefits", *Proc. International Dedicated Conference on MECHATRONICS in conjunction with the 31st ISATA: International Symposium on Automotive Technology and Automation*, Duesseldorf, Germany, 2nd - 5th June 1998, Paper No. 98AE017, pp. 271-278.
179. FIJALKOWSKI B (1999): A family of driving, braking, steering, absorbing, rolling and throttling controls -'X-By-Wire Automotive Control. *Proc. Programme Track on Auto-motive Electronics and New Products, in conjunction with the 32nd ISATA: Inter-national Symposium on Automotive Technology and Automation - Advances in Automotive and Transportation Technology and Practice for the 21st Century*, Vienna, Austria, 14th - 18th June 1999, Paper No. 99AE016, pp. 287-294.
180. FIJALKOWSKI B (2000): Advanced Chassis Engineering. *World Market Series Business Briefing* – Global Automotive Manufacturing & Technology, World Markets Research Centre, August 2000, pp. 109-116.

181. FISCH A, J NIKITCZUK, B WEINBERG, J MELLI-HUBER, C MAVROIDIS AND C WAMPLER (2003): Development of an electro-rheological fluidical actuator and haptic systems for vehicular instrument control. *Proceedings of IMECE2003 – 2003 ASME International Mechanical Engineering Congress and Exposition*, November 15 – 21, 2003 – Washington, D.C., USA, pp. 1-10.
182. FISCHER D AND R ISERMANN (2004): Mechatronic semi-active and active vehicle suspensions. *Control Engineering Practice*, Vol.12, 2004, pp.1353 – 1367.
183. FOAG W (1989): A practical concept for passenger car active suspension with preview. *Proc. IMechE*, Vol. 203, 1989.
184. FODOR M AND R REDFIELD (1995): Resistance Control, Semi-Active Damping Performance. *Advanced Automotive Technologies*, American Society of Automotive Engineers (ASME), Dynamic Systems and Control Division, DSC, Vol.56, 1995, pp. 161-169.
185. FOLONARI CV (2006): Nanotechnology for Automotive. *Presentation to Centro Richerche FIAT,* Torino, Italy, 2006, ss. 1-19.
186. FORTUNECITY (2005: Active Suspension. *Fortune City*, 2005. Available online at http:// fortunecity.com/ .
187. FREESCALE (2005): Semi-Active Suspension. *Freescale™ Semiconductor, Inc.* 2004 – 2005. Available online at http://www.freescale.com/ .
188. GAO J, NJ LEIGHTON AND C MORGAN (1996): Modeling and Open-Loop Performance of a Novel Active Suspension for a Road Vehicle. *Proc. UKACC International Conference on Control 1996*, Vol.2, No. 427, Sept. 1996, pp. 1160-1165.
189. GASPAR P, I. SZASZI AND J BOKOR (2003): Design of robust controllers for active vehicle suspension using the mixed μ-synthesis. *Vehicle System Dynamics*, Vol. 40, No. 4, 2003, pp.193–228.
190. GARRET TG, G CHEN, FY CHENG AND W HUEBNER (2001): Experimental characterization of piezoelectric friction dampers. *Proceedings of the SPIE*, New Port Beach, CA, 2001, pp. 405-415.
191. GAVIN H, H HOAGG AND M DOBOSSY (2001): Optimal Design of MR Dampers. *Proceedings of the U.S. – Japan Workshop on Smart Structures for Improved Seismic Performance in Urban Regions*, Seattle, WA, 2001, pp. 225-236.
192. GEHM R (2001): Delphi Improves Cadillac's Ride. *Automotive Engineering International (AEI)*, Vol. 109, No. 10, October 2001, pp. 32-33. Available online at http:// www.sae.org/automag/techbrief/10-2001/index.htm .
193. GENTA G (1997): *Motor Vehicle Dynamics*, Modeling and Simulation. WorldScientific, 1997.
194. GIACOMIN J (1991): Neural Network Simulation of an Automotive Shock Absorber. *Engineering Applications of Artificial Intelligence*, Vol. 4, No. 1, 1991, pp. 59-64.
195. GILBERT R AND M JACKSON (2002): Magnetic Ride Control. *TECHlink*, GM, January 2002, Vol. 4, No. 1, pp. 1-4, Available online at http://service.gm.com .

196. GILLESPIE TD (1992): *Fundamentals of Vehicle Dynamics*. Society of Automotive Engineers, Inc. Warrendale, Pa., 1992.
197. GILLESPIE TD, N ESLAMINASAB, T KIM AND MF GOLNARAGHI (2004A): Evaluation of Existing Active Valve Technologies For Use in Semi-Active Shock Absorbers. *Technical Report Submitted to General Kinetics Engineering Corporation,* Ontario, June 30, 2004.
198. GILLESPIE, TD, N ESLAMINASAB, T KIM AND MF GOLNARAGHI (2004B): Concept Development of a Magnetorheological Piloted Valve. *Technical Report Submitted to General Kinetics Engineering Corporation,* Ontario, August 31, 2004.
199. GILLIOMEE CL AND PS ELS (1998): Semi-active hydropneumatic spring and damper system. *Journal of Terramechanics*, Vol. 35, 1998, pp. 109-117.
200. GILSDORF H-J, A THOMÄ AND M MÜNSTER (2006): Electro-mechanically Actuated Systems for Roll and Body Control. *ZF Antriebs- und Fahrwerktechnik,* PPT Presentation S-DDA, 2/05.2006, ZF Sachs AG, ZF Friedrichshafen AG.
201. GIUA A, A SAVASTANO, C SEATZU AND G USAI (1998): Approximation of an Optimal Gain Switching Active Law with a Semiactive Suspension. *IEEE Conference on Control Applications -- Proceedings*, Vol. 1, 1998, Paper No. 98CH36104, pp. 248-252.
202. GIUA A, C SEATZU AND G USAI (1999): Semi-Active Suspension Design with Optimal Gain Switching Target. *Vehicle System Dynamics*, Vol. 31, No. 4, 1999, pp. 213-232.
203. GIUA A, SEATZU AND G USAI (2000): A Mixed Suspension System for a Half-Car Vehicle Model. *Dynamics and Control*, Vol. 10, No. 4, December 2000, pp. 375-397.
204. GIUA A, M MELAS, C SEATZU AND G USAI (2004A): Design of a Predictive Semiactive Suspension System. *Vehicle System Dynamics*, Vol. 41 No. 4, 2004, pp. 277-300.
205. GIUA A, M MELAS AND C SEATZU (2004B): Design of a control law for a semiactive suspension system using a solenoid valve damper. *Proc. 2004 IEEE Conference on Control Applications*, Taipei, Taiwan, September 2004, pp. 1-14.
206. GOBBI M AND G MASTIMU (2001): Analytical Description and Optimization of the Dynamic Behaviour of Passively Suspended Road Vehicles. *Journal of Sound and Vibrations*, Vol. 254, No. 3, 2001, pp. 457-481.
207. GOLNARAGHI MF (2003A): Smart Sensing Systems. Presented to *Ford Canada President*, March 4, 2003.
208. GOLNARAGHI MF (2003B): Advanced Mechatronics Vehicle Systems. Presented to the *GM Technical Center*, Warren Michigan, April 10, 2003.
209. GOLNARAGHI MF (2003C): Smart Vehicle Systems. Presented to *Daimler Chrysler*, December 5, 2003.
210. GOLNARAGHI MF (2004): Smart Mechatronics & Sensing Systems for Automotive Applications. Presented to *MARK IV Corporation*, January 20, 2004.

211. GONCALVES FD (2001): Dynamic Analysis of Semi-Active Control Techniques for Vehicle Applications. M.S. Thesis, Virginia Polytechnic Institute and State University, Blacksburg, VA, August 2001.
212. GONCALVES FD AND M AHMADIAN (2003A): A hybrid control policy for semi-active vehicle suspensions. *Shock and Vibration*, Vol. 10, No.1, 2003, pp. 59-69.
213. GONCALVES FD, J-H KOO AND M AHMADIAN (2003): Experimental Approach for Finding the Response Time of MR Dampers for Vehicle Applications. *Proceedings of DETC'03: ASME 2003 Design Engineering Technical Conferences and Computer and Information in Engineering Conference*, Chicago, Illinoius, USA, September 2-6, 2003, pp. 1-6.
214. GONCALVES FD AND M AHMADIAN (2003): Experimental investigation of the energy dissipated in a magneto-rheological damper for semi-active vehicle suspensions. *Proceedings of IMECE'03: ASME 2003 International Mechanical Engineering Congress and RD&D Expo,* Washington, DC, USA, November 15-212. 2003, 9 p.
215. GOODSELL D (1989): *Dictionary of Automotive Engineering*. Butterworths, London, 1989, 182 p.
216. GOPALASAMY S AND JK HEDRICK (1994): Tracking Nonlinear Non-minimum Phase System Using Sliding Control. *International Journal of Control*, Vol. 57, No.5, 1994, 1141-1158.
217. GOPALASAMY S, JK HEDRICK, C OSORIO AND R RAJAMANI (1997): Model Predictive Control for Active Suspensions – Controller Design and Experimental Study. *Trans. of ASME, J. of Dynamic Systems and Control*, Vol. 61, 1997, pp/ 725-733.
218. GORAN MB. BI BACHRACH AND RE SMITH (1992): The design and development of a broad bandwidth active suspension concept car. *Proc. ImechE*, 1992, pp. 231-252.
219. GORDANINEJAD F AND DG BREESE (1999): Heating of Magnetorheological Fluid Dampers. *Journal of Intelligent Materials, Systems and Structures*, Vol. 10, No. 8, 1999, pp. 634-645.
220. GORDANINEJAD F AND DG BREESE (2000): Magneto-Rheological Fluid Dampers. *U.S. Patent,* No. 6,019,201, 2000.
221. GORDANINEJAD F AND SP KELSO (2000): Magneto-Rheological Fluid Shock Absorbers for HMMWV. *Proceeding of SPIE Conference on Smart Materials and Structures*, Newport Beach, CA, March 2000.
222. GORDANINEJAD F AND SP KELSO (2001): Fail-Safe Magneto-Rheological Fluid Dampers for Off-Highway, High-Payload Vehicles. *Journal of Intelligent Material System and Structures*, Vol. 11. No. 5, 2001, pp. 395-406.
223. GORDON TJ, C MARCH AND MG MILSTED (1991): A comparison of adaptive LQG and nonlinear controllers for vehicle active suspension systems. *Vehicle System Dynamics*, Vol. 20, 1991, pp. 321-340.
224. GORDON TJ AND MC BEST (1994): Dynamic Optimization of Nonlinear Semi-Active Suspension Controllers, *IEE Conference Publication*, Published by IEE, Vol. 1, No. 389, 1994, pp. 332-337.

225. GORDON J (1999): Understanding Electronic Suspension Systems. *Motor Age*, Feb, 1999.
226. GÖRING E, EC VON GLASNER, R POVEL AND P SCHÜTZNER (1993): Intelligent suspension systems for commercial vehicles. *Proc. Int. Cong. MV2*, Active Control in Mechanical Engineering, Lyon, France, June 1993, pp. 1-12.
227. GREEN D AND J BUSH (2003): Active Pneumatic Suspension System. *Project*, 1993, Electrical & Computer Engineering, Bradley University. Available online at http://cegt201.bradley.edu/projects/proj1993/apss/shtml/
228. GRIMA M AND C RENOU (1993): Modelization of semiactive suspensions. *Proc. Int. Cong. MV2*, Active Control in Mechanical Engineering, Lyon, France, Vol. 1, June 1993 (In French).
229. GRUNDLER D (1997): *Multilevel Fuzzy Process Control Optimized by Genetic Algorithm*. Ph.D. Thesis, University of Zagreb, Croatia, 1997.
230. GUY Y, MB LIZELL AND MW KERASTAS (1988): A solenoid-actuated pilot valve in a semi-active damping system. *SAE Technical Paper Series,* 1988, Paper 881139.
231. HAC A (1985): Suspension optimisation of a 2-DOF vehicle model using a stochastic optimal control technique. *Journal of Sound and Vibration*, Vol. 100, No. 3, 1985, pp. 343-357.
232. HAC A (1986): Stochastic optimal control of vehicle with elastic body and active suspension. *ASME Journal Dynamic Systems, Measurement, and Control*, Vol. 108, 1986, pp. 106-110.
233. HAC A (1992): Optimal linear preview control of active vehicle suspension. *Vehicle System Dynamics*, Vol. 21, 1992, pp. 167-195.
234. HAC A AND I YOUN (1992): Optimal Semi-Active Suspension with Preview Based on a Quarter Car Model. *Journal of Vibrations and Acoustics,* Stress Reliability Design, Vol. 114, No, 1, 1992, pp. 84-92.
235. HAC A AND I YOUN (1993): Optimal design of active and semi-active suspensions including time delays and preview. *Journal of Vibration and Acoustics*, Vol. 115, 1993, pp. 498-508.
236. HAC A AND AV FRATINI JR (1999): Estimation of Limit Cycles Due to Signal Estimation in Semi-Active Suspensions. *SAE Technical Paper Series*, 1999, Paper 1999-01-0728, pp. 1-7 -- Reprint From: Steering and Suspension Technology Symposium 1999 (SP-1438, *International Congress and Exposition,* Detroit, Michigan, March 1-4, 1999.
237. HAC A (2002): Rollover Stability Index Including Effects of Suspension Design. *SAE Technical Paper Series,* 2002, Paper No. 2002-01-0965.
238. HAGET KH ET AL. (1990): Continuous adjustable shock absorbers for rapid-acting ride control systems (RCS). *XXIII FISITA Congress*, Torino, Italy, 1990, # 905125.
239. HAGIWARA T, SA PANFILOV, SV ULYANOV AND I KURAWAKI (2000): Intelligent robust control suspension system based on soft computing. *Proceeding of the 4th International Conference on Application of Fuzzy Systems and Soft Computing*, 2000, pp. 180-189.

240. HAGIWARA T, S PANFILOV, K TAKAHASHI AND O DIAMANTE (2002): Application of smart control suspension system based on soft computing to a passenger car. *2002 ISMA -- International Conference on Noise and Vibration Engineering*, III Active suspensions, September 16 – 18, 2002, Katholieke Universiteit Leuven, Belgium.
241. HAGIWARA T, SA PANFILOV, SV ULYANOV, K TAKAHASHI AND O DIAMANTE (2003): An Application of a smart control suspension system for a passenger car based on soft computing. *Yamaha Motor Technical Review*, Technical Papers and Articles, 2003.01.05, pp. 1-10.
242. HAMMOND JK AND RF HARRISON (1981): Non-stationary response of vehicle on rough ground – a state space approach. *Journal of Dynamic System Measurement and Control*, Vol. 103, 1981, pp. 243-250.
243. HARDER G (1998): Recurrent RBF Networks for Suspension System Modelling and Wear Diagnosis of a Damper. *IEEE International Conference on Neural Networks*, Vol. 3, No. 3, 1998, pp. 2441-2446.
244. HASHIYAMA T, T FURUHASHI AND Y UCHIKAWA (1995A): Study on Finding Fuzzy Rules for Semi-Active Suspension Controllers with Genetic Algorithm. *Proceedings of the IEEE Conference on Evolutionary Computation*, Vo. 1, 1995, pp. 279-282.
245. HASHIYAMA T, T FURUHASHI AND Y UCHIKAWA (1995B): Fuzzy Controllers for Semi-Active Suspension System Generated through Genetic Algorithms. *Proceedings of the IEEE International Conference Syst Man Cybern*, Vol. 95, No. 5, 1995, pp. 4361-4366.
246. HASHIYAMA T, T FURUHASHI AND Y UCHIKAWA (1995C): On Finding Fuzzy Rules and Selecting Input Variables for Semi-Active Suspension Control Using Genetic Algorithm. *Proceedings of the 11th Fuzzy System Symposium*, 1995, pp. 225-228.
247. HATHAWAY KB AND AE CLARK (1993): Magnetostrictive Materials. *MRS Bulletin*, Vol. XVIII, No. 4, April 1993.
248. HAYAKAWA K ET AL. (1993): Robust H-infinity output feedback control of decoupled automobile active suspension system. *Proc. CDC*, 1993.
249. HAYAKAWA K, K MATSUMOTO, M YAMASHITA, Y SUZUKI, K FUJIMORI AND H KIMURA (1999): Robust H_1-output feedback control of decoupled automobile active suspension systems. *IEEE Transactions on Automatic Control*, Vol. 44, 1999, pp.392–396.
250. HEDRICK JK (1973): Some optimal control techniques applicable to suspension system design. *ASME*, 1973, Paper 73-ICT-55, pp. 1-12.
251. HEDRICK JK AND DN WORMELY (1975): Active Suspension for Ground Support Transportation – A Review. *ASME-AMD*, Vol. 15, 1975, pp. 21-40.
252. HEDRICK JK (1981): Railway vehicle active suspensions. *Vehicle System Dynamics,* **10**, 1981, pp. 267-283.
253. HEDRICK JK AND T BUTSUEN (1988): Invariant properties of automotive suspensions. *Proc. IMechE Conference on Advanced Suspensions*, London, UK, 1988, pp. 35-42.

254. HEDRICK JK, N MEHTA AND C OSORIO (1996): Simulation Models for Active Suspension Development. *Interim Report,* PAAE07-96-C-X007, March 1996.
255. HENNECKE D, B JORDAN UND U OEHNER (1987): Elektronische Dämpfer-Kontrolle – eine vollautomatisch adaptive Dämpfkraftverstellung für den BMW 635 CS9, *ATZ Automobiltechnische Zeitschrift*, 89, 1987, Nr 9.
256. HENNECKE D, P BAUER, B JORDAN AND E WALEK (1990): Further market-oriented development of adaptive damper force control. *XXIII FISITA Congress*, Torino, Italy, 1990, Paper 905143.
257. HENNECKE D, P BAUER, B JORDAN AND E WALEK (1990): EDCIII – The new variable damper system for BMW's top models – a further development of our adaptive, frequency-dependent damper control. *SAE Technical Paper Series,* 1990, Paper 900662.
258. HENNING H (1990): Citroen SM V6. *ATZ 92*, 1990, pp. 23-27.
259. HEO S, K PARK AND S HWANG (2000): Performance and Design Consideration for Continuously Controlled Semi-Active Suspension Systems. *International Journal of Vehicle Design*, Vol. 23, No. 3, 2000, pp. 376-389.
260. HERRERA F, M LOZANO AND JL VERDEGAY (1994): Generating Fuzzy Rules from Examples using Genetic Algorithms. *Proceedings Fifth International Conference of Information Processing and Management of Uncertainty in Knowledge-Based Systems.* Paris, July 4-8, 1994, pp. 675-680.
261. HIEMENZ GJ, YT CHOI AND NM WERELEY (2000): Seismic Control of Civil Structures Utilizing Semi-Active Bracing Systems. *Proceedings of SPIE*, Vol. 3988, 2000, pp. 217-228.
262. HITACHI (2004): Semi-Active Suspension. *Automotive Systems,* 2004. Available online at http://www.hitachi.co.jp/Div/apd/en/products/dcs/dcs_004.html .
263. HOLDMAN P AND M HOLLE (1999): Possibilities to Improve the Ride and Handling Performance of Delivery Trucks by Modern Mechatronic System. *JSAE Review*, Vol. 20, No. 4, 1999, pp. 505-510.
264. HOOGTERP FB, NL SAXON AND PJ SCHIHL (1993): Semiactive Suspension for Military Vehicles. *SAE Technical Paper Series*, 1993, Paper No. 930847.
265. HOOGTERP FB (1995): Active suspension technology for combat vehicles. *SPIE Proceedings*, Vol. CR59, Paper No. CR59-06, 04/17 – 04/24, 1005, Orlando, FL, USA.
266. HOU B, F GONCALVES, C SANDU AND M AHMADIAN (2004): Dynamic Simulation of a Full Vehicle with Magneto-Rheological Dampers, *Proceedings of the ASME IMECE 2004, 6th Annual Symposium on Advanced Vehicle Technology,* Paper No. IMECE2004-59768, November 14-19, 2004, Anaheim, California.
267. HOUSNER GW, LA BERGMAN, TK CAUGHEY, AG CHASSIAKOS, RO CLAUS, SF MASRI, RE SKELTON, TT SOONG, BF SPENCER JR, AND JTP YAO (1997): Structural Control: Past Present, and Future. *Journal of Engineering Mechanics*, ASCE, **123**, 1997, pp. 897–971.

268. HROVAT D, DL MARGOLIS AND M HUBBARD (1988): An approach toward the optimal semi-active suspension. *Transactions of the ASME,* Vol. 110, Sept. 1988, pp. 288.
269. HROVAT D (1990): Optimal active suspension structures for quarter-car vehicle models. *Automatica,* Vol. 26, 5, 1990, pp. 845-860.
270. HROVAT D (1991): Optimal active suspensions for 3D vehicle models. *Proceedings of 1991 American Control Conference*, 1991, pp. 1534-1541.
271. HROVAT D (1993): Application of Optimal Control to Advanced Automotive Suspension Design. *ASME Journal of Dynamic System Measurement and Control*, Vol. 115, No. 2B, 1993, pp. 328-342.
272. HROVAT D (1997): Survey of Advanced Suspension Developments and Related Optimal Control Applications. *Automatica*, Vol. 33, No. 10, 1997, pp. 1781-1817.
273. HUANG S-J AND W-CH LIN (2003): A Self-Organizing Fuzzy Controller for an Active Suspension System. *Journal of Vibration and Control*, Vol. 9, No. 9, 2003, pp. 1023-1040.
274. HUISMAN RGM (1994): *A controller and observer for active suspension with preview.* Ph.D. Dissertation. Department of Mechanical Engineering, Eindhoven University of Technology, 1994.
275. HWANG S, S HEO, H KIM AND K LEE (1997): Vehicle Dynamic Analysis and Evaluation of Continuously Controlled Semi-Active Suspensions Using Hardware-in-the-Loop Simulation. *Vehicle System Dynamics*, Vol. 27, No. 5-6, June 1997, pp. 423-434.
276. HYVÄRINEN J-K (2004): *The Improvement of Full Vehicle Semi-Active Suspension through Kinematical Model.* Academic Dissertation, Faculty of Technology, University of Oulu, Finland, December 11th 2004, Oulu University Press, Oulu 2004, 159 p.
277. I-CAR (1990): Air Suspension Systems. *I-CAR Advantage Online*™, *Technical Information for the Collision Industry*, Vol. III, No. 5 September-October 1990, pp.1-6. Available online at http://www.i-car.com/pdf/program_support/advantage/1990/sepoct 190.pdf .
278. I-CAR (1991): Computer Controlled Suspensions – Part 2. *I-CAR Advantage Online*™, *Technical Information for the Collision Industry*, Vol. I, No. 1, January-February 1991, pp. 6-9. Available online at http://www.i-car.com/pdf/program_support/advantage/1991/janfeb91.pdf .
279. IIYAMA F, K SUNAKODA AND K SUZUKI (1998): Hybrid Control of Seismic Isolation System Using Mechatro Damper. *Proceedings of the ASME/JSME Joint Pressure Vessels and Piping Conference*, San Diego, CA, Vol. 379, 1998, pp. 157-161.
280. IKENAGA S, FL LEWIS, J CAMPOS AND L DAVIS (2000): Active Suspension Control of Ground Vehicle based on a Full-Vehicle Model. *Proceedings of the American Control Conference (ACC)*, Chicago, Illinois, USA, June 2000.

281. IRIE N, Y SHIBAHATA, H ITO AND T UNO (1986): HICAS – Improvement of vehicle stability and controllability by rear suspension steering characteristics. *Proceedings of the FISITA 21st Congress*, 1986, Paper 865114, pp. 2.81-2.88.
282. ISO 5008 --E (1979): *Agricultural wheeled tractors and field machinery* – Measurement of whole-body vibration of the operator.
283. ISO 2631 -- E (1985): *Mechanical vibration and shock.* International Standard Organisation (ISO) 1990, pp. 481-495.
284. ISO 8855 --E/F (1991): *Road vehicles* – Vehicle dynamics and road-holding ability – Vocabulary. International Standard Organisation (ISO).
285. ISO 2631-1 (1997): *Mechanical Vibration and Shock -- Evaluation of Human Exposure to Whole-Body Vibration*, Part 1: General requirements, International Standard Organisation (ISO), 1997, 19 p.
286. ISOBE O, T KAWABE, Y WATANABE, Y MIYASATO AND S HANBA (1996): Semi-active suspension system for heavy duty vehicles using a sliding mode control theory, *JSAE Review*, Vol. 17, No. 4, 1996, p. 444.
287. IVERS DE AND LR MILLER (1991): Semiactive Suspension Technology: An Evolutionary View. *DE-40, Book No. Hoo719-1991*, 1991, pp, 327-346.
288. JAMEI M (2003): *Symbiotic Evolution-Based Design of Fuzzy Inference Systems with Application to Active Suspension Systems.* Ph.D. Thesis, The University of Sheffield, Sheffield, UK, 2003.
289. JENNINGS J (2005): A Bose Suspension? *Automobile Magazine*, News, 2005.
290. JEONG SG, IS KIM, KS YOON, JN LEE, JI BAC AND MH LEE (2000): Robust H_∞-controller design for performance improvement of a semi-active suspension system. *Proc. 2000 IEEE International Symposium on Industrial Electronics*, Vol. 2, 2000, pp. 706-709.
291. JOHNSON ER, MIS ELMASRY (2003): *Parametric frequency domain identification using variable stiffness and damping devices.* Project 01-10, Metrans Research Project, Final Report, June 2003, pp. i-viii; 1-45.
292. JOSHI CHH (2000): Compact Magnetostrictive Actuators and Linear Motors. *Actuator 2000 Conference*, Bremen, Germany, June 2000. Available online at: http://www. EnergenInc.com .
293. JOSHI CHH AND ES BOBROV (2000): *Compact, Efficient Deformable Mirrors for Space-borne Telescopes.* Energen Inc., Bedford, Massachusetts, 2000.
294. JORDAN TC AND MT SHAW (1989): Electrorheology. *IEEE Transactions on Electrical Insulation*, 1989, 24(5), pp. 849-878.
295. JUNG-SHAN L AND I KANELLAKOPOULOS (1997): Nonlinear design of active suspensions. *IEEE Control Systems Magazine,* Vol. 17, 1997, pp.45–59.
296. KALAUGHER L ED. (2003): Nanoparticles go against the flow in electrorheology break-through. *The world service for nanotechnologies*, nanotech.web.org, IoP Publishers of the Journal *Nanotechnology*, 8 October 2003.
297. KANARACHOS A, D KOULOCHERIS AND H VRAZOPOULOS (2000): Optimisation of fuzzy logic controller using genetic algorithms. *8th IEEE Mediterranean Conference on Control and Automation,* July 17-19, 2000, University of Patras, Rio, Greece.

298. ANNAN S, HM URAS AND HM AKTAN (1995): Active control of building seismic response by energy dissipation. *Earthquake Engineering and Structural Dynamics*, Vol. 24, 1995, pp. 747-759.
299. KARNOPP DC AND AK TRIKHA (1969): Comparative study of optimization techniques for shock and vibration isolation. Trans. ASME, Journal of Engineering for Industry, Series B, Vol. 91, No. 4, 1969, pp.1128-1132.
300. KARNOPP DC (1973): Active and passive isolation of random vibration. *ASME Journal of Applied Mechanics*, Vol. 1, 1973, pp. 64-86.
301. KARNOPP DC, MJ CROSBY AND RA HARWOOD (1974): Vibration Control Using Semi-Active Force Generators. *Journal of Engineering for Industry, Transactions of the ASME*, Vol. 96, Ser. B, No. 2, May 1974, pp. 619-626.
302. KARNOPP DC (1983): Active damping in road vehicle suspension systems. *Vehicle System Dynamics,* Vol. 12, 1983, pp. 291-316.
303. KARNOPP DC AND DL MARGOLIS (1984): Adaptive suspension concepts for road vehicles. *Vehicle System Dynamics*, Vol. 13, 1984, pp. 145-160.
304. KARNOPP DC (1987): Active suspension based on fast load levelers. *Vehicle System Dynamics*, Vol. 16, 1987, pp. 355-380.
305. KARNOPP DC (1989): Analytical results for optimum actively damped suspension under random excitation. *Journal of Acoustic, Stress and Reliability in Design*, Vol. 111, 1989, pp. 278-283.
306. KARNOPP DC AND G HEESS (1991): Electronically Controllable Vehicle Suspensions. *Vehicle System Dynamics*, Vol. 20, No. 3-4, 1991, pp. 207-217.
307. KARNOPP DC (1992): Active and semi-active suspensions. *Kia Academic Seminar on the Automotive Technology for Safety and Environment*, Vol. 1, 1992, pp. 123-147.
308. KARNOPP DC (1995): Active and Semi-Active Vibration Isolation. *Journal of Mechanical Design*, Vol. 117, 1995, pp. 177-185.
309. KARR CL (1992): Design of Adaptive Fuzzy Logic Controller Using a Genetic Algorithm. *Proceedings of the 4th International Conference on Genetic Algorithms,* 1992, pp. 450-457.
310. KASHANI R AND JE STRELOW (1999): Fuzzy logic active and semi-active control of off-road vehicle suspensions. *Vehicle System Dynamics*, Vol. 32, 1999, pp. 409-420.
311. KASPRZAK EM AND DL MILLIKEN (2000): MRA Vehicle Dynamics Simulation – Matlab®/Simulink®. *Proceedings of Automotive Dynamics and Stability Conference*, May 15-17, 2000, SAE International, Warrendale, USA, Paper No. 2000-01-1624, 8 p.
312. KEEBLER J (1995): Ford Kills Active-Suspension Program. *Automotive News*, January 23, 1995, p. 18.
313. KELLET M (2001): An Electronic Controller for Adaptive Suspension. *IEE Review*, November 1998.
314. KELSO SP (2001): Experimental Characterization of Commercially Practical Magneto-rheological Fluid Damper Technology. *Proceedings of SPIE*, 2001, Vol. 4332, pp. 292-299.

315. KIM H AND YS YOON (1995): Semi-Active Suspension with Preview using a Frequency Shape Performance Index. *Vehicle System Dynamics*, Vol. 24, 1995, pp. 759-780.
316. KIM E (1996): Nonlinear indirect adaptive control of a quarter car active suspension. *Proceedings of the 1996 IEEE International Conference on Control Applications*, September 1996, pp. 61-66.
317. KINETIC (2005): Kinetic Technology, an overview. *Kinetic™ Suspension Technology*, 2005. Available online at http://www.kinetic.au.com/techno.htm .
318. KIRSNER S (2004): Beyond speakers: Bose turns to potholes, *The Boston Globe*, June 21, 2004.
319. KITCHING KJ, DJ COLE AND D CEBON (2000): Performance of Semi-Active Damper for Heavy Vehicles. *ASME Journal of Dynamic Systems,* Measurement and Control, Vol. 122, 2000, pp. 498-506.
320. KNAAP ACM VAN DER (1991): *Design of a Low Power Anti-Roll/Pitch System for a Passenger Car*, Delft University of Technology, Vehicle Research Laboratory, Report No. 89.3.VT.2628, 1991 (In Dutch).
321. KNAAP ACM VAN DER, PJTH VENHOVENS AND HB PACEJKA (1994): Evaluation and Practical Implementation of a Low Power Attitude and Vibration Control System, *Proceedings of the International Symposium on Advanced Vehicle Control 1994 (AVEC'94),* Tsukuba, Japan, JSAE 9438466, October 1994.
322. KNAAP ACM VAN DER (1995): *Delft Active Suspension*, PhD Thesis, Delft University of Technology, Vehicle Research Laboratory, November 1995.
323. KNOWLES (2002): Road Suspension System. *Knowles. CM-Ch-08.gxd*, 9/14/02, pp. 202-223.
324. KNUTSON D (2002): *Selection and Design of Electrohydraulic Valves for Electronically-Controlled Automotive Suspension Systems*. Applied Power Inc., 2002.
325. KOBORI T, AND M TAKAHASHI (1993): Seismic Response Controlled Structure with Active Variable Stiffness System. *Earthquake Engineering and Structural Dynamics*, **22**(12), 1993,pp. 925–941.
326. KOO JH, M AHMADIAN, M SETAREH AND T MURRAY (2003A): An Experimental Evaluation of Magneto-Rheological Dampers for Semi-Active Tuned Vibration Absorbers. *SPIE 2003 Smart Structures and Materials/NDE,* March 2-6, 2003, San Diego, CA, USA.
327. KOO JH, M AHMADIAN, M SETAREH AND T MURRAY (2003B): Robustness Analysis of Semi-Active Tuned Vibration Absorbers with Magneto-Rheological Dampers: An Experimental Study. *ASME 2003 International Mechanical Engineering Congress and Exposition,* November 16-21, 2003, Washington, DC. USA.
328. KOO J-H, DF GONCALVES AND M AHMADIAN (2005): Investigation of the response time of magnetorheological fluid dampers. Advanced Vehicle Dynamics, Laboratory, Virginia Tech, Blacksburg, VA 24061-0238, 2005, 9 p.

329. KORTŰM W, M VALASEK, Z SIKA, W SCHWARTZ, P STEINBAUER AND O VACULIN (2002): Semi-Active Damping in Automotive Systems: Design-by-Simulation. *IJVD-Sonderheft*, Vol. 28, 2002, pp.103-120.
330. KOSLIK B (1997): Optimal Active Suspension of Vehicles Using an On_Line Dynamic Programming Method. *ISMP97: International Symposium on Mathematical Programming*, Lausanne, EPFL, August 24-29, 1997. Available online at http://dmawww.epfl.ch/roso.mosaic/ismp97/ismp_sess 3 _F031.html .
331. KOSLIK B, G RILL, O VON STRYK AND DE ZAMPIERI (1998): Active Suspension Design for a Tractor by Optimal Control Methods. *Mathematische Modellierung, Simulation und Verifikation in materialorientierten Prozessen und intelligenten Systemen*, Sonderfor-schungsbereich 438: TU-München, Universität Augsburg, 1998, Preprint SFB-438-9801, 23 p.
332. KOWAL J, M SZYMKAT AND T UHL (1995): Synthesis and analysis of active suspension control. *ACTIVE 95 Proceedings of the 1995 International Symposium on Active Control of Sound and Vibration*, Newport Beach, CA, July 06-08, 1995, p. 0067.
333. KOWALICK T (2007): IEEE P1616 Work Plan (2nd deaft). *Presentation at Crash Zone Technologies CzT*, 24 March 2007, ss. 1-27.
334. KOWALSKI M (2006): *Optymalizacja wymiarowa wybranych mechanizmow wielowahaczowych zawieszen samochodow*. Rozprawa doktorska (Ph.D. Dissertation). Politechnika Krakowska im. Tadeusza Kosciuszki, Krakow, March 29, 2006 (In Polish).
335. KRASNICKI EJ (1980): Comparison of analytical and experimental results for a semi-active vibration isolator. *Journal of Sound and Vibration*, September 1980, pp. 69-76.
336. KRTOLICA R AND D HROVAT (1992): Optimal active suspension control based on a half-car model: an analytical solution. *IEEE Transactions on Automatic Control*. Vol. 37, Issue 4, 1992, pp. 528-532.
337. KRUCZEK A, J STŘIBRSKÝ, J HONCŮ AND K HYNIOVÁ (2003): H_∞ control of an automotive active suspension. *Proceeding of Control '03*, Strbske Pleso, Slovakia, 2003, pp. 33.1-33.7.
338. KURIMOTO M AND T YOSHIMURA (1998): Active suspension of passenger cars using sliding mode controller (Based on reduced models). *International Journal of Vehicle Design*, Vol. 19, No. 4, 1998, pp. 402-414.
339. LACHEISSERIE T DE (1993): *Magnetostriction:* Theory and Applications of Magneto-elasticity, CRC Press Inc., 1993.
340. LACHNER R (1997): Collision Avoidance as Differential Game: Real-Time Approximation of Optimal Strategies using Higher Derivatives of the Value Function. *ISMP97: International Symposium on Mathematical Programming*, Lausanne, EPFL, August 24-29,, 1997. Available online at http://dma www.epfl.ch/roso.mosaic/ismp97/ismp_abs_304.html .
341. LAI C-Y AND WH LIAO (2002): Vibration Control of a Suspension System via a Magneto-rheological Fluid Damper. *Journal of Vibration and Control*, Vol. 8, No. 4 (2002), pp. 527-547.

342. LANE JS, AA FERRI AND BS HECK (1992): Vibration control using semi-active friction damping. *Proceedings of the Winter Annual Meeting of ASME*, New York, NY, Vol, 49, 1992, pp. 165-171.
343. LANG HH (1977): *A Study of the Characteristics of Hydraulic Dampers at High Stroking Frequency*. Ph.D. Thesis, University of Michigan, USA, 1977.
344. LANGLOIS RC AND RJ ANDERSON (1995): Preview Control Algorithms for the Active Suspension of an Active Suspension. *Vehicle System Dynamics*, Vol. 24, No. 1, January 1995, pp. 65-97.
345. LAUWERYS C, J SWERES AND P SAS (2002): Linear control of car suspension using nonlinear control. *2002 ISMA -- International Conference on Noise and Vibration Engineering*, III Active suspensions, September 16 – 18, 2002, Katholieke Universitieit Leuven, Belgium.
346. LEBEDEV AV, EA LOMONOVA, PG VAN LEUVEN AND J STEINBERG (2004): Analysis and initial synthesis of a novel linear actuator with active magnetic suspension, *IEEE IAS*, 05/2004, pp. 2111-2118.
347. LEE C-R, J-W KIM, JO HALLQUIST, Y ZHANG AND AD FARAHANI (1997): Validation of a FEA tire model for vehicle dynamic analysis and full vehicle real proving ground simulations. *SAE International Congress and Exposition*, February 24-27, 1997, SAE International, Warrendale, USA, Paper No. 971100, pp. 1-8.
348. LEIGHTON NJ AND J PULLEN (1994): A novel active suspension system for automotive applicaction. *Proc Instn Mech Engrs*, 1994.
349. LENOCH P (2005): Millen Works' active damper suspension system demonstrates dramatic mobility improvements for the U.S. Army Stryker Armoured Combat Vehicle. *MILLENWORKS -- For Immediate Release*, February 3, 2005. Available online at http://www.millenworks.com/Stryker_Test.pdf .
350. LI W, G YAO, G CHEN, S YEO AND F YAP (2000): Testing and steady state modelling of a linear MR damper under sinusoidal loading. *Smart Materials and Structures*, Vol. 9, 2000, pp. 95-102.
351. LI TH AND K-Y PIN (2000): Evolutionary Algorithms for Passive Suspension Systems. *JSME International Journal,* Ser. C, Vol. 43, No. 3, 2000, pp. 537 -544.
352. LI Y, F MARCASSA, R HOROWITZ, R OBOE AND R EVANS (2003): Track-Following Control with Active Vibration Damping of a PZT-Actuated Suspension Dual-Stage Servo System. *Proceedings of the American Control Conference*, Denver, Colorado, June 4-5, 2003, pp. 2553-2559.
353. LIAO W AND C LAI (2002): Harmonic analysis of a magneto-rheological damper for vibration control. *Smart Materials and Structures*, Vol. 11, 2002, pp. 288-296.
354. LIEH J (1991): Control of Vibrations in Elastic Vehicles Using Saturation Nonlinear Semiactive Dampers. *Transactions of the ASME*, American Society of Mechanical Engineers (ASME), Paper No. 91-WA-DSC-4, pp. 1-6.

355. LIEH J (1993): Effect of Bandwidth of Semiactive Dampers on Vehicle Ride. *Journal of Dynamic Systems Measurement & Control -- Transactions of the ASME*, Vol 115, No. 3, September 1993, pp. 571-575.
356. LIEH J (1997): Semiactive and Active Suspensions for Vehicle Ride Control Using Velocity Feedback. *JVC Journal of Vibration & Control*, Vol.3, No. 2, May 1997, pp. 201-212.
357. LIEH J AND WJ LI (1997): Adaptive Fuzzy Control of Vehicle Semi-Active Suspensions. *ASME Dynamic Systems and Control Division, DSC*, Vol. 61, 1997, Fairfield, NJ, pp. 293-297.
358. LIN Y AND Y ZHANG (1989): Suspension Optimization by Frequency Domain Equivalent Optimal Control Algorithm. *Journal of Sound and Vibration*, Vol. 133, No. 2, 1989, pp. 239-249.
359. LIN YJ. YQ LU AND J PADOVAN (1993): Fuzzy logic control of vehicle suspension systems. *International Journal of Vehicle Design*, Vol. 14, No. 6, 1993, pp. 457-470.
360. LINDHOLM M (2003): *Side wind compensation using active suspension*. MSc thesis, Linköpings Universitet, 2003.
361. LINDLER JE AND NM WERELEY (1999): Analysis and testing of electrorheological bypass damper/ *Proceedings of SPIE*, Vol. 3327, 1999, pp. 226-241.
362. LINDLER JE, GA DIMOCK AND NM WERELEY (2000): Design of Magneto-rheological Automotive Shock Absorber. *Proceedings. of SPIE*, 2000, Vol. 3985, pp426-437.
363. LIU Y, F GORDANINEJAD, CA EVRENSEL, U DOGUER, M-S YEO, ES KARAKAS AND A FUCHS (2004): Temperature dependent skyhook control of HMMWV suspension using a failsafe magneto-rheological damper. *Project, U.S. Army Research Office*, University of Nevada, Reno, NV 89557, USA, 2004, pp. 1-9.
364. LIZELL M (1990): Dynamic levelling for ground vehicles. Doctoral Thesis, Norsteds tryckeri, Stockholm 1990, 113 p.
365. LONGORIA RG (2004): Problem 1: The Bose® Active Suspension. *ME383Q – Modeling of Physical Systems – Quiz 1*, October 14, 2004, Department of Mechanical Engineering, The University of Texas at Austin.
366. LORD (1998): MRF-132LD Fluid Specifications. *Lord Corporation*, 1998.
367. LORD (2002): Basic Electromagnet controlled MR Fluid Valve. Available online at http:// www.lord.com/~/MRFluidValve.eng note.pdf .
368. LORD (2003): MR Brake, MRB-2107-3. *Lord Corporation Product Bulletin*, Cary, NC, 2003.
369. LORD (2005): Lord Corporation. Available online at http://www.mrfluid.com
370. LOTUS (2005): Feature -- The Future Role of Vehicle Dynamics. *Proactive – The official industry newsletter of Lotus Engineering*, Issue 6, February 2005, pp. 15-16, Available online at http://www.just-auto.com .
371. LOUAM N, DA WILSON AND RS SHARP (1988): Optimal control of a vehicle suspension incorporating the time delay between front and rear wheel inputs. *Vehicle System Dynamics*, Vol. 17, 1988, pp. 317-336.

372. LU J AND M DEPOYSTER (2002): Multi objective optimal suspension control to achieve integrated ride and handling performance. *IEEE Transactions on Control Systems Technology*, Vol.10, November 2002.
373. MANZO L AND T TIRELLI, A (1994): *Controllo di un sistema di sospensioni per autovetture da competizione: modelli ed approcci progettuali.* Degree Thesis, University of Brescia, 1994 (in Italian).
374. MARGOLIS D (1982): The Response of Active and Semi-Active Suspensions to Realistic Feedback Signals. *Vehicle System Dynamics*, Vol. 11, 1982, pp. 267-282.
375. MARGOLIS DL (1983): Semi-active control of wheel hop in ground vehicles. *Vehicle System Dynamics*, Vol. 12, 1983, pp. 317-330.
376. MARGOLIS D AND CM NOBLES (1991): Semi-active heave and roll control for large off-road vehicles. *Proceedings of International Truck & Bus Meeting and Exposition* - Commercial Vehicle Suspensions, Steering Systems and Traction (SP-892), November 18-21, 1991, SAE International, Warrendale, USA, Paper No. 912672, pp. 25-34.
377. MARTINEZ V (2004): Conceptual Development of Advanced Suspension (ADSS). *FISITA 2004: World Automotive Congress*, 23-27 May 2004, Barcelona, Spain, Paper F2004 F175.
378. MATSCHINSKY W (1999): *Road Vehicle Suspensions.* Professional Engineering Publishing Ltd, 1999, 360 p.
379. MATCHINSKY W (2000): *Road Vehicle Suspensions.* Edmundsbury Press Ltd., Suffolk, UK, 2000, 359 p.
380. MAVROIDIS C, Y BAR-COHEN AND M BOUZIT (2005): Chapter 19 – Haptic Interfaces Using Electrorheological Fluids. Rutgers University, Jet Propulsion Laboratory, pp. 1-27.
381. MCCORMICK (2004): Shooting for a smoother ride behind the wheel. *The Detroit News AUTOS Insider*, September 5, 2004.
382. MEAD S (1999): First Drive: 2000 Mercedes-Benz CL500. *Edmunds Reviews*, Road Test: First Drive Test, Date posted: 01-01-1999. Available online at: http://www.edmunds.com/reviews/roadtest/firstdrive/ 43954/ article.html .
383. MEILLAUD L (2005): Tommorow's suspension. *The ViaMichelin Newsletter*, New technology and mobility, Magazine -- 01/04/05. Available online at http://www.via michelin.co.uk/viamichelin/gbr/htm/tech_suspension.htm .
384. MELAS M (2002): *Nonlinear semiactive suspension design.* Laurea Thesis, University of Cagliari, Italy, October 2002 (in Italian).
385. MEMER S (1999): Suspension III: Active Suspension. *Edmunds Reviews*, Tech Center, Date posted: 01-01-1999. Available online at: http://www.emunds.com/zipcode/new?tid=edmunds.h.new.1
386. METZ D AND J MADDOCK (1986): Optimal ride height and pitch control for championship racecars. *Automatica*, Vol. 22, No. 5, 1986.
387. MILECKI A (2001): Investigation and control of magneto–rheological fluid dampers. *International Journal of Machine Tools & Manufacture*, Vol. 41, 2001, pp. 379–391.

388. MILLENWORKS (2005A): DARPA -- Servovalve Optimized Active Damper Suspension (SOADS). *MILLENWORKS -- Printer Friendly Version*, 2005. Available online at http://www.millenworks.com/Soads.htm .
389. MILLENWORKS (2005B): TACOM -- Magnetorheological (MR) damper design MagnetoRheological fluid Optimized Active Damper Suspension (MROADS). *MILLEN WORKS -- Printer Friendly Version*, 2005. Available online at http://www.millenworks. com/Mroads.htm .
390. MILLER LR AND CM NOBLES (1990): Methods for Eliminating Jerk and Noise in Semi-Active Suspensions. *SAE (Society of Automotive Engineers) Transactions*, Vol. 99, Sect. 2, 1990, pp. 943-951.
391. MILLIKEN WF AND DL MILLIKEN (1995): *Race car vehicle dynamics*. SAE International, Warrendale, USA, 1995, 890 p.
392. MILLIKEN WF AND DL MILLIKEN (2002): *CHASSIS DESIGN -- Principles and Analysis*, Professional Engineering Publishing Ltd, February 2002, 656 p.
393. MORAN A AND M NAGAI (1994): Optimal active control of nonlinear vehicle suspensions using neural networks, *JSME International Journal*, Vol. 37, No. 4, 1994, pp. 707-718.
394. MOTTA DS, DE ZAMPIERI AND AK PEREIRA (2000): Optimization of a Vehicle Suspension Using a Semi-Active Damper. *SAE Technical Paper Series*, 2000-01-3304 E, ISSN 0148-7191, 2000.
395. MUCKA P (2000): The influence of quarter-car model parameters on the quality of active suspension. *Journal of Mechanical Engineering*, Vol. 51, N0. 3, 2000, p. 174 (in Slovak).
396. MUCKA P (2002): Active suspension of a heavy-vehicle driven axle. *Journal of Mechanical Engineering*, Vol. 53, No. 3, 2002, p. 153 (In Slovak).
397. NABAGLO T (2006): Controller of Magneto-Rheological Semi-Active Car Suspension. *Proceedings of the 2006 SAE Automotive Dynamics Stability and Controls Conference and Exhibition*, February 14-16, 2006, Novi, MI, USA, P-395, Paper No. 2006-01-1969, pp. 119-128.
398. NABAGLO T (2006): *Synthesis of an Automobile Semi-Active Suspension Control System with Magneto-Rheological Elements (Synteza ukladu sterowania semiaktywnego zawieszenia samochodu z elementami magnetoreologicznymi)*. PhD Thesis, Cracow University of Technology (POLITECHNIKA KRAKOWSKA*), 2007, 187 p. (In Polish).*
399. *NAGAI M, A MORAN AND S TANAKA (1989): Optimal Active Secondary Suspension Applied to Magnetically Levitated Vehicle Systems. Proceedings of the 28th SICE Annual Conference*, Vol. II, July 1989, pp. 1396, 1167 -1170.
400. NAGAI M (1993): Recent Researches on Active Suspensions for Ground Vehicles. *JSME International Journal*, Series C, Vol. 36, No. 2, 1993, pp. 161-170.
401. NAGAI M AND T HASEGAWA (1997): Vibration isolation analysis and semi-active control of vehicles with connected front and rear suspension dampers. *JSAE Review*, Vol. 18, Issue 1, 1997, pp. 45-50.

402. NAGARAJAIAH S, AND X MA (1996): System Identification Study of a 1:10 Scale Steel Model Using Earthquake Simulator, *Proc. of Eng. Mech. Conf.*, ASCE, **2**, 1996, pp. 764–767.
403. NAKAI H, K YOSHIDA, S OHSAKU AND Y MOTOZONO (1997): Design of Practical Observer for Semiactive Suspensions. *Nippon Kikai Gakkai Ronbunshu, C Hen,* Vol. 63, No. 615, November 1997, pp. 3898-3904.
404. NAKANO K AND Y SUDA (2004): Self-Powered Active Vibration Control for Truck Suspensions. *FISITA 2004: World Automotive Congress*, 23-27 May 2004, Barcelona, Spain, Paper F2004F394.
405. NARAGHI M AND E NAJAF ZADEH (2001): Effect of heavy vehicle suspension designs on dynamic road loading – A comparative study. *Proceedings of International Truck and Bus Meeting* – Truck and Bus Chassis, Suspension, Stability and Handling (SP-1651), November 12-14, 2001, SAE International, Warrendale, PA, Paper No. 2001-011-2766. 7 p.
406. NELL S AND JL STEYN (1998): An alternative control strategy for semi-active dampers on off-road vehicles. *Journal of Terramechanics*, Vol. 35, No. 1, 1998, pp. 25-40.
407. NESTICO V (1995): Hybrid electric drive active suspension HMMWV. *SPIE Proceedings*, Vol. CR59, Paper No. CR59-05, 04/17 -- 04/24, 1995, Orlando, FL, USA.
408. NEUHAUS P (1992): *Design of a Nonlinear Controller for Active Suspension.* Master Thesis, Vehicle Dynamics Laboratory, University of California at Berkeley, July 1992.
409. NICOLAS CF, J LANDALUZE, E CASTILLO, M GASTON AND R REYERO (1997): Application of Fuzzy Logic Control to the Design of Semi-Active Suspension Systems. *IEEE International Conference on Fuzzy Systems*, Vol. 2, 1997, Paper No. 97CB36032, pp. 987-993.
410. NOVAK R (2005): 'Stiff' fluid could soon put the brakes on cars. *New Scientist,* 11 October 2005, p. 23. Available online at http://www.new scientist.com .
411. NWAGBOSO CHO ED. (1993): *Road Vehicle Automation.* PENTECH PRESS Publishers: London, 1993, 309 p.
412. NWAGBOSO CHO ED. (1997): *Advanced Vehicle and Infrastructure Systems: Computer Applications, Control, and Automation.* John Wiley, New York, 1997, 502 p.
413. OLLEY M (1934): Independent Wheel Suspension -- Its Whys and Wherefores. *SAE Journal*, Vol. 34, No. 3, 1934, pp. 73-81.
414. OLLEY M (1946): *Road Manners of the Modern Car*. Institution of Automobile Engineers, 1946, pp. 147-182.
415. OSORIO C, S GOPALASAMY AND J HEDRICK (1999): Force Tracking Control for Electro-hydraulic Active Suspensions Using Output Redefinition. *Proceedings of the ASME Winter Annual Meeting*, Nashville, TN, 1999.
416. PACEJKA HB (2000): *Tire and Vehicle Dynamics*. SAE International, Warrendale, PA, USA, 2000.

417. PALKOVICS L AND PJTH VENHOVENS (1992): Investigation of stability and possible chaotic motions in the controlled wheel suspension systems. *Vehicle System Dynamics*, Vol. 21, 1992, pp. 269-296.
418. PALKOVICS L AND M EL-GINDY (1993): Examination of Different Control Strategies on Heavy-Vehicle Performance. *Advanced Automotive Technologies*, American Society of Mechanical Engineers (ASME), Dynamic Systems and Control Division, DSC, Vol. 52, 1993, pp. 349-360.
419. PANG L, GM KAMATH AND NM WERELEY (1998): Dynamic characterization and analysis of magnetorheological damper behavior. *Proceedings of SPIE*, Vol. 3327, 1998, pp. 284-302.
420. PATEL A AND JF DUNNE (2000): Neural Network Modelling of Frequency Dependent Automotive Dampers. *Structural Dynamics: Recent Advances, Proceedings of the 7th International Conference,* ISVR Southampton, Vol. 1, 2000, pp. 501-513.
421. PATEL A AND JF DUNNE (2001): Physical Versus Neural Network Models of Automotive Dampers. *Proceedings of the 1st Vehicle Technology Conference*, (ImechE), University of Sussex, 2001, pp. 241-251.
422. PATEL A AND JF DUNNE (2002): Neural network modelling of suspension dampers for variable temperature operation. Available online at http://www.sussex.ac.uk/automotive/ tvt2002/17_patel.pdf , 2002.
423. PATTEN WN, S JINGHUI, L GUANGJUN, J KUEHN AND G SONG (1999): Field Test of an Intelligent Stiffener for Bridges at The I-35 Walnut Creek Bridge. *Earthquake Engineering and Structural Dynamics,* Vol. 28, 1999, pp. 109–126.
424. PENG H, R STRATHEARN AND AG ULSOY (1997): A Novel Active Suspension Design Technique - Simulation and Experimental Results. *Proceedings of the 1997 American Control Conference*, Albuqueque, New Mexico, June 4-6, 1997.
425. PETEK NK, DJ ROMSTADT, MB LIZELL AND TR WEYENBERG (1995): Demonstration of an automotive semi-active suspension using electrorheological fluid. *Proceedings of International Congress and Exposition* -- New Developments in Vehicle Dynamics, Simulation and Suspension Systems (SP -1074), February 27 -- March 2, 1995, SAE International, Warrendale, USA, Paper No. 950586, pp. 237-242.
426. PHULE P AND J GINDER (1998): The materials science of field-responsive fluids. *MRS Bulletin*, August 1998, pp. 19-21.
427. PINKOS A, E SHTARKMAN AND T FITZGERALD (1993): Actively Damped Passenger Car Suspension System with Low Voltage Electro-Rheological Magnetic Fluid. *Vehicle Suspension and Steering Systems SAE Special Publications*, No. 952, 1993, pp. 87-93.
428. POYNOR JC (2000): *Innovative Designs for Magneto-Rheological Dampers*, Master Thesis, Virginia Polytechnic Institute and State University, Blacksburg, VA, 2001.
429. PREM H (1987): A laser-based highway-speed road profile measuring system. *Proc. 10th IAVSD*, 1987.

430. PROKOP G AND RS SHARP (1995): Performance Enhancement of Limited Bandwidth Active Automotive Suspensions by Road Preview. *IEE Proc. – Control Theory*, A, Vol. 142, No. 2, 1995, pp. 140-148.
431. RABINOW J (1948): The Magnetic Fluid Clutch. *AIEE Trans.*, Vol. 67, 1948, pp. 1308–1315.
432. RADATEC (2002): Precision Terrain Radar doe Automotive Active Suspension Control. *Applications*, Radatec Inc. 2002. Available online at http://www.radatec.com/radatec/ applications.asp .
433. RAJU GV AND S NARAYANAN (1991): Optimal estimation and control of non-stationary response of a two-degree-of-freedom vehicle model. *Journal of Sound and Vibration*, Vol. 149, 1991, pp. 413-428.
434. RAJAMANI R (1993): *Observers for Nonlinear Systems, with Application to Automotive Active Suspensions*. Ph.D. Dissertation, The University of California at Berkeley, Berkeley, CA, 1993.
435. RAJAMANI R AND JK HEDRICK (1994): Performance of active automotive suspensions with hydraulic actuators: theory and experiment. *Proceedings of the 1994 American Control Conference*, June 1994, pp. 1214-1218.
436. RAJAMANI R (2005): *Vehicle Dynamics and Control*. Springer Verlag, New York, 2005.
437. REDFIELD RC AND DC KARNOPP (1988): Optimal Performance of Variable Component Suspensions. *Vehicle System Dynamics*, Vol. 29, 1988, pp. 231 -253.
438. REDFIELD RC (1991): Performance of Low-Bandwidth, Semi-Active Damping Concepts for Suspension Control. *Vehicle System Dynamics*, Vol. 20, No. 5, 1991, pp. 245-267.
439. REIMPELL J, H STOLL AND JW BETZLER (2001): *The Automotive Chassis*. 2nd Edition. Butterworth Heinmann. 2001.
440. RETTIG U AND O VON STRYK (2001): Numerical Optimal Control Strategies for Semi-Active Vehicle Suspension with Electrorheological Fluid Dampers. *Proceedings of the Workshop" Fast solution of discretized optimisation problems*, International Series on Numerical Mathematics, Technische Universität München, Technische Universität Darmstadt, (Birkhauser) Germany, 2001, pp. 221-241.
441. RETTIG U AND O VON STRYK (2005): Optimal and Robust Damping Control for Semi-Active Vehicle Suspension. *ENOC 2005*, Eindhoven, The Netherlands, 7-12 August 2005, pp. 1-10.
442. RICHERZHAGEN M (2004): Auslegung der elastokinematik unter anwendung regelungs technischermethoden. *Steuerung und Regelung von Fahrzeugen und Motoren–AUTOREG 2004*, Düsseldorf, Germany, 2004.
443. RILL G (1983): The influence of Correlated Random Road Excitation Processes on Vehicle Vibration. *Proc. of the 8th IAVSD Symposium*, 1983.
444. RMSV (2004): Advanced Active Damper Designed by Rod Millen Special vehicles Survives Gruelling Military Endurance Testing. *Rod Millen Special Vehicles, For Immediate Release*, October 1, 2004. Available online at http://www.rodmillen.com .

445. ROBERTI V, B OUYAHIA AND A DEVALLET (1993): Oleopneumatic suspension with preview semi-active control law. *Proc. Int5. Cong. MV2*, Active Control in Mechanical Engineering, Lyon, France, Vol. 1, June 1993.
446. ROBERTSON A.J (1992): *Vehicle design.* MSc Lecture Notes, Cranfield Institute of Technology, 1992. Robertson, A.J.: "Vehicle design", MSc lecture notes, Cranfield Inst. of Technology, 1992.
447. ROBSON JD AND CJ DODDS (197): The Response of Vehicle Component to Random Road Surface Undulation. *Proceedings of the 13th FISITA Congress*, 1970.
448. ROH HS AND Y PARK (1999): Stochastic Optimal Preview Control of an Active Vehicle Suspension. *Journal of Sound and Vibration,* Vol. 220, No. 2, 1999, pp. 313-330.
449. ROUKIEH S AND A TITLI (1993): Design of Active and Semiactive Automotive Suspension Using Fuzzy Logic, *IFAC 12th World Congress*, Sydney, Australia, Vol. 2, 1993, pp. 253-257.
450. SACHS (2003A): Sachs AG's www-pages. 22.07.2003, Available online at http://www.sachs.de/oxx_medien/8_462_8_9_12_00000000000000.html
451. SACHS (2003B): Sachs AG's www-pages. 22.07.2003, Available online at http://www.sachs.de/oxx_medien/8/461.pdf .
452. SAE J6A (1965): *Ride and Vibration Data Manual.* SAE J6a, Society of Automotive Engineers, Warrendale, PA, December 1965 (see Appendix B).
453. SAE J670E (1978): *Vehicle Dynamics Terminology.* SAE J670e, Society of Automotive Engineers, Warrendale, PA, 1978 (see Appendix A).
454. SAMMLER D (2001): *Modeling and control of suspension systems for road vehicles*. Ph.D. Thesis, Lab. of Automatic Control of Grenoble, France, November 2001 (in French).
455. SAMMLER D, O SENAME AND L DUGARD (2000): H_∞ control of active vehicle suspensions, *Proceedings of the 2000 IEEE International Conference on Control Applications*, September 25-27, 2000, pp. 976-981.
456. SAMMLER D, O SENAME AND L DUGARD (2002): Commande par placement de poles de suspensions automobiles. *Conf. Int. Francophone d'Automatique*, Nantes, France, July 2002 (in French).
457. SAMPSON DJM (2000): *Active Roll Control of Articulated Heavy Vehicles.* Ph.D. Dissertation, Churchill College, University of Cambridge, September 2000. 277 p. Available online at http://david.sampson.id.au/vehicle/arcoahv.pdf .
458. SASSI S, K CHERIF, L MEZGHANI, M THOMAS AND A KONTRANC (2005): An innovative magnetorheological damper for automotive suspension from design to experimental characterization. *Smart Material Structures*, Vol. 14, 2005, pp. 811-822.
459. SAXON NL, WR MELDRUM JR AND TK BONTE (1998): *Semiactive Suspension: A Field Testing Case Study. Developments in Tire, Wheel, Steering and Suspension Technology.* SAE Special Publications, Vol. 1338, February 1998, pp. 187-194.

460. SCHÖNER HP (1999): 42 V: New Prospects for Mechatronic Systems for Chassis and Engine Technology. *DaimlerChrysler Presentation*, 8 Aachener Kolloquium Fahrzeug-Motorentechnik 1999, 5. October 1999, ss. 1-25.
461. SCHULDINER H (2005): Engineering innovations highlight Porsche 911. *Kane County Chronicle – Online*, 2005.
462. SCHWARZ R AND P RIETH (2004): Global chassis control – system vernetzung im Fahrwerk. *Steuerung und Regelung von Fahrzeugen und Motoren–AUTOREG 2004*, Düsseldorf, Germany, 2004.
463. SEIFFERT U, AND P WALZER (1991): *Automotive Technology of the Future.* VDI-Verlag GmbH, 1991, 251 p.
464. SENSORSMAG (2001): *Magnetostrictive position sensors enter the automotive market.* Sensorsmag, December 2001, http://www.sensorsmag.com/articles/1201/26/main.shtml .
465. SHARIATI A, DRIVER TAGHIRAD AND A FATCHI (2004): Decentralized Robust H_∞ Controller Design for a Half-Car Active Suspension System. *Control 2004*, University of Bath, UK, September 2004, Paper ID-216, pp. 1-5.
466. SHARP RS AND SA HASSAN (1984): The fundamentals of passive automotive suspension system design. *Proceeding of the Society of Environmental Engineers Conference on Dynamics in Automotive Engineering*, Cranfield Institute of Technology, UK, 1984, pp. 104-115.
467. SHARP RS AND SA HASSAN (1986A): The Relative Performance Capabilities of Passive, Active and Semi-Active Car Suspension Systems. *Proc. ImechE*, Vol. 200, No. D3. 1986, pp. 219-228.
468. SHARP RS AND SA HASSAN (1986B): An evaluation of passive automotive suspension systems with variable stiffness and damping parameters. *Vehicle System Dynamics*, Vol 15, No. 6, 1986, pp. 335-350.
469. SHARP RS AND SA HASSAN (1986C): The Relative Performance Capabilities of Passive, Active and Semi-Active Car Suspension Systems. *Proc. IMechE*, Part D, Vol. 203, No. 3, 1986, pp. 219-228.
470. SHARP RS AND SA HASSAN (1987): Performance and design considerations for dissipative semi-active suspension systems for automobiles. *Proc. IMechE*, Vol. 201, No. D2, 1986, pp. 149-153.
471. SHARP RS AND DA CROLLA (1987A): Road Vehicle Suspension System Design – A Review. *Vehicle System Dynamics*, Vol. 16, 1987, pp. 167-192.
472. SHARP RS AND DA CROLLA (1987B): Intelligent Suspensions of Road Vehicles – Current and Future Developments. *Proc. International Conference on New Developments in Power Train and Chassis Engineering.* Düsseldorf, Germany, 1987, pp. 579-601.
473. SHARP RS AND DA WILSON (1990): On Control Law for Vehicle Suspension Accounting for Input Correlations. *Vehicle System Dynamics,* Vol. 19, 1990, pp. 353-363.
474. SHARP RS (1998): Variable geometry active suspension for cars. *Computing and Control Engineering Journal,* Vol. 9, Issue 5, 1998, pp. 217-222.
475. SHEN X AND H PENG (2003): Analysis of Active Suspension Systems with Hydraulic Actuators. *Proceedings of the 2003 IAVSD Conference*, Atsugi, Japan, August 2003, pp. 1-10.

476. SHEN Y, MF GOLNARAGHI AND GR HEPPLER (2004A): Semi-active suspension control with a magnetorheological damper. *Proceedings: 7th CanSmart Meeting International Workshop Smart Materials & Structures,* 21-22 October, 2004, Montreal, Quebec, Canada, pp.143-152.
477. SHEN Y, MF GOLNARAGHI, GR HEPPLER (2004B): Experimental Research and Modeling of Magnetorheological Elastomers. *Journal of Intelligent Material Systems and Structures,* Vol. 15, Issue 1, 2004, pp. 27-35(9).
478. SHEN Y, MF GOLNARAGHI AND GR HEPPLER (2005): Analytical and Experimental Study of response of a Suspension System with a Magnetorheological Damper. *International Journal of Intelligent Materials* and *Structures,* Vol. 16, Issue 02, 2005, pp. 135-148.
479. SHI XD, J WANG AND LF QIAN (2004): Study on Semi-Active Suspension Controller Model of the Quarter Car with Magneto-Rheological Dampers by Using Fuzzy Reasoning Based on Lyapunov Stability Theory. *ELECTRORHEOLOGICAL FLUIDS AND MAGNETO-RHEOLOGICAL SUSPENSIONS (ERMR 2004) Proceedings of the Ninth International Conference*, Beijing, China, 29 August – 3 September 2004, p.749.
480. SHIOZAKI M, S KAMIYA, M KUROYANAGI, K MATSUI AND R KIZU (1991): High speed control of damping force using piezoelectric elements. *Proceedings of International Congress and Exposition* – Vehicle Dynamics and Electronic Controlled Suspensions (SP-861), February 25 – March 1, 1991, SAE International, Warrendale, USA, Paper No. 901661, pp. 148-154.
481. SIMON DE AND M AHMADIAN (1999): Application of Magneto-Rheological Dampers for Heavy Truck Suspensions. *Proceedings of the 32nd International Symposium on Automotive Technology and Automation (ISATA*, Vienna, Austria, 1999.
482. SIMON DE (2001): *An Investigation of the Effectiveness of Skyhook Suspensions for Controlling Roll Dynamics of Sport Utility Vehicles Using Magneto-Rheological Dampers.* Ph.D. Dissertation, Virginia Polytechnic Institute and State University, Blacksburg, Virginia, November 28, 2001 229 p.
483. SIMON DE AND M AHMADIAN (2002): An alternative semiactive control method for sport utility vehicles. *Journal of Automobile Engineering*, Vol. 216, No. 2, 2002, pp. 125-139.
484. SIMON DE (2005): Ch2.pdf, 2005. Available online at http://scholar.lib.vt.edu/theses/available/etd91898203427/unrestricted/ch2.PDF .
485. SIMS ND, DJ PEEL, R STANWAY, AR JOHNSON AND WA BULLOUGH (2000): The electro-rheological long-stroke damper; a new modelling technique with experimental validation. *Journal of Sound and Vibration*, Vol. 229, Issue 2, 2000, pp. 207-227.
486. SINCEBAUGH P AND W GREEN (1996): A Neural Network Based Diagnostic Test System for Armoured Vehicle Shock Absorbers. *Expert Systems with Applications*, Vol.11, No. 2, 1996, pp. 237-244.
487. SMAKMAN H (2000): *Integration of Slip Control With Active Suspension for Improved Lateral Vehicle: Dynamics*. Maschinenwesen. Herbert Utz Verlag – Wissenschaft, 2000.

488. SMITH MC, KM GRIGORIADIS AND RE SKELTON (1992): The Optimal Mix of Passive and Active Control in Structures. *Journal of Guidance, Control and Dynamics*, Vol. 15, No. 4, 1992, pp. 912-919.
489. SMITH MC (1993): Achievable dynamic response for automotive active suspensions. 1993.
490. SMITH MC AND GW WALKER (1997): Performance Limitations for Active and Passive Suspensions. *Proceedings of the European Control Conference*, 1997.
491. SMITH MC AND GW WALKER (2000): Limitations and Constraints for Active and Passive Suspensions: A Mechanical Multi-Port Approach. *Vehicle Dynamics*, Vol. 33, No. 3, 2000, pp. 137-168.
492. SMITH M AND F WANG (2002): Controller parameterization for disturbance response decoupling: application to vehicle active suspension control. *IEEE Transactions on Control Systems Technology*, Vol. 10, May 2002. Vol. 10, May 2002.
493. SNYDER R AND N WERELEY (1999): Characterization of a magneto-rheological fluid damper using a quasi-steady model. *Proceedings of SPIE*, Vol 3668, 1999, pp. 507-519.
494. SOBCZYK K, DB MACVEAN AND JD ROBSON (1997): Response to profile imposed excitation with randomly varying traversal velocity. *Journal of Sound and Vibration*, Vol. 52, 1977, pp. 37-49.
495. SOHN HC, KS HONG AND JK HEDRICK (2000): Semi-Active Control of the MacPherson Suspension System Hardware-in-the-Loop Simulations. *Proceedings of the 2000 IEEE International Conference on Control Applications*, Vol. 1, 2000, Piscataway, NJ, pp. 982-987.
496. SOLIMAN A (2004): A Comparison of the Performance and Power Requirements of Three Semi-Active Suspension Control Laws. *FISITA 2004: World Automotive Congress*, 23-27 May 2004, Barcelona, Spain, Paper F2004F001.
497. SONG X (1999): *Design of adaptive vibration control systems with application of magneto-rheological dampers.* PhD Dissertation, Virginia Polytechnic Institute and State University, Blacksburg, VA, 1999.
498. SONG X, M AHMADIAN, S SOUTHWARD AND LR MILLER (2005): An Adaptive Semiactive Control Algorithm for Magnetorheological Suspension Systems. *Journal of Vibration and Acoustics, Transactions of the ASME,* October 2005. Vol. 127. pp. 493-502.
499. SOONG TT, AND GF DARGUSH (1997): *Passive Energy Dissipation Systems in Structural Engineering*, Wiley & Sons, Chichester, England.
500. SORGE K AND H WILHELM (2001): Integration of Damper Control Algorithm into ADAMS/Car Full Vehicle Model. *16th European Mechanical Dynamics User Conference,* Bechtesgarden, Germany, November 2001.
501. SORSCHE JH, K ENCKE AND K BAUER (1974): Some Aspects of Suspension and Steering Design for Modern Cars. *SAE Technical Paper Series,* 1974, Paper 741039, 9 p.

502. SPENCER BF JR., SJ DYKE, MK SAIN AND JD CARLSON (1997): Phenomenological Model for Magnetorheological Dampers. *Journal of Engineering Mechanics*, ASCE, **123**(3), 1997, pp. 230-238.
503. SPENTZAS CN (1993): Optimization of Vehicle Ride Characteristics by Means of Box's Method. *International Journal of Vehicle Design*, Vol. 14, 1993, pp. 539-551.
504. SPROSTON JL, R STANWAY, EW WILLIAMS AND S RIGBY (1994): The electrorheological automotive engine mount. *Journal of Electrostatics*. Vol. 32, 1994, pp. 253-259.
505. STAMMERS CW AND T SIRETANU (1998): Vibration control of machines by use of semi-active dry-friction damping. *Journal of Sound and Vibration*, Vol. 209, Issue 4, 1998, pp. 671-684.
506. STANWAY R, JL SPROSTON AND AK EL-WAHED (1996): Application of electro-rheological fluids in vibration control: a survey. *Smart Materials and Structures*, Vol. 5, No. 4, 1996, pp. 464-482.
507. STRATHEARN RR (1996): *Active suspension design and evaluation using a quarter car test rig*. M.S. Thesis, University of Michigan, 1996.
508. STŘIBRSKÝ A, K HYNIOVÁ AND J HONCŮ (2000): Using Fuzzy Logic to Control Active Suspension of Vehicles. *Intelligent Systems in Practice*, Luhačovice, Slovakia, 2000, pp. 41-49.
509. STŘIBRSKÝ A, K HYNIOVÁ AND J HONCŮ (2002): Limitations in Active Suspension Systems. *CTU REPORTS – Proceedings of Workshop 2002*, 2002, Vol. 6, 212 p.
510. STRYK O VON (1997): Optimal Guidance of Full Car Simulations in Real-Time. *ISMP97: International Symposium on Mathematical Programming*, Lausanne, EPFL, August 24-29, 1997. Available online at http://dmawww.epfl.ch/roso.mosaic/ismp97/ismp_abs_280. html .
511. STRYK O VON (2004): Computational Engineering at TUD: Education and Research Across Disciplines. *Presentation to CE*, Technische Universitat Darmstadt. Blacksburg, Feb. 12, 2004, ss. 1-71. Available online at http://www.sim.informatik.tu-darmstadt.de/pers/stryk.html .
512. SU Y, J YUE, H JIANG AND R PENG (2004): Research on Control Character of Pneumatic Servo System with MR. *ELECTRORHEOLOGICAL FLUIDS AND MAGNETO-RHEOLOGICAL SUSPENSIONS (ERMR 2004) Proceedings of the Ninth International Conference*, Beijing, China, 29 August – 3 September 2004, p. 790.
513. SWEVERS J, P SAS AND CH LAUWERYS (2005): Control of a semi-active suspension system for a passenger car. *Noise and Vibration Research Group*, Katholieke Universiteit Leuven, 08/07/2005. Available online at http://www.kuleuven.be/mod/other/ topic_03 _05.en.html .
514. SWRI (1997): Innovative vehicle suspension system wins R&D 100 award. *Southwest Research Institute (SwRI) News*, August 22, 1997. Available online at http://www. swri.org/9what/releases/1997/rd.htm .

515. SUGASAWA F ET AL. (1985): Electronically controlled shock absorber system used as a road sensor which utilizes supersonic waves. *SAE Technical Paper Series,* 1985, Paper 851652.
516. TAGHIRAT DRIVER AND E ESMAILZADEH (1998): Automobile Passenger Comfort Assured through LQG/LQR Active Suspension. *Journal of Vibration Control*, Vol. 4, No. 5, 1998, pp. 603-618.
517. TAKANO A (2004): Showa Presents Electromagnetic Suspension. *Tech-On*, NikkeiBP, October 21, 2004. Available online at http://nikkeibp.co.jp/ English/NEWS_EN/20051021/109955/?SS=imgview_e&FD=1121857826 .
518. TAMBOLI JA AND SG JOSHI (1999): Optimum design of a passive suspension system of a vehicle subjected to actual random road excitations. *Journal of Sound and Vibration*, Vol. 219, No.2 1999, pp. 193-205. Article No. jsvi, 1998, 1882. Available online at http://www.idealibrary.com .
519. TAN XB (2004): Smart Materials and Microsystems. *Presentation at Michigan State University*, Smart MicroSystems Laboratory, November 16, 2004, ss. 1-17. Available online at: http://www.egr.msu.edu/`xbtan .
520. TANAHASHI H, K SHINDO, T NOGAMI AND T OONUMA (1987): Toyota electronic modulated air suspension for the 1986 Soarer. *SAE Technical Paper Series,* 1987, Paper 870541.
521. TAO R ED. (1999): *Proceedings of the Seventh International Conference on ER Fluids and MR Suspensions*. Honolulu, Hawaii, July 19-23, World Scientific Publishing Company.
522. TEN (2004): Tenneco Automotive to Supply Innovative Electronic Suspension System on New Audi AS. *Tenneco Automotive Inc. Corporate News* November 17, 2004. Available online at http://www.tenneco-automotive.com/news/ .
523. TEN (2005): Tenneco to supply innovative electronic suspension system on new Audi A6. *AUTOMOTIVE BUSINESS Review Online*, Spring 2005. Available online at http://www.automotive-business-review.com/article_news.asp/guid=2E2D1088-FA18-86-88E C-C8D566E6277 .
524. THOMPSON AG (1976): An active suspension with optimal linear state feedback. *Vehicle System Dynamics*, Vol. 5, 1976, pp. 187-203.
525. THOMPSON AG (1984): Optimal and suboptimal linear active suspensions for road vehicles. *Vehicle System Dynamics*, Vol. 13, 1984, pp. 61-72.
526. THOMPSON AG, BR DAVIS AND FJM SALZBORN (1984): Active suspensions with vibration absorbers and optimal output feedback control. *SAE Technical Paper Series*, 1984, Paper 841253.
527. THOMPSON AG AMD PM CHAPLIN (1996): Force Control in Electrohydtaulic Active Suspensions. *Vehicle System Dynamics*, Vol. 25, 1996, pp. 185-202.
528. THOMPSON AG AND CEM PEARCE (1998): Physically Realisable Feedback Controls for a Fully Active Preview Suspension Applied to a Half-Car Model. *Vehicle System Dynamics*, Vol. 30, 1998, pp. 17-35.
529. THOMPSON AG AND BR DAVIS (2005): Computation of the rms state variables and control forces in a half-car model with preview active suspension using spectral decomposition methods. *Journal of Sound and Vibration*, Vol. 285, No. 3, 2005, pp. 571-583.

530. TING C, T LI AND F KUNG (1995): Design of fuzzy controller for active suspension system. *Mechatronics*, Vol. 5, No. 4, 1995, pp.365–383.
531. TOBATA H, N FUKUSHIMA AND T KIMURA (1992): Advanced control method of active suspension. *AVEC92*, Yokohama. Japan, 1992, Paper 923023.
532. TOMIZUKA M (1976): Optimum linear preview control with application to vehicle suspension -- Revisited. *ASME Journal of Dynamic Systems*, Measurement and Control, Vol. 98. 1976, pp. 309-315.
533. TRIANTAFYLLIDIS N AND S KANKANALA (2005): On finitely strained magneto-rheological elastomers. Department of aerospace Engineering, The University of Michigan, Ann Arbor, MI 48109.
534. TREVETT NRC (2002): *X-by-Wire, New Technologies for 42V Bus -- Automobile of the Future*. Submitted in Partial Fulfillment of the requirements for Graduation with Honors from the South Carolina Honors College, April, 2002
535. TRUSCOTT AJ AND PE WELLSTEAD (1994): Adaptive ride control in active suspension systems. *Control Systems Centre Reports*, No. 798, 1994, University of Manchester, 43 p. Available online at http://www.csc.umist.ac.uk/CSCReports/798.htm .
536. TRW (2005): Active Suspension Control Systems -- ASCS. *Chassis Systems* – Body Control Systems, TRW Automotive, 2005.
537. TSENG HE AND JK HEDRICK (1994): Semi-Active Control Laws -- Optimal and Sub-Optimal. *Vehicle System Dynamics*, Vol. 23, 1994, pp. 545-549.
538. TSUTSUMI Y, H SATO, H KAWAGUCHI AND M HIROSE (1990): Development of Piezo TEMS (Toyota Electronic Modulated Suspension). *SAE Technical Paper Series,* 1990, Paper 901745.
539. TYLIKOWSKI A (1995): Active stabilization of beam vibrations parametrically excited by wide-band Gaussian force. *ACTIVE 95 Proceedings of the 1995 International Symposium on Active Control of Sound and Vibration*, Newport Beach, CA, July 06-08, 1995, p. 0091
540. UEKI N, J KUBO, T TAKAYAMA, I KANARI AND M UCHIYAMA (2004): Vehicle Dynamics Electric Control Systems for Safe Driving, *Hitachi Review*, Vol. 53 (2004), No. 4, pp. 222-226.
541. ULSOY AG, D HROVAT AND T TSENG (1994): Stability robustness of LQ and LQG active suspension. *ASME Journal of Dynamic Systems*, Measurement and Control, Vol. 116, 1994, pp. 123-131.
542. UNSAL M (2002): *Force control of a new semi-active piezoelectric-based friction damper.* M.S. Thesis, University of Florida, 2002, pp. i-xiii; 1-88.
543. UNSAL M, C NIEZRECKI AND C CRANE III (2004): Two Semi-Active Approaches for Vibration Isolation: Piezoelectric Friction Damper and Magneto-rheological Damper. *US DOE Grant Number DE-FG04086NE37967*, Department of Mechanical and Aerospace Engineering, University of Florida, Gainsville, FL, 2004, pp. 1-6.
544. UOM (1994): *Electrorheology for smart automotive suspensions.* Final Report, The University of Michigan Transportation Research Institute, June 1994.

545. URSU I, F URSU, T SIRETANU AND CW STAMMERS (2000): Artificial intelligence base synthesis of semiactive suspension system. *The Shock and Vibration Digest*, Vol. 32, No. 1, January 2000, pp. 3-10.
546. UT-CEM (2004): Electronically Controlled Active Suspension System (ECASS). *Ground Combat Element*, 18 December 2004, The University of Texas at Austin – Center for Electro Magnetics (UN-CEM), POC: (703) 432-0453, p. V-GCE-6.
547. VALÁŠEK M, M NOVAK, Z SIKA AND O VACULIN (1997): Extended groundhook – new concept of semi-active control of truck's suspension. *Vehicle System Dynamics*, Vol. 27, No. 5-6, 1997, pp. 289-303.
548. VALÁŠEK M, M NOVAK, Z SIKA AND O VACULIN (1998): Development of semi-active road-friendly truck suspension. *Control Engineering Practice*, Vol. 6, 1998, pp. 735-744.
549. VANHEES G AND M MAES (2002): Vehicle suspension characterization by using simulation on a 4 poster test rig. *2002 ISMA -- International Conference on Noise and Vibration Engineering*, III Active suspensions, September 16 – 18, 2002, Katholieke Universitieit Leuven, Belgium.
550. VELARDOCCHIA M AND A SORNIOTTI (2004): Development of an Active Suspension Control Strategy and its Integration with Vehicle Dynamics Control. *FISITA 2004: World Automotive Congress*, 23-27 May 2004, Barcelona, Spain, Paper F2004F254.
551. VENHOVENS PJTH, KNAAP ACM VAN DER AND HB PACEJKA (1992): Semi-Active Vibration and Attitude Control. *Proceedings of the International Symposium on Advanced Vehicle Control 1992 (AVEC'92),* Yokohama, Japan, JSAE 923032, September 1992.
552. VENHOVENS PJTH, ACM VAN DER KNAPP AND HB PACEJKA (1993): Semi-Active Attitude and Vibration Control. *Vehicle System Dynamics*, Vol. 22, No. 5-6, September-November 1993, pp. 359-381.
553. VENHOVENS PJTH (1993): *Optimal Control of Vehicle Suspensions*, PhD Thesis, Delft University of Technology, Vehicle Research Laboratory, December 1993.
554. VENHOVENS PJTH, ACM VAN DER KNAAP, AR SAVKOOR AND AJJ VAN DER WEIDEN (1994): Semi-Active Control of Vibration and Attitude of Vehicles. *Vehicle System Dynamics*, Vol. 23, Suppl., 1994, pp. 522-540.
555. VENHOVENS PJTH AND ACM VAN DER KNAAP (1995): Delft Active Suspension (DAS), Background Theory and Physical Realization, *Hand-out of the presentation on the Seminar Smart Vehicles*, 13-16 February 1995.
556. VETTURI D (1993): *Aspetti del comportamento dinamico e modellizzazione di un ammortizzatore per vetture da competizione*, Degree Thesis, University of Brescia, 1993 (In Italian).
557. VILLEGAS-RAMOS C (*2004): Influence of an active suspension system on the lateral dynamics of a passenger vehicle.* MSc thesis, CINVESTAV-IPN, Mexicocity, 2004.

558. VILLEGAS C AND R SHORTEN (2005): *Complex Embedded Automotive Control Systems (CEACS), Public State of the Art of Integrated Chassis* Control, Deliverable D2, STREP project 004175 CEMACS, DaimlerChrysler, SINTEF, Glasgow University, Hamilton Institute, Lund University, February 2005, pp. 1-31
559. VOY C (1988): *Die frequenzmodulierte Dämpfung von Fahrzeugschwingungen. VDI-BeRICHTE 699 Berechnung im Automobilbau.* VDI Gesellschaft Fahrzeugtechnik, Düsseldorf, VDI-Verlag, 1988, S. 93-120.
560. WAL M VAN DE, P PHILIPS AND AG DE JAGER (1998): Actuator and sensor selection for an active vehicle suspension aimed at robust performance. *Int. J. of Control*, Vol. 70, 1*998, pp. 703-720.*
561. *WALKER GW (1997A): Constraints upon the Achieveable Performance of Vehicle Suspension Systems.* Ph.D. Thesis, University of Cambridge, Department of Engineering, 1997
562. WALKER GW (1997B): An Introduction to Active Suspension Systems. University of Cambridge, Department of Engineering, 1997. Available online at http://www.control.eng.cam.ac.uk/gww/what_is_active.html .
563. WALLENTOWITZ H (2002): *Vertikal-Querdynamik von Kraftfahrzeugen.* Forschungs-gesellschaft Kraftfahrwesen Aachen mbH, 2002.
564. WANG D, I HAGIWARA AND G ZHONGYANG (2001): Polytopic modelling and state observer synthesis for suspension with variant damping and stiffness. *Proceeding of SAE Noise and Vibration Conference and Exposition*, April 30 – May 3, 2001, SAE International, Warrendale, USA, Paper No. 2001 -01-1579, 6 p.
565. WANG F-CH AND MC SMITH (2001): Active and Passive Suspension Control for Vehicle Dive and Squat. *Nonlinear and Adaptive Control Network (NACO 2), Workshop on Automotive Control*, Lund, Sweden, May 18-10, 2001.
566. WANG J AND J CHEN (2004): Energy Efficiency Analysis and Optimal Control of Car Active Suspension Systems. *Electrical Drives, Power and Control Group*, The University of Liverpool, Department of Electrical Engineering and Electronics, 17 May 2004. Available online at http://www.liv.ac.uk/EEE/research/edpc/project9.htm .
567. WARDLAW CH (1999): Full Test: 1999 Land Rover Discovery Series II, *Edmunds Reviews*, Road Test: Full Test, Date posted: 01-01-1999. Available at http://www.edmunds. com/reviews/roadtests/roadtest44309/article.html .
568. WATANABE Y AND RS SHARP (1999): Mechanical and control design of a variable geometry active suspension system. *Vehicle System Dynamics*, Vol. 32, Issue 2/3, 1999, pp. 217-235.
569. WEBER F AND G FELTRIN (2002): Cable vibration mitigation using controlled magneto-rheological fluid dampers. *EMPA Activities 2002*: Materials and Systems for Civil Engineering, Report on technical and scientific activities. Swiss Federal Laboratories for Materials Testing and Research, An institution of the ETH Domain, pp. 36-37.

570. WEEKS DA, JH BENO, AM GUENIN AND DA BRESIE (2000): Electromechanical Active Suspension Demonstration for Off-Road Vehicles. *SAE Technical Paper Series*, 2000, Paper 00-01-0102.
571. WEN W, X HUANG, S YANG, K LU AND P SHENG (2005): The giant electrorheological effect in suspension of nanoparticles. *Nature Materials*, Vol. 2, 2003, pp. 727-730.
572. WIKIPEDIA (2005): Active Body Control. *Wikipedia, the free encyclopedia*, Category: Automotive suspension technology, 2005. Available online at http://en.wikipedia.org/ wiki/Active_Body_Control.htm .
573. WILDE JR (2005): *Experimental Evaluation and ADAMS Simulation of the Kinetic™ Suspension System.* The Ohio State University, 2005.
574. WILDE JR, GJ HEYDINGER, DA GUENTHER, T MALLIN AND AM DEVENISH (2005): Experimental Evaluation of Fishhook Maneuver Performance of a Kinetic™ Suspension System. *SAE Technical Paper Series*, 2005, Paper No. 2005-01-0392.
575. WILLIAMS RA (1992): Automotive active suspensions. *IEE Colloquium on Active Suspension technology for Automotive and Railway Applications*, 1992.
576. WILSON DA, RS SHARP AND SA HASSAN (1986): The application of linear optimal control theory to the design of automotive suspension. *Vehicle System Dynamics*, Vol. 15, 1986, pp. 105-118.
577. WINSLOW WM (1949): Induced Vibration of Suspensions. *J. Applied Physics*, 20, 1949, pp. 1137–1140.
578. WONG JY (1993): *Theory of Ground Vehicles*, John Wiley & Sons, New York 2001.
579. WRIGHT PG (1982): The influence of aerodynamics on the design of Formula One racing cars. *International Journal of Vehicle Design*, 1982.
580. WRIGHT PG (1984): The application of active suspension to high performance road vehicle. *Proc. IMechE*, C239/84, 1984.
581. WYCZALEK FA (2001): Hybrid Electric Vehicles -- Year 2000 Status. IEEE Aerospace and Electronic Systems Magazine, Vol. 16, 2001, pp. 15-19.
582. XIA P (2003): An inverse model of MR damper using optimal neural network and system identification. *Journal of Sound and Vibration*, Vol. 266, 2003, pp. 1009–1023.
583. XU L AND B YAO (1999): Output Feedback Adaptive Robust Control of Uncertain Linear Systems with Large Disturbances. *Proceedings of the 1999 American Control Conference*, June 2-4, 1999, San Diego CA, USA, pp. 556-560.
584. YAGIZ N, V OZBULUR, A DERDIYOK AND N INANC (1997): Sliding Modes Control of Vehicle Suspension Systems. *ESM'97 European Simulation Multi Conference*, Istanbul, Turkey. June 1997.
585. YAGIZ N AND I YÜKSEK (2001): Robust Control of Active Suspensions Using Sliding Modes. Turk J Engin Environ Sci, Vol. 25, 2001, pp. 79-87.

586. YAHAYA MS, JHS OSMAN, M RUDDIN AND A GHANI (2003): Active Suspension Control: Performance Comparison Using Proportional-Integral Sliding Mode and Linear Quadratic Regulator Methods. *Proc. of the 2003 IEEE Conference on Control Applications -- CCA 2003*, Control Systems Society, June 23-25, 2003, Istanbul, Turkey, Paper No. CF-003430. Available online at http://www.mecha.ee.boun.edu.tr/cca2003 .
587. YANG JN, JC WU AND Z LI (1996): Control of Seismic-Excited Buildings Using Active variable Stiffness Systems. *Engineering Structures*, Vol. 18,No. 8, 1996, pp. 589–596.
588. YANG G (2001): *Large-Scale Magnetorheological Fluid Damper for Vibration Mitigation: Modeling, Testing and Contro*l. Ph.D. Dissertation, University of Notre Dame, 2001.
589. YANG G, JD CARLSON, MK SAIN AND BF SPENCER (2002): Large-scale MR fluid dampers: Modeling and dynamic performance considerations. *Engineering Structures*, 2002, Vol. 24, No. 3, pp. 309-323. Available online at http://cee.uiuc.edu/sstl/gyang2/ Ch2.pdf .
590. YAMAKADA M AND Y KADOMUKAI (1994): A jerk sensor and its application to vehicle motion control system. *Proceedings for the Dedicated Conferences on Mechatronics and Supercomputing Applications in the Transportation Industries in conjunction with the 27th ISATA*, Aachen, Germany, 1994.
591. YAMASHITA M, K FUJIMORI, K HAYAKAWA AND H KIMURA (1994): Application of H_∞ Control to Active Suspension Systems. *Automatica*, 1994, Vo. 30, No. 11, pp. 1717-1729.
592. YANG G, BF SPENCER, JD CARLSON AND MK SAIN (2002): Large-scale MR fluid dampers: modelling and dynamic performance considerations. *Engineering Structures*, Vol. 24, 2002, pp. 309-323.
593. YAO B, F BU, J REEDY AND GTC CHIU (1999): Adaptive Robust Motion Control of Single Rod Hydraulic Actuators: Theory and Experiments. *Proceedings of the 1999 American Control Conference*, June 2-4, 1999, San Diego, CA, USA, pp. 759-763.
594. YAO GZ, FF YAP, G CHEN, WH LI AND SH YEO (2002): MR damper and its application for semi-active control of vehicle suspension system. *Mechatronics*, Vol. 12, Issue 7, 2002, pp. 963-973.
595. YEH EC AND YJ TSAO (1994): A fuzzy preview control scheme of active suspension for rough road. *International Journal of Vehicle Design*, Vol. 15, No.1/2, 1994, pp. 166-180.
596. YEO MS, HG LEE AND MC KIM (2002): *Journal of Intelligent Material Systems and Structures*, Vol. 13, No. 7-8, 2002, pp. 485-489
597. YESTER J AND J SUN (1993): Design of Automatic Tuning of Fuzzy Logic Control for an Active Suspension. *IFAC World Congress*. Sydney, Australia, Vol. 2, 1993, pp. 151-153.
598. YI K, M WANGELIN AND K HEDRICK (1992): Dynamic tire force control by semi-active suspension. *ASME Dynamic Systems and Control Division DSC,* Transportation Systems, Vol. 44, 1992, pp. 299-310.

599. YI K AND K HEDRICK (1993): Dynamic Tire Force Control by Semiactive Suspensions. *Journal of Dynamic Systems Measurement & Control – Transactions of the ASME*, Vol. 115, No. 3, September 1993, pp. 465-474.
600. YI K AND BK SONG (1999): Observer design for semi-active suspension control. *Vehicle System Dynamics*, Vol. 32, 1999, pp. 129-148.
601. YOKOYA Y, R KIZU, H KAWAGUCHI, K OHASHI AND K OHNO (1990): Integrated control system between active control suspension and four wheel steering for the 1989 CELICA. *SAE Technical Paper Series*, 1990, Paper 901748, pp. 87-102.
602. YOKOTA Y, K ASAMI AND T HAJIMA (1984): Toyota electronic modulated suspension system for the 1983 Soarer. *SAE Technical Paper Series,* 1984, Paper 840341.
603. YOSHIMURA T AND M SUGIMOTO (1990): Active suspension for a vehicle travelling on flexible beams with an irregular surface. *Journal of Sound and Vibration*, Vol. 138, No. 3, 1990, pp. 433-445.
604. YOSHIMURA T AND K EDOKORO (1993): An active suspension model for rail/vehicle systems with preview and stochastic optimal control. *Journal of Sound and Vibration*, Vol. 166, 1993, pp. 507-519.
605. YOSHIMURA T, Y ISARI, Q LI AND J HINO (1997A): Active suspension of motor coaches using skyhook damper and fuzzy logic controls. *Control Engineering Practice*, Vol. 5, No. 2, 1997, pp. 175-184.
606. YOSHIMURA T, K NAKAMINAMI AND J HINO (1997B): A semi-active suspension with dynamic absorbers of ground vehicles using fuzzy reasoning. *International Journal of Vehicle Design*, Vol. 18, No. 1, 1997, pp. 19-34.
607. YOSHIMURA T, T HIWA, M KURIMOTO AND J HINO (2001): Construction of an active suspension system of a quarter car model using the concept of sliding mode control. *Journal of Sound and Vibration*, Vol. 239, No. 2, pp. 187-199.
608. YOSHIMURA T, T HIWA, M KURIMOTO AND J HINO (2003): Active suspension of a one-wheel car model using fuzzy reasoning and compensators. *International Journal of Vehicle Autonomous Systems*, Vol. 1, No. 2. 2003, pp. 196 -205.
609. YOSHIMURA T AND K WATANABE (2003): Active suspension of a full car model using fuzzy reasoning based on single input rule modules with dynamic absorbers. *International Journal of Vehicle Design*, Vol. 31, No. 1, 2003, pp. 22-40.
610. YOSHIMURA T AND A TAKAGI (2004): Pneumatic active suspension for a one-wheel car model using fuzzy reasoning and a disturbance observer. *J. of Zhejiang Univ Sci*, Vol 5, No. 9, September 2004, pp. 1060-1068. Available online at http://www.zju.edu.cn/jzus/2004/0409/040907.pdf .
611. YOSHIOKA H, JC RAMALLO AND BF SPENCER (2002): "Smart" Base Isolation Strategies Employing Magnetorheological Dampers. *Journal of Engineering Mechanics*, Vol. 128, No. 5, May 2002, pp. 540-551.
612. YU F AND DA CROLLA (1998): Analysis on Benefits of an Adaptive Kalman Filter Active Vehicle Suspension. *SAE Technical Paper Series*, 1998, Paper No. 981120, 7 p. Avail-able online at http://www.paper.edu.cn .

613. YU M, CR LIAO, XM DONG, WM CHEN AND ZSH LI (2004): Road Testing of Automotive MR Shock Absorber. *ELECTRORHEOLOGICAL FLUIDS AND MAGNETO-RHEOLOGICAL SUSPENSIONS (ERMR 2004) Proceedings of the Ninth International Conference*, Beijing, China, 29 August – 3 September 2004, p. 694.
614. YUE C, T BUTSUEN AND JK HEDRICK (1988): Alternative control laws for automotive active suspensions. *American Control Conference*, 1988.
615. YUE C AND T BUTSUEN (1989): Alternative Control Laws for Automotive Active Suspension. *Journal of Dynamic Systems, Measurements and Control*, 1989, pp. 286-291.
616. YUKSEK I AND F KAYA (1995): Vibration Optimization of Vehicle Systems. *ASME Structural Dynamics and Vibration*, PD-70, 1995, pp. 217-221.
617. ZHANG Y AND A ALLEYNE (2002): A Practical and Effective Approach to Active Suspension Control. *Proceedings of the 6th International Symposium on Advanced Vehicle Control*, Hiroshima, Japan, September 2002.
618. ZHANG Y AND A ALLEYNE (2003): A New Approach to Half-Car Active Suspension Control. *Proceedings of the 2003 American Control Conference*, Denver, Colorado.
619. ZHANG J-Q AND N-L LU (2004): Analysis of Vibration Control of Tracked Vehicles Suspension Using Magnetorheological Suspensions Damper. *ELECTRORHEOLOGICAL FLUIDS AND MAGNETO-RHEOLOGICAL SUSPENSIONS (ERMR 2004) Proceedings of the Ninth International Conference*, Beijing, China, 29 Aug. – 3 Sept. 2004, p. G783.
620. ZHUANG D, X SHEN AND F YU (2004): Study on the Band-Limited Active Vehicle Suspension Based on the Co-Simulation Technology. *FISITA 2004: World Automotive Congress*, 23-27 May 2004, Barcelona, Spain, Paper F2004F283.
621. ZIPSER L, L RICHTER AND U LANGE (2001): Magnetorheological Fluid for Actuators. *Sensors and Actuators*, 2001, pp. 318-325.
622. ZUO L AND SA NAYFEH (2002): Design of Passive Mechanical Systems via Decentralized Control Techniques. *Proc. of the 43rd AIAA/ASME/ASCE/AHS/ASC Structures, Structural Dynamics and Materials Conference*, April 2002, AIAA 2002-1282.
623. ZUO L AND SA NAYFEH (2003A): Structured H_2 Optimization of Vehicle Suspensions Based on Multi-Wheel Models. *Vehicle System Dynamics*. Vol. 40, No. 5, 2003, pp. 351-371,
624. ZUO L AND SA NAYFEH (2003B): Low-Order Continuous Time Filters for Approximation of the ISO 2631-1 Human Vibration Sensitivity Weightings. *Journal of Sound and Vibration*, 2003.

Acronyms

Acronyms

A&J	Acceleration-and-jerk
ABC	Active body control
ABS	Anti-lock braking system
ABW	Absorb-by-wire
AC	Alternating current; active control
ACC	Adaptive cruise control
ACE	Active cornering enhancement
ACT	Active control technology
ADAMS	Automatic dynamics analysis of mechanical systems
ADC	Active dynamic control
A/D	Analogue/digital
AE	All-electric
AECV	All-electric combat vehicle
AEIV	All-electric intelligent vehicle
AEV	All-electric vehicle
AF	All-fluidical
AFS	Active front steering
AI	Artificial intelligence
AP	All-pneumatical
ARC	Adaptive robust control; active roll control
ARP	Adaptive ride control; active rollover protection
ARS	Active roll stabilisation
ASC	Active suspension controller
ASIC	Application specific integrated circuit
ASIM	Application specific integrated matrixer
ATC	Adaptive traction control
AV	Automotive vehicle
AVA	Active vibration absorber
AV/driver	Automotive vehicle/human driver
AW	Active wheel
AWA	All-wheel absorbed
AWB	All-wheel-braked; all-wheel-brake
AWD	All-wheel driven
AWS	All-wheel steered
BBW	Brake-by-wire
BMW	*Bayerische Motoren Werke (Bawarian Motor Works)*
CCD	Controller and control display
CCV	Control configured vehicle
CCVRTD	Chassis continuously variable real-time damping
C&HS	Comfort-and-handling stability

CH-E	Chemo-electrical
CoG	Centre-of-gravity
COTS	Commercial off-the-shelf
CSU	Central storage unit
CVD	Continuously variable damper
CVDA	Continuously variable-damping-force shock absorber
CVRSS	Continuously variable road-sensing suspension
CVT	Continuously variable transmission
DBW	Drive-by-wire
DC	Direct current; decision coverage
DD	Dynamic drive
DoF	Degrees-of-freedom
DSC	Dynamic stability control
DSP	Digital signal processor
DYC	Direct yaw-moment control
E&IN	Energy-and-information network
EBA	Electronic brake assist
EBD	Electronic brake-force distribution
ECC	Electrically-heated catalytic converter
ECE	External combustion engine
ECI&C	Entertainment, communication, information and control
ECU	Electronic control unit
E-CH	Electro-chemical
EDC	Electronic damping control; electronic damper control
EED	Electrical energy distribution
E-F-M	Electro-fluido-mechanical
EMB	Electro-mechanical brake
EMU	Engine management unit
E-M	Electro-mechanical
E-M-F	Electro-mechano-fluidical
E-M-P	Electro-mechano-pneumatical
ESP	Electronic stability control
ER	Electro-rheological
ERF	Electro-rheological fluid
FA	Front axle
FBW	Fly-by-wire
FCU	Fail-consistent unit
FECU	Fluidical-and-electronic control unit
FF	Force-feedback
FL	Fuzzy logic
F-M	Fluido-mechanical
FMEA	Failure-mode effect analysis
F-P-M	Fluido-pneumo-mechanical
FO	Fibre optic
FTU	Fault-tolerant unit
FWA	Front-wheel absorbed

FWD	Front-wheel driven; front-wheel-drive
FWS	Front-wheel steered
GA	Genetic algorithm
GER	Giant-electro-rheological
GERF	Giant-electro-rheological fluid
HAC	Hill-start assist control
HD	Human driver
HE	Hybrid-electric
HM	Hybrid-mechanical
HMMWV	High mobility multi-purpose wheeled vehicle
H-M	Human-machine
HW	Hand wheel
HWA	Hand-wheel angle
H&TD	Human and/or telerobotic driver
ICE	Internal combustion engine
ICR	Instantaneous centre of rotation
ICV	Infantry carrier vehicle
I/O	Inlet/outlet; input/output
IPM	Interior permanent magnet
IR	Infra-red
ISO	International Organization for Standardization
LC	Level control
LQG	Linear-quadratic-Gaussian
LSD	Limited slip differential
LSI	Leaf Spring Institute
IVDL	In-vehicle data link
IVS	In-vehicle sensor
L	Laser
LQR	Linear-quadratic regulator
MARC	Modular-adaptive-robust-control
MBD	Model-based design
MC	Modified condition
MCU	Microcontroller unit
M-E	Mechano-electrical
M-F	Mechano-fluidical
MMI	Multi-media interface
M-M	Mechano-mechanical
MOSFET	Metal-oxide semiconductor field-effect
M-P	Mechano-pneumatical
MR	Magneto-rheological
MRE	Magneto-rheological elastomer
MRF	Magneto-rheological fluid
MROADS	Magneto-rheological optimised damper suspension
MV	Minimum variance
MWD	Middle-wheel-drive
NF	Neuro-fuzzy

NMRF	Nano-magneto-rheological fluid
NN	Neural network
NVH	Noise, vibration, and harshness
OEM	Original equipment manufacturer
PCG	Production code generation
PF	Piezoelectric friction
PG	Parallelogram
PI	Proportional-integral
PID	Proportional-integral-derivative
P-M	Pneumo-mechanical
PS	Power steering
PSD	Power spectral density
PWM	Pulse width modulator
QCF	Quadratic cost functional
QoS	Quality of service
RBW	Ride-by-wire
RC&RH	Ride control and road handling
R&D	Research-and-development
RFS	Reverse function stabiliser
RMD	Roll moment distribution
RMS	Root-mean-square
R&P	Rack-and pinion
RTD	Real-time damping
RWA	Rear-wheel absorbed
RWD	Rear-wheel-driven; rear-wheel-drive
RWS	Rear-wheel steered
SASE	Self-adaptive symbiotic evaluation
SBW	Steer-by-wire
SCM	Software configuration management
SCS	Stability control system
SDoF	Single-degree-of-freedom
SEC	Specific energy consumption
SFC	Specific fuel consumption
SIL	Safety integrity level
SLA	Short-long arm
SM&GW	Steered, motorised and/or generatorised wheel
SOADS	Servo-valve optimised active damper suspension
SOSE	Self-organising symbiotic evaluation
SRF	Stimuli responsive fluid
SSA	State-switching absorber
SUV	Smart utility vehicle
TCS	Traction control system
TD	Torque distribution
TDMA	Time division multiple access
TTA	Time-triggered architecture
TTP	Time-triggered protocol

TV	Tracked vehicle
TVA	Tuned vibration absorber
US	Ultra-sonic
USPA	Ultrasonic parking assistance
VDC	Vehicle dynamics control
VH	Vehicle handling
VIP	Vehicle intrusion protection
VMFC	Virtual model following control
VO	Variable orifice
VR	Vehicle ride
VRPS	Vehicle-velocity-responsive power steering
VSC	Vehicle skid control
VTD	Variable traction distribution
VVT	Variable valve timing
V&V	Verification and validation
WDT	Watch-dog timer
WV	Wheeled vehicle
XBW	X-by-wire
3D	Three dimensions
2-DoF	Two-degrees-of-freedom
4-DoF	Four-degrees-of-freedom
7-DoF	Seven-degrees-of-freedom
8-DoF	Eight-degrees-of-freedom
4WA	Four-wheel absorbed; four-wheel-absorb
4WB	Four-wheel braked; four-wheel-brake
2WD	Two-wheel driven, two-wheel-drive
4WD	Four-wheel driven; four-wheel-drive
2WS	Two-wheel steered; two-wheel-steering
4WS	Four-wheel steered; four-wheel-steering

Nomenclature

Nomenclature

A – Steady state gain in yaw angular velocity: [s^{-1}]
α_x – Longitudinal acceleration of the centre of gravity (barycentre): [$m\ s^{-1}$]
α_y - Lateral acceleration of the centre of gravity (barycentre): [$m\ s^{-1}$]
B_f – Steady state gain vehicle sideslip angle for front wheels: [s]
B_r – Steady state gain vehicle sideslip angle for rear wheels: [s]
b_f - Tread of the front wheels: [m]
b_r - Tread of the rear wheels: [m]
f_n – Natural frequency ($f_n = \omega_n / 2\pi$): [s^{-1}] or [Hz]
I_z – Moment of the inertia of vehicle: [$kg\ m^2$]
i_s - Steering ratio in over all
K – Stability factor: [$s^2 m^{-2}$]
K_f– Equivalent cornering power of front: [N] or [$N\ rad^{-1}$]
K_r – Equivalent cornering power of rear: [N] or [$N\ rad^{-1}$]
k - Steering angle ratio: [unitless]
l - Wheelbase: [m]
l_f – Distance between centre of gravity (barycentre) and front wheel shaft: [m]
l_r – Distance between centre of gravity (barycentre) and rear wheel shaft: [m]
m – Mass of the vehicle: [kg]
s - Complex frequency ($s = \mathrm{Re}(s) + \mathrm{Im}(s) = \sigma + j\omega$) or *Laplace* numerator: [s^{-1}]
M_{Zf} - Inner-front-wheel yaw torque around the vertical axis z: [Nm]
M_{Zr} - Inner-rear-wheel yaw torque around the vertical axis z: [Nm]
V – Vehicle velocity $\left(V = \sqrt{v_x^2 + v_y^2}\right)$: [$m\ s^{-1}$]
v_x – Longitudinal velocity of the centre of gravity (barycentre): [$m\ s^{-1}$]
v_y – Lateral velocity of the centre of gravity (barycentre): [$m\ s^{-1}$]
δ_f – Front sideslip angle: [rad]
δ_r – Rear sideslip angle:[rad]
β - Sideslip angle of the centre of gravity (barycentre): [rad]
δ_f - Front wheel steering angle (average of left and right front wheels): [m]
δ_r - Rear wheel steering angle (average of left and right rear wheels): [m]
δ_H - Steering wheel angle: [rad]
ϕ - Roll angle: [rad]
$\dot{\phi}$ - Roll rate: [$rad\ s^{-1}$]
T_f – Time constant for front sideslip angle: [s]
T_r – Time constant for rear sideslip angle: [s]
τ_f – Time constant for front yaw angular velocity: [s]
τ_r – Time constant for rear yaw angular velocity: [s]

σ - *Neperian* frequency $\left(\sigma = \text{Re}(s)\right)$: [$s^{-1}$]

ω – Angular frequency $\left(\omega = 2\pi f\right)$: [rad s^{-1}]

ω_n – Response angular frequency $\left(\omega_n = 2\pi f_n\right)$: [rad s^{-1}]

ω_y – Yaw angular velocity (frequency): [rad s^{-1}]

ψ - Yaw angle: [rad]

$\dot{\psi}$ - Yaw angular velocity (frequency): [rad s^{-1}]

$\ddot{\psi}$ - Yaw angular acceleration: [rad s^{-2}]

ζ - Damping rate

$\zeta\,\omega_n$ – : [s^{-1}]

$\zeta\,\omega_n\,V$ – Steering capacity: [m s^{-2}]

Index

Index

Printed by Publishers' Graphics LLC
MLSI130426.15.16.339